T0255033

RIEMANNIAN GEOMETRY

A Modern Introduction

Second Edition

This book provides an introduction to Riemannian geometry, the geometry of curved spaces, for use in a graduate course. Requiring only an understanding of differentiable manifolds, the book covers the introductory ideas of Riemannian geometry, followed by a selection of more specialized topics. Also featured are Notes and Exercises for each chapter to develop and enrich the reader's appreciation of the subject. This second edition has a clearer treatment of many topics from the first edition, with new proofs of some theorems. Also a new chapter on the Riemannian geometry of surfaces has been added.

The main themes here are the effect of curvature on the usual notions of classical Euclidean geometry, and the new notions and ideas motivated by curvature itself. Among the classical topics shown in a new setting is isoperimetric inequalities – the interplay of volume of sets and the areas of their boundaries – in curved space. Completely new themes created by curvature include the classical Rauch comparison theorem and its consequences in geometry and topology, and the interaction of microscopic behavior of the geometry with the macroscopic structure of the space.

Isaac Chavel is Professor of Mathematics at The City College of the City University of New York. He received his Ph.D. in Mathematics from Yeshiva University under the direction of Professor Harry E. Rauch. He has published in international journals in the areas of differential geometry and partial differential equations, especially the Laplace and heat operators on Riemannian manifolds. His other books include *Eigenvalues in Riemannian Geometry* (1984) and *Isoperimetric Inequalities: Differential Geometric and Analytic Perspectives* (2001). He has been teaching at The City College of the City University of New York since 1970, and he has been a member of the doctoral program of the City University of New York since 1976. He is a member of the American Mathematical Society.

CAMBRIDGE STUDIES IN ADVANCED MATHEMATICS

Already published

17 W. Dicks & M. Dunwoody *Groups acting on graphs*
18 L.J. Corwin & F.P. Greenleaf *Representations of nilpotent Lie groups and their applications*
19 R. Fritsch & R. Piccinini *Cellular structures in topology*
20 H. Klingen *Introductory lectures on Siegel modular forms*
21 P. Koosis *The logarithmic integral II*
22 M.J. Collins *Representations and characters of finite groups*
24 H. Kunita *Stochastic flows and stochastic differential equations*
25 P. Wojtaszczyk *Banach spaces for analysis*
26 J.E. Gilbert & M.A.M. Murray *Clifford algebras and Dirac operators in harmonic analysis*
27 A. Fröhlich & M.J. Taylor *Algebraic number theory*
28 K. Goebel & W.A. Kirk *Topics in metric fixed point theory*
29 J.F. Humphreys *Reflection groups and Coxeter groups*
30 D.J. Benson *Representations and cohomology I*
31 D.J. Benson *Representations and cohomology II*
32 C. Allday & V. Puppe *Cohomological methods in transformation groups*
33 C. Soulé et al. *Lectures on Arakelov geometry*
34 A. Ambrosetti & G. Prodi *A primer of nonlinear analysis*
35 J. Palis & F. Takens *Hyperbolicity, stability and chaos at homoclinic bifurcations*
37 Y. Meyer *Wavelets and operators I*
38 C. Weibel *An introduction to homological algebra*
39 W. Bruns & J. Herzog *Cohen–Macaulay rings*
40 V. Snaith *Explicit Brauer induction*
41 G. Laumon *Cohomology of Drinfeld modular varieties I*
42 E.B. Davies *Spectral theory and differential operators*
43 J. Diestel, H. Jarchow, & A. Tonge *Absolutely summing operators*
44 P. Mattila *Geometry of sets and measures in Euclidean spaces*
45 R. Pinsky *Positive harmonic functions and diffusion*
46 G. Tenenbaum *Introduction to analytic and probabilistic number theory*
47 C. Peskine *An algebraic introduction to complex projective geometry*
48 Y. Meyer & R. Coifman *Wavelets*
49 R. Stanley *Enumerative combinatorics I*
50 I. Porteous *Clifford algebras and the classical groups*
51 M. Audin *Spinning tops*
52 V. Jurdjevic *Geometric control theory*
53 H. Volklein *Groups as Galois groups*
54 J. Le Potier *Lectures on vector bundles*
55 D. Bump *Automorphic forms and representations*
56 G. Laumon *Cohomology of Drinfeld modular varieties II*
57 D.M. Clark & B.A. Davey *Natural dualities for the working algebraist*
58 J. McCleary *A user's guide to spectral sequences II*
59 P. Taylor *Practical foundations of mathematics*
60 M.P. Brodmann & R.Y. Sharp *Local cohomology*
61 J.D. Dixon et al. *Analytic pro-p groups*
62 R. Stanley *Enumerative combinatorics II*
63 R.M. Dudley *Uniform central limit theorems*
64 J. Jost & X. Li-Jost *Calculus of variations*
65 A.J. Berrick & M.E. Keating *An introduction to rings and modules*
66 S. Morosawa *Holomorphic dynamics*
67 A.J. Berrick & M.E. Keating *Categories and modules with K-theory in view*
68 K. Sato *Levy processes and infinitely divisible distributions*
69 H. Hida *Modular forms and Galois cohomology*
70 R. Iorio & V. Iorio *Fourier analysis and partial differential equations*
71 R. Blei *Analysis in integer and fractional dimensions*
72 F. Borceaux & G. Janelidze *Galois theories*
73 B. Bollobás *Random graphs*
74 R.M. Dudley *Real analysis and probability*
75 T. Sheil-Small *Complex polynomials*

(continued on overleaf)

Series list (*continued*)

76 C. Voisin *Hodge theory and complex algebraic geometry, I*
77 C. Voisin *Hodge theory and complex algebraic geometry, II*
78 V. Paulsen *Completely bounded maps and operator algebras*
79 F. Gesztesy & H. Holden *Soliton equations and their algebro-geometric solutions*
81 S. Mukai *An Introduction to invariants and moduli*
82 G. Tourlakis *Lectures in logic and set theory I*
83 G. Tourlakis *Lectures in logic and set theory II*
84 R. Bailey *Association schemes*
85 J. Carlson, S. Müller-Stach, & C. Peters *Period mappings and period domains*
86 J. Duistermaat & J. Kolk *Multidimensional real analysis I*
87 J. Duistermaat & J. Kolk *Multidimensional real analysis II*
89 M. Golumbic & A. Trenk *Tolerance graphs*
90 L. Harper *Global methods for combinatorial isoperimetric problems*
91 I. Moerdijk & J. Mrcun *Introduction to foliations and lie groupoids*
92 J. Kollar, K. Smith & A. Corti *Rational and nearly rational varieties*
93 D. Applebaum *Lévy processes and stochastic calculus*
95 M. Schechter *An introduction to nonlinear analysis*

RIEMANNIAN GEOMETRY

A Modern Introduction

Second Edition

ISAAC CHAVEL

Department of Mathematics
The City College of the
City University of New York

CAMBRIDGE
UNIVERSITY PRESS

32 Avenue of the Americas, New York NY 10013-2473, USA

Cambridge University Press is part of the University of Cambridge.

It furthers the University's mission by disseminating knowledge in the pursuit of education, learning and research at the highest international levels of excellence.

www.cambridge.org
Information on this title: www.cambridge.org/9780521619547

First published 1994
Second edition published 2006

A catalogue record for this publication is available from the British Library

Library of Congress Cataloguing in Publication data

Chavel, Isaac.
Riemannian geometry : a modern introduction / Isaac Chavel. – 2nd ed.
p. cm. – (Cambridge studies in advanced mathematics; 98)
Includes bibliographical references and indexes.
ISBN-13: 978-0-251-85368-2 (alk. paper)
ISBN-10: 0-521-85368-0 (alk. paper)
ISBN-13: 978-0-521-61954-7 (pbk. : alk. paper)
ISBN-10: 0-521-61954-8 (pbk. : alk. paper)
1. Geometry, Riemannian. I. Title. II. Series.
QA649.C45 2006
516.3'73 – dc22 2005029338

ISBN 978-0-521-61954-7 Paperback

for
HARRY ERNEST RAUCH
(1925–1979)

Contents

Preface to the Second Edition

In this second edition, the first order of business has been to correct mistakes, mathematical and typographical, large and small, and clarify a number of arguments that were unclear or given short shrift the first time round. I can only hope that, in this process, and in the process of changes and additions described below, I have not introduced any new errors.

I have added some proofs of theorems, and sketches to some of the exercises, that were originally left completely to the reader in the first edition. I have added some new notes and exercises as well.

In the text itself, I have made a few changes. I added a chapter with topics from surfaces, immediately following the chapter on coverings (Chapter IV). The chapter (Chapter V) now includes the Gauss–Bonnet theorem; but, it also contains topics of current interest, showing that the Riemannian geometry of surfaces is alive and well, and is a constant testing ground, as well as a source, of new ideas. As it contained the introduction to the isoperimetric problem in Riemannian manifolds, presenting the Bol–Fiala inequalities, and the Benjamini–Cao solution of the isoperimetric problem on the paraboloid of revolution, I thought it best to follow the chapter with isoperimetric inequalities in the classical constant curvature space forms (Chapter VI).

This last chapter (Chapter VI) is a bit different from what I presented in the first edition. New proofs were given for the isoperimetric problem in Euclidean space, with the famous proof by M. Gromov, using Stokes' theorem, now appearing in my other book *Isoperimetric Inequalities: Differential Geometric and Analytic Perspectives* (2001). The Brunn–Minkowski inequalities in hyperbolic space and the sphere were redone, hopefully improving on the first presentation.

Chapter VI is followed by the original (now Chapter VII) on the kinematic density, with little change. I was sorely tempted to include the Burago–Ivanov solution to the E. Hopf conjecture that metrics on the torus, of *all* dimensions, without conjugate points are flat. But, such an undertaking would have taken the discussion too far afield. This chapter is then followed by the one on isoperimetric inequalities in general Riemannian manifolds, and the chapter on the Rauch comparison theorem and its consequences.

Beyond the Notes and Exercises sections that conclude each chapter, the reader is highly recommended to M. Berger's recent survey *A Panoramic View of Riemannian Geometry* (2003), preceded by his preparatory essay *Riemannian Geometry During the Second Half of the Twentieth Century* (2000). Just about every page of this introduction to Riemannian geometry could have contained references to Berger's surveys for further background and future work.

It is a pleasure to thank the readers of my first edition for their warm reception of the book and for the helpful criticisms – both in pointing out errors and in suggesting improvements. I should add that I found the reviews very helpful, and I am grateful for the effort that went into them. I hope this edition merits the effort they invested.

ISAAC CHAVEL
Riverdale, New York
February 2005

Preface

My goals in this book on Riemannian geometry are essentially the same as those that guided me in my *Eigenvalues in Riemannian Geometry* (1984): to introduce the subject, to coherently present a number of its basic techniques and results with a mind to future work, and to present some of the results that are attractive in their own right. This book differs from *Eigenvalues* in that it starts at a more basic level. Therefore, it must present a broader view of the ideas from which all the various directions emerge. At the same time, other treatments of Riemannian geometry are available at varying levels and interests, so I need not introduce everything. I have, therefore, attempted a viable introduction to Riemannian geometry for a very broad group of students, with emphases and developments in areas not covered by other books.

My treatment presupposes an introductory course on manifolds, the construction of associated tensor bundles, and Stokes' theorem. When necessary, I recall the facts and/or refer to the literature in which these matters are discussed in detail.

I have not hesitated to prove theorems more than once, with different points of view and arguments. Similarly, I often prove weaker versions of a result and then follow with the stronger version (instead of just subsuming the former result under the latter). The variety of levels, ideas, and approaches is a hallmark of mathematics; and an introductory treatment should display this variety as part of the development of broad technique and as part of the aesthetic appreciation of the mathematical endeavor.

I am confident that a short course could be easily crafted from Chapters I to IV and VII (the second edition: Chapter IX), and a more ambitious course from the remaining material. Every chapter of the book features a Notes and Exercises section. These sections cover (i) references to earlier literature and to other results; (ii) "toes in the water" introductions to topics emerging from the ideas presented in the main body of the text; and (iii) examples and applications. The Notes and Exercises sections of the first four chapters are quite extensive.

These sections in the later chapters are not as ambitious as those in the first four, since the first four chapters are genuinely introductory.

The Notes and Exercises sections are organized loosely under subheadings of topics. These are not to be taken too literally; rather, they attempt to restrain the variety of material in these sections from becoming chaotic.

A submotif in the Notes and Exercises sections is the method of calculation with moving frames, even though the method is not used extensively in the main body of the text itself. Besides the obvious claim that such calculations should be included in an introductory treatment, I had in mind a quiet tribute to the late William F. Pohl. I learned the magic long ago of *repère mobile* from Bill Pohl at the University of Minnesota. I can still see his full frame at the blackboard, extending his arm gracefully in front of him, moving his hands descriptively with his fingers playing the role of the frame vectors, and declaring that ${\omega_2}^3 = 0$ since the frame vector field e_2 did not turn in the direction of e_3.[1]

It is a pleasure to thank my colleagues and friends for their contributions to my work, in general, and to this book, in particular. P. Buser provided me with some helpful discussions and read portions of the work. So did J. Dodziuk and E. A. Feldman. Finally, I wish to thank the geometers of the doctoral program of the City University of New York, namely J. Dodziuk, L. Karp, B. Randol, R. Sacksteder, J. Velling, and Edgar A. Feldman – who have provided, over the years, all sorts of help, mathematical stimulation and insight, and scientific partnership. Their contribution permeates all the pages of this book.

ISAAC CHAVEL
Riverdale, New York
July 1992

[1] My memory hits the mark. As soon as I mentioned to Ed Feldman that I put moving frames in the book because of Bill Pohl, Ed performed an imitation of the grand gesture that was Bill's trademark.

I
Riemannian Manifolds

One cannot start discussing Riemannian geometry without mention of the classics. By "the classics," we refer to the essays of C. F. Gauss (1825, 1827) and B. Riemann (1854), to G. Darboux's summary treatise (1894) of the work of the nineteenth century (and beginning of the twentieth), and to E. Cartan's lectures (1946) in which the method of moving frames became a powerful exciting tool of differential geometry.

Nor may one forget to recommend to the reader the delightful discussion of differential geometry in D. Hilbert–S. Cohn-Vossen (1952).

H. Hopf's notes (1946, 1956) remain eminently readable. A very helpful collection of more current introductory essays is the *MAA Studies* volume edited by S. S. Chern (1989).

In addition, one should refer to the "introductory" five-volume opus of M. Spivak (1970) – wherein the practice of differential geometry is presented in loving detail.

Most recently, one has a definitive overview of the subject at the end of the twentieth century by M. Berger (2003).

Our treatment here is mostly inspired by, and follows in many respects, J. Milnor's elegant and exceptionally clear lecture notes Milnor (1963).[1]

A short summary of the progression of ideas of this chapter is as follows.

Whereas one has, given a differentiable manifold, a natural differentiation of functions on the manifold, one does not have a naturally determined method of differentiation of vector fields on the manifold. Therefore, one considers all possibilities of such differentiation – connections on the manifold. Once one actually picks such a differentiation procedure (i.e., a connection), one determines differentiation of vector fields along paths in the manifold. In particular,

[1] See Note 1 in §I.9.

one has an acceleration vector field (the derivative of the velocity vector field) associated with each C^2 path in the manifold. "Straight lines," usually referred to as *geodesics*, are then the paths in the manifold for which the acceleration is zero – they are the collection of paths describing the "law of inertia" for the manifold with the given connection.

The exponential map (the name inspired by analogy to the exponential map in Lie Theory) then provides a map from the tangent space of any given point of the manifold to the manifold itself, in which lines emanating from the origin of the tangent space are mapped to geodesics in the manifold itself emanating from the point in question. It is in this context that we introduce the torsion and curvature tensors of a connection. For the torsion and curvature tensors arise from the linearization of the differential equations of geodesics; therefore, they will ultimately play a role in studying the differential of the exponential map – the precise role to be explicated in detail in later chapters.

Next, we introduce the Riemannian metrics, the ability to calculate the length of paths in the manifold and to calculate angles of tangent vectors in the same tangent space of the manifold. Again, the specification of the Riemannian metric is not uniquely determined. However, once one has such a metric, one automatically has a preferred connection associated with it. It will always be assumed, unless some explicit comment is made to the contrary, that this connection – the *Levi-Civita connection* – is the one under consideration when examining a given Riemannian metric.

The ability to determine lengths of paths in the manifold then induces a natural metric space structure on the manifold. Namely, the distance between any two points of the manifold is the infimum of the length of all paths connecting the two points. One has the classical theorems that (i) if a path between two points has length equal to the distance between them, then the path may be reparameterized to be a geodesic, and (ii) given any point in the manifold, the point has a neighborhood for which there is one and only one distance minimizing geodesic connecting the original point to any other point in the neighborhood. (Actually, more is true – see §I.6.) This development of ideas concludes (§I.7) with the full characterization of the completeness of the metric structure of the Riemannian metric in terms of the infinite extendability of the geodesics of the Riemannian metric.

The chapter closes with a discussion of calculations using moving frames. We do not really present any new material; rather, we revisit some of the previous calculations with a new tool to be developed in its own right and to be used later on.

§I.1. Connections

We refer the reader to Narasimhan (1968) and Warner (1971) for background on differentiable manifolds. Needless to say, these are not the only possible quality choices.

For a path $\omega(t)$ in a manifold M, we let $\omega'(t)$ denote the velocity vector of ω at $\omega(t)$. When the manifold is $\mathbb{R}^n$, we will distinguish between the velocity vector and the derivative of the vector valued function, when necessary.

Unless otherwise stated, either explicitly or by unequivocal context, all our manifolds are C^∞, Hausdorff, with countable base, and connected. Unless otherwise indicated, *differentiable* means C^∞. When speaking of manifolds that possess boundary, our use of the word "manifold" (nearly) always refers to the interior. In particular, our compact manifolds are without boundary.

Let M be an n–dimensional differentiable manifold, with tangent bundle TM and associated natural projection $\pi : TM \to M$. For any $p \in M$, we let M_p denote the tangent space to M at p. We denote the collection of C^ℓ, $\ell = 0, 1, \ldots, \infty$, vector fields on M by $\Gamma^\ell(TM)$.

If $\phi : M \to N$ is a differentiable map from the manifold M to the manifold N, we let $\phi_* : TM \to TN$ denote the induced bundle map (in local coordinates the Jacobian linear transformation) linear on each fiber. We also let ϕ^* denote the pullback maps of the associated cotangent bundles.

Definition. A *connection on M* is a map $\nabla : TM \times \Gamma^1(TM) \to TM$, which we write as $\nabla_\xi Y$ instead of $\nabla(\xi, Y)$, with the following properties: First we require that $\nabla_\xi Y$ be in the same tangent space as ξ, and that for $\alpha, \beta \in \mathbb{R}$, $p \in M, \xi, \eta \in M_p, Y \in \Gamma^1(TM)$,

$$\nabla_{\alpha\xi+\beta\eta} Y = \alpha\nabla_\xi Y + \beta\nabla_\eta Y.$$

Second, we require that for $p \in M, \xi \in M_p, Y, Y_1, Y_2 \in \Gamma^1(TM), f \in C^1(M)$, we shall have

$$\nabla_\xi(Y_1 + Y_2) = \nabla_\xi Y_1 + \nabla_\xi Y_2,$$
$$\nabla_\xi(fY) = (\xi f)Y_{|p} + f(p)\nabla_\xi Y.$$

Finally, we require that ∇ be smooth in the following sense: if $X, Y \in \Gamma^\infty(TM)$, then $\nabla_X Y \in \Gamma^\infty(TM)$.

The example that motivates the above definition is, naturally, $\mathbb{R}^n$. We let $\Im_p : \mathbb{R}^n \to (\mathbb{R}^n)_p$ be the natural identification of $\mathbb{R}^n$ with the (abstract) tangent space to $\mathbb{R}^n$ at any $p \in \mathbb{R}^n$. For the natural basis $\{\mathfrak{e}_1, \ldots, \mathfrak{e}_n\}$ of $\mathbb{R}^n$, the natural basis of $(\mathbb{R}^n)_p$ determined by the chart consisting of the identity map of $\mathbb{R}^n$ to itself is given by

$$\partial_{j|p} = \Im_p \mathfrak{e}_j$$

for $j = 1, \ldots, n$. Let Y be a differentiable vector field on $\mathbb{R}^n$ such that $Y = \Sigma_j \eta^j \partial_j$. Then, the *standard connection on* $\mathbb{R}^n$ is given by

$$\nabla_\xi Y = \sum_{j=1}^{n} (\xi \eta^j) \partial_j. \tag{I.1.1}$$

One easily checks that the requirements of the definition of a connection are satisfied.

A more explicit geometric expression for the standard connection on $\mathbb{R}^n$ is given as follows: Given $\xi \in (\mathbb{R}^n)_p$, let $\omega : (-\epsilon, \epsilon) \to \mathbb{R}^n \in C^1$ be a path in $\mathbb{R}^n$ with $\omega(0) = p$, $\omega'(0) = \xi$. Then one verifies that

$$\nabla_\xi Y = \lim_{t \to 0} \frac{\Im_p \circ \Im_{\omega(t)}{}^{-1} Y_{|\omega(t)} - Y_{|p}}{t}. \tag{I.1.2}$$

Thus, the natural identification of the tangent spaces $(\mathbb{R}^n)_p$ and $(\mathbb{R}^n)_q$ via the map $\Im_q \circ \Im_p{}^{-1}$, for any p, q in $\mathbb{R}^n$, is that which allows for the natural differentiation of vector fields on $\mathbb{R}^n$. In an abstract differentiable manifold, no such natural identification exists, a priori. Therefore, it must be postulated in advance. However, it is far more natural to postulate the differentiation of vector fields first, and to then investigate the resultant identification of tangent spaces at different points of the manifold. See §I.2.

Let M be our differentiable manifold with connection ∇. We note that, for $p \in M, \xi \in M_p, \nabla_\xi Y$ is uniquely determined by the restriction of Y to any open set U containing p. To see this, fix $p \in M, \xi \in M_p$, and an open neighborhood U of p.

We first show that $Y|U = 0$ implies $\nabla_\xi Y = 0$. Pick a differentiable function $f : M \to \mathbb{R}$ such that $f(p) = 0$ and $f|M \backslash U = 1$. Then, $fY = Y$ and

$$\nabla_\xi Y = \nabla_\xi (fY) = (\xi f) Y_{|p} + f(p) \nabla_\xi Y,$$

both terms of which vanish at p. We conclude that, if two vector fields agree on all of U, then so do their covariant derivatives.

We may proceed conversely, namely, if Y is given as defined only on U, then pick open V relatively compact in U and $\phi : M \to [0, 1]$ differentiable with compact support such that $\phi|V = 1$ and $\operatorname{supp} \phi \subseteq U$. Then define $\overline{Y} \in \Gamma^1(TM)$ by setting $\overline{Y} = \phi Y$ on U, and $\overline{Y} = 0$ on $M \backslash U$; and finally define

$$\nabla_\xi Y = \nabla_\xi \overline{Y}.$$

Then, $\nabla_\xi Y$ is well-defined, that is, it is independent of the choice of extension of Y to $\overline{Y} \in \Gamma^1(TM)$. Thus, we may effectively calculate ∇ by restricting the vector fields in question to, for example, the domain of a chart.

We now show more, namely, that to calculate $\nabla_\xi Y$, for given $Y \in \Gamma^1(TM)$, we need only know Y restricted to a path through $p = \pi(\xi)$ with velocity vector at p equal to ξ. Indeed, let $x : U \to \mathbb{R}^n$ be a chart about p, and ξ given by

$$\xi = \sum_j \xi^j \partial_{j|p}.$$

Then $\nabla_\xi Y$ is given by

$$\nabla_\xi Y = \sum_j \xi^j \nabla_{\partial_{j|p}} Y.$$

Also, one has the functions $\eta^j : U \to \mathbb{R}$, $j = 1, \ldots, n$, such that

$$Y|U = \sum_j \eta^j \partial_j.$$

Now, there exist functions $\Gamma_{jk}{}^\ell : U \to \mathbb{R}$, $j, k, \ell = 1, \ldots, n$, referred to as *Christoffel symbols*, such that

$$\nabla_{\partial_k} \partial_j = \sum_\ell \Gamma_{jk}{}^\ell \partial_\ell \tag{I.1.3}$$

on U. We then have

$$\begin{aligned}
\nabla_\xi Y &= \sum_k \xi^k \nabla_{\partial_{k|p}} Y \\
&= \sum_k \xi^k \nabla_{\partial_{k|p}} \left(\sum_j \eta^j \partial_j \right) \\
&= \sum_k \xi^k \left\{ \sum_j (\partial_k \eta^j)(p) \partial_{j|p} + \sum_{j,\ell} \eta^j(p) \Gamma_{jk}{}^\ell(p) \partial_{\ell|p} \right\} \\
&= \sum_\ell \left\{ \sum_k \xi^k (\partial_k \eta^\ell)(p) + \sum_{j,k} \Gamma_{jk}{}^\ell(p) \eta^j(p) \xi^k \right\} \partial_{\ell|p},
\end{aligned}$$

that is,

$$\nabla_\xi Y = \sum_\ell \left\{ \sum_k \xi^k (\partial_k \eta^\ell)(p) + \sum_{j,k} \Gamma_{jk}{}^\ell(p) \eta^j(p) \xi^k \right\} \partial_{\ell|p}. \tag{I.1.4}$$

In particular, if $\omega : (\alpha, \beta) \to M$ is differentiable such that $t_0 \in (\alpha, \beta)$, $\omega(t_0) = p$, $\omega'(t_0) = \xi$, one then has

$$\nabla_\xi Y = \sum_\ell \left\{ (\eta^\ell \circ \omega)'(t_0) + \sum_{j,k} \Gamma_{jk}{}^\ell(p)\eta^j(p)\xi^k \right\} \partial_{\ell|p},$$

which was our claim.

Next, we note that the choice of connection on M is highly undetermined. Given any chart $x : U \to \mathbb{R}^n$, then any choice of n^3 functions $\Gamma_{jk}{}^\ell : U \to \mathbb{R} \in C^\infty$ determine a local connection on U, via the equations (I.1.3) and (I.1.4). One can then create global connections on M from local ones, by using a partition of unity.

Finally, we note the change of variable formula for the Christoffel symbols. Given two charts $x : U \to \mathbb{R}^n$, $y : U \to \mathbb{R}^n$, on M, with respective Christoffel symbols ${}_x\Gamma_{jk}{}^\ell$, ${}_y\Gamma_{st}{}^r$, then one verifies by direct calculation

(I.1.5)
$$\sum_\ell {}_x\Gamma_{jk}{}^\ell \frac{\partial(y^r \circ x^{-1})}{\partial x^\ell} = \frac{\partial^2(y^r \circ x^{-1})}{\partial x^j \partial x^k} + \sum_{s,t} \frac{\partial(y^s \circ x^{-1})}{\partial x^j} \frac{\partial(y^t \circ x^{-1})}{\partial x^k} {}_y\Gamma_{st}{}^r.$$

Definition. Let $\omega : (\alpha, \beta) \to M$ be a C^1 path in M. We define a *vector field along the path* ω to be a map $X : (\alpha, \beta) \to TM$, such that $\pi \circ X = \omega$, that is, $X(t) \in M_{\omega(t)}$ for all t. (Note that in such a situation, we do not necessarily obtain a vector field on the image of ω in M since it is possible, for example, that $t_1, t_2 \in (\alpha, \beta)$, $t_1 \neq t_2$, $\omega(t_1) = \omega(t_2)$, but $X(t_1) \neq X(t_2)$.)

We define the *derivative of* X *along* ω, $\nabla_t X$, as follows: Assume $x : U \to \mathbb{R}^n$ is a chart containing $\omega((\alpha, \beta))$ and define $\Gamma_{jk}{}^\ell$ as in (I.1.3). Also set

$$\omega^j = x^j \circ \omega, \quad j = 1, \ldots, n,$$

write X as

$$X = \sum_j \xi^j (\partial_j \circ \omega),$$

and finally, define

(I.1.6)
$$\nabla_t X = \sum_\ell \left\{ (\xi^\ell)' + \sum_{j,k} (\Gamma_{jk}{}^\ell \circ \omega)\xi^j (\omega^k)' \right\} (\partial_\ell \circ \omega),$$

for $X \in C^1$.

One checks, using (I.1.5), that the definition (I.1.6) is independent of the choice of chart on M and thereby obtains a well-defined vector field $\nabla_t X$ along ω even if the image of ω is not contained in the domain of one chart on M. Also,

$$\nabla_t(X_1 + X_2) = \nabla_t X_1 + \nabla_t X_2, \tag{I.1.7}$$

$$\nabla_t(fX) = f'X + f\nabla_t X, \tag{I.1.8}$$

for all vector fields X, X_1, X_2 along ω, and $f : (\alpha, \beta) \to \mathbb{R} \in C^1$.

One can now use the above to consider a more general situation, namely,

Definition. Let N, M be differentiable manifolds, $\phi : N \to M$ differentiable. Then, define a *vector field X along ϕ* to be a map $X : N \to TM$ satisfying $\pi \circ X = \phi$, that is, $X(q) \in M_{\phi(q)}$ for all $q \in N$.

If X is a differentiable vector field along ϕ, $q \in N$, $\xi \in N_q$, and ∇ a connection on M, define the *derivative of X along ϕ in the direction ξ*, $\nabla_\xi X$, as follows: Let $\omega : (-\epsilon, \epsilon) \to N$ be any differentiable path for which $\omega(0) = q$, $\omega'(0) = \xi$, and let $Y : (-\epsilon, \epsilon) \to TM$ be the vector field along $\phi \circ \omega$ given by

$$Y = X \circ \omega.$$

Define $\nabla_\xi X$ by

$$\nabla_\xi X = (\nabla_t Y)(0).$$

$\nabla_\xi X$ is seen to be independent of the choice of ω, and is therefore well-defined and satisfies

$$\nabla_\xi(X_1 + X_2) = \nabla_\xi X_1 + \nabla_\xi X_2,$$
$$\nabla_\xi(fX) = (\xi f)X_{|p} + f(p)\nabla_\xi X,$$

where X, X_1, X_2 are differentiable vector fields along ϕ and $f : N \to \mathbb{R}$ is differentiable.

§I.2. Parallel Translation of Vector Fields

Let M be a given differentiable manifold with connection ∇.

Definition. Let $\omega : (\alpha, \beta) \to M$ be a C^1 path in M. We say that a vector field X along ω is *parallel along ω* if

$$\nabla_t X = 0$$

on all of (α, β).

By (I.1.7), (I.1.8) one has that, given ω, ∇_t is a linear operator on vector fields along ω; thus, the set of parallel vector fields along ω is a vector space over $\mathbb{R}$. From (I.1.6), one has (via the theory of linear ordinary differential equations), to each $t_0 \in (\alpha, \beta)$, $\xi \in M_{\omega(t_0)}$, the existence of a unique parallel vector field X along ω satisfying $X(t_0) = \xi$. In particular, the space of parallel vector fields along ω is finite dimensional and has dimension equal to that of M.

Thus, we can construct isomorphisms between the tangent spaces to M at different points of ω, namely, let $t_1, t_2 \in (\alpha, \beta)$, and for $\xi \in M_{\omega(t_1)}$ let X_ξ be the parallel vector field along ω satisfying $X_\xi(t_1) = \xi$. Now set

$$\tau_{t_1,t_2}(\xi) = X_\xi(t_2).$$

Then, τ_{t_1,t_2} is a linear isomorphism of $M_{\omega(t_1)}$ onto $M_{\omega(t_2)}$ and is called *parallel translation along ω from $M_{\omega(t_1)}$ to $M_{\omega(t_2)}$*.

Theorem I.2.1. *Let $\omega : (\alpha, \beta) \to M$ be a differentiable path, X a differentiable vector field along ω, and $t_0 \in (\alpha, \beta)$. Then,*

$$(\nabla_t X)(t_0) = \lim_{t \to t_0} \frac{\tau_{t,t_0}(X(t)) - X(t_0)}{t - t_0}. \tag{I.2.1}$$

Proof. Let $E_1(t), \ldots, E_n(t)$ be n parallel vector fields along ω, which are pointwise linearly independent (of course, as soon as they are linearly independent at one point, they are linearly independent at all points), $n = \dim M$; then, there exist functions $\xi^j : (\alpha, \beta) \to \mathbb{R}$, $j = 1, \ldots, n$ such that

$$X(t) = \sum_{j=1}^{n} \xi^j(t) E_j(t)$$

on (α, β). One now calculates explicitly both sides of (I.2.1) and the result follows. ∎

Remark I.2.1. The reader is invited to compare (I.2.1) with (I.1.2). Note that the identification of tangent spaces $\tau_{t_1,t_2}(\xi)$ depends on the path ω connecting $\omega(t_1)$ to $\omega(t_2)$. A local calculation shows that if parallel translation of vector fields is independent of the choice of path connecting any two given points in M, then the curvature tensor of ∇ – to be defined below in §I.4 – vanishes identically on M. Almost needless to say, if parallel translation of vector fields on M were independent of the choice of path connecting any two given points in M, then one could construct, at will, n linearly independent (over $\mathbb{R}$) nonvanishing vector fields on M – a global *topological* restriction on M. See, also, Remark I.5.2.

§I.3. Geodesics and the Exponential Map

We are still with our differentiable manifold M and connection ∇.

Definition. A path $\omega : (\alpha, \beta) \to M \in C^\ell$, $\ell \geq 2$, is called is called a *geodesic* if

$$\nabla_t \omega' = 0. \tag{I.3.1}$$

on all of (α, β).

To write the equation for a geodesic in a chart, we let $x : U \to \mathbb{R}^n$ be the chart, set $\omega^j = x^j \circ \omega$, $j = 1, \ldots, n$, and let $\Gamma_{jk}{}^\ell$, $j, k, \ell = 1, \ldots, n$ be given by (I.1.3). Then, (I.1.6) implies that (I.3.1) reads as

$$(\omega^\ell)'' + \sum_{j,k} (\Gamma_{jk}{}^\ell \circ \omega)(\omega^j)'(\omega^k)' = 0. \tag{I.3.2}$$

We now exhibit the second-order system (I.3.2) as a first-order system on TM. With the projection $\pi : TM \to M$ and chart $x : U \to \mathbb{R}^n$, we associate the natural chart $Q : \pi^{-1}[U] \to \mathbb{R}^{2n} = \mathbb{R}^n \times \mathbb{R}^n$ by $Q(\xi) = (q(\xi), \dot{q}(\xi))$, where

$$q = x \circ \pi, \quad \dot{q}(\xi) = \xi x$$

(where by ξx we mean $\xi^j = \xi x^j$, $j = 1, \ldots, n$). Thus,

$$\xi = \sum_j \dot{q}^j(\xi) \partial_{j|\pi(\xi)}.$$

We find it convenient to write the basis of tangent spaces to TM at points of $\pi^{-1}[U]$ by $\{\partial/\partial q^1, \ldots, \partial/\partial q^n, \partial/\partial \dot{q}^1, \ldots, \partial/\partial \dot{q}^n\}$. One immediately has

$$\pi_*(\partial/\partial q^j) = \partial/\partial x^j, \quad \pi_*(\partial/\partial \dot{q}^j) = 0.$$

The differential equation (I.3.2) can then be written as a first-order equation in $\pi^{-1}[U]$:

$$(q^\ell)' = \dot{q}^\ell, \tag{I.3.3}$$

$$(\dot{q}^\ell)' = -\sum_{j,k} (\Gamma_{jk}{}^\ell \circ \pi) \dot{q}^j \dot{q}^k. \tag{I.3.4}$$

The solutions to (I.3.3), (I.3.4) are therefore integral curves of the vector field $\mathcal{G}$ on $\pi^{-1}[U]$ given by

$$\mathcal{G} = \sum_\ell \left\{ \dot{q}^\ell \frac{\partial}{\partial q^\ell} - \sum_{j,k} (\Gamma_{jk}{}^\ell \circ \pi) \dot{q}^j \dot{q}^k \frac{\partial}{\partial \dot{q}^\ell} \right\}.$$

Since the geodesic equations are independent of the choice of coordinates on M, we conclude that $\mathcal{G}$ defines a global vector field on TM.

Definition. The maximal flow of $\mathcal{G}$ is called the *geodesic flow*.

One easily has the following:

Theorem I.3.1. *Let $\Omega : (\alpha, \beta) \to TM$ be an integral curve of $\mathcal{G}$, and $\omega = \pi \circ \Omega$. Then,*

$$\omega' = \Omega \tag{I.3.5}$$

and ω is a geodesic in M. Conversely, given a geodesic $\omega : (\alpha, \beta) \to M$ and Ω defined by (I.3.5), *then Ω is an integral curve of $\mathcal{G}$.*

Thus, if $\varphi(t, \xi)$ denotes the maximal flow of $\mathcal{G}$ on TM, where $t \in \mathbb{R}, \xi \in TM$, then

$$\gamma_\xi(t) := \pi \circ \varphi(t, \xi)$$

is the unique maximal (relative to its domain in $\mathbb{R}$) geodesic in M satisfying

$$\gamma_\xi(0) = \pi(\xi), \quad \gamma_\xi{}'(0) = \xi.$$

Of course,

$$\gamma_\xi{}'(t) = \varphi(t, \xi).$$

In particular, $\gamma_\xi(t)$ depends differentiably (i.e., C^∞) on t and ξ.

Finally, if I_ξ is the maximal interval on which γ_ξ is defined, then for any $\alpha \in \mathbb{R}, \alpha \neq 0$ we have

$$I_{\alpha\xi} = (1/\alpha)I_\xi, \quad \gamma_{\alpha\xi}(t) = \gamma_\xi(\alpha t), \tag{I.3.6}$$

where if $I_\xi = (\beta_1, \beta_2)$ then $(1/\alpha)I_\xi := (\beta_1/\alpha, \beta_2/\alpha)$ when $\alpha > 0$, and $(1/\alpha)$ $I_\xi := (\beta_2/\alpha, \beta_1/\alpha)$ when $\alpha < 0$.

Remark I.3.1. We note that (I.3.5) – the coordinate-free version of (I.3.3), (I.3.4) – is the heart of a coordinate-free definition of a second-order differential equation on a manifold. Namely, we say that *a vector field $\mathfrak{X}$ on TM determines a second-order ordinary differential equation on M* if

$$\pi_*(\mathfrak{X}_{|\xi}) = \xi$$

for all $\xi \in TM$, that is, if we consider $\mathfrak{X} : TM \to TTM$ as a section on TM into TTM then

$$\pi_* \mathfrak{X} = \mathrm{id}_{TM}. \tag{I.3.7}$$

Thus, if Ω denotes an integral curve of $\mathfrak{X}$ on TM, and $\omega = \pi \circ \Omega$ is the projection of Ω to M, then the validity of (I.3.5) for all integral curves Ω is equivalent to (I.3.7).

Definition. By the *zero section in* TM, we mean the vector field $\mathfrak{o} \in \Gamma(TM)$ where $\mathfrak{o}_{|p}$ is the zero vector in M_p.

Note that $\mathfrak{o}$ is an imbedding of M in TM.

Theorem I.3.2. *Let $\mathcal{T}M$ be the subset of TM defined by*

$$\mathcal{T}M = \left\{\xi \in TM : 1 \in I_\xi\right\}.$$

Then $\mathcal{T}M$ is open and is starlike with respect to $\mathfrak{o} \in \Gamma(TM)$, that is, for any $\xi \in \mathcal{T}M$ we have $\alpha\xi \in \mathcal{T}M$ for all $\alpha \in [0, 1]$.

If we define $\exp : \mathcal{T}M \to M$ *by*

$$\exp \xi = \gamma_\xi(1),$$

then exp *is differentiable, and* exp *has maximal rank on $\mathfrak{o}(M) \subseteq TM$.*

If for any given $p \in M$ we define $\exp_p : M_p \cap \mathcal{T}M \to M$ *by*

$$\exp_p = \exp | M_p \cap \mathcal{T}M,$$

then $\exp_p$ *is differentiable, and* $\exp_p$ *has maximal rank at $0 \in M_p$.*

Finally, the map $\pi \times \exp : \mathcal{T}M \to M \times M$ given by

$$(\pi \times \exp)(\xi) = (\pi(\xi), \exp \xi)$$

is differentiable and of maximal rank ($=$ 2dim M) on $\mathfrak{o}(M) \subseteq TM$. Thus, there exists a neighborhood W of $\mathfrak{o}(M)$ in TM such that $(\pi \times \exp)|W$ is a diffeomorphism of W onto its image, an open subset of $M \times M$.

Proof. That $\mathcal{T}M$ is open is a consequence of the fact that the domain of the maximal flow is open. That $\mathcal{T}M$ is starlike follows from (I.3.6).

We shall calculate $(\pi \times \exp)_*$ on $\mathfrak{o}(M)$, since this calculation contains that of $\exp_*$ on $\mathfrak{o}(M)$, and $(\exp_p)_{*|0}$.

Fix $p \in M$, and a chart $x : U \to \mathbb{R}^n$. On $\pi^{-1}[U]$ we have the chart $Q : \pi^{-1}[U] \to \mathbb{R}^n \times \mathbb{R}^n$ as discussed previously; and on $U \times U$ we have the chart $y : U \times U \to \mathbb{R}^n \times \mathbb{R}^n$ given by $y(p_1, p_2) = (x(p_1), x(p_2))$. Then

$(\pi \times \exp)^{-1}[U \times U]$ is open in TM and contains $\mathfrak{o}(U)$. So our goal is to calculate the rank of the Jacobian matrix of

$$z := y \circ (\pi \times \exp) \circ Q^{-1} : Q(\pi^{-1}[U]) \to \mathbb{R}^{2n}$$

on $Q(\mathfrak{o}(U))$.

Let $\pi_1 : \mathbb{R}^n \times \mathbb{R}^n \to \mathbb{R}^n$ denote the projection onto the first factor. Then,

$$x \circ \pi \circ Q^{-1} = \pi_1;$$

so it remains to calculate $(\exp_p)_{*|0}$.

For the calculation that follows, it will be more convenient to denote the origin of M_p by o, and the origin of $\mathbb{R}$ by 0.

Let $\Im_o : M_p \to (M_p)_o$ be the canonical identification of M_p with *its* tangent space at the origin o. We shall show that

$$(\exp_p)_{*|o} \circ \Im_o = \mathrm{id}_{M_p}.$$

Indeed, given $\xi \in M_p$ consider the linear path ω in M_p given by $\omega(t) = t\xi$. Then

$$\omega'(0) = \Im_o \xi = \omega_{*|0} \frac{\partial}{\partial t},$$

which implies

$$\begin{aligned}
(\exp)_{*|o} \circ \Im_o \xi &= (\exp)_{*|o} \circ \omega_{*|0} \frac{\partial}{\partial t} \\
&= (\exp \circ \omega)_{*|0} \frac{\partial}{\partial t} \\
&= \left\{ \frac{d}{dt}((\exp \circ \omega)(t)) \right\}_{t=0} \\
&= \gamma_\xi{}'(0) \\
&= \xi,
\end{aligned}$$

which is the claim.

The final statement of the theorem follows from the calculation just done, and the following fact:

If M, N are Hausdorff spaces, M locally compact with countable basis, $\phi : M \to N$ a local homeomorphism, A closed in M, and $\phi|A$ is one-to-one, then there exists a neighborhood V of A such that $\phi|V$ is a homeomorphism. ∎

Definition. We refer to $\exp : TM \to M$ as the *exponential map*.

§I.4. The Torsion and Curvature Tensors

M is our differentiable manifold with connection ∇.

Definition. For any $p \in M$, we shall define the multilinear maps

$$T : M_p \times M_p \to M_p, \quad R : M_p \times M_p \times M_p \to M_p,$$

henceforth referred to as the *torsion* and *curvature tensors of* ∇, respectively, as follows: For $\xi, \eta, \zeta \in M_p$, let X, Y, Z be extensions of ξ, η, ζ, respectively, to vector fields on a neighborhood of p. Then define

$$T(\xi, \eta) = \nabla_\eta X - \nabla_\xi Y - [Y, X]_{|p}, \tag{I.4.1}$$

$$R(\xi, \eta)\zeta = \nabla_\eta \nabla_X Z - \nabla_\xi \nabla_Y Z - \nabla_{[Y,X]_{|p}} Z. \tag{I.4.2}$$

In the above, and in all that follows, [,] denotes the Lie bracket of vector fields.

To show that T and R are well-defined (i.e., they are independent of the extensions of ξ, η, ζ), it is best to change the point of view and consider T and R as defined on the vector fields X, Y, Z. We first show that T and R are multilinear (with respect to X, Y, Z) over functions on M, that is, for example,

$$T(X, Y) = \nabla_Y X - \nabla_X Y - [Y, X]$$

satisfies

$$T\left(\sum_{i=1}^{2} f_i X_i, \sum_{j=1}^{2} g_j Y_j\right) = \sum_{i,j=1}^{2} f_i g_j T(X_i, Y_j).$$

Certainly, $T(X, Y) = -T(Y, X)$; so we only have to show that T is linear over functions with respect to the first variable. Clearly, $T(X, Y)$ is additive with respect to X, and

$$\begin{aligned} T(fX, Y) &= \nabla_Y(fX) - \nabla_{fX} Y - [Y, fX] \\ &= (Yf)X + f\nabla_Y X - f\nabla_X Y - \{(Yf)X + f[Y, X]\} \\ &= fT(X, Y). \end{aligned}$$

So T is linear in X, hence multilinear in X and Y, over functions. But then, in any coordinate chart $x : U \to \mathbb{R}^n$, we have for

$$X = \sum_j \xi^j \frac{\partial}{\partial x^j}, \qquad Y = \sum_k \eta^k \frac{\partial}{\partial x^k},$$

the calculation

$$T(X, Y) = \sum_{j,k} \xi^j \eta^k T\left(\frac{\partial}{\partial x^j}, \frac{\partial}{\partial x^k}\right).$$

So $T(X, Y)_{|p}$ is determined exclusively by the values of $X_{|p}$ and $Y_{|p}$. The discussion for the curvature tensor field R is similar.

As mentioned, one has

$$T(\xi, \eta) + T(\eta, \xi) = 0. \tag{I.4.3}$$

Also,

$$R(\xi, \eta)\zeta + R(\eta, \xi)\zeta = 0. \tag{I.4.4}$$

Remark I.4.1. The above equations (I.4.3) and (I.4.4) are obvious consequences of the definitions (I.4.1) and (I.4.2). For more involved calculations, such as (I.4.5) immediately below, it helps to note that, for any chart on M, the Lie bracket of coordinate vector fields associated with the chart vanishes identically on its domain. Since for calculating identities for tensor fields it suffices to verify the results for coordinate vector fields of a chart, one has a considerable simplification of the resulting calculations, by verifying the identities for coordinate vector fields of a chart.

If we are given that $T = 0$ on all of M then direct calculation yields the *first Bianchi identity*

$$R(\xi, \eta)\zeta + R(\zeta, \xi)\eta + R(\eta, \zeta)\xi = 0. \tag{I.4.5}$$

To prove (I.4.5), it suffices to check that, in any chart $x : U \to \mathbb{R}^n$, we have

$$R(\partial_j, \partial_k)\partial_\ell + R(\partial_\ell, \partial_j)\partial_k + R(\partial_k, \partial_\ell)\partial_j = 0.$$

Well, since $T = 0$, we have $\nabla_{\partial_j}\partial_k = \nabla_{\partial_k}\partial_j$ for all j, k, which implies

$$\begin{aligned}
&R(\partial_j, \partial_k)\partial_\ell + R(\partial_\ell, \partial_j)\partial_k + R(\partial_k, \partial_\ell)\partial_j \\
&\quad = \nabla_{\partial_k}\nabla_{\partial_j}\partial_\ell - \nabla_{\partial_j}\nabla_{\partial_k}\partial_\ell + \nabla_{\partial_j}\nabla_{\partial_\ell}\partial_k - \nabla_{\partial_\ell}\nabla_{\partial_j}\partial_k + \nabla_{\partial_\ell}\nabla_{\partial_k}\partial_j - \nabla_{\partial_k}\nabla_{\partial_\ell}\partial_j \\
&\quad = \nabla_{\partial_k}(\nabla_{\partial_j}\partial_\ell - \nabla_{\partial_\ell}\partial_j) + \nabla_{\partial_j}(\nabla_{\partial_\ell}\partial_k - \nabla_{\partial_k}\partial_\ell) + \nabla_{\partial_\ell}(\nabla_{\partial_k}\partial_j - \nabla_{\partial_j}\partial_k) \\
&\quad = 0.
\end{aligned}$$

We leave to the reader to verify that if $x : u \to \mathbb{R}^n$ is a chart on M, and the functions ${\Gamma_{jk}}^\ell : U \to \mathbb{R}$, $j, k, \ell = 1, \ldots, n$, are given by (I.1.3), then for

$$T(\partial_j, \partial_k) := \sum_\ell {T_{jk}}^\ell \partial_\ell$$

we have

$$T_{jk}{}^{\ell} = \Gamma_{jk}{}^{\ell} - \Gamma_{kj}{}^{\ell},$$

and for

$$R(\partial_j, \partial_k)\partial_i := \sum_{\ell} R_{ijk}{}^{\ell}\partial_\ell$$

we have

$$R_{ijk}{}^{\ell} = \partial_k\Gamma_{ij}{}^{\ell} - \partial_j\Gamma_{ik}{}^{\ell} + \sum_r \left\{\Gamma_{ij}{}^{r}\Gamma_{rk}{}^{\ell} - \Gamma_{ik}{}^{r}\Gamma_{rj}{}^{\ell}\right\}. \tag{I.4.6}$$

Both the torsion and curvature tensors appear in *Jacobi's equations of geodesic deviation*, that is, the linearized geodesic equations: Let $\epsilon_0 > 0$, $v : (\alpha, \beta) \times (-\epsilon_0, \epsilon_0) \to M$ be differentiable such that for each ϵ in $(-\epsilon_0, \epsilon_0)$ the path $\omega_\epsilon : (\alpha, \beta) \to M$ given by $\omega_\epsilon(t) = v(t, \epsilon)$ is a geodesic. The coordinate vector fields along v will be written as

$$\partial_t v = v_* \partial_t, \qquad \partial_\epsilon v = v_* \partial_\epsilon,$$

and differentiation of vector fields along v with respect to the directions ∂_t, ∂_ϵ by ∇_t, ∇_ϵ respectively. One has, since $[\partial_t, \partial_\epsilon] = 0$,

$$\nabla_\epsilon \partial_t v - \nabla_t \partial_\epsilon v = T(\partial_t v, \partial_\epsilon v), \tag{I.4.7}$$

$$\nabla_\epsilon \nabla_t - \nabla_t \nabla_\epsilon = R(\partial_t v, \partial_\epsilon v). \tag{I.4.8}$$

Theorem I.4.1. (C. F. Jacobi (1836)) *Given the above, we have*

$$0 = \nabla_t{}^2 \partial_\epsilon v + \nabla_t(T(\partial_t v, \partial_\epsilon v)) + R(\partial_t v, \partial_\epsilon v)\partial_t v. \tag{I.4.9}$$

Proof. We are given on all of $(\alpha, \beta) \times (-\epsilon_0, \epsilon_0)$ that

$$0 = \nabla_t \partial_t v,$$

which implies

$$\begin{aligned} 0 &= \nabla_\epsilon \nabla_t \partial_t v \\ &= \nabla_t \nabla_\epsilon \partial_t v + R(\partial_t v, \partial_\epsilon v)\partial_t v \\ &= \nabla_t \{\nabla_t \partial_\epsilon v + T(\partial_t v, \partial_\epsilon v)\} + R(\partial_t v, \partial_\epsilon v)\partial_t v \\ &= \nabla_t{}^2 \partial_\epsilon v + \nabla_t(T(\partial_t v, \partial_\epsilon v)) + R(\partial_t v, \partial_\epsilon v)\partial_t v, \end{aligned}$$

which yields the claim. ■

§I.5. Riemannian Metrics

Definition. Given a differentiable manifold M, define a *Riemannian metric* g *on* M to be a mapping that associates with each $p \in M$ an inner product $g_p : M_p \times M_p \to \mathbb{R}$ satisfying the following differentiability property: If U is any open set in M and X, Y are differentiable vector fields on U, then the function $g(X, Y) : U \to \mathbb{R}$ given by

$$g(X, Y)(p) = g_p(X_{|p}, Y_{|p})$$

is differentiable on U.

By a *Riemannian manifold* we mean a differentiable manifold with a given Riemannian metric.

Definition. Let M, N be differentiable manifolds, h a Riemannian metric on N, $\phi : M \to N$ differentiable, and $M_0 = \{p \in M : \phi_{*|p} \text{ is one-to-one}\}$. Of course M_0 is a, possibly empty, open submanifold of M. The *pull-back* ϕ^*h *of* h is defined to be the Riemannian metric on M_0 given by

$$(\phi^*h)(\xi, \eta) = h(\phi_*\xi, \phi_*\eta) \tag{I.5.1}$$

where $\xi, \eta \in M_p$, $p \in M_0$.

If $p \in M\backslash M_0$, then (I.5.1) defines a symmetric bilinear form on M_p, but the form is only nonnegative.

Should M also be a Riemannian manifold with Riemannian metric g, then we say that ϕ is a *local isometry of* M_0 *into* N if $g = \phi^*h$ on M_0. If M is connected, then $g = \phi^*h$ also implies that $M_0 = M$, that is, ϕ is a *Riemannian immersion*. If ϕ is an imbedding satisfying $g = \phi^*h$ then we call ϕ an *isometry of* M *into* N.

An *isometry of* M is a diffeomorphism of M onto itself that is an isometry.

When there is only one Riemannian metric under consideration, we usually write $\langle\ ,\ \rangle$ for $g(\ ,\ )$.

Theorem I.5.1. (T. Levi-Civita (1929)) *If M is a Riemannian manifold, then there exists a unique connection ∇ (henceforth called the* Levi-Civita connection) *for which*

$$\nabla_X Y = \nabla_Y X + [X, Y], \tag{I.5.2}$$

$$X\langle Y, Z\rangle = \langle\nabla_X Y, Z\rangle + \langle Y, \nabla_X Z\rangle, \tag{I.5.3}$$

for all differentiable vector fields $X, Y, Z \in \Gamma(TM)$.

Proof. Since at each point the inner product is a nondegenerate bilinear form, to calculate $\nabla_X Y$ it suffices to calculate $\langle \nabla_X Y, Z\rangle$ for any $X, Y, Z \in \Gamma(TM)$. To this end we have, using (I.5.2), (I.5.3) alternatively,

$$\begin{aligned}
\langle \nabla_X Y, Z\rangle &= X\langle Y, Z\rangle - \langle Y, \nabla_X Z\rangle \\
&= X\langle Y, Z\rangle - \langle Y, \nabla_Z X\rangle - \langle Y, [X, Z]\rangle \\
&= X\langle Y, Z\rangle - Z\langle Y, X\rangle + \langle \nabla_Z Y, X\rangle - \langle Y, [X, Z]\rangle \\
&= X\langle Y, Z\rangle - Z\langle Y, X\rangle + \langle \nabla_Y Z, X\rangle + \langle [Z, Y], X\rangle - \langle Y, [X, Z]\rangle \\
&= X\langle Y, Z\rangle - Z\langle Y, X\rangle + Y\langle Z, X\rangle - \langle Z, \nabla_Y X\rangle + \langle [Z, Y], X\rangle \\
&\quad - \langle Y, [X, Z]\rangle \\
&= X\langle Y, Z\rangle - Z\langle Y, X\rangle + Y\langle Z, X\rangle - \langle Z, \nabla_X Y\rangle \\
&\quad - \langle Z, [Y, X]\rangle + \langle [Z, Y], X\rangle - \langle Y, [X, Z]\rangle.
\end{aligned}$$

Therefore, we have

$$\begin{aligned}
\text{(I.5.4)} \qquad \langle \nabla_X Y, Z\rangle = (1/2)\{&X\langle Y, Z\rangle + Y\langle Z, X\rangle - Z\langle X, Y\rangle \\
&- \langle X, [Y, Z]\rangle - \langle Y, [X, Z]\rangle - \langle Z, [Y, X]\rangle\},
\end{aligned}$$

that is, if we are given the restrictions (I.5.2), (I.5.3), then we have the explicit calculation of $\langle \nabla_X Y, Z\rangle$ – thus $\nabla_X Y$ is uniquely determined. To establish the existence of ∇, one takes (I.5.4) to define ∇, and verifies directly that ∇ indeed defines a connection satisfying (I.5.2), (I.5.3). ■

Remark I.5.1. Note that (I.5.2) says that the torsion tensor of the Levi–Civita connection vanishes on all of M.

Remark I.5.2. Before proceeding with the further development of ideas and results, it is worth illustrating the point made in Remark I.2.1, namely, that parallel translation of vectors along paths might very well depend on the choice of path, equivalently, that parallel translation around a closed path might have distinct initial and terminal vectors. Consider, for simplicity, the 2–sphere $\mathbb{S}^2$ of radius 1 in $\mathbb{R}^3$. Its standard metric is induced by restricting the standard metric of $\mathbb{R}^3$ to the tangent bundle of $\mathbb{S}^2$. We shall show in §II.3 that the geodesics of $\mathbb{S}^2$ are the "great circles" of the sphere, that is, those circles obtained as the intersection of $\mathbb{S}^2$ with any 2–plane in $\mathbb{R}^3$ passing through the center of $\mathbb{S}^2$.

To illustrate parallel translation (see Figure I.1), consider a unit tangent vector ξ at the north pole P. Consider the geodesic γ_ξ determined by ξ, and parallel translate ξ along γ_ξ to obtain $\eta = {\gamma_\xi}'(\pi/2)$ on the equator. Rotate η through $\pi/2$ radians to obtain the unit tangent vector ζ tangent to the equator. Parallel translate η along γ_ζ (i.e., along the equator) to obtain the tangent vector σ_α at

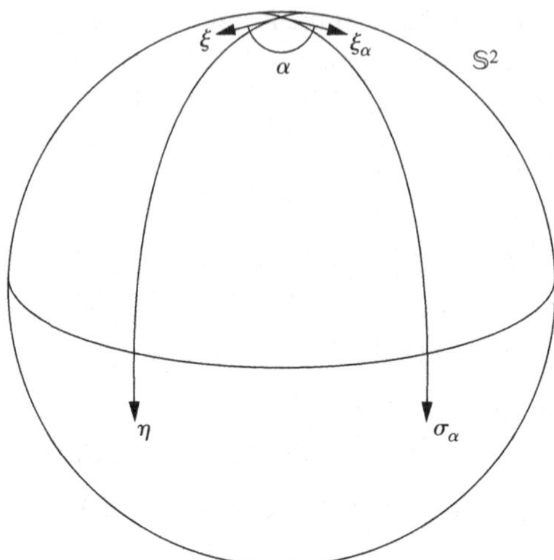

Figure I.1. Parallel translation on $\mathbb{S}^2$.

$\gamma_\zeta(\alpha)$. Since (I.5.3) implies that lengths and angles are preserved by parallel translation, then σ_α is a unit tangent vector, perpendicular to $\gamma_\zeta{}'(\alpha)$, and pointing toward the south pole. Finally, parallel translate σ_α along $\gamma_{-\sigma_\alpha}$, back to the unit tangent vector $\xi_\alpha = -\gamma_{-\sigma_\alpha}{}'(\pi/2)$ at the north pole P. Then, for $\alpha \in (0, 2\pi)$, we certainly have $\xi_\alpha \neq \xi$.

Let $x : U \to \mathbb{R}^n$ be a chart on the Riemannian manifold M. Then, for each $p \in U$, the matrix $G(p)$ given by

(I.5.5) $$G(p) = (g_{ij}(p)), \quad g_{ij}(p) = \langle \partial_i|_p, \partial_j|_p \rangle,$$

is positive definite symmetric, and the functions $g_{ij} : U \to \mathbb{R}$, $i, j = 1, \ldots, n$, are C^∞ on U. Conversely, given any n–dimensional manifold M, and a chart $x : U \to \mathbb{R}^n$ on M, one determines a local Riemannian metric on U by specifying a C^∞ function from U to the space of $n \times n$ positive definite symmetric matrices. Since the $n \times n$ positive definite symmetric matrices form a cone in the space of all $n \times n$ matrices, one may use partitions of unity to pass from the existence of local Riemannian metrics to global ones.

Given the matrix function G on U, we denote the inverse G^{-1} by

(I.5.6) $$G^{-1} = (g^{ij}).$$

Let ∇ be the Levi–Civita connection of the Riemannian metric, and $\Gamma_{jk}{}^\ell$ given by (I.1.3). Then (I.5.2) becomes

(I.5.7) $$\Gamma_{jk}{}^\ell = \Gamma_{kj}{}^\ell,$$

and (I.5.4) becomes

$$\Gamma_{jk}{}^{\ell} = \frac{1}{2}\sum_{r} g^{\ell r}\left\{\partial_j g_{rk} + \partial_k g_{jr} - \partial_r g_{jk}\right\}. \tag{I.5.8}$$

Finally, we note that, for formal calculations, one uses the classical expression for the Riemannian metric

$$ds^2 = \sum_{j,k} g_{jk}(x)\,dx^j dx^k,$$

which on $\mathbb{R}^n$ becomes

$$ds^2 = |dx|^2.$$

§I.6. The Metric Space Structure

Let M, N be differentiable manifolds and A an arbitrary set in M. Recall that a map $\phi : A \to N$ is C^k on A, $k \geq 1$, if there exist an open set U such that $A \subseteq U \subseteq M$ and a map $\hat{\phi} : U \to N \in C^k$ satisfying $\hat{\phi}|A = \phi$.

Definition. For a given differentiable manifold M, $k = 1, \ldots, \infty$, we let D^k denote the collection of all maps ω from closed intervals of $\mathbb{R}$ into M that are continuous and piecewise C^k, that is, ω is given by $\omega : [\alpha, \beta] \to M \in C^0$ and there exist $\alpha = t_0 < t_1 < \ldots < t_\ell = \beta$ such that $\omega|[t_{j-1}, t_j] \in C^k$ for $j = 1, \ldots, \ell$.

Let M be a Riemannian manifold. For any $\xi \in TM$, define the *length of* ξ, $|\xi|$, by

$$|\xi| = \langle \xi, \xi \rangle^{1/2}.$$

For any path $\omega : [\alpha, \beta] \to M \in D^1$ define the *length of* ω, $\ell(\omega)$ by

$$\ell(\omega) = \int_\alpha^\beta |\omega'(t)|\,dt.$$

For M connected (our usual assumption), $p, q \in M$, define the *distance between* p *and* q, $d(p, q)$, by

$$d(p, q) = \inf_\omega \ell(\omega),$$

where ω ranges over all $\omega : [\alpha, \beta] \to M \in D^1$ satisfying $\omega(\alpha) = p, \omega(\beta) = q$. One immediately verifies that

$$\begin{aligned} &d(p, q) = d(q, p),\\ &d(p, q) \geq 0,\\ &d(p, q) \leq d(p, r) + d(r, q), \end{aligned}$$

for all $p, q, r \in M$. Thus, to show that $d(\ ,\ )$ turns M into a metric space, it remains to show that if p and q are distinct points of M, then $d(p, q) > 0$. We give two proofs of this fact. The first is short – it reduces the problem to that of Euclidean space, which we may take as known. The second approach is far more detailed, in that it recaptures those features of the Euclidean case that are pertinent to the matter.

The first proof goes as follows (see Hopf–Rinow (1931, p. 213)): Assume M is n–dimensional, and (henceforth) let $\mathbb{B}(x; r)$ denote the open ball in $\mathbb{R}^n$, centered at x, with radius r. Given $p \in M$, let $x : U \to \mathbb{R}^n$ be a chart on M, $p \in U$. Then there exists $r > 0$ for which

$$\mathbb{B}(x(p); r) \subset\subset x(U),$$

which determines the existence of a constant $\lambda > 0$ such that, for all $\xi \in T(x^{-1}[\mathbb{B}(x(p); r)])$,

$$\xi = \sum_j \xi^j \frac{\partial}{\partial x^j},$$

we have

$$|\xi| \geq \lambda \sqrt{\sum_j (\xi^j)^2}.$$

So, on $\mathbb{B}(x(p); r)$, the Riemannian lengths are uniformly bounded below by the corresponding Euclidean lengths. Therefore, for $q \in x^{-1}[\mathbb{B}(x(p); r)]$, we have

$$d(p, q) \geq \lambda |x(p) - x(q)| > 0.$$

For $q \in M \setminus x^{-1}[\mathbb{B}(x(p); r)]$, we obviously have

$$d(p, q) \geq \lambda r.$$

So $p \neq q$ implies $d(p, q) > 0$.

For the second proof, recall that, for any $\xi \in TM$, the path

$$\gamma_\xi(t) = \exp t\xi \tag{I.6.1}$$

is the unique geodesic in M satisfying

$$\gamma_\xi(0) = \pi(\xi), \qquad \gamma_\xi{}'(0) = \xi, \tag{I.6.2}$$

where $\pi : TM \to M$ is the projection map. Note that an immediate consequence of (I.5.3), extended to differentiation of vector fields along paths, is that

$$|\gamma_\xi{}'(t)| = |\xi| \tag{I.6.3}$$

for all t for which γ_ξ is defined.

Also recall that if $\mathfrak{o} : M \to TM$ is the zero section of M in TM, and $\mathcal{T}M$ is the domain of the exponential map exp, then by Theorem I.3.2 there exists an open set W in TM such that $\mathfrak{o}(M) \subseteq W \subseteq \mathcal{T}M$, and $\pi \times \exp : W \to M \times M$ is a diffeomorphism of W onto its image, an open subset of $M \times M$.

Definition. For any $q \in M$, we let

$$\mathsf{B}(q;\epsilon) = \{\xi \in M_q : |\xi| < \epsilon\}, \qquad \mathsf{B}_q = \mathsf{B}(q;1),$$

and for any $V \subseteq M$ we write

$$\mathsf{B}(V;\epsilon) = \bigcup_{q\in V} \mathsf{B}(q;\epsilon).$$

Similarly, we define

$$\mathsf{S}(q;\epsilon) = \{\xi \in M_q : |\xi| = \epsilon\}, \qquad \mathsf{S}_q = \mathsf{S}(q;1),$$
$$\mathsf{S}(V;\epsilon) = \bigcup_{q\in V} \mathsf{S}(q;\epsilon).$$

Theorem I.6.1. *For each $p \in M$, there exists $\epsilon > 0$ and a neighborhood U of p in M such that*

(i) *any two points of U are joined by a unique geodesic in M of length $< \epsilon$;*
(ii) *the geodesic depends differentiably on its endpoints; and*
(iii) *for each $q \in U$, $\exp_q$ maps $\mathsf{B}(q;\epsilon)$ diffeomorphically onto an open set in M.*

Proof. Let W be the open set in $\mathcal{T}M$ described above. Then, for any $p \in M$, there exists a neighborhood V of p in M and an $\epsilon > 0$ such that $\mathsf{B}(V;\epsilon) \subseteq W$. Next, there exists an open neighborhood U of p in M such that $U \times U \subseteq (\pi \times \exp)(\mathsf{B}(V;\epsilon))$. One now checks that U and ϵ will do the job. ■

Theorem I.6.2. *Let $\epsilon > 0$ and U be given as in Theorem* I.6.1, *$p,q \in U$, and $\gamma : [0,1] \to M$ the geodesic of length less than ϵ satisfying $\gamma(0) = p$, $\gamma(1) = q$. Let $\omega : [0,1] \to M \in D^1$ be any path satisfying $\omega(0) = p$, $\omega(1) = q$. Then*

$$\ell(\gamma) \le \ell(\omega) \tag{I.6.4}$$

with equality only if $\gamma([0,1]) = \omega([0,1])$.

Proof. We first require the following:

Lemma I.6.1. (Gauss's lemma (1825, p. 107), (1827, p. 24)) *Let $p \in M$ and $\mathsf{B}(p;\delta_0) \subseteq \mathcal{T}M$. Then, for any $t \in (0,\delta_0)$, $\xi \in \mathsf{S}_p$, and $\zeta \in (\mathsf{S}(p;t))_{t\xi}$, we have*

$$\langle \gamma_\xi{}'(t), (\exp_p)_{*|t\xi}\zeta\rangle = 0. \tag{I.6.5}$$

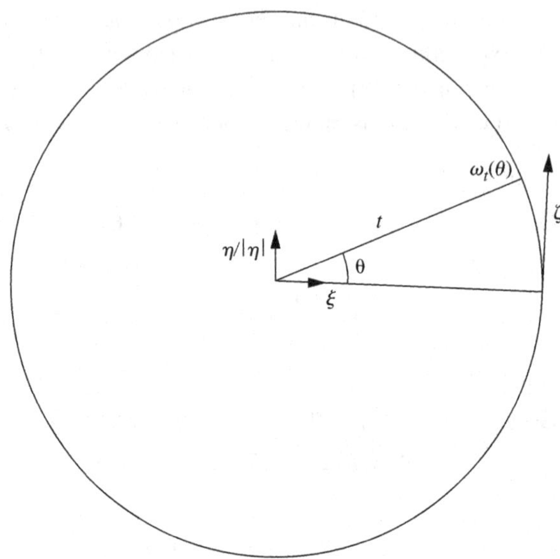

Figure I.2. The geodesic variation in Gauss' lemma.

Proof. Fix $\xi \in \mathsf{S}_p$, and let

$$\xi^{\perp} = \{\eta \in M_p : \langle \eta, \xi \rangle = 0\}.$$

Recall that $\mathfrak{I}_{t\xi} : M_p \to (M_p)_{t\xi}$ denotes the canonical isomorphism. For fixed $\eta \in \xi^{\perp}$, $t > 0$, consider the path in M_p given by

$$\omega_t(\theta) = t\{(\cos|\eta|\theta)\xi + (\sin|\eta|\theta)\eta/|\eta|\}.$$

(See Figure I.2.) Then,

$$\omega_t(0) = t\xi, \qquad \omega_t{}'(0) = t\mathfrak{I}_{t\xi}\eta \in (\mathsf{S}(p;t))_{t\xi}.$$

Thus, the map

$$\eta \mapsto t\mathfrak{I}_{t\xi}\eta$$

is an isomorphism of $\xi^{\perp}$ onto $(\mathsf{S}(p;t))_{t\xi}$.

In particular, if $\zeta \in (\mathsf{S}(p;t))_{t\xi}$, let $\eta \in \xi^{\perp}$ be given by

$$\eta = t^{-1}\mathfrak{I}_{t\xi}{}^{-1}\zeta,$$

and consider

$$v(t,\theta) = \exp \omega_t(\theta);$$

then, for $\partial_t v = v_*\partial_t$, $\partial_\theta v = v_*\partial_\theta$, we have

$$\partial_t v(t,0) = \gamma_\xi{}'(t), \qquad \partial_\theta v(t,0) = (\exp)_{*|t\xi}\zeta.$$

Thus, (I.6.5) is equivalent to showing

$$\langle \partial_t v, \partial_\theta v\rangle = 0. \tag{I.6.6}$$

First note that

$$|\partial_t v| = 1, \qquad \nabla_t \partial_t v = 0,$$

and that by (I.5.2), we have

$$\nabla_t \partial_\theta v = \nabla_\theta \partial_t v.$$

Thus, by (I.5.3), (I.6.3),

$$\begin{aligned}
\partial_t \langle \partial_t v, \partial_\theta v\rangle &= \langle \nabla_t \partial_t v, \partial_\theta v\rangle + \langle \partial_t v, \nabla_t \partial_\theta v\rangle \\
&= \langle \partial_t v, \nabla_t \partial_\theta v\rangle \\
&= \langle \partial_t v, \nabla_\theta \partial_t v\rangle \\
&= (1/2)\partial_\theta \langle \partial_t v, \partial_t v\rangle \\
&= 0.
\end{aligned}$$

So, for each fixed θ, $\langle \partial_t v, \partial_\theta v\rangle$ is constant along the geodesic determined by the unit vector $t^{-1}\omega_t(\theta)$. To evaluate the constant at $t = 0$, we have $\partial_\theta v(0,\theta) = 0$ for all θ. Thus, (I.6.6) is valid for all (t,θ). ■

Lemma I.6.2. *Let $\sigma : [\alpha, \beta] \to U\backslash\{p\} \in D^1$. Then, $\sigma(\tau)$ may be written as*

$$\sigma(\tau) = \exp t(\tau)\xi(\tau),$$

where $t : [\alpha, \beta] \to (0, \epsilon)$, $\xi : [\alpha, \beta] \to \mathsf{S}_p$. We then have

$$\ell(\sigma) \geq |t(\beta) - t(\alpha)|, \tag{I.6.7}$$

with equality if and only if $t(\tau)$ is monotone and $\xi(\tau)$ is constant.

Proof. The functions $t(\tau)$, $\xi(\tau)$ are defined by setting

$$\begin{aligned}
t(\tau) &= |(\exp|\mathsf{B}(p;\epsilon))^{-1}\sigma(\tau)|, \\
\xi(\tau) &= (t(\tau))^{-1}(\exp|\mathsf{B}(p;\epsilon))^{-1}\sigma(\tau).
\end{aligned}$$

To verify (I.6.7), introduce *geodesic spherical coordinates about p* by defining $V : [0,\epsilon) \times \mathsf{S}_p \to M$ by

$$V(t,\xi) = \exp t\xi.$$

As usual, set $\partial_t V = V_*(\partial_t)$ and $(\partial_\xi V)\eta = V_*(\eta)$ for $\eta \in (\mathsf{S}_p)_\xi$. Then, for $(t, \xi) \in [0, \epsilon) \times \mathsf{S}_p$, we have

$$\partial_\xi V(t, \xi) = t(\exp_p)_{*|t\xi} \mathfrak{I}_{t\xi},$$

as we argued previously in the proof of Gauss's lemma. Since $\exp |\mathsf{B}(p; \epsilon)$ is a diffeomorphism, so is $V|(0, \epsilon) \times \mathsf{S}_p$.

We now have

$$\sigma(\tau) = V(t(\tau), \xi(\tau)),$$

$$\sigma' = t'\partial_t V + (\partial_\xi V)\xi'.$$

But (I.6.3), (I.6.6) imply

$$|\sigma'|^2 = (t')^2 + |(\partial_\xi V)\xi'|^2 \geq (t')^2.$$

Thus,

$$\ell(\sigma) = \int_\alpha^\beta |\sigma'| \geq \int_\alpha^\beta |t'| \geq |\int_\alpha^\beta t'| = |t(\beta) - t(\alpha)|,$$

which is the inequality (I.6.7). The case of equality is handled easily. ∎

Proof of Theorem I.6.2. Pick $\epsilon_0 > 0$ sufficiently small so that $(\epsilon_0 + 1)\ell(\gamma) < \epsilon$; then, γ can be extended to a geodesic $\widehat{\gamma} : [-\epsilon_0, 1] \to U$. Then, $\ell(\widehat{\gamma}) = (\epsilon_0 + 1)\ell(\gamma)$; also, ϵ, U satisfy Theorem I.6.1 with respect to $\widehat{\gamma}(-\epsilon_0)$.

Let G be the image under exp of the open annulus in $M_{\widehat{\gamma}(-\epsilon_0)}$ given by

$$\mathsf{B}(\widehat{\gamma}(-\epsilon_0); \ell(\widehat{\gamma})) \backslash \overline{\mathsf{B}}(\widehat{\gamma}(-\epsilon_0); \epsilon_0),$$

that is,

$$G = \exp \mathsf{B}(\widehat{\gamma}(-\epsilon_0); \ell(\widehat{\gamma})) \backslash \overline{\mathsf{B}}(\widehat{\gamma}(-\epsilon_0); \epsilon_0).$$

Then, $\{\tau : \omega(\tau) \in G\}$ is open in $\mathbb{R}$ and is given by a countable disjoint union of open intervals $\{(\alpha_j, \beta_j) : j = 1, \ldots\}$. Applying Lemma I.6.2 to each $[\alpha_j, \beta_j]$, we have

$$\ell(\omega) \geq \sum_j \ell(\omega|[\alpha_j, \beta_j]) \geq \sum_j |t_j(\beta_j) - t_j(\alpha_j)|,$$

where t_j is the function $t(\tau)$, above, for each $\omega|[\alpha_j, \beta_j]$. Each summand in the last term is either equal to 0 or $\ell(\gamma)$ and, at least one of the summands is $\ell(\gamma)$.

This implies the inequality (I.6.4). Equality in (I.6.4) implies there is precisely one interval in G, and the image of ω is that of γ by Lemma I.6.2. ■

Corollary I.6.1. *If $p \neq q$, then $d(p, q) > 0$. Thus, $d : M \times M \to \mathbb{R}$ given above turns M into a metric space. For $p \in M$ and $\delta > 0$, we write*

$$B(p; \delta) := \{q \in M : d(p, q) < \delta\},$$
$$S(p; \delta) := \{q \in M : d(p, q) = \delta\}.$$

Then, for a given $p \in M$ and $\epsilon > 0$ determined in Theorem I.6.1, *we have for all $\delta \in (0, \epsilon)$*

$$B(p; \delta) = \exp \mathsf{B}(p; \delta),$$
$$S(p; \delta) = \exp \mathsf{S}(p; \delta)$$

diffeomorphically – thus the metric space topology coincides with the topology of M possessed by the differentiable structure.

Corollary I.6.2. *Let $\omega : [0, \ell] \to M \in D^1$, $|\omega'| = 1$ when ω' exists, have the property that $\ell = d(\omega(0), \omega(\ell))$. Then, ω, is a geodesic.*

Proof. First, note that $\ell = d(\omega(0), \omega(\ell))$ implies that $\delta = d(\omega(\tau), \omega(\tau + \delta))$ for all $\tau, \tau + \delta \in [0, \ell]$. Otherwise, the triangle inequality would imply $\ell < d(\omega(0), \omega(\ell))$.

Next, it suffices to show that, for every $\tau \in (0, \ell)$, there exists $\delta > 0$ such that $\omega|[\tau, \tau + \delta]$ is a geodesic. To prove the existence of δ, given $\omega(\tau)$, let ϵ be given by Theorem I.6.1 for $p = \omega(\tau)$, and set $\delta = \min\{\epsilon/2, \ell - \tau\}$. Then, there exists a unique geodesic $\gamma : [\tau, \tau + \delta] \to M$, $|\gamma'| = 1$, from $p = \omega(\tau) = \gamma(\tau)$ to $\gamma(\tau + \delta) = \omega(\tau + \delta)$ of length δ. This implies $\delta = \ell(\omega|[\tau, \tau + \delta]) \geq \ell(\gamma|[\tau, \tau + \delta]) = \delta$. Therefore, $\omega|[\tau, \tau + \delta] = \gamma|[\tau, \tau + \delta]$. Since both paths are parametrized with respect to arc length based at p, they must be identical. ■

Corollary I.6.3. *Given a compact set $K \subseteq M$, there exists a $\delta > 0$ so that any two points of K with the distance less than δ are joined by a unique geodesic of length less than δ. This geodesic minimizes distance between the two points and depends differentiably on the endpoints.*

Proof. To each point $p \in K$, construct $U = U(p)$ and $\epsilon = \epsilon(p)$ as in Theorem I.6.1 – in particular, $B(p; \epsilon(p)) \supseteq U(p)$ for all $p \in K$. If our corollary is false, then to each $\delta > 0$, there exist points $p, q \in K$ with $d(p, q) < \delta$

and for which there exists no p' such that $p, q \in U(p')$. One then obtains a sequence $(p_n, q_n) \in K \times K$ with $d(p_n, q_n) \to 0$ such that $\{p_n, q_n\} \not\subset U(p')$ for any $p' \in K$. Now one easily uses the compactness of K to obtain a contradiction. ■

§I.7. Geodesics and Completeness

Definition. Let M be a Riemannian manifold. We say that M is *geodesically complete* if for every $\xi \in TM$, the geodesic γ_ξ is defined on all of $\mathbb{R}$, that is, if exp is defined on all of TM.

Theorem I.7.1. (H. Hopf & W. Rinow (1931, pp. 216ff)) *If M is connected and geodesically complete, then any two points of M can be joined by a minimal geodesic.*

Proof. (G. de Rham (1952))[2] The idea of the proof is to lay one's hands on a candidate geodesic at the very outset. It goes as follows:

Let $p, q \in M$, $d(p,q) = \delta > 0$, and let $\epsilon = \epsilon(p) > 0$ as given in Theorem I.6.1. If $\delta < \epsilon$, then there is nothing to prove; so assume $\delta \geq \epsilon$ and fix $\delta_0 \in (0, \epsilon)$. Then, there exists $p_0 \in S(p; \delta_0)$ such that

$$d(p_0, q) = d(S(p; \delta_0), q),$$

that is, p_0 is the point on $S(p; \delta_0)$ closest to q. Our candidate geodesic γ_ξ, therefore, is given by ξ the unit vector in M_p determined by p and p_0, that is,

$$\xi = (1/\delta_0)(\exp|\mathsf{B}(p;\epsilon))^{-1}(p_0).$$

(See Figure I.3.) We shall prove for all $t \in [\delta_0, \delta]$ that

$$d(\gamma_\xi(t), q) = \delta - t. \tag{I.7.1}$$

This will certainly prove the theorem.

First, we note that (I.7.1) is true for $t = \delta_0$. Indeed,

$$\begin{aligned} \delta &= d(p, q) \\ &\leq d(p, p_0) + d(p_0, q) \\ &= \delta_0 + d(p_0, q). \end{aligned}$$

[2] See Note I.8 in §I.9.

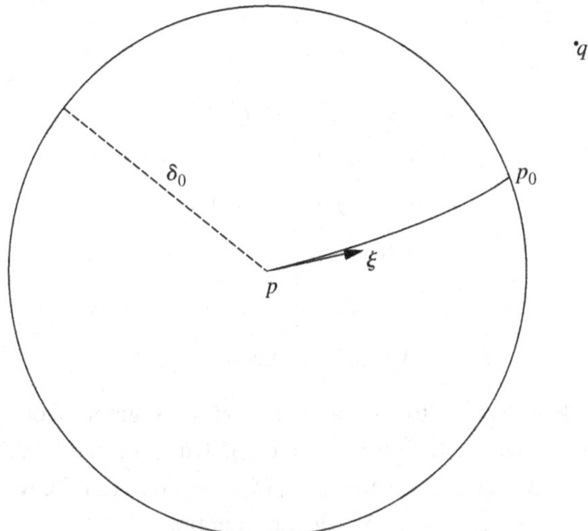

Figure I.3. Picking a candidate geodesic.

Furthermore, for every path $\omega : [0, 1] \to M \in D^1$ satisfying $\omega(0) = p, \omega(1) = q$, there exists $\alpha \in (0, 1)$ such that $\omega(\alpha) \in S(p; \delta_0)$. Then,

$$\begin{aligned}\ell(\omega) &= \ell(\omega|[0, \alpha]) + \ell(\omega|[\alpha, 1]) \\ &\geq \delta_0 + d(\omega(\alpha), q) \\ &\geq \delta_0 + d(p_0, q) \\ &\geq \delta,\end{aligned}$$

by the above argument. But, $\delta = \inf \ell(\omega)$ over ω just described. Thus, we have $\delta = \delta_0 + d(p_0, q)$ and (I.7.1) is valid for $t = \delta_0$.

Note that (I.7.1) implies $d(p, \gamma_\xi(t)) = t$.

Now let

$$\delta_1 = \max \{t \in [\delta_0, \delta] : \text{(I.7.1) is valid for } t\}.$$

Our claim, of course, is that $\delta_1 = \delta$. So we assume $\delta_1 < \delta$ and obtain a contradiction. Set $p_1 = \gamma_\xi(\delta_1)$ and pick δ_2 to satisfy $0 < \delta_2 < \min \{\epsilon(p_1), \delta - \delta_1\}$.

Let $p_2 \in S(p_1; \delta_2)$ satisfy $d(p_2, q) = d(S(p_1; \delta_2), q)$. Then, by our previous argument, using p_1 for p and p_2 for p_0 and δ_2 for δ_0, we have

$$\delta_2 + d(p_2, q) = d(p_1, q) = \delta - \delta_1. \tag{I.7.2}$$

Thus,

$$\begin{aligned} d(p,q) &= \delta \\ &= \delta_1 + \delta_2 + d(p_2,q) \\ &= d(p,p_1) + d(p_1,p_2) + d(p_2,q) \\ &\geq d(p,p_2) + d(p_2,q) \\ &\geq d(p,q), \end{aligned}$$

which implies

$$d(p,p_2) = d(p,p_1) + d(p_1,p_2).$$

Thus, the path γ_ξ from p to p_1, and the geodesic segment from p_1 to p_2 in $B(p_1;\epsilon(p_1))$ minimize arc length from p to p_2. Thus, the path is differentiable and $p_2 = \gamma_\xi(\delta_1+\delta_2)$. If we insert $p_2 = \gamma_\xi(\delta_1+\delta_2)$ into (I.7.2), we obtain that (I.7.1) is valid for $t = \delta_1 + \delta_2 > \delta_1$ – a contradiction. ∎

Corollary I.7.1. (H. Hopf & W. Rinow (1931)) *If M is geodesically complete, then every closed and bounded subset is compact. As a consequence, M is a complete metric space.*

Proof. Let $E \subseteq M$ be closed and bounded, $p \in M$, and $\delta = \sup\{d(p,q) : q \in E\} < +\infty$. Then,

$$B(p;\delta) = \exp \mathsf{B}(p;\delta), \quad S(p;\delta) \subseteq \exp \mathsf{S}(p;\delta) \subseteq \overline{B(p;\delta)}$$

by the argument of Corollary I.6.1. Thus, $E \subseteq \overline{B(p;\delta)} = \exp \overline{\mathsf{B}(p;\delta)}$, which is compact. Thus, E is compact. ∎

Theorem I.7.2. (H. Hopf & W. Rinow (1931)) *If M is a complete metric space then M is geodesically complete.*

Proof. Assume M is a complete metric space, but there exists $\xi \in TM$ such that its maximal interval I_ξ for the integral curve of the geodesic flow through ξ is not all of $\mathbb{R}$. Then, $-\infty < \alpha := \inf I_\xi$ and/or $\sup I_\xi := \beta < +\infty$. Assume the first case.

Pick $t_n \downarrow \alpha$ as $n \uparrow \infty$. Then, (t_n) is a Cauchy sequence and for n, m we have

$$d(\gamma_\xi(t_n), \gamma_\xi(t_m)) \leq \ell(\gamma_\xi|[t_n,t_m]) = |\xi||t_n - t_m|;$$

thus, $\gamma_\xi(t_n)$ is also a Cauchy sequence in M and $\gamma_\xi(t_n) \to q$ for some $q \in M$, as $n \to \infty$.

But, for a relatively compact neighborhood $U = U(q)$ of q, we have $\{\zeta \in \pi^{-1}[\overline{U}] : |\zeta| = |\xi|\}$ is compact in TM. Thus, we have a subsequence (τ_k) of

(t_n), and $\eta \in M_q$ for which $\gamma_\xi'(\tau_k) \to \eta$ as $k \to \infty$. If we set

$$\gamma_\eta(t) = \exp t\eta,$$

then for all k we have

$$\gamma_\eta(\tau_k - \alpha) = \gamma_\xi(\tau_k)$$

by the uniqueness of integral curves of the geodesic flow. Then, $\alpha \neq \inf I_\xi$ – a contradiction. ∎

Corollary I.7.2. *If M has a point p for which* exp *is defined on all of M_p, that is, $\mathcal{T}M \cap M_p = M_p$, then M is a complete metric space, and therefore* exp *is defined on all of TM.*

Proof. The argument of Theorem I.7.1 shows that minimizing geodesics exist between p and any other point of M. But, then one can argue as in Corollary I.7.1, followed by Theorem I.7.2. ∎

§I.8. Calculations with Moving Frames

In this section, we do not derive new material; rather, we are more interested in viewing some of our previous calculations from the perspective of moving frames. We then hope to use these calculations in some of our future work.

In what follows, for ξ, η in a real vector space E, and α, β in its dual space E^*, our definition of the wedge product will be given by

$$(\alpha \wedge \beta)(\xi, \eta) = \alpha(\xi)\beta(\eta) - \alpha(\eta)\beta(\xi). \tag{I.8.1}$$

Also, for a differentiable manifold M, differentiable vector fields X, Y, and a differentiable 1–form ω on M, our normalization of the exterior derivative $d\omega$ will be given by

$$d\omega(X, Y) = X(\omega(Y)) - Y(\omega(X)) - \omega([X, Y]). \tag{I.8.2}$$

Now, let M have a connection ∇, with associated torsion and curvature tensors.

Definition. The *covariant differentiation of* 1*–forms* is defined by

$$(\nabla_X \omega)(Y) = X(\omega(Y)) - \omega(\nabla_X Y), \tag{I.8.3}$$

that is,

$$X(\omega(Y)) = (\nabla_X \omega)(Y) + \omega(\nabla_X Y) \tag{I.8.4}$$

(one checks that $\nabla_X \omega$ is indeed a 1–form).

Note that

$$d\omega(X, Y) = (\nabla_X \omega)(Y) - (\nabla_Y \omega)(X) - \omega(T(X, Y)),$$

where T denotes the torsion tensor of ∇.

Let U be open in M and $\{e_1, \ldots, e_n\}$ differentiable vector fields on U, which are pointwise linearly independent. We refer to $\{e_1, \ldots, e_n\}$ as a *moving frame* on U. We let $\{\omega^1, \ldots, \omega^n\}$ denote the dual coframe field.

Then, for any tangent vector ξ in TU, one has an expansion of $\nabla_\xi e_k$ in terms of the basis $\{e_1, \ldots, e_n\}$, namely,

$$\nabla_\xi e_j = \sum_k {\omega_j}^k(\xi) e_k.$$

Since $\nabla_\xi e_j$ is linear in ξ, the collection $\{{\omega_j}^k : j, k = 1, \ldots, n\}$ forms a matrix of differentiable 1–forms, loosely referred to as the *connection 1–forms*.

Then, for any $\xi \in TU$, we have, by (I.8.4),

$$\begin{aligned} 0 &= \xi(\omega^\ell(e_j)) \\ &= (\nabla_\xi \omega^\ell)(e_j) + \omega^\ell(\nabla_\xi e_j) \\ &= (\nabla_\xi \omega^\ell)(e_j) + {\omega_j}^\ell(\xi). \end{aligned}$$

Therefore,

$$\nabla_\xi \omega^\ell = \sum_j (\nabla_\xi \omega^\ell)(e_j)\omega^j = -\sum_j ({\omega_j}^\ell(\xi))\omega^j. \tag{I.8.5}$$

We then have

$$\begin{aligned} d\omega^j(X, Y) &- \sum_k (\omega^k \wedge {\omega_k}^j)(X, Y) \\ &= X(\omega^j(Y)) - Y(\omega^j(X)) - \omega^j([X, Y]) \\ &\quad - \sum_k \{\omega^k(X){\omega_k}^j(Y) - \omega^k(Y){\omega_k}^j(X)\} \\ &= X(\omega^j(Y)) + \sum_k \omega^k(Y){\omega_k}^j(X) \\ &\quad - Y(\omega^j(X)) - \sum_k \omega^k(X){\omega_k}^j(Y) - \omega^j([X, Y]) \\ &= X(\omega^j(Y)) - (\nabla_X \omega^j)(Y) \\ &\quad - Y(\omega^j(X)) + (\nabla_Y \omega^j)(X) - \omega^j([X, Y]) \\ &= \omega^j(\nabla_X Y - \nabla_Y X - [X, Y]) \\ &= -\omega^j(T(X, Y)) \end{aligned}$$

(the first equality follows from (I.8.1) and (I.8.2), the second from rearranging terms, the third from (I.8.5), the fourth from (I.8.3), and the fifth from the definition of the torsion tensor). We therefore conclude

$$d\omega^j = \sum_k \omega^k \wedge \omega_k{}^j - \omega^j(T), \tag{I.8.6}$$

where we think of T as a 2–form with values in the tangent bundle, and therefore $\omega^j(T)$ is a 2–form on U.

We now consider, by (I.8.2)

$$d\omega_j{}^k(X, Y) = X(\omega_j{}^k(Y)) - Y(\omega_j{}^k(X)) - \omega_j{}^k([X, Y]).$$

To this end, we note

$$\begin{aligned}\nabla_Y \nabla_X e_j &= \nabla_Y \sum_k \omega_j{}^k(X) e_k \\ &= \sum_k \left\{Y(\omega_j{}^k(X))e_k + \omega_j{}^k(X)\nabla_Y e_k\right\} \\ &= \sum_k \left\{Y(\omega_j{}^k(X)) + \sum_\ell \omega_j{}^\ell(X)\omega_\ell{}^k(Y)\right\} e_k,\end{aligned}$$

from which we have

$$\begin{aligned}&\nabla_Y \nabla_X e_j - \nabla_X \nabla_Y e_j - \nabla_{[Y,X]} e_j \\ &= \sum_k \left\{Y(\omega_j{}^k(X)) - X(\omega_j{}^k(Y)) + \sum_\ell (\omega_j{}^\ell \wedge \omega_\ell{}^k)(X, Y) - \omega_j{}^k([Y, X])\right\} e_k \\ &= \sum_k \left\{d\omega_j{}^k(Y, X) + \sum_\ell \omega_j{}^\ell \wedge \omega_\ell{}^k(X, Y)\right\} e_k.\end{aligned}$$

We conclude that if $\Omega_j{}^k$ denotes the 2–form given by

$$\Omega_j{}^k(X, Y) = \omega^k(R(X, Y)e_j),$$

then

$$d\omega_j{}^k = \sum_\ell \omega_j{}^\ell \wedge \omega_\ell{}^k - \Omega_j{}^k. \tag{I.8.7}$$

For Riemannian manifolds, we summarize the discussion in the following

Theorem I.8.1. *If M is a Riemannian manifold, and the frame $\{e_1, \ldots, e_n\}$ is picked to be orthonormal at every point of U, with dual coframe $\{\omega^1, \ldots, \omega^n\}$, then*

$$d\omega^j = \sum_k \omega^k \wedge \omega_k{}^j. \tag{I.8.8}$$

where

(I.8.9) $$\omega_j{}^k(\xi) = \langle \nabla_\xi e_j, e_k \rangle, \qquad \omega_j{}^k = -\omega_k{}^j.$$

For the curvature forms, we have

(I.8.10) $$d\omega_j{}^k = \sum_\ell \omega_j{}^\ell \wedge \omega_\ell{}^k - \Omega_j{}^k,$$

where

(I.8.11) $$\Omega_j{}^k(X,Y) = \langle R(X,Y)e_j, e_k \rangle, \qquad \Omega_j{}^k = -\Omega_k{}^j,$$

which implies the skew-symmetry of $\langle R(X,Y)Z, W \rangle$ *in* Z *and* W. (See (II.1.4).)

§I.9. Notes and Exercises

Bibliographic Sampler

Note I.1. One must start by referring to de Rham (1955) for manifolds, differential forms, de Rham cohomology and theorems, and the Laplacian on differential forms.

For more up-to-date treatment of differential topology, see Milnor (1965) and Hirsch (1976).

One can construct an ambitious shopping list of books with which to start differential, and, more particularly, Riemannian geometry. Many of the more introductory books start with surfaces in Euclidean space and then deal with the exclusively intrinsic Riemannian geometry of surfaces. We refer to some of them in Note II.2. Here, in addition to the references cited in the introduction to this chapter, we just note the influential two-volume treatise of Kobayashi–Nomizu (1969) on the general foundations – from connections through Riemannian metrics through curvature through the first level of specializations of areas in differential geometry, for example, curvature and geodesics, homogeneous spaces, Kähler manifolds, etc.

For works more exclusively devoted to intrinsic Riemannian geometry see, for example, Cheeger–Ebin (1975), do Carmo (1992), Gallot–Hulin–Lafontaine (1987), Gromoll–Klingenberg–Meyer (1968), Klingenberg (1982), Lang (1995), Lee (1997), O'Neill (1983), and Petersen (1998) – just to name a few!

Lie Brackets of Vector Fields

Note I.2. It may be helpful to remind the reader that, if X, Y are vector fields on a differentiable manifold M, then their *Lie bracket* $[X, Y]$ can be defined by

$$[X,Y]f = X(Yf) - Y(Xf)$$

for any differentiable function on M.

One checks that if $x : U \to \mathbb{R}^n$ is a chart on M, and X and Y are given by

$$X = \sum_j \xi^j \frac{\partial}{\partial x^j}, \qquad Y = \sum_k \eta^k \frac{\partial}{\partial x^k},$$

then $[X, Y]$ is given by

$$[X, Y] = \sum_{k,j} \left\{ \xi^k \frac{\partial \eta^j}{\partial x^k} - \eta^k \frac{\partial \xi^j}{\partial x^k} \right\} \frac{\partial}{\partial x^j}.$$

Another formula for $[X, Y]$ is given by the Lie derivative of Y with respect to X, that is, one lets φ_t denote the 1–parameter flow on M determined by X, and defines

$$\mathcal{L}_X Y := \frac{d}{dt} \left\{ (\varphi_{-t})_* Y_{|\varphi_t} \right\} \Big|_{t=0}.$$

Then,

$$\mathcal{L}_X Y = [X, Y].$$

Connections and Covariant Differentiation

Note I.3. The coordinate-free definition of connections presented in §I.1 is originally due to Koszul, who communicated it to Nomizu (1954).

Note I.4. In nearly all of our discussions, we started with global coordinate-free definitions of the concepts and then gave a calculation in local coordinates of the objects defined. However, in the "old" days, the method ran the other way; that is, one started with an expression in local coordinates and then established the global character of the notions so defined. For a nice treatment of this approach, we recommend Laugwitz (1966). Two vestiges of this approach are left in our treatment of (i) differentiation of vector fields along a path and (ii) the resulting definition of the vector field of the geodesic flow. For (i), an intrinsic a priori global definition is given in Gromoll–Klingenberg–Meyer (1968), the approach originally due to Dombrowski (1962). One can find a similar treatment in Besse (1987), Klingenberg (1982), and also an intrinsic definition of (ii). Finally, one can find a global characterization of the geodesic flow, via analytical mechanics, in §V.1.

For the general covariant differentiation of tensor fields, one proceeds as follows: Recall that given a finite dimensional vector space F over $\mathbb{R}$, the

(r, s)–tensor space $F^{\otimes r,s}$ may be considered as the collection of multilinear maps

$$A : (F^*)^r \times F^s \to \mathbb{R}.$$

Given the n–dimensional manifold M with connection ∇, one extends the covariant differentiation to general (r, s)–tensor fields over M, namely, given the (r, s)–tensor field A, we define its *covariant differential* ∇A to be the $(r, s + 1)$–tensor field given by

$$\begin{aligned}
&(\nabla A)(\theta^1, \dots, \theta^r; X_0, X_1, \dots, X_r) \\
&\quad := (\nabla_{X_0} A)(\theta^1, \dots, \theta^r; X_1, \dots, X_s) \\
&\quad := X_0(A(\theta^1, \dots, \theta^r; X_1, \dots, X_s)) \\
&\qquad - \sum_{j=1}^{r} A(\theta^1, \dots, \nabla_{X_0}\theta^j, \dots, \theta^r; X_1, \dots, X_s) \\
&\qquad - \sum_{k=1}^{s} A(\theta^1, \dots, \theta^r; X_1, \dots, \nabla_{X_0} X_k, \dots, X_s),
\end{aligned}$$

where $\theta^1, \dots, \theta^r$ are 1–forms on M, and $X_0, \dots, X_s$ are vector fields on M.

Exercise I.1. Show that, for any $p \in M$, the value of ∇A at p only depends on the values of A on a neighborhood of p, and on the values of $\theta^1, \dots, \theta^r$, $X_0, \dots, X_s$ at p.

If M is Riemannian and $r = 1$, it is occasionally easier to view to A as a section $\mathfrak{A}$ in the bundle Hom $(TM^{\otimes s}, TM)$ (the identification of the two viewpoints is achieved through the canonical identification of F with $(F^*)^*$), in which case one defines

$$\begin{aligned}
(\nabla \mathfrak{A})(X_0, X_1, \dots, X_s) &:= (\nabla_{X_0} \mathfrak{A})(X_1, \dots, X_s) \\
&:= \nabla_{X_0}(\mathfrak{A}(X_1, \dots, X_s)) \\
&\quad - \sum_{k=1}^{s} \mathfrak{A}(X_1, \dots, \nabla_{X_0} X_k, \dots, X_s),
\end{aligned}$$

where $X_0, \dots, X_s$ are vector fields on M.

Note I.5. For more on the covariant differentiation of general tensor fields see, for example, Klingenberg (1982) and O'Neill (1983).

The Second Bianchi Identity

Exercise I.2. Suppose we are given the connection ∇ with torsion and curvature tensors T and R, respectively. Prove *Bianchi's second identity*, that is, show that $T = 0$ on M implies

$$(\nabla_\zeta R)(\xi, \eta) + (\nabla_\eta R)(\zeta, \xi) + (\nabla_\xi R)(\eta, \zeta) = 0$$

for all ξ, η, ζ in M_p, $p \in M$.

Covariant Differentiation in Vector Bundles

Let M be our manifold, and E a vector bundle over M, with fiber space the finite-dimensional vector space F over $\mathbb{R}$.

Definition. A *connection on E* is a map $\nabla : TM \times \Gamma^1(E) \to E$ (where $\Gamma^1(E)$ denotes the C^1 sections in E), which we write as $\nabla_\xi Y$ instead of $\nabla(\xi, Y)$, with the following properties: First, we require that $\nabla_\xi Y$ be in the same fiber as ξ, and that for $\alpha, \beta \in \mathbb{R}$, $p \in M$, $\xi, \eta \in M_p$, $Y \in \Gamma^1(E)$,

$$\nabla_{\alpha\xi+\beta\eta} Y = \alpha \nabla_\xi Y + \beta \nabla_\eta Y.$$

Second, we require that, for $p \in M$, $\xi \in M_p$, $Y, Y_1, Y_2 \in \Gamma^1(E)$, $f \in C^1(M)$, we shall have

$$\nabla_\xi (Y_1 + Y_2) = \nabla_\xi Y_1 + \nabla_\xi Y_2,$$

$$\nabla_\xi (fY) = (\xi f) Y_{|p} + f(p) \nabla_\xi Y.$$

Finally, we require that ∇ be smooth in the following sense: if $X \in \Gamma^\infty(TM)$, $Y \in \Gamma^\infty(E)$, then $\nabla_X Y \in \Gamma^\infty(E)$.

One has the same local character of the covariant differentiation as that described for $E = TM$, with corresponding notions of parallel translation along paths in M.

Similarly, one has the curvature tensor defined as follows: For any $p \in M$ we define the multilinear map henceforth referred to as the *curvature tensor of* ∇, by: For $\xi, \eta \in M_p$, $\zeta \in E_p$ (where E_p denotes the fiber over p), let X, Y be extensions of ξ, η, respectively, to vector fields, and Z an extension of ζ to a section in E, on a neighborhood of p. Then define

$$R(\xi, \eta)\zeta = \nabla_\eta \nabla_X Z - \nabla_\xi \nabla_Y Z - \nabla_{[Y,X]_{|p}} Z.$$

For an example, see Exercise II.11.

Coordinate Characterization of $T = 0$

Exercise I.3. Prove that, for any connection ∇ on a manifold, the torsion tensor T vanishes at a point p if and only if there exists a chart $y : V \to \mathbb{R}^n$ for which the Christoffel symbols of the connection vanish at p.

Gradients

Let $f : M \to \mathbb{R}$ be a differentiable function on the Riemannian manifold M.

Definition. The *gradient vector field of f on M*, grad f, is defined by

$$\operatorname{grad} f = \theta^{-1}(df),$$

where df denotes the differential of f, and $\theta : TM \to TM^*$ denotes the natural bundle isomorphism given by

$$\theta(\xi)(\eta) = \langle \xi, \eta \rangle,$$

for all $p \in M$ and $\xi,\ \eta \in M_p$.

Exercise I.4. Assume that

$$|\operatorname{grad} f| = 1$$

on all of M. Show that the integral curves of grad f are geodesics.

Exercise I.5. (See Davies (1987, pp. 325–326)) Show that, for any Riemannian manifold M, the distance function may be given analytically by

$$d(x, y) = \sup\,\{|\psi(x) - \psi(y)| : \psi \in C^\infty, |\operatorname{grad} \psi| \le 1\},$$

that is, where ψ varies over smooth functions for which $|\operatorname{grad} \psi| \le 1$ on all of M.

On Theorem I.6.1

Note I.6. See Hopf–Rinow (1931, p. 219), where they refer this result to O. Bolza (1909, §33). A stronger theorem, that the unique geodesic of length $< \epsilon$ is contained in U itself, was proved by Whitehead (1932). See our discussion in §VII.6.

The Hopf–Rinow Theorems

Note I.7. A precursor to the Hopf–Rinow theorems is in the work of Koebe. See the references and comments in the original paper, Hopf–Rinow (1931).

Note I.8. The strength of de Rham's proof of Theorem I.7.1 lies in its immediate selection of a candidate geodesic to minimize distance between two points. If, however, one tries to carry out the argument for a class of closed loops, for example, if one attempts to minimize the length of loops in a nontrivial free homotopy class of a compact Riemannian manifold, then one has nowhere to aim. So one must return to the original argument of Hopf–Rinow (1931). Namely, one must deal directly with a sequence of loops whose lengths approach the infimum of all such possible lengths. See our discussion in the proof of Theorem IV.12.

Also note that the hypothesis here in Theorem I.7.1 is geodesic completeness. If one assumes instead, metric completeness of the Riemannian metric, then one has an easy direct proof of the existence, given any two points, of a minimizing geodesic joining the two points. The argument is direct variational, and uses the Arzela–Ascoli theorem. (See the sketch to Exercise I.7.)

Exercise I.6. Let M be a complete noncompact Riemannian manifold. Show that to each $p \in M$, there exists a *geodesic ray* emanating from p, that is, there exists a geodesic $\gamma : [0, +\infty) \to M$ such that $\gamma(0) = p$, and $d(p, \gamma(t)) = t$ for all $t > 0$.

Note I.9. The definition of the metric structure on Riemannian manifolds, and the Hopf–Rinow theorems have been the subject of refinement, axiomatization, and development (see W. Rinow (1961)). In what follows, we introduce some of the work of M. Gromov (1981). A subsequent English edition (more than just a translation) appeared in Gromov (1999). An excellent introduction to the field is Burago–Burago–Ivanov (2001).

Length Spaces

The standard approach to lengths and distances that we presented starts with the definition of lengths of a specified collection of paths (namely, D^1), from which is created a distance function, which is then shown to determine the same topology as that with which we started. A more abstract approach is to consider an arbitrary set X with a *length structure*, that is, to each closed interval $I \subseteq \mathbb{R}$

is associated a collection of maps $\mathcal{C}_I$ from I to X, the resulting collection

$$\mathcal{C} = \cup_I \mathcal{C}_I,$$

and a *length function*

$$\ell : \mathcal{C} \to [-\infty, +\infty]$$

satisfying:

(a) $\ell(f) \geq 0$ for all $f \in \mathcal{C}$;

(b) $\ell(f) = 0$ if and only if f is the constant map;

(c) if $I \subseteq J$ then the restriction of any $f \in \mathcal{C}_J$ to the interval I is an element of $\mathcal{C}_I$;

(d) if $f \in \mathcal{C}_{[a,b]}$ and $g \in \mathcal{C}_{[b,c]}$, with $f(b) = g(b)$, then $f \cdot g \in \mathcal{C}_{[a,c]}$ with

$$\ell(f \cdot g) = \ell(f) + \ell(g);$$

(e) if $\phi : I \to J$ is a homeomorphism, and $f \in \mathcal{C}_J$, then $f \circ \phi \in \mathcal{C}_I$ and

$$\ell(f \circ \phi) = \ell(f);$$

(f) if $I = [a, b]$, $f \in \mathcal{C}_I$, then the map

$$t \mapsto \ell(f|[a, t])$$

is continuous. One then naturally proposes a *length distance $d_\ell(x, y)$ between any two points x, y* to be given by

$$d_\ell(x, y) = \inf\{\ell(f) : f \in \mathcal{C},\ f \text{ joins } x \text{ to } y\}.$$

(To ultimately speak of *all* maps $I \to X$ for all I, one sets $\ell(f) = +\infty$ for any $f : I \to X \notin \mathcal{C}_I$.)

Note I.10. If X is given to be a topological space, then the paths in $\mathcal{C}$ are restricted to be continuous and are assumed to satisfy: Given any $x \in X$, and any open neighborhood U_x of x, the length of paths connecting x to points in $X \setminus U_x$ is bounded away from 0 (Burago–Burago–Ivanov (2001, p. 27)).

Also (see p. 28), if X is Hausdorff, then d_ℓ is indeed a metric (not just nonnegative) – although points in the same component might have infinite distance between them.

To construct a natural example of a length structure, simply stand the discussion on its head, namely, start with a metric space X with distance function d. The collection of paths $\mathcal{C}$ is to consist of all continuous maps of closed intervals to

X; and for any continuous $f : [a, b] \to X$, define the length of f by

$$\ell_d(f) = \sup \sum_{j=1}^{k} d(f(t_{j-1}), f(t_j)),$$

where the supremum is taken over all partitions $[a = t_0 \leq t_1 \leq \ldots \leq t_k = b]$ of the interval $[a, b]$. Thus, $\mathcal{C}$ consists of all continuous maps of closed intervals in $\mathbb{R}$ to X.

We then have two "cycles": First, start with a length structure ℓ, determine a length distance function d_ℓ, from which one determines a length structure $\tilde{\ell}$. Second, start with a metric space with distance function d, determine the length structure ℓ_d, which then determines a new length distance function $\tilde{d}$. The first two questions, of course, are: When is $\ell = \tilde{\ell}$, and when is $d = \tilde{d}$? Here are some results:

Start with the length structure ℓ. Show that, if for every closed interval I, the length function ℓ restricted to $\mathcal{C}_I$ is lower semicontinuous in the compact–open topology of $\mathcal{C}_I$, then $\ell = \tilde{\ell}$.

Proposition A. *Start with the metric d, and consider the following two properties of d:*
1°: For any $x, y \in X$ and $\epsilon > 0$, there exists $z \in X$ such that

$$\sup \{d(x, z), d(z, y)\} \leq \frac{d(x, y)}{2} + \epsilon.$$

2°: For any $x, y \in X$ and $r_1, r_2 > 0$ satisfying

$$r_1 + r_2 \leq d(x, y),$$

one has

$$d(\overline{B(x; r_1)}, \overline{B(y; r_2)}) \leq d(x, y) - r_1 - r_2.$$

Then, the two properties are equivalent. Also, $d = \tilde{d}$ implies that these properties are valid.

Conversely, if the metric d is complete, with these properties valid, show that $d = \tilde{d}$.

Definition. The metric space (X, d) is called a *length space* if $d = \tilde{d}$.

Proposition B. *If X is a complete locally compact length space, then closed and bounded sets are compact.*

For the details, see Burago–Burago–Ivanov (2001, §2.3–4).

One can now consider geodesics without ordinary differential equations. Namely,

Definition. Let X be a length space with respect to the metric d. A *minimizing geodesic* is a path $f : I \to X$ such that

$$d(f(t), f(t')) = |t - t'|$$

for all $t, t' \in I$.

The path $f : I \to X$ is a *geodesic* if the restriction of f to all sufficiently small intervals I is always a minimizing geodesic.

Exercise I.7. If X is a complete locally compact length space, then any two points are joined by a minimizing geodesic. (See Remark IV.5.1.)

One has corresponding Hopf–Rinow theorems; see Burago–Burago–Ivanov (2001, pp. 51ff).

Exercise I.8. Let M be a Riemannian manifold. Then, the *Riemannian length structure* is given by all maps $f : I \to M \in D^1$, and the usual length function. Show that the induced metric d – the Riemannian distance function – satisfies 1° of Proposition A. In particular, if M is Riemannian complete, it is a length space.

Continue with M a Riemannian manifold with standard Riemannian length structure ℓ and associated distance function d. We wish the length structure ℓ_d induced on *all* continuous maps of compact intervals into M to be an extension of ℓ from D^1 maps to C^0 maps. Note that one cannot invoke the result just prior to Proposition A. Why?

Exercise I.9. Give an easy proof that

$$\ell_d(\omega) = \int_a^b |\omega'|\, dt,$$

when ω is a D^1 path in a Riemannian manifold.

A more general result is:

Theorem (See Rinow (1961, p. 106ff)) *If $f : [a, b] \to X$ is Lipschitz, then*

$$\ell_d(f) = \int_a^b |f'|\, dt$$

(note that, by Rademacher's theorem – see Chavel (2001, p. 20ff) – f' exists for almost all t in $[a, b]$). For the proof, see Burago–Burago–Ivanov (2001, §2.7).

Moving Frames

Note I.11. The method of moving frames goes back at least to G. Darboux's (1898) triply orthogonal systems, was turned into a modern tool in differential geometry in the works of É. Cartan (1946), and widely disseminated in the 1950s and 1960s in the various mimeographed lecture notes of S.S. Chern. An attractive entrée into the method can be found in Flanders (1963).

Hessians

Definition. Given a function f on a Riemannian manifold M, its first covariant derivative is, simply, its differential df.

The *Hessian of* f, Hess f, is defined to be the second covariant derivative of f, that is, ∇df. So,

$$(\mathrm{Hess}\, f)(\xi, \eta) = \xi(df(Y)) - (df)(\nabla_\xi Y),$$

where Y is any extension of η.

Exercise I.10. Prove

(a) Hess f is symmetric in ξ, η;
(b) the $(1, 1)$–tensor field associated with the $(0, 2)$–field Hess f is given by

$$\xi \mapsto \nabla_\xi \,\mathrm{grad}\, f;$$

(c) a function with positive definite Hessian has no local maxima.

Exercise I.11. So, for a moving orthonormal frame $\{e_1, \ldots, e_n\}$ on M with dual coframe $\{\omega^1, \ldots, \omega^n\}$, we may write

$$\nabla f = df = \sum_j F_j \omega^j.$$

Prove

(a) that

$$\mathrm{grad}\, f = \sum_j F_j e_j;$$

(b) that

$$\nabla \omega^j = -\sum_k {\omega_k}^j \otimes \omega^k;$$

(c) that Hess f is given by

$$\begin{aligned}\text{Hess}\, f = \nabla df &= \sum_j \left\{ dF_j - \sum_k {\omega_j}^k F_k \right\} \otimes \omega^j \\ &:= \sum_{j,k} F_{jk}\, \omega^j \otimes \omega^k;\end{aligned}$$

(d) F_{jk} is symmetric in j, k.
(e) that

$$\begin{aligned}& d\left\{ \sum_j (-1)^{j-1} F F_j\, \omega^1 \wedge \cdots \wedge \widehat{\omega^j} \wedge \cdots \wedge \omega^n \right\} \\ &= \left\{ \sum_j {F_j}^2 + F F_{jj} \right\} \omega^1 \wedge \cdots \wedge \omega^n\end{aligned}$$

(we use the $\widehat{\ }$ to indicate that the term in question is missing).

Moving Frames in Euclidean Space

Consider Euclidean space $\mathbb{R}^m$ with its standard Riemannian metric and associated Levi-Civita connection (I.1.1). Then, certainly, both the torsion and curvature tensors vanish identically on $\mathbb{R}^m$. One can recapture the fact with moving frames as follows:

Let $\{\mathfrak{e}_1, \ldots, \mathfrak{e}_m\}$ denote the standard basis of $\mathbb{R}^m$, and let $\mathfrak{I}_p : \mathbb{R}^m \to (\mathbb{R}^m)_p$ the standard identification of $\mathbb{R}^m$ with the tangent space of any of its points p. Set

$$(E_A)_{|\mathbf{x}} = \mathfrak{I}_{\mathbf{x}} \mathfrak{e}_A, \qquad A = 1, \ldots, m, \qquad \mathbf{x} \in \mathbb{R}^m,$$

and let e_A denote a moving frame on $\mathbb{R}^m$; so one has a matrix function (from a neighborhood in $\mathbb{R}^m$ to the orthogonal matrices) ${S_A}^B$ for which

$$\mathbf{e}_A = \sum_B {S_A}^B E_B.$$

Set

$$T = S^{-1} = S^* \qquad \text{(where * denotes transpose).}$$

Exercise I.12.

(a) Let

$$\mathbf{x} = \sum_A x^A \mathbf{e}_A.$$

So the coordinates $\{x^A\}$ are functions on $\mathbb{R}^n$. Check that the 1–forms $\{dx^A\}$ are dual to the frame $\{E_A\}$.

(b) Prove that, for the moving coframe $\{\theta^1, \ldots, \theta^n\}$ dual to $\{e_1, \ldots, e_n\}$, we have

$$(\theta^A)_{|\mathbf{x}} = \sum_B dx^B {T_B}^A.$$

(c) Prove that the connection 1–forms $\{{\theta_A}^B\}$ are given by

$${\theta_A}^B = \sum_C d{S_A}^C {T_C}^B.$$

(d) Prove

$$d\theta^A = \sum_B \theta^B \wedge {\theta_B}^A$$

$$d{\theta_A}^B = \sum_C {\theta_A}^C \wedge {\theta_C}^B.$$

Note I.12. One can substitute (a) and (b) into (c) to verify it, but a more natural way is to think of $\mathbf{x}$ and e_A as functions on $\mathbb{R}^n$ with values in a fixed vector space. There, one can use

$$0 = d^2\mathbf{x} = d^2\mathbf{e}_A,$$

where d denotes exterior differentiation.

Examples

We first mention some obvious examples of Riemannian manifolds, namely,

- imbedded submanifolds of Euclidean space, wherein the Riemannian metric of the Euclidean space is restricted to the submanifold (after all, this is how the subject was created)
- the classical examples of constant curvature, the Euclidean space, sphere, and hyperbolic space of all dimensions ≥ 2.

Both types of examples are discussed in Chapter II.

A more unusual example is given by

Example I.9.1. Consider $\mathbb{R}^{n+1}$ with a Minkowski quadratic form $\mathcal{M}$ given by

$$\mathcal{M}((x,\tau)) = |x|^2 - \tau^2, \qquad x \in \mathcal{R}^n, \quad \tau \in \mathbb{R},$$

where $|x|$ denotes the usual norm on $\mathbb{R}^n$. For any $\rho > 0$, consider the submanifold M in $\mathbb{R}^{n+1}$ (consisting of two components) given by

$$|x|^2 - \tau^2 = -\rho^2.$$

Show that the restriction of the quadratic form $\mathcal{M}$ to the submanifold M is positive definite and therefore determines a Riemannian metric on M.

It will turn out that this example has constant sectional curvature $= -1/\rho^2$. See Exercise II.15.

Example I.9.2. Given Riemannian manifolds (M, g), (N, h), one naturally considers the *product Riemannian metric* on the product manifold $M \times N$, as follows: For $x \in M$, $y \in N$ the tangent space $(M \times N)_{(x,y)}$ is canonically isomorphic to $M_x \oplus N_y$. For vectors $\xi, \eta \in M_x$, $\zeta, \nu \in N_y$, we define the inner product of $\xi \oplus \zeta$ and $\eta \oplus \nu$ by

$$(g,h)(\xi \oplus \zeta, \eta \oplus \nu) = g(\xi, \eta) + h(\zeta, \nu).$$

One can easily follow through with the calculation of the curvature tensor of the Levi–Civita connection. (Which method is easier? Local coordinates, or moving frames?)

Example I.9.3. Given Riemannian manifolds M and N, a smooth map $\pi : M \to N$ is a submersion if π_* maps M_x onto $N_{\pi(x)}$ for all $x \in M$. If π_* maps the orthogonal complement of the kernel of π_*, $(\ker \pi_*)^\perp$, isometrically onto $N_{\pi(x)}$ for all x, then π is called a *Riemannian submersion.*

Of course, the projection of a Riemannian product onto either of its factors is a Riemannian submersion.

If $\pi : M \to N$ is a submersion, then for any $y \in N$, the preimage $\pi^{-1}[y]$ is a submanifold of dimension equal to $\dim M - \dim N$. Of course, if $\dim M = \dim N$ then $\pi^{-1}[y]$ is a discrete collection of points. Such examples include Riemannian coverings, discussed at length in Chapter IV.

Suppose we are given a Riemannian submersion $\pi : M \to N$. Then, with each $p \in M$ is associated an orthogonal decomposition of the tangent space

$$M_p = H_p \oplus V_p,$$

where V_p *the vertical subspace* is the tangent space to the submanifold π^{-1} $[\pi(p)]$ at p, and H_p *the horizontal subspace* is the orthogonal complement of

V_p in M_p. With each $q \in N$, $\xi \in N_q$, and $p \in \pi^{-1}[q]$, we associate a unique *horizontal lift* $\overline{\xi} \in H_p$ satisfying

$$\pi_* \overline{\xi} = \xi.$$

Exercise I.13.

(a) Show that if T, S are vertical vector fields, and X is a horizontal vector field, on M, then

$$\langle [T, S], X \rangle = 0.$$

(b) Show that if X, Y are horizontal vector fields, and T is a vertical vector field, on M then, for any p in M, $\langle [X, Y], T \rangle(p)$ depends only on the values of X, Y, T at the point p.

Definition. If X is a vector field on M, and X is a vector field on N such that

$$\pi_* X = \mathsf{X},$$

then we say that X *is* π*–related to* X. We also say that a vector field X *on* M *projects to* N if there exists a vector field X on N that is π–related to X.

(c) Show that if X, Y are π–related to X, Y, respectively, then $[X, Y]$ is π–related to $[\mathsf{X}, \mathsf{Y}]$.

(d) Show that if X, Y, Z are horizontal vector fields on M, all of which π–project, and T is a vertical vector field on M then

$$\langle [X, Y], Z \rangle = \langle [\pi_* X, \pi_* Y], \pi_* Z \rangle, \qquad \langle [X, T], Z \rangle = 0.$$

Exercise I.14. Here, we introduce coordinate systems adapted to the considerations. Let $n = \dim M$, $k = \dim N$, $\ell = n - k$. Let

$$\pi_k : \mathbb{R}^k \times \mathbb{R}^\ell \to \mathbb{R}^k$$

denote the projection onto the first factor.

(a) Show, by the implicit function theorem, that, for each $p \in M$, there exist neighborhoods U of p and $V = \pi(U)$ of $q = \pi(p)$, and charts

$$x : U \to \mathbb{R}^n, \qquad y : V \to \mathbb{R}^k$$

such that

$$x(p) = 0, \qquad \pi_k \circ x = y \circ \pi.$$

Also, show that, for any vector field X on N, there exists a smooth horizontal lift of X to M, that is, a vector field X on M π–related to X.

(b) Show that charts x and y can also be chosen so that the submanifold Q given by

$$x^\alpha = 0, \qquad \alpha = k+1, \ldots, n,$$

is horizontal at p.

(c) Consider the section $\mu : V \to Q \subseteq U$ given by

$$x \circ \mu = (y, 0), \qquad 0 \in \mathbb{R}^\ell.$$

Show

$$\mu^* dx^r = dy^r, \qquad r = 1, \ldots, k.$$

Exercise I.15. Now for moving frames. Suppose we are given the orthonormal frame field $\{E_r : r = 1, \ldots, k\}$ with dual coframe field $\{\theta^r\}$ on $V \subseteq N$. We then associate with this frame field the horizontal lifts $\{e_r\}$ on U of $\{E_r\}$ on V. Complete this collection of vector fields to an orthonormal frame field $\{e_A : A = 1, \ldots, n\}$ on U, with dual coframe field $\{\omega^A\}$. In particular, $\{e_\alpha : \alpha = k+1, \ldots, n\}$ are vertical vector fields.

(a) Show that

$$\theta^s = \mu^* \omega^s \text{ on } V, \qquad \mu^* \omega^\alpha = 0 \text{ at } q = \pi(p).$$

We let $\omega_B{}^A$, $\theta_s{}^r$ denote the connection 1–forms on U, V of the respective frame fields. Then,

$$d\omega^A = \sum_B \omega^B \wedge \omega_B{}^A, \qquad d\theta^r = \sum_s \theta^s \wedge \theta_s{}^r,$$

with

$$\omega_B{}^A = -\omega_A{}^B, \qquad \theta_s{}^r = -\theta_s{}^r.$$

We set

$$\omega_B{}^A = \sum_C \Gamma_{BC}{}^A \omega^C, \qquad \Gamma_{BC}{}^A = \langle \nabla_{e_C} e_B, e_A \rangle.$$

(b) Show

$$\Gamma_{rs}{}^\alpha = -\Gamma_{\alpha s}{}^r = -\Gamma_{s\alpha}{}^r = \frac{1}{2}\langle [e_s, e_r], e_\alpha \rangle.$$

(c) Show

$$\sum_{\alpha,\beta} \Gamma_{\alpha\beta}{}^r \omega^\alpha \wedge \omega^\beta = 0.$$

(d) Show

$$d\theta^r = \sum_s \theta^s \wedge \mu^* \left(\omega_s{}^r - \sum_\alpha \Gamma_{\alpha s}{}^r \omega^\alpha \right),$$

and therefore

$$\theta_s{}^r = \mu^* \left(\omega_s{}^r - \sum_\alpha \Gamma_{\alpha s}{}^r \omega^\alpha \right) \quad \text{on } V, \qquad \theta_s{}^r = \mu^* \omega_s{}^r \quad \text{at } q.$$

(e) Conclude that, in general, for horizontal vector fields X, Y on M, which are π–related to vector fields X, Y on N, we have

$$\nabla_X Y = \overline{\nabla_{\mathsf{X}} \mathsf{Y}} + \frac{1}{2}[X, Y]^V,$$

where $\overline{\nabla_{\mathsf{X}} \mathsf{Y}}$ denotes the horizontal lift of $\nabla_{\mathsf{X}} \mathsf{Y}$ (we use the same ∇ for both Levi-Civita connections), and the superscript V denotes the vertical component. In particular, a horizontal path in M is a geodesic if and only if its image in N is a geodesic.

Exercise I.16. We now relate the "horizontal" curvature of M to the curvature of N. We continue with our calculations in moving frames. Let $\Omega_B{}^A, \Theta_s{}^r$ denote the curvature 2–forms on U, V, respectively. Then

$$d\omega_B{}^A = \sum_C \omega_B{}^C \wedge \omega_C{}^A - \Omega_B{}^A, \qquad d\theta_s{}^r = \sum_t \theta_s{}^t \wedge \theta_t{}^r - \Theta_s{}^r.$$

(a) Show that at $q = \pi(p)$, we have

$$\Theta_s{}^r = \mu^* \Omega_s{}^r - \sum_{\alpha, u, v} \mu^* \left\{ \Gamma_{rs}{}^\alpha \Gamma_{uv}{}^\alpha + \frac{1}{2} \left[\Gamma_{sv}{}^\alpha \Gamma_{ru}{}^\alpha - \Gamma_{su}{}^\alpha \Gamma_{rv}{}^\alpha \right] \right\} \theta^u \wedge \theta^v.$$

(b) Show that, for any horizontal vectors $\xi, \eta \in H_p$, we have for the curvature tensors R, R of M, N, respectively,

$$\langle \mathsf{R}(\pi_*\xi, \pi_*\eta)\pi_*\xi, \pi_*\eta \rangle = \langle R(\xi, \eta)\xi, \eta \rangle + \frac{3}{4} |[\xi, \eta]^V|.$$

Of course, $[\xi, \eta]^V$ is well-defined, by Exercise I.13(b). (See §II.1 for specific interest in the values of $\langle \mathsf{R}(\pi_*\xi, \pi_*\eta)\pi_*\xi, \pi_*\eta \rangle$.)

Note I.13. The first differential geometric calculations associated to submersions were developed by B. O'Neill (1966b).

Example I.9.4. We now give an extremely barebones discussion of *Lie groups*, just enough to do some calculations for invariant Riemannian metrics on them.

The reader can start with Warner (1971, Chapter III) and progress from there into the subject. Short discussions emphasizing our interests here, and our discussion below of Riemannian homogeneous spaces, can be found in Chavel (1970, 1972), Cheeger–Ebin (1975), and Milnor (1963). Results go back to E. Cartan (1927) and, later on, to K. Nomizu (1954).

Recall that a Lie group is an n–dimensional, real analytic manifold G whose elements possess a multiplication, for which the map $x, y \mapsto x \cdot y^{-1}$ of $G \times G \to G$ is analytic.

For each $g \in G$, the maps $L_g, R_g : G \to G$ given by

$$L_g(g') = g \cdot g', \qquad R_g(g') = g' \cdot g,$$

are analytic diffeomorphisms of G, and are referred to as the *left* and *right translations of* G, respectively. A vector field X on G is said to be *left-invariant* if for every $g, h \in G$ we have

$$X_{|g \cdot h} = (L_g)_*(X_{|h}).$$

Since L_g is always a diffeomorphism, $(L_g)_*$ is always nonsingular. The set of left-invariant vector fields on G form a real n–dimensional vector space, since any left-invariant vector field X on G is of the form

$$X_{|g} = (L_g)_*\xi,$$

where ξ is a fixed element of the tangent space to G at the identity e. Conversely, for any given $\xi \in G_e$, the vector field X on G is left-invariant. One checks that for left-invariant vector fields X, Y on G, the vector field $[X, Y]$ is left-invariant. Hence, the set of left-invariant vector fields on G form an n–dimensional Lie algebra $\mathfrak{g}$.

A 1*–parameter subgroup* is an analytic homomorphism $\gamma : \mathbb{R} \to G$. It is known that any left-invariant vector field X is complete, that is, its integral curves are defined over the whole real line; that the integral curve γ of $X \in \mathfrak{g}$ through the identity e of G is a 1–parameter subgroup of G; and that all other integral curves of X are left translates of γ.

We may identify G_e with $\mathfrak{g}$, by our remarks. The *exponential map* $\exp : \mathfrak{g} \to G$ is defined by

$$\exp \xi = \gamma_\xi(1),$$

where γ_ξ is the 1–parameter subgroup with velocity vector ξ at e. The 1–parameter subgroup γ_ξ is then given by

$$\gamma_\xi(t) = \exp t\xi.$$

As in the case of Riemannian manifolds, there exists a starlike neighborhood $\mathcal{W}$ of the origin of $\mathfrak{g}$ such that $\exp|\mathcal{W}$ is a diffeomorphism of $\mathcal{W}$ onto a neighborhood of e in G. (We shall be careful with notation when the two exponential maps coexist in the same discussion.)

Note that one always has a natural collection of Riemannian metrics on Lie groups. Indeed, given any basis $\{e_1, \ldots, e_n\}$ of G_e, one declares them to be orthonormal, and then uses left translation to declare the associated left-invariant vector fields $X_1, \ldots, X_n$ (satisfying $X_{j|e} = e_j$) orthonormal at every point of G.

(1) The most basic example of a Lie group is $\mathcal{GL}(V)$, the *general linear group* of a finite dimensional vector space, that is, $\mathcal{GL}(V)$ consists of all nonsingular linear transformations of V to itself. The group multiplication is given by the composition of elements of $\mathcal{GL}(V)$ as mappings, and the analytic manifold structure is given by viewing $\mathcal{GL}(V)$ as an open submanifold of the space $\mathcal{L}(V)$ of *all* linear transformations of V, this latter space identified (after a choice of basis of V) with $\mathbb{R}^{n^2}$, where $n = \dim V$. Given any $A \in \mathcal{L}(V)$ one sees that, in the topology of $\mathcal{L}(V)$, the series

$$e^{tA} := \sum_{k=0}^{\infty} t^k A^k / k!$$

converges (uniformly for bounded t), and is a 1–parameter subgroup of $\mathcal{GL}(V)$. One concludes that the Lie algebra of $\mathcal{GL}(V)$ is $\mathfrak{g} = \mathcal{L}(V)$, and that the Lie multiplication in $\mathfrak{g}$ is given by

$$[A, B] = AB - BA.$$

(2) Once one has a Lie group G, one furthers the collection of examples by considering Lie subgroups. A subgroup H of G is called a *Lie subgroup of* G if H is a 1–1 immersed submanifold of G. Certainly, a Lie subgroup is itself a Lie group (note: its topology might not coincide with relative topology, since it is only immersed), with its Lie algebra $\mathfrak{h}$ a Lie subalgebra of $\mathfrak{g}$. Also, one knows that if G is a Lie group, and H an (abstract) subgroup of G which is also a closed subset of G, then there exists a unique analytic structure on H such that H is a Lie group. Standard examples of subgroups of $\mathcal{GL}(V)$ are (i) the *special linear group*, $\mathcal{SL}(V)$, consisting of those elements of $\mathcal{GL}(V)$ with determinant equal to 1, (ii) the *orthogonal group*, $\mathcal{O}(V)$, consisting of those elements of $\mathcal{GL}(V)$ that preserve a given inner product on V.

Exercise I.17.

(a) Let V be a real n–dimensional vector space, $q \in V$, A a linear transformation of V, and $a_{A;q} : V \to V$ given by $a_{A;q}(p) = Ap + q$. Check that

$$(a_{A;q})_{*|p} = \mathfrak{I}_{a_{A;q}(p)} A \mathfrak{I}_p{}^{-1}.$$

In particular, if V is an inner product space and A an orthogonal transformation, then $a_{A;q}$ is an isometry of V for every $q \in V$.

(b) Show that the Lie algebra of $\mathcal{SL}(V)$ consists of linear transformations with trace equal to 0 (see Proposition II.8.2), and the Lie algebra of $\mathcal{O}(V)$ consists of the skew-symmetric linear transformations of V.

(3) Here, we consider Lie groups with bi-invariant metrics, namely we consider Lie groups that carry a Riemannian metric such that the metric is invariant relative to both left and right translations by elements of G. Before considering any details of this situation, we first comment on inner automorphisms.

For every $g \in G$, the map of G to itself, inn_g, given by

$$\operatorname{inn}_g(g') = g \cdot g' \cdot g^{-1}$$

is an automorphism of G that fixes the identity (referred to as an *inner* automorphism). In particular $(\operatorname{inn}_g)_*$ maps G_e to itself (nonsingular!), thereby determining the *representation* $g \mapsto (\operatorname{inn}_g)_*$ *of* G *acting on* $\mathfrak{g}$. It is common to write $\operatorname{Ad} g$ for $(\operatorname{inn}_g)_*$. So $\operatorname{Ad} : G \to \mathcal{GL}(\mathfrak{g})$ is an analytic homomorphism, *the adjoint representation of* G *in* $\mathcal{GL}(\mathfrak{g})$. It is well-known that for $\xi \in \mathfrak{g}$, and $\operatorname{ad} : \mathfrak{g} \to \mathcal{L}(\mathfrak{g})$ given by

$$(\operatorname{ad} \xi)(\eta) = [\xi, \eta],$$

we have

$$\operatorname{Ad} \exp t\xi = e^{t \operatorname{ad} \xi}.$$

Exercise I.18. If G possesses a bi-invariant Riemannian metric, show that for every $\xi, \eta, \zeta \in \mathfrak{g}$ we have

$$\langle [\zeta, \xi], \eta \rangle + \langle \xi, [\zeta, \eta] \rangle = 0.$$

Exercise I.19.

(a) Show that if G is any Lie group with left-invariant vector fields $\{e_1, \ldots, e_n\}$, $n = \dim G$, with dual 1–forms $\{\omega^1, \ldots, \omega^n\}$, then there exist constants ${C_{jk}}^i$ such that

$$[e_j, e_k] = \sum_i {C_{jk}}^i e_i, \qquad d\omega^i = -\frac{1}{2} \sum_{j,k} {C_{jk}}^i \, \omega^j \wedge \omega^k,$$

$$\sum_\ell {C_{jk}}^\ell {C_{\ell i}}^r + {C_{ij}}^\ell {C_{\ell k}}^r + {C_{ki}}^\ell {C_{\ell j}}^r = 0.$$

(b) Assume that G possesses a bi-invariant Riemannian metric relative to which $\{e_1, \ldots, e_n\}$ are orthonormal. Show that

$$C_{jk}{}^{i} + C_{ji}{}^{k} = 0.$$

(c) Continue our assumptions as in (b). Show that connection forms of the Levi–Civita connection are given, relative to the frame $\{e_1, \ldots, e_n\}$, by

$$\omega_j{}^i = \frac{1}{2}\sum_k C_{kj}{}^i \omega^k;$$

so for left-invariant vector fields X, Y on G, we have

$$\nabla_X Y = [X, Y]/2.$$

Also show that the curvature 2–forms are given by

$$\Omega_j{}^i = \frac{1}{4}\sum_{\ell,r,s} C_{js}{}^{\ell} C_{\ell r}{}^{i}\, \omega^r \wedge \omega^s;$$

so

$$R(\xi,\eta)\zeta = \frac{1}{4}\{[[\zeta,\eta],\xi] - [[\zeta,\xi],\eta]\} \quad \text{and} \quad \langle R(\xi,\eta)\xi,\eta\rangle = \frac{1}{4}\|[\xi,\eta]\|^2$$

for all $\xi, \eta \in \mathfrak{g}$.

Example I.9.5. We now consider calculations associated to Riemannian metrics on homogeneous spaces. Let G be a Lie group, H a closed subgroup, consider the set of cosets $\{gH : g \in G\}$, which we denote by G/H, and let $\pi : G \to G/H$ denote the projection $\pi(g) = gH$. Then, the quotient topology on G/H induced by π will be the unique topology on G/H such that π is continuous and open. Since H is closed, G/H is a Hausdorff space. Let $\mathfrak{g}$, $\mathfrak{h}$ denote the Lie algebras of G, H, respectively, $\mathfrak{m}$ a complementary subspace of $\mathfrak{h}$ in $\mathfrak{g}$, and $\psi = \exp|\mathfrak{m}$. Then, G/H has a unique analytic structure for which the following are true: (a) Let $\mathcal{W}$ be the neighborhood of 0 in G_e on which $\exp|\mathcal{W}$ is a diffeomorphism (as above). Then, there exists a neighborhood $\mathcal{U}$ of 0 in $\mathfrak{m}$, such that $\mathcal{U} \subseteq \mathfrak{m} \cap \mathcal{W}$, and such that ψ maps it diffeomorphically onto its image Q in G (with relative topology) and $\pi \circ \psi$ maps $\mathcal{U}$ diffeomorphically onto a neighborhood V of $o = \pi(H)$ on G/H. In particular, π is a submersion, and there exists a local C^∞ section $\mu : V \to Q \subseteq G$ such that

$$\mu(\pi(\exp\xi)) = \exp\xi$$

for all $\xi \in \mathcal{U}$. (b) If to each $g \in G$ we assign the *left translation of* G/H, $\tau_g : G/H \to G/H$ by

$$\tau_g(g'H) = (gg')H,$$

then G is a *transitive Lie transformation group of the homogeneous space* G/H, that is,

(i) $\tau_{g_1g_2} = \tau_{g_1} \circ \tau_{g_2}$ for all $g_1, g_2 \in G$;

(ii) for all $x \in G/H$, $g \in G$ the map $G \times G/H \to G/H$ given by $(g, x) \mapsto \tau_g(x)$ is differentiable,

(iii) the collection of elements of G that leave the point $p = gH$ fixed, the *isotropy group of p*, is given by gHg^{-1},

(iv) and for any $x_1, x_2 \in G/H$, there exists an element $g \in G$ such that $\tau_g(x_1) = x_2$.

Note that $\pi_* : \mathfrak{m} \to (G/H)_o$ is an isomorphism, and we may henceforth identify $\mathfrak{m}$ with the tangent space to G/H at o. Certainly, for any $h \in H$, we have $\tau_h(o) = o$. One checks that if $\mathrm{Ad}\, h : \mathfrak{g} \to \mathfrak{g}$ leaves $\mathfrak{m}$ invariant, then $(\tau_h)_{*|o} : \mathfrak{m} \to \mathfrak{m}$ is given by

$$(\tau_h)_{*|o} = \mathrm{Ad}\, h|\mathfrak{m}.$$

We say that G/H is *reductive* if $\mathrm{Ad}\, h$ leaves $\mathfrak{m}$ invariant, for all $h \in \mathfrak{h}$. Naturally we have, under all circumstances,

$$[\mathfrak{h}, \mathfrak{h}] \subseteq \mathfrak{h}.$$

We now assume G/H to be *Riemannian homogeneous,* that is, G/H is endowed with a Riemannian metric relative to which τ_g is an isometry, for every $g \in G$. One can now think of the inner product as also existing on $\mathfrak{m}$. We also assume that G/H is reductive. Thus, we have

$$[\mathfrak{h}, \mathfrak{m}] \subseteq \mathfrak{m}, \quad \text{and} \quad \langle [\varpi, \xi], \eta\rangle + \langle \xi, [\varpi, \eta]\rangle = 0$$

for all $\varpi \in \mathfrak{h}$, $\xi, \eta \in \mathfrak{m}$.

To calculate the curvature and geodesics of G/H, we use a variant of the argument for Riemannian submersions. We restrict ourselves to *naturally reductive* Riemannian metrics, that is, in addition to the above, we assume that

$$\langle [\zeta, \xi]_{\mathfrak{m}}, \eta\rangle + \langle \xi, [\zeta, \eta]_{\mathfrak{m}}\rangle = 0$$

for all $\xi, \eta, \zeta \in \mathfrak{m}$, whereby the subscript $\mathfrak{m}$ we mean projection onto $\mathfrak{m}$.

Assume that $\dim G = n$, $\dim H = \ell$, $\dim G/H = k$, with $k + \ell = n$. Let $\{e_1, \ldots, e_k\}$ be an orthonormal basis of $\mathfrak{m}$, and $\{e_{k+1}, \ldots, e_n\}$ a basis of $\mathfrak{h}$, and determine the associated left-invariant vector fields $\{X_1, \ldots, X_n\}$ on G. Let $\mathcal{U}$

be the neighborhood of 0 in $\mathfrak{m}$, V be the neighborhood of $o = \pi(H)$ in G/H described above, and define the orthonormal frame field $\{E_1, \ldots, E_k\}$ on V by

$$E_{j|\pi(\exp \xi)} = (\tau_{\exp \xi})_* e_j, \qquad j = 1, \ldots, k, \qquad \xi \in \mathfrak{m}.$$

Now imitate the calculations given for Riemannian submersions (the projection π of G to G/H *is* a submersion) to:

Exercise I.20.

(a) Show that G/H is complete, and for every $\xi \in \mathfrak{m}$, we have

$$\mathrm{Exp}_o \xi = \pi(\exp \xi),$$

where (just here) Exp_o denotes the Riemannian exponential map of G/H at o.

(b) Show that if $\xi, \eta \in \mathfrak{m}$ and if $Y(t)$ is the vector field along the geodesic $\gamma(t) = \pi(\exp \xi)$ given by

$$Y(t) = (\tau_{\exp t\xi})_* \eta.$$

Then

$$\nabla_t Y(t) = \frac{1}{2}(\tau_{\exp t\xi})_* [\xi, \eta]_{\mathfrak{m}}.$$

(c) Show that the curvature at $o = \pi(H)$ is given by

$$R(\xi, \eta)\zeta = [[\xi, \eta]_{\mathfrak{h}}, \zeta] + \frac{1}{2}[[\xi, \eta]_{\mathfrak{m}}, \zeta]_{\mathfrak{m}} + \frac{1}{4}[[\zeta, \xi]_{\mathfrak{m}}, \eta]_{\mathfrak{m}} + \frac{1}{4}[[\eta, \zeta]_{\mathfrak{m}}\xi]_{\mathfrak{m}}$$

for all $\xi, \eta, \zeta \in \mathfrak{m}$.

(d) Assume that G possesses a bi-invariant Riemannian metric, with $\mathfrak{m}$ the orthogonal complement of $\mathfrak{h}$ in $\mathfrak{g}$. Then $\mathrm{Ad}\, h$ leaves $\mathfrak{m}$ invariant, for all $h \in \mathfrak{h}$, and there therefore exists a naturally reductive Riemannian metric on G/H such that the projection π is a Riemannian submersion. Prove that

$$\langle R(\xi, \eta)\xi, \eta\rangle = |[\xi, \eta]_{\mathfrak{h}}|^2 + \frac{1}{4}|[\xi, \eta]_{\mathfrak{m}}|^2$$

for all $\xi, \eta \in \mathfrak{m}$.

Exercise I.21. (Hopf fibration of $\mathbb{S}^3$) We let $\mathbf{1}, \mathbf{i}, \mathbf{j}, \mathbf{k}$ denote the standard basis of $\mathbb{R}^4$. Beyond the vector space structure of $\mathbb{R}^4$, we define a multiplication of elements of $\mathbb{R}^4$, where the multiplication of the natural basis is given by

$$\mathbf{1i} = \mathbf{i} = \mathbf{i1}, \;\; \mathbf{1j} = \mathbf{j} = \mathbf{j1}, \;\; \mathbf{1k} = \mathbf{k} = \mathbf{k1}, \qquad \mathbf{i}^2 = \mathbf{j}^2 = \mathbf{k}^2 = -\mathbf{1},$$

and

$$\mathbf{ij} = \mathbf{k} = -\mathbf{ji}, \quad \mathbf{jk} = \mathbf{i} = -\mathbf{kj}, \quad \mathbf{ki} = \mathbf{j} = -\mathbf{ik}.$$

With this bilinear multiplication, $\mathbb{R}^4$ becomes an algebra, the *quaternions*.

With each element

$$\mathbf{x} = \alpha\mathbf{1} + \beta\mathbf{i} + \gamma\mathbf{j} + \delta\mathbf{k}$$

we associate its *conjugate*

$$\overline{\mathbf{x}} = \alpha\mathbf{1} - \beta\mathbf{i} - \gamma\mathbf{j} - \delta\mathbf{k},$$

and its *norm* $|\mathbf{x}|$ defined by

$$|\mathbf{x}|^2 := \mathbf{x}\overline{\mathbf{x}} = \overline{\mathbf{x}}\mathbf{x}$$

with associated bilinear form

$$\langle \mathbf{x}|\mathbf{y}\rangle = \frac{1}{2}(\overline{\mathbf{x}}\mathbf{y} + \overline{\mathbf{y}}\mathbf{x}).$$

Note that $\overline{\mathbf{x}\mathbf{y}} = \overline{\mathbf{y}}\,\overline{\mathbf{x}}$, which implies

$$|\overline{\mathbf{x}}|^2 = |\mathbf{x}|^2, \qquad |\mathbf{x}\mathbf{y}| = |\mathbf{x}||\mathbf{y}|.$$

So

$$|\mathbf{x}| = 1 \qquad \Rightarrow \qquad |\mathbf{x}\mathbf{y}| = |\mathbf{y}\mathbf{x}| = |\mathbf{y}|.$$

We conclude that the unit quaternions, $\mathbb{S}^3$, is a compact Lie group under the quaternionic multiplication.

Also, for any $\mathbf{x} \neq 0$, we have

$$\mathbf{x}^{-1} = \frac{\overline{\mathbf{x}}}{|\mathbf{x}|^2}.$$

Since $\mathbf{1}$ is the identity element of the unit quaternions, the basis of the tangent space to $\mathbb{S}^3$ at $\mathbf{1}$ can be thought of as given by

$$\mathbf{i},\ \ \mathbf{j},\ \ \mathbf{k}.$$

More precisely, it is given by

$$\Im_{\mathbf{1}}\mathbf{i},\ \ \Im_{\mathbf{1}}\mathbf{j},\ \ \Im_{\mathbf{1}}\mathbf{k}.$$

(a) Show that if $\boldsymbol{\xi}$ is a linear combination of $\mathbf{i}, \mathbf{j}, \mathbf{k}$, then (i) $\boldsymbol{\xi}^2 = -|\boldsymbol{\xi}|^2$, and (ii) the 1–parameter subgroup in $\mathbb{S}^3$, $\gamma(t) = \exp t\boldsymbol{\xi}$, is given by

$$\exp t\boldsymbol{\xi} = (\cos |\boldsymbol{\xi}|t)\mathbf{1} + (\sin |\boldsymbol{\xi}|t)\frac{\boldsymbol{\xi}}{|\boldsymbol{\xi}|}.$$

(b) Show that the Lie algebra is given by

$$[\mathbf{i}, \mathbf{j}] = 2\mathbf{k}, \quad [\mathbf{j}, \mathbf{k}] = 2\mathbf{i}, \quad [\mathbf{k}, \mathbf{i}] = 2\mathbf{j}.$$

(c) Declare the basis $\{\mathbf{i}, \mathbf{j}, \mathbf{k}\}$ of $\mathbb{S}^3$ to be orthonormal in the tangent space to $\mathbb{S}^3$ at $\mathbf{1}$ ($\mathbb{S}^3$ does not have a Riemannian metric, yet), and use left-invariance to define a Riemannian metric on $\mathbb{S}^3$. Show that the Riemannian metric is bi-invariant and has sectional curvature identically equal to 1 (see §II.1 for the definition of sectional curvature).

(d) Let H denote the Lie subgroup

$$H = \mathbb{S}^1 = \{\cos\theta\mathbf{1} + \sin\theta\mathbf{i} : \theta \in \mathbb{R}\}.$$

Show that G/H is the 2–sphere in $\mathbb{R}^3$ with constant sectional curvature equal to 4.

Exercise I.22. Given the orthogonal group $\mathcal{O}(n)$ acting on $\mathbb{R}^n$.

(a) Show that the bilinear form given by

$$\langle A, B\rangle = -\frac{1}{2}\operatorname{tr} AB$$

for $A, B \in \mathfrak{g} = \mathfrak{o}(n)$, where $\mathfrak{o}(n)$ is the Lie algebra of skew-symmetric linear transformations of $\mathbb{R}^n$, is a bi-invariant Riemannian metric on $G = \mathcal{O}(n)$.

(b) Calculate the Riemannian metric and curvature of the *Grassmann manifold of k–planes in* $\mathbb{R}^n$,

$$\mathcal{O}(n)/\{\mathcal{O}(k) \times \mathcal{O}(n-k)\}.$$

II
Riemannian Curvature

In this chapter, we begin to consider the invariant that truly characterizes differential geometry – curvature. The original formal definition of the curvature tensor, in §I.4, gives little hint to its profound geometric meaning; nevertheless, we indicated there the direction in which we are most interested in studying curvature – Jacobi's equation. In the Riemannian case, the torsion tensor of the Levi-Civita connection vanishes identically, so the curvature is the exclusive influence in studying the behavior of geodesics neighboring a given geodesic.

This, of course, is not the historical origin of curvature. In the beginning of differential geometry (i.e., in the beginning of the nineteenth century), it was viewed from the perspective of immediate human experience. Namely, the curvature of a curve attempted to measure the deviation of a curve in a plane or in space from being a straight line, and the various studies of curvature of a surface situated in space attempted to express how the surface deviated from being a plane in space. C. F. Gauss (1825, 1827) was the first to realize that one aspect of curvature, what we refer to as the Gauss curvature,[1] did not depend on how the surface is situated in Euclidean space; that if the surface was bent – that is, deformed in such a manner as to preserve the measurement of lengths and angles *in* the surface – then while some curvatures were changed, other curvatures (namely, the Gauss curvature) were left invariant under the bending. This very fact created the distinction between intrinsic and extrinsic properties of the surface: The intrinsic properties are invariant under bending of the surface, and, therefore, belong to the geometric study of the surface for itself – independent of how the surface is visualized as situated in Euclidean space. The extrinsic properties are those properties, once the original surface with its intrinsic geometry is given, which describe the particular details of how

[1] R. Osserman (1990) has noted that the definition and first study of the Gauss curvature goes back to O. Rodrigues (1816).

the surface is situated in Euclidean space. It studies precisely those properties that change under the bending.

Both studies of submanifolds – the intrinsic and extrinsic – are alive and well. Any particular focus on one of them will indicate the point of view with which one looks at curvature. Subsequent experience (of the last 150 years) has shown that the notion of curvature is so rich that, even within each of the categories of intrinsic and extrinsic, one still constantly chooses, and thereby refines, the view with which one studies the curvature of a Riemannian manifold.

Here, our view, by and large, is in the intrinsic category. Within this category, the original most striking perspective of the study of curvature is through the celebrated Gauss–Bonnet theorem and formula (see §V.2). Nevertheless, as mentioned, we are mainly interested in studying the curvature from the perspective of the "straight lines" of the Riemannian manifold – the geodesics.

The flow of ideas of the chapter is as follows:

In the first three sections, we present the basic notions and facts of the curvature of the Levi-Civita connection of a given Riemannian manifold. We recapture Gauss' original calculation, relating (it is no longer a discovery) the intrinsic curvature of a submanifold to the curvature of the ambient manifold and the extrinsic geometry of the imbedding of the submanifold. We also describe, in some detail, the model spaces of constant sectional curvature. These are the spaces that represent the first level of study beyond Euclidean space, and it is by reference to these spaces that the general Riemannian manifolds are studied.

We then, in the succeeding sections, study the "local" theory of geodesics. The word "local" in this context has at least two meanings. In Chapter I, it referred to the fact that, given a point in the Riemannian manifold, one can find a sufficiently small neighborhood of that point for which the shortest path, from the point in question to any other point in the neighborhood, is given by a unique length minimizing geodesic segment joining the two points. Here, in Chapter II, by "local" we mean something else: Given a geodesic emanating from a point, we wish to know for how long the geodesic minimizes the length of all paths joining the initial point of the geodesic to the point in question on the geodesic. The first level of study (which we initiate in this chapter) is to study the competing length of curves which are "close" to the original geodesic in which we are interested. This is what we mean by "local." It is these questions that lead to the detailed study of Jacobi's equations and the associated Jacobi criteria, and the study of the role played by curvature in these phenomena. At the end of the chapter, we bring these notions back to the exponential map, in that the study of one-parameter families of geodesics is essentially equivalent to the study of the differential of the exponential map.

§II.1. The Riemann Sectional Curvature

Unless otherwise noted, when given a Riemannian metric, we only use the Levi-Civita connection.

So, M is our Riemannian manifold with Levi-Civita connection ∇. Recall that, for $X, Y, Z, W \in \Gamma(TM)$, we have

$$R(X, Y)Z = \nabla_Y \nabla_X Z - \nabla_X \nabla_Y Z - \nabla_{[Y,X]} Z,$$

where R is the curvature tensor of ∇. Of course we have

$$R(X, Y)Z + R(Y, X)Z = 0; \tag{II.1.1}$$

and since the torsion of ∇ vanishes identically, we also have

$$R(X, Y)Z + R(Z, X)Y + R(Y, Z)X = 0. \tag{II.1.2}$$

One now establishes, using (I.5.2), (I.5.3), and Remark I.3.1,

$$\langle R(X, Y)Z, W\rangle - \langle R(Z, W)X, Y\rangle = 0, \tag{II.1.3}$$

$$\langle R(X, Y)Z, W\rangle + \langle R(X, Y)W, Z\rangle = 0. \tag{II.1.4}$$

Since (II.1.1) implies that the curvature tensor vanishes identically for $\dim M = 1$, all discussions concerning the curvature tensor will assume $\dim M \geq 2$.

Proposition II.1.1. *For $p \in M$, $\xi, \eta \in M_p$ define*

$$\mathbf{k}(\xi, \eta) = \langle R(\xi, \eta)\xi, \eta\rangle.$$

Then, by (II.1.1)–(II.1.4), *we have for any $\xi, \eta, \zeta, \mu \in M_p$*

$$\langle R(\xi, \eta)\zeta, \mu\rangle = \frac{1}{6} \frac{\partial^2}{\partial s \partial t}\bigg|_{s=t=0} \{\mathbf{k}(\xi + s\zeta, \eta + t\mu) - \mathbf{k}(\xi + s\mu, \eta + t\zeta)\}.$$

Thus, the function $\mathbf{k} : M_p \times M_p \to \mathbb{R}$ and the properties (II.1.1)–(II.1.4) *completely determine $R : M_p \times M_p \times M_p \to M_p$.*

Proof. Direct calculation as for (II.1.3), (II.1.4).

Proposition II.1.2. *For $p \in M$, $\xi, \eta, \zeta \in M_p$, define*

$$R_1(\xi, \eta)\zeta = \langle \xi, \zeta\rangle\eta - \langle \eta, \zeta\rangle\xi, \qquad \mathbf{k}_1(\xi, \eta) = \langle R_1(\xi, \eta)\xi, \eta\rangle.$$

Then

$$\mathbf{k}_1(\xi, \eta) = \langle \xi, \xi \rangle \langle \eta, \eta \rangle - \langle \xi, \eta \rangle^2,$$

and R_1 *satisfies the axioms* (II.1.1)–(II.1.4). *Furthermore, if* ξ, η *are linearly independent tangent vectors in* M_p, *then*

$$\mathcal{K}(\xi, \eta) := \frac{\mathbf{k}(\xi, \eta)}{\mathbf{k}_1(\xi, \eta)} = \frac{\langle R(\xi, \eta)\xi, \eta \rangle}{|\xi|^2|\eta|^2 - \langle \xi, \eta \rangle^2}$$

is well-defined and only depends on the 2–dimensional subspace determined by ξ *and* η.

Proof. All of the claims, except for the last, are straightforward exercises. For the last claim, it suffices to note that if $\alpha, \beta, \gamma, \delta$ are real numbers and ξ, η linearly independent tangents vectors in M_p, then

$$\langle R(\alpha\xi + \beta\eta, \gamma\xi + \delta\eta)(\alpha\xi + \beta\eta), \gamma\xi + \delta\eta \rangle = (\alpha\delta - \beta\gamma)^2 \langle R(\xi, \eta)\xi, \eta \rangle,$$

and

$$\begin{aligned} &|\alpha\xi + \beta\eta|^2 |\gamma\xi + \delta\eta|^2 - \langle \alpha\xi + \beta\eta, \gamma\xi + \delta\eta \rangle \\ &\quad = (\alpha\delta - \beta\gamma)^2 \{|\xi|^2|\eta|^2 - \langle \xi, \eta \rangle\}, \end{aligned}$$

which implies the claim. ∎

Definition. We refer to $\mathcal{K}(\xi, \eta)$ as the *sectional curvature of the 2–section determined by* ξ, η. We note that, if G_2 is the complete collection of all 2–dimensional spaces tangent to M, then G_2 can be provided a differentiable structure in a natural manner, and $\mathcal{K} : G_2 \to \mathbb{R}$ will then be differentiable.

Theorem II.1.1. *If* $\dim M = 2$, *then* $G_2 = M$; $\mathcal{K}$ *is called* the Gauss curvature of M. *For* $p \in M$, $\xi, \eta, \zeta \in M_p$, *we have*

$$R(\xi, \eta)\zeta = \mathcal{K}(p) R_1(\xi, \eta)\zeta. \tag{II.1.5}$$

Proof. We only have to deal with the last claim. Let $\{e_1, e_2\}$ be an orthonormal basis of M_p. One easily sees that we only have to verify (II.1.5) for $\xi = \zeta = e_1$, $\eta = e_2$.

To this end, map $\mathsf{R} : M_p \to M_p$ by

$$\mathsf{R}(\xi) = R(e_1, \xi)e_1.$$

Then, by (II.1.3), R is self-adjoint and therefore diagonalizable. Since e_1 is an eigenvector of R with eigenvalue 0, we have that e_2 is an eigenvalue of

R with eigenvalue

$$\langle \mathsf{R}e_2, e_2 \rangle = \langle R(e_1, e_2)e_1, e_2 \rangle = \mathcal{K}(p).$$

Thus,

$$R(e_1, e_2)e_1 = \mathcal{K}(p)e_2 = \mathcal{K}(p)R_1(e_1, e_2)e_1,$$

and the claim is proven. ∎

Definition. For $p \in M$, we define the *Ricci curvature tensor* $\mathrm{Ric} : M_p \times M_p \to \mathbb{R}$ by

$$\text{(II.1.6)} \qquad \mathrm{Ric}\,(\xi, \eta) = \text{ trace } (\zeta \mapsto R(\xi, \zeta)\eta),$$

and the *scalar curvature* S is the trace of Ric with respect to the Riemannian metric.

In particular, we have for any orthonormal basis of M_p, $\{e_1, \ldots, e_n\}$,

$$\text{(II.1.7)} \qquad \mathrm{Ric}\,(\xi, \eta) = \sum_{j=1}^{n} \langle R(\xi, e_j)\eta, e_j \rangle.$$

Thus, Ric is a symmetric bilinear form on M_p. To calculate its associated quadratic form, pick $\{e_1, \ldots, e_n\}$ so that $e_n = \xi/|\xi|$; then

$$\mathrm{Ric}\,(\xi, \xi) = \left\{ \sum_{j=1}^{n-1} \mathcal{K}(e_j, \xi) \right\} |\xi|^2.$$

For any orthonormal basis $\{e_1, \ldots, e_n\}$ of M_p, we have

$$S = \sum_{j \neq k;\, j,k=1}^{n} \mathcal{K}(e_j, e_k).$$

§II.2. Riemannian Submanifolds

We are given Riemannian manifolds $M, \overline{M}$ with respective Riemannian metrics $g, \overline{g}$ and an isometric imbedding φ of M in $\overline{M}$. Thus, $\dim M \le \dim \overline{M}$; should the dimensions be equal, M will be an open submanifold of $\overline{M}$. Since φ is a diffeomorphism, for any differentiable vector field $X \in \Gamma(TM)$, we have $\varphi_* X$, a well-defined vector field on $\varphi(M)$, that is, $\varphi_* X \in \Gamma(\varphi_* TM)$. One easily verifies that

$$\varphi_*[X, Y] = [\varphi_* X, \varphi_* Y]$$

for all $X, Y \in \Gamma(TM)$. Since φ is an isometry, we have by definition

$$\overline{g}(\varphi_*\xi, \varphi_*\eta) = g(\xi, \eta)$$

for all $\xi, \eta \in M_p$, $p \in M$. For the rest of this section, we assume $\dim M < \dim \overline{M}$.

Thus, in what follows, nothing is lost in assuming that $M \subseteq \overline{M}$ and φ is the inclusion map. For any $p \in M$, we let $M_p{}^\perp$ denote the orthogonal complement of M_p in $\overline{M}_p$. For any $p \in M, \overline{\xi} \in \overline{M}_p$, we shall denote the projection of $\overline{\xi}$ onto M_p by $\overline{\xi}^T$, and the projection of $\overline{\xi}$ onto $M_p{}^\perp$ by $\overline{\xi}^N$. The *normal bundle of M in $T\overline{M}$* is defined by

$$\nu M = \bigcup_{p \in M} M_p{}^\perp,$$

and has a natural differentiable structure such that the inclusion of νM in $T\overline{M}$, the projection of $T\overline{M}$ to νM, and the projection of νM to M are all differentiable. Also, we let $\Gamma(\nu M)$ denote the differentiable sections of νM, that is, those differentiable maps of M into νM such that the image of any point p is an element of $M_p{}^\perp$.

Proposition II.2.1. *Let $\nabla, \overline{\nabla}$ be the respective Levi-Civita connections of $g, \overline{g}$. Then, for any $p \in M$, $\xi \in M_p$, and $Y \in \Gamma(TM)$, we have*

$$\nabla_\xi Y = (\overline{\nabla}_\xi Y)^T. \tag{II.2.1}$$

Furthermore, to each $p \in M$, there exists a symmetric bilinear map $\mathfrak{B} : M_p \times M_p \to M_p{}^\perp$ such that, for any $\xi, \eta \in M_p$, $Y \in \Gamma(TM)$ satisfying $Y_{|p} = \eta$, we have

$$\mathfrak{B}(\xi, \eta) = (\overline{\nabla}_\xi Y)^N. \tag{II.2.2}$$

$\mathfrak{B}$ is called the second fundamental form of M in $\overline{M}$. *If to each $v \in M_p{}^\perp$, we let $\mathfrak{b}_v : M_p \times M_p \to \mathbb{R}$ be the bilinear form defined by*

$$\mathfrak{b}_v(\xi, \eta) = \langle \mathfrak{B}(\xi, \eta), v \rangle, \tag{II.2.3}$$

then the self-adjoint linear transformation $\mathfrak{A}^v : M_p \to M_p$ determined by

$$\mathfrak{b}_v(\xi, \eta) = \langle \mathfrak{A}^v \xi, \eta \rangle$$

is given by

$$\mathfrak{A}^v \xi = -(\overline{\nabla}_\xi V)^T, \tag{II.2.4}$$

where V is any extension of v to an element of $\Gamma(\nu M)$.

Proof. To establish (II.2.1), one first decomposes $\overline{\nabla}_\xi Y$, for $\xi \in M, Y \in \Gamma(TM)$ by

$$\overline{\nabla}_\xi Y = D_\xi Y + \mathfrak{B}(\xi, Y),$$

where $D_\xi Y = (\overline{\nabla}_\xi Y)^T$, and $\mathfrak{B}(\xi, Y) = (\overline{\nabla}_\xi Y)^N$ as in (II.2.2). Next, one verifies directly that D is a connection on M with no torsion (i.e., it satisfies (I.5.2)) and preserves the inner product (i.e., it satisfies (I.5.3)). Thus, by Theorem I.5.1, D is the Levi-Civita connection of M, and (II.2.1) follows.

For the symmetry of $\mathfrak{B}$, we have for any $X, Y \in \Gamma(TM)$,

$$\mathfrak{B}(X, Y) - \mathfrak{B}(Y, X) = \{\overline{\nabla}_X Y - \overline{\nabla}_Y X\}^N = [X, Y]^N = 0,$$

since X, Y are tangent to M. That $(\mathfrak{B}(X, Y))_{|p}$ depends only on $X_{|p}, Y_{|p}$ follows from the symmetry of $\mathfrak{B}$ and (II.2.2).

Finally, we wish to show that if $\xi, \eta \in M_p$, $v \in {M_p}^\perp$ and V is any extension of v in $\Gamma(\nu M)$, then

$$\langle \mathfrak{B}(\xi, \eta), v\rangle = -\langle(\overline{\nabla}_\xi V)^T, \eta\rangle.$$

To do so extend η to a vector field $Y \in \Gamma(TM)$. Then

$$\begin{aligned} -\langle(\overline{\nabla}_\xi V)^T, \eta\rangle &= -\langle\overline{\nabla}_\xi V, \eta\rangle = -\langle\overline{\nabla}_\xi V, Y\rangle \\ &= -\xi\langle V, Y\rangle + \langle V, \overline{\nabla}_\xi Y\rangle = \langle v, \mathfrak{B}(\xi, \eta)\rangle. \quad \blacksquare \end{aligned}$$

Remark II.2.1. If we refer to the second fundamental form, we should have already described the first fundamental form. The *first fundamental form* is just the restriction of the Riemannian metric of $\overline{M}$ to the immersed submanifold M.

The map $\mathfrak{A}^v$ is commonly referred to as the *Weingarten map*.

The *principal curvatures of M in $\overline{M}$ at $p \in M$, relative to the normal direction v*, are the eigenvalues of the second fundamental form $\mathfrak{b}_v$ relative to the first fundamental form – equivalently, the eigenvalues of $\mathfrak{A}^v$. The associated eigenvectors are referred to as the *principal directions*.

Theorem II.2.1. *If R and $\overline{R}$ denote the respective curvature tensors of M and $\overline{M}$, then for any $p \in M$, $\xi, \eta, \zeta, \mu \in M_p$ we have*

$$R(\xi, \eta)\zeta = (\overline{R}(\xi, \eta)\zeta)^T + \mathfrak{A}^{\mathfrak{B}(\xi,\zeta)}\eta - \mathfrak{A}^{\mathfrak{B}(\eta,\zeta)}\xi, \tag{II.2.5}$$

$$\langle R(\xi, \eta)\zeta, \mu\rangle = \langle\overline{R}(\xi, \eta)\zeta, \mu\rangle + \langle\mathfrak{B}(\xi, \zeta), \mathfrak{B}(\eta, \mu)\rangle - \langle\mathfrak{B}(\xi, \mu), \mathfrak{B}(\eta, \zeta)\rangle.$$

In particular, if $\mathcal{K}$, $\overline{\mathcal{K}}$ denote the respective sectional curvatures of M, $\overline{M}$, then for linearly independent ξ, η we have

$$\mathcal{K}(\xi,\eta) = \overline{\mathcal{K}}(\xi,\eta) + \frac{\langle \mathfrak{B}(\xi,\xi), \mathfrak{B}(\eta,\eta)\rangle - |\mathfrak{B}(\xi,\eta)|^2}{|\xi|^2|\eta|^2 - \langle \xi,\eta\rangle^2}. \tag{II.2.6}$$

Proof. Direct calculation, using (II.2.1)–(II.2.4) and (I.4.2), (I.5.2), (I.5.3). ∎

The Second Fundamental Form Via Moving Frames

Now assume $\overline{M}$ is m–dimensional, and M n–dimensional with $n < m$. For $p \in M$, one may pick a neighborhood U of p in $\overline{M}$, with orthonormal frame field $\{\overline{e}_1, \ldots, \overline{e}_m\}$, and dual coframe $\{\overline{\omega}^1, \ldots, \overline{\omega}^m\}$. We then have equations (I.8.8), (I.8.9), and (I.8.10), (I.8.11), namely,

$$d\overline{\omega}^A = \sum_B \overline{\omega}^B \wedge \overline{\omega}_B{}^A, \qquad \overline{\omega}_A{}^B = -\overline{\omega}_B{}^A, \tag{II.2.7}$$

and

$$d\overline{\omega}_A{}^B = \sum_C \overline{\omega}_A{}^C \wedge \overline{\omega}_C{}^B - \overline{\Omega}_A{}^B, \tag{II.2.8}$$

where

$$\overline{\Omega}_A{}^B(X,Y) = \langle \overline{R}(X,Y)\overline{e}_A, \overline{e}_B\rangle = -\overline{\Omega}_B{}^A(X,Y). \tag{II.2.9}$$

For calculations relating M to $\overline{M}$, we may pick $\{\overline{e}_1, \ldots, \overline{e}_m\}$ so that $\{\overline{e}_1, \ldots, \overline{e}_n\}$ are tangent to M at all points of $M \cap U$, and $\{\overline{e}_{n+1}, \ldots, \overline{e}_m\}$ are normal to M at all points of $M \cap U$. (See Figure II.1.)

In the calculation that follows, we let $A, B, \ldots$ range over $1, \ldots, m$; $j, k, \ldots$ range over $1, \ldots, n$; and $\alpha, \beta, \ldots$ range over $n+1, \ldots, m$. We also let

$$e_A = \overline{e}_A|M, \qquad \omega^A = \overline{\omega}^A|M, \qquad \omega_B{}^A = \overline{\omega}_B{}^A|M.$$

Then, on M, we have

$$\omega^\alpha = 0,$$

which implies

$$d\omega^j = \sum_k \omega^k \wedge \omega_k{}^j, \tag{II.2.10}$$

which is the analogue of (II.2.1); and

$$0 = d\omega^\alpha = \sum_j \omega^j \wedge \omega_j{}^\alpha, \tag{II.2.11}$$

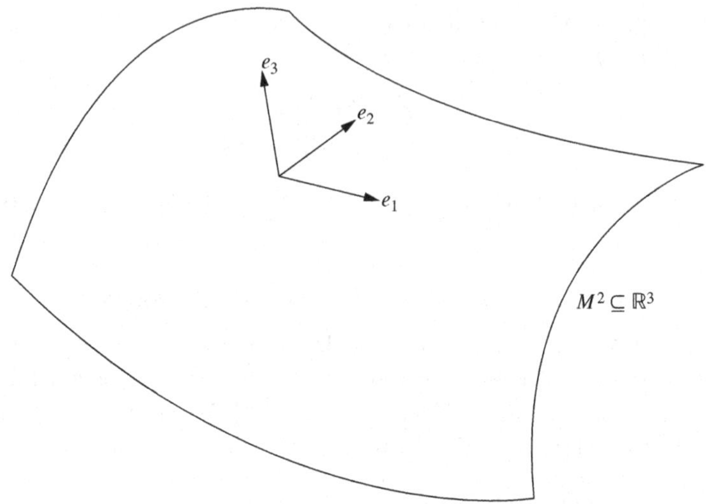

Figure II.1. Moving frame.

which implies

$$\omega_j{}^\alpha = \sum_k h_{jk}{}^\alpha \omega^k, \qquad h_{jk}{}^\alpha = h_{kj}{}^\alpha. \tag{II.2.12}$$

The equation (II.2.12) then encapsulates the description of the second fundamental form in Proposition II.2.1.

The Gauss equations of Theorem II.2.1 are obtained as follows: Let $\overline{\Omega}_A{}^B$ denote the matrix curvature 2–form of $\overline{M}$, as above, and $\Omega_j{}^k$ the matrix curvature 2–form of M. Then,

$$d\omega_j{}^k = \sum_\ell \omega_j{}^\ell \wedge \omega_\ell{}^k + \sum_\alpha \omega_j{}^\alpha \wedge \omega_\alpha{}^k - (\overline{\Omega}|M)_j{}^k,$$

which implies

$$\Omega_j{}^k = (\overline{\Omega}|M)_j{}^k - \sum_\alpha \omega_j{}^\alpha \wedge \omega_\alpha{}^k, \tag{II.2.13}$$

which is, using (II.2.12), a rewrite of Theorem II.2.1.

§II.3. Spaces of Constant Sectional Curvature

Let M be a Riemannian manifold of dimension ≥ 2, $\mathcal{K}$ the Riemann sectional curvature of 2–dimensional spaces tangent to M.

Definition. We say that M *has constant sectional curvature* $\kappa, \kappa \in \mathbb{R}$, if $\mathcal{K}(\sigma) = \kappa$ for all 2–sections σ.

From Propositions II.1.1 and II.1.2, we have immediately:

Proposition II.3.1. *M has constant sectional curvature κ if and only if for any $p \in M$, $\xi, \eta, \zeta \in M_p$ we have*

$$R(\xi, \eta)\zeta = \kappa\{\langle \xi, \zeta \rangle \eta - \langle \zeta, \eta \rangle \xi\}. \tag{II.3.1}$$

Euclidean Space

Let M be $\mathbb{R}^n$ with its usual inner product $(\xi, \eta) \mapsto \xi \cdot \eta$. As in §I.1, we let $\mathfrak{I}_p : \mathbb{R}^n \to (\mathbb{R}^n)_p$ be the canonical isomorphism associated to each $p \in \mathbb{R}^n$. The *standard Riemannian metric on* $\mathbb{R}^n$ will be defined by

$$\langle \xi, \eta \rangle = \mathfrak{I}_p{}^{-1}\xi \cdot \mathfrak{I}_p{}^{-1}\eta$$

for $p \in \mathbb{R}^n$, $\xi, \eta \in (\mathbb{R}^n)_p$. A straightforward calculation shows that the Levi-Civita connection on $\mathbb{R}^n$ given by Theorem I.5.1 is the standard connection on $\mathbb{R}^n$ given by (I.1.1). One easily sees that the Riemann curvature tensor vanishes identically. Thus, $\mathbb{R}^n$ with its standard Riemannian metric is *flat*, that is, it has constant sectional curvature equal to 0.

The straight lines of $\mathbb{R}^n$ are easily seen to be its geodesics; since they are infinitely extendible in both directions, $\mathbb{R}^n$ is complete.

For comparison to later considerations, we write the metric of $\mathbb{R}^n$ in spherical coordinates, namely, for $x \in \mathbb{R}^n$ set

$$x = t\xi, \qquad t > 0,\ \xi \in \mathbb{S}^{n-1},$$

where $\mathbb{S}^{n-1}$ denotes the unit $(n-1)$–sphere in $\mathbb{R}^n$. Then,

$$dx = (dt)\xi + t\,d\xi,$$

and since

$$|\xi|^2 = 1, \qquad \xi \cdot d\xi = 0$$

(we are giving the phenomena on which the Gauss lemma is predicated), we have

$$|dx|^2 = dt^2 + t^2|d\xi|^2, \tag{II.3.2}$$

where $|d\xi|^2$ denotes the induced Riemannian metric on $\mathbb{S}^{n-1}$.

We now consider $\mathbb{S}^n(\rho)$, the sphere in $\mathbb{R}^{n+1}$ of radius $\rho > 0$, and show that $\mathbb{S}^n(\rho)$ has constant sectional curvature $1/\rho^2$. Here, $\mathbb{S}^n(\rho)$ has the Riemannian

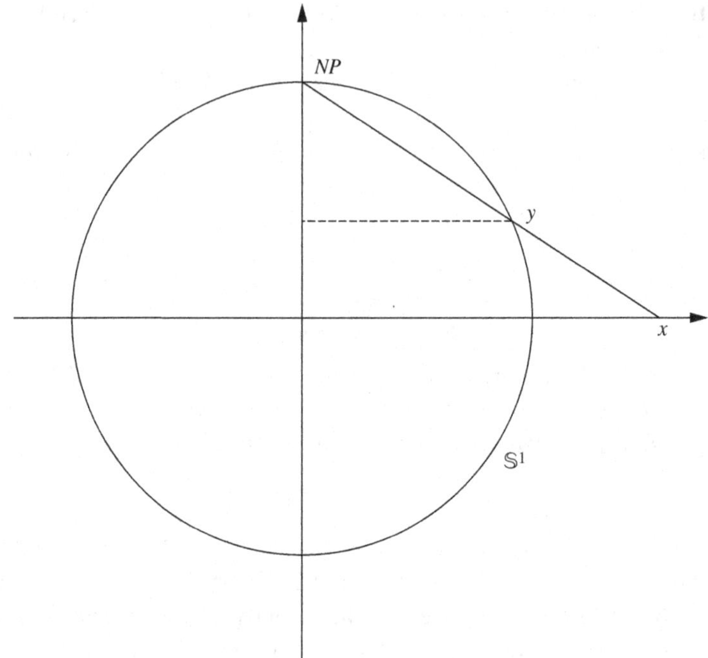

Figure II.2. 1–Dimensional stereographic projection.

metric for which the inclusion of $\mathbb{S}^n(\rho)$ in $\mathbb{R}^{n+1}$ is an isometry. Three methods are available.

Spheres: The First Method

The first approach consists of using coordinates on $\mathbb{S}^n(\rho)$, given for example, by stereographic projection, and carrying out the calculation explicitly.

Namely, the stereographic projection fixes the "north pole" NP of $\mathbb{S}^n(\rho)$, and its equatorial n–dimensional hyperplane $\mathbb{R}^n$ in $\mathbb{R}^{n+1}$. Then, with every point $y \in \mathbb{S}^n(\rho)$, one associates $x \in \mathbb{R}^n$ to be the point of intersection of the line in $\mathbb{R}^{n+1}$, determined by NP and y, with the equatorial hyperplane $\mathbb{R}^n$. (See Figure II.2.) To carry out the associated calculations, let $y = (y^1, \ldots, y^{n+1})$, $n \geq 1$, denote any point on $\mathbb{S}^n(\rho) \subset \mathbb{R}^{n+1}$, and $(x^1, \ldots, x^n)$ its image under stereographic projection from the north pole to

$$\mathbb{R}^n = \{z \in \mathbb{R}^{n+1} : z^{n+1} = 0\}.$$

It is standard that for $j = 1, \ldots, n$, we have

$$y^j = x^j(\rho - y^{n+1})/\rho,$$

from which one derives

$$y^{n+1} = \frac{-\rho^2 + |x|^2}{\rho^2 + |x|^2}\rho,$$

$$(y^1, \ldots, y^n) := \overline{y} = \frac{2\rho^2 x}{\rho^2 + |x|^2}.$$

We claim that

$$g_{jk} = \left\langle \frac{\partial}{\partial x^j}, \frac{\partial}{\partial x^k} \right\rangle = \frac{4\delta_{jk}}{\{1 + |x|^2/\rho^2\}^2}. \tag{II.3.3}$$

The simplest way to derive (II.3.3) is to compute formally:

$$d\overline{y} = 2\rho^2 \frac{\{\rho^2 + |x|^2\}dx - 2x(x \cdot dx)}{\{\rho^2 + |x|^2\}^2},$$

$$|d\overline{y}|^2 = 4\rho^4 \frac{\{\rho^2 + |x|^2\}^2 |dx|^2 - 4\rho^2 (x \cdot dx)^2}{\{\rho^2 + |x|^2\}^4},$$

$$dy^{n+1} = 4\rho^3 \frac{x \cdot dx}{\{\rho^2 + |x|^2\}^2}, \qquad (dy^{n+1})^2 = 16\rho^6 \frac{(x \cdot dx)^2}{\{\rho^2 + |x|^2\}^4},$$

$$ds^2 = \sum_{a=1}^{n+1} (dy^\alpha)^2 = 4\rho^4 \frac{|dx|^2}{\{\rho^2 + |x|^2\}^2}, \tag{II.3.4}$$

which is (II.3.3).

Therefore, (II.3.3) and (I.5.8) imply

$$\Gamma_{jk}{}^{\ell} = \frac{-2}{\rho^2 + |x|^2} \{\delta_{\ell k} x^j + \delta_{j\ell} x^k - \delta_{jk} x^{\ell}\}.$$

Therefore, for $n \geq 2$, one now uses (I.4.6) to verify, by a long and tedious calculation, that $\mathbb{S}^n(\rho)$ has constant sectional curvature $1/\rho^2$.

Spheres: The Second Method

The second approach is to use the apparatus of §II.2 to calculate the sectional curvatures of $\mathbb{S}^n(\rho)$ in terms of those of $\mathbb{R}^{n+1}$ and the second fundamental form of $\mathbb{S}^n(\rho)$ in $\mathbb{R}^{n+1}$.

It goes as follows: For any $q \in \mathbb{S}^n(\rho) \subseteq \mathbb{R}^{n+1}$, the exterior unit normal vector ν at q is given by

$$\nu = (1/\rho)\mathfrak{I}_q q;$$

more informally,

$$\nu = q/\rho,$$

which implies

$$dv = (1/\rho)dq;$$

more precisely, (II.2.4) becomes

$$\mathfrak{A}^{v} = -(1/\rho)I. \tag{II.3.5}$$

For $n \geq 2$, one now can easily use (II.3.5) to show that the images of geodesics on $\mathbb{S}^n(\rho)$ are obtained by intersecting $\mathbb{S}^n(\rho)$ with 2–planes through the center of $\mathbb{S}^n(\rho)$. Also, one substitutes (II.3.5) into (II.2.6) to verify that sectional curvatures of $\mathbb{S}^n(\rho)$ are equal to $1/\rho^2$.

The third approach is via Jacobi's equation, relating the curvature of $\mathbb{S}^n(\rho)$ with its geodesics. First,

Some Generalities about Isometries

Let M be a connected Riemannian manifold and $\Phi : M \to M$ be a local isometry, that is, Φ is C^∞, and for all $p \in M, \xi, \eta \in M_p$ we have

$$\langle \Phi_*\xi, \Phi_*\eta \rangle = \langle \xi, \eta \rangle.$$

If Φ is a global isometry, that is, Φ is a diffeomorphism in addition to being a local isometry, then one verifies that Φ preserves the distance function, that is,

$$d(\Phi(p), \Phi(q)) = d(p, q)$$

for all $p, q \in M$. One also verifies that, when Φ is a local isometry, Φ preserves the Levi-Civita connection; in particular, one has that Φ preserves geodesics, that is,

$$\Phi(\exp \xi) = \exp \Phi_*\xi \tag{II.3.6}$$

for all $\xi \in \mathcal{T}M$ the domain of exp in TM. Also, Φ preserves sectional curvature, that is, if $p \in M, \xi, \eta \in M_p$ are linearly independent, then

$$K(\Phi_*\xi, \Phi_*\eta) = K(\xi, \eta).$$

Finally, if M_1 is a submanifold of M, and Φ is an isometry of M satisfying $\Phi(M_1) \subseteq M_1$, then $\Phi|M_1$ is an isometry.

Let V be a real n–dimensional vector space, $q \in V$, A a linear transformation of V, and $a_{A;q} : V \to V$ given by $a_{A;q}(p) = Ap + q$. Then (Exercise I.I.17),

$$(a_{A;q})_{*|p} = \mathfrak{I}_{a_{A;q}(p)} A \mathfrak{I}_p{}^{-1}.$$

In particular, if V is an inner product space and A an orthogonal transformation, then $a_{A;q}$ is an isometry of V for every $q \in V$.

Spheres: The Third Method

Let $V = \mathbb{R}^{n+1}, n \geq 2$, with its standard Riemannian metric, and let $\mathcal{O}(n+1)$ be the group of orthogonal transformations of $\mathbb{R}^{n+1}$. Then, for each $A \in \mathcal{O}(n+1)$, $\rho > 0$, we have $A|\mathbb{S}^n(\rho)$ is an isometry of $\mathbb{S}^n(\rho)$ to itself. Also, one easily sees that for $p \in \mathbb{S}^n(\rho)$

$$(\mathbb{S}^n(\rho))_p = \mathfrak{I}_p(p^\perp),$$

where $p^\perp$ is the orthogonal complement of the span of p in $\mathbb{R}^{n+1}$. From these two facts, one easily has that for $p, q \in \mathbb{S}^n(\rho)$, ξ_1, ξ_2 orthonormal in $(\mathbb{S}^n(\rho))_p$ and η_1, η_2 orthonormal in $(\mathbb{S}^n(\rho))_q$, there exists $A \in \mathcal{O}(n+1)$ such that $A(p) = q$, $A_*\xi_j = \eta_j$, $j = 1, 2$. Thus, $\mathbb{S}^n(\rho)$ has constant sectional curvature. The only question is: What is the constant?

We first consider the geodesics of $\mathbb{S}^n(\rho)$. (See Milnor (1963, p. 65).) By Corollary I.6.3, there exists $\delta > 0$ so that for any $p, q \in \mathbb{S}^n(\rho)$ satisfying $d(p, q) < \delta$, there exists a unique geodesic γ: $[0, d(p, q)] \to \mathbb{S}^n(\rho)$, $|\gamma'| = 1$, such that $\gamma(0) = p$, $\gamma(d(p, q)) = q$. Pick such a p, q, $p \neq q$, let σ be the 2–plane through the origin of $\mathbb{R}^{n+1}$ spanned by $\{p, q\}$, and let $A : \mathbb{R}^{n+1} \to \mathbb{R}^{n+1}$ be the reflection of $\mathbb{R}^{n+1}$ through σ, that is, if $\{\mathfrak{e}_1, \ldots, \mathfrak{e}_{n+1}\}$ is an orthonormal basis of $\mathbb{R}^{n+1}$ such that span $\{\mathfrak{e}_1, \mathfrak{e}_2\} = \sigma$ then A is determined by

$$A\mathfrak{e}_1 = \mathfrak{e}_1, \qquad A\mathfrak{e}_2 = \mathfrak{e}_2,$$

and

$$A\mathfrak{e}_\alpha = -\mathfrak{e}_\alpha, \qquad \alpha = 3, \ldots, n+1.$$

Then, $A \in \mathcal{O}(n+1)$ with fixed-point set equal to σ.

Let $\Sigma = \sigma \cap \mathbb{S}^n(\rho)$ be the great circle through p, q. Then $A|\mathbb{S}^n(\rho)$ is an isometry with fixed-point set equal to Σ. Now, $A(p) = p$, $A(q) = q$ implies that A takes any minimizing geodesic from p to q to a minimizing geodesic from p to q. But there is only one such geodesic: γ. Thus, $A_*\gamma'(0) = \gamma'(0)$, which implies $A\gamma(t) = \gamma(t)$, that is, γ is contained in Σ. So, the great circles are the geodesics of $\mathbb{S}^n(\rho)$.

We now show that $\mathbb{S}^n(\rho)$ has constant sectional curvature equal to $1/\rho^2$.

We first argue informally: Let $\mathbf{v}_0 \in \mathbb{S}^n(\rho)$ (it is helpful to think of $\mathbf{v}_0$ as the north pole), and let ξ vary over unit vectors in $(\mathbb{S}^n(\rho))_{\mathbf{v}_0}$. We think of any such ξ as an element of $\mathbb{S}^n \cap \mathbf{v}_0{}^\perp$. So, write $y \in \mathbb{S}^n(\rho)$ as

$$y = (\cos t/\rho)\mathbf{v}_0 + (\rho \sin t/\rho)\xi,$$

$$\text{(II.3.7)} \qquad dy = \left\{\frac{-\sin t/\rho}{\rho}\mathbf{v}_0 + (\cos t/\rho)\xi\right\} dt + \rho(\sin t/\rho)\, d\xi.$$

Since, for each fixed ξ, y as a function of t describes the geodesic $\gamma_\xi(t)$, then the Jacobi field of the variation of geodesics, determined by $d\xi$, is given by

$$Y(t) = \rho(\sin t/\rho)\, d\xi$$

with $d\xi$ an $\mathbb{S}^n(\rho)$–parallel tangent vector field along $\gamma_\xi(t)$. If we substitute $Y(t)$ into Jacobi's equation (I.4.9) (with $T = 0$), we obtain $\kappa = 1/\rho^2$.

More precisely, let $\mathbf{v_o} \in \mathbb{S}^n(\rho), \xi, \eta \in (\mathbb{S}^n(\rho))_{\mathbf{v_o}}$ be orthonormal, and consider the geodesic variation v given by

$$\begin{aligned} v(t, \epsilon) &= \exp_{\mathbf{v_o}} \{(\cos \epsilon)\xi + (\sin \epsilon)\eta\}t \\ &= (\cos t/\rho)\mathbf{v_o} + (\sin t/\rho)\mathfrak{I}_{\mathbf{v_o}}{}^{-1} \rho\{(\cos \epsilon)\xi + (\sin \epsilon)\eta\}. \end{aligned}$$

Let γ be the geodesic given by

$$\gamma(t) = v(t, 0) = (\cos t/\rho)\mathbf{v_o} + (\sin t/\rho)\mathfrak{I}_{\mathbf{v_o}}{}^{-1} \rho\xi,$$

and $Y(t)$ the vector field along γ given by

$$Y(t) = (\partial_\epsilon v)(t, 0) = \mathfrak{I}_{\gamma(t)}(\sin t/\rho)\mathfrak{I}_{\mathbf{v_o}}{}^{-1} \rho\eta.$$

So, if we let $e(t)$ be the $\mathbb{S}^n(\rho)$–parallel vector field along γ satisfying $e(0) = \eta$, that is,

$$e(t) = \mathfrak{I}_{\gamma(t)}\mathfrak{I}_{\mathbf{v_o}}{}^{-1}\eta,$$

then

$$Y(t) = \rho(\sin t/\rho)e(t).$$

Since v is a geodesic variation, Y satisfies Jacobi's equation (I.4.9)

$$\nabla_t{}^2 Y + R(\gamma', Y)\gamma' = 0;$$

note that in the Riemannian case, $T = 0$. Thus,

$$\begin{aligned} -(1/\rho)(\sin t/\rho)e &= \nabla_t{}^2 Y \\ &= -R(\gamma', Y)\gamma' \\ &= -\rho(\sin t/\rho)R(\gamma', e)\gamma'. \end{aligned}$$

In particular, at $t = 0$, we obtain

$$\rho R(\xi, \eta)\xi = \eta/\rho,$$

which implies

$$\mathcal{K}(\xi, \eta) = \langle R(\xi, \eta)\xi, \eta\rangle = 1/\rho^2,$$

which is the claim. ∎

Remark II.3.1. It may be worth commenting on the approach of the last method. Its appeal is extremely natural, in that the curvature tensor is created to describe the linearization of the equations for geodesics. Since, in general, the geodesics of a Riemannian manifold are unknown, one tries at least to study how geodesics behave "near" a given one. So one uses knowledge of the curvature to inform about geodesics. But our situation with the sphere is quite the reverse, namely, we know all the geodesics at the outset. So, we certainly know the local behavior of geodesics, which implies that we know the curvature tensor. This is what drives the third method.

Note that we have, from (II.3.7),

$$ds^2 = |dy|^2 = dt^2 + \rho^2(\sin^2 t/\rho)|d\xi|^2. \tag{II.3.8}$$

Finally, we note that, from the above formal calculation, one can reclaim the Riemannian metric for stereographic projection as follows: Let $x \in \mathbb{R}^n$,

$$x = r\xi, \qquad r > 0,\ \xi \in \mathbb{S}^{n-1},$$

$$r = \rho \tan t/2\rho, \qquad t \in [0, \pi\rho].$$

Then, one verifies directly that

$$ds^2 = \frac{4|dx|^2}{\{1 + |x|^2/\rho^2\}^2} = dt^2 + \rho^2(\sin^2 t/\rho)|d\xi|^2.$$

Hyperbolic Space

To describe the model space of constant sectional curvature equal to -1, we work with two models: (i) $\mathbb{B}^n$ the unit disk in $\mathbb{R}^n$ with radius 1, and (ii) $\mathbb{R}^n_+$ the upper half-space of $\mathbb{R}^n$.

The first step is to give an identification of $\mathbb{R}^n_+$ with $\mathbb{B}^n$. To this end, set

$$y = x + (1/2 - 2x^n)\mathfrak{e}_n,$$

where $\mathfrak{e}_n$ denotes the n-th element of the natural basis of $\mathbb{R}^n$. Then, $x \mapsto y$ takes $\mathbb{R}^n_+$ to $\{y^n < 1/2\}$. Now map $\{y^n < 1/2\}$ to $\mathbb{B}^n$ by

$$z = \mathfrak{e}_n + (y - \mathfrak{e}_n)|y - \mathfrak{e}_n|^{-2}.$$

This provides a diffeomorphism of $\mathbb{R}^n_+$ to $\mathbb{B}^n$.

For the Riemann metric on $\mathbb{B}^n$, choose:

$$ds^2 := \frac{4|dz|^2}{\{1 - |z|^2\}^2}. \tag{II.3.9}$$

To write the induced Riemannian metric on $\mathbb{R}^n_+$, we have

$$\begin{aligned} dz &= |y - \mathfrak{e}_n|^{-2}\,dy - 2|y - \mathfrak{e}_n|^{-4}\{(y - \mathfrak{e}_n)\cdot dy\}(y - \mathfrak{e}_n), \\ |dz|^2 &= |y - \mathfrak{e}_n|^{-4}|dy|^2, \\ |z|^2 &= 1 + 2\mathfrak{e}_n \cdot (y - \mathfrak{e}_n)|y - \mathfrak{e}_n|^{-2} + |y - \mathfrak{e}_n|^{-2}, \\ 1 - |z|^2 &= \{1 - 2y^n\}|y - \mathfrak{e}_n|^{-2}, \end{aligned}$$

which implies

$$ds^2 = \frac{4|dz|^2}{\{1 - |z|^2\}^2} = \frac{4|dy|^2}{\{1 - 2y^n\}^2}.$$

Next,

$$|dy| = |dx|, \qquad 2y^n - 1 = -2x^n,$$

which implies

$$ds^2 = \frac{4|dy|^2}{\{1 - 2y^n\}^2} = \frac{|dx|^2}{\{x^n\}^2}.$$

So, in $\mathbb{R}^n_+$ the metric is written as

$$ds^2 = \frac{|dx|^2}{\{x^n\}^2}. \tag{II.3.10}$$

To show that the sectional curvature $\mathcal{K}$ is identically equal to -1, we work in $\mathbb{R}^n_+$. Then, (I.5.5) implies

$$g_{ij} = \frac{\delta_{ij}}{(x^n)^2},$$

which implies, by (I.5.8),

$$\Gamma_{jk}{}^{\ell} = -(x^n)^{-1}\{\delta_{jn}\delta_{\ell k} + \delta_{kn}\delta_{j\ell} - \delta_{\ell n}\delta_{jk}\}.$$

Therefore, for $\alpha, \beta = 1, \ldots, n-1$, we have

$$\nabla_{\partial_\alpha}\partial_\beta = (x^n)^{-1}\delta_{\alpha\beta}\partial_n, \quad \nabla_{\partial_n}\partial_\beta = -(x^n)^{-1}\partial_\beta, \quad \nabla_{\partial_n}\partial_n = -(x^n)^{-1}\partial_n,$$
$$\nabla_{\partial_\beta}\nabla_{\partial_\alpha}\partial_\beta = -(x^n)^{-2}\delta_{\alpha\beta}\partial_\beta, \qquad \nabla_{\partial_\alpha}\nabla_{\partial_\beta}\partial_\beta = -(x^n)^{-2}\partial_\alpha.$$

Therefore, we have, by (I.4.6), for $\alpha \neq \beta$

$$R(\partial_\beta, \partial_\alpha)\partial_\beta = -(x^n)^{-2}\partial_\alpha,$$

which implies, from the definition in Proposition II.1.2, that

$$\mathcal{K}(\partial_\alpha, \partial_\beta) = -1.$$

Also, we similarly have

$$R(\partial_n, \partial_\alpha)\partial_n = -(x^n)^{-2}\partial_\alpha;$$

so

$$\mathcal{K}(\partial_\alpha, \partial_n) = -1.$$

Thus, our space – referred to as *hyperbolic space* – has constant sectional curvature equal to -1.

We note that, if we start with the n–disk $\mathbb{B}^n(\rho)$ in $\mathbb{R}^n$, endowed with metric

$$ds^2 = \frac{4|dz|^2}{\{1 - |z|^2/\rho^2\}^2},$$

then the sectional curvature becomes the constant $-1/\rho^2$. Furthermore, if we substitute

$$z = r\xi, \qquad r > 0,\ \xi \in \mathbb{S}^{n-1},$$
$$r = \rho \tanh t/2\rho,$$

then we obtain

$$ds^2 = dt^2 + \rho^2(\sinh^2 t/\rho)|d\xi|^2. \tag{II.3.11}$$

One easily sees that, in this model, the geodesics emanating from the origin are given by straight lines emanating from origin, and their length to the boundary $\mathbb{S}^{n-1}$ is infinite. So hyperbolic space is Riemannian complete.

Remark II.3.2. A particularly clear and elegant discussion of the Riemannian geometry of hyperbolic space can be found in B. Randol's Chapter 11 of Chavel (1984). See Note II.5 and II.6 in §II.9 for further references.

§II.4. First and Second Variations of Arc Length

M is our Riemannian manifold.

Definition. We are given $\omega : [\alpha, \beta] \to M \in D^\infty$ (see §I.6). A *variation* v *of* ω is a continuous mapping $v : [\alpha, \beta] \times (-\epsilon_0, \epsilon_0) \to M \in D^\infty$ for some $\epsilon_0 > 0$, for which

$$\omega(t) = v(t, 0)$$

for all $t \in [\alpha, \beta]$, and such that there exists a subdivision $[\alpha = t_0 < \ldots < t_k = \beta]$ of $[\alpha, \beta]$ for which $v|[t_{j-1}, t_j] \times (-\epsilon_0, \epsilon_0) \in C^\infty$ for each $j = 1, \ldots, k$.

If v *fixes endpoints,* that is, v also satisfies

$$\omega(\alpha) = v(\alpha, \epsilon), \qquad \omega(\beta) = v(\beta, \epsilon)$$

for all ϵ in $(-\epsilon_0, \epsilon_0)$, then we say that v is a *homotopy of* ω.

We refer to v as a *smooth variation* if v is differentiable on all of $[\alpha, \beta] \times (-\epsilon_0, \epsilon_0)$.

We call v a *geodesic variation of* ω if for every ϵ in $(-\epsilon_0, \epsilon_0)$ the path $\omega_\epsilon : [\alpha, \beta] \to M$ given by

$$\omega_\epsilon(t) = v(t, \epsilon)$$

is a geodesic.

We write $\partial_t v$, $\partial_\epsilon v$ for $v_*(\partial_t)$, $v_*(\partial_\epsilon)$, respectively, and denote differentiation of vector fields along v with respect to $\partial_t v$, $\partial_\epsilon v$ by ∇_t, ∇_ϵ, respectively.

For a geodesic variation v, we have (see (I.4.7), (I.4.8))

$$\nabla_\epsilon \partial_t v - \nabla_t \partial_\epsilon v = 0,$$

since the Levi-Civita connection has no torsion, and

$$\nabla_\epsilon \nabla_t - \nabla_t \nabla_\epsilon = R(\partial_t v, \partial_\epsilon v).$$

Theorem II.4.1. (The first variation of arc length) *Assume $\omega : [\alpha, \beta] \to M$ is differentiable, and v a differentiable variation of ω. For each ϵ in $(-\epsilon_0, \epsilon_0)$, let $L(\epsilon)$ be the length of ω_ϵ, namely,*

$$L(\epsilon) = \int_\alpha^\beta |\partial_t v(t, \epsilon)|\, dt.$$

Then, L is differentiable and

$$\text{(II.4.1)} \qquad \frac{dL}{d\epsilon} = \left\langle \frac{\partial v}{\partial \epsilon}, \frac{\partial v}{\partial t} \middle/ \left| \frac{\partial v}{\partial t} \right| \right\rangle \Bigg|_\alpha^\beta - \int_\alpha^\beta \langle \partial_\epsilon v, \nabla_t(\partial_t v / |\partial_t v|) \rangle \, dt.$$

In particular, if ω is parameterized with respect to arc length, that is, $|\omega'| = 1$, and we set

$$Y(t) = (\partial_\epsilon v)(t, 0),$$

then

$$\text{(II.4.2)} \qquad \frac{dL}{d\epsilon}(0) = \langle Y, \omega' \rangle \Big|_\alpha^\beta - \int_\alpha^\beta \langle Y, \nabla_t \omega' \rangle \, dt.$$

Proof. We have

$$\begin{aligned}
\partial_\epsilon L &= \partial_\epsilon \int_\alpha^\beta |\partial_t v|\,dt \\
&= \int_\alpha^\beta \partial_\epsilon(\langle \partial_t v, \partial_t v\rangle^{1/2})\,dt \\
&= \int_\alpha^\beta |\partial_t v|^{-1}\langle \nabla_\epsilon \partial_t v, \partial_t v\rangle\,dt \\
&= \int_\alpha^\beta \langle \nabla_t \partial_\epsilon v, \partial_t v/|\partial_t v|\rangle\,dt \\
&= \int_\alpha^\beta \{\partial_t \langle \partial_\epsilon v, \partial_t v/|\partial_t v|\rangle - \langle \partial_\epsilon v, \nabla_t(\partial_t v/|\partial_t v|)\rangle\}\,dt \\
&= \left\langle \frac{\partial v}{\partial \epsilon}, \frac{\partial v}{\partial t} \Big/ \left|\frac{\partial v}{\partial t}\right|\right\rangle\Bigg|_\alpha^\beta - \int_\alpha^\beta \langle \partial_\epsilon v, \nabla_t(\partial_t v/|\partial_t v|)\rangle\,dt,
\end{aligned}$$

which is (II.4.1). Equation (II.4.2) follows immediately. ■

Lemma II.4.1. *If $\omega : [\alpha, \beta] \to M \in D^\infty$ and Y is a D^∞ vector field along ω, then there exists a variation v of ω for which $Y = \partial_\epsilon v|_{\epsilon=0}$.*

Proof. Simply set $v(t, \epsilon) = \exp \epsilon Y(t)$. ■

Theorem II.4.2. *A path $\omega : [\alpha, \beta] \to M$, $|\omega'| = 1$, is a geodesic if and only if*

(II.4.3) $$L'(0) = 0$$

for every homotopy v of ω.

Proof. For any vector field Z along ω, set

$$\Delta Z(t_0) = \lim_{t \downarrow t_0} Z(t) - \lim_{t \uparrow t_0} Z(t)$$

for any $t_0 \in (\alpha, \beta)$.

If $\omega \in D^\infty$, then for any homotopy v of ω, we have

(II.4.4) $$L'(0) = -\sum_t \langle Y, \Delta\omega'\rangle - \int_\alpha^\beta \langle Y, \nabla_t \omega'\rangle\,dt.$$

Thus, if ω is a geodesic, then (II.4.3) is valid for every homotopy v of ω. If, on the other hand, (II.4.3) is valid for all homotopies v of ω, then

$$0 = -\sum_t \langle Y, \Delta\omega'\rangle - \int_\alpha^\beta \langle Y, \nabla_t \omega'\rangle\,dt$$

for every D^∞ vector field Y along ω vanishing at $t = \alpha, \beta$. Let $t_1 < \ldots < t_{k-1}$ denote the discontinuities of ω' and assume that there exists $t_0 \in (\alpha, \beta)\backslash\{t_1 \ldots, t_{k-1}\}$ such that $(\nabla_t \omega')(t_0) \neq 0$. Pick $\delta_1 > 0$ so that $\{t : |t - t_0| < \delta_1\} \subset (\alpha, \beta)\backslash \{t_1, \ldots, t_{k-1}\}$, and let $Z(t)$ be the parallel vector field on $\{t : |t - t_0| < \delta_1\}$ satisfying

$$Z(t_0) = (\nabla_t \omega')(t_0).$$

Then there exists $\delta_2 \in (0, \delta_1)$ such that $\langle Z, \nabla_t \omega' \rangle > 0$ on $\{t : |t - t_0| < \delta_2\}$. Finally pick $\varphi : [\alpha, \beta] \to [0, +\infty) \in C^\infty$ such that $\varphi(t_0) > 0$, with $\operatorname{supp} \varphi$ contained in $\{t : |t - t_0| < \delta_2\}$, and set $Y = \varphi Z$. Then, the right-hand side of (II.4.4) will be strictly negative – a contradiction.

Thus, $\nabla_t \omega' = 0$ on $(\alpha, \beta)\backslash\{t_1, \ldots, t_{k-1}\}$, and ω is at least piecewise geodesic. It remains to show that $\omega \in C^1$ on $[\alpha, \beta]$. For, if so, then the differential equations of geodesics in local coordinates (I.3.2) will then imply that $\omega \in C^\infty$, that is, ω is a geodesic. Well, our assumption now is that

$$0 = \sum_{t_j} \langle Y, \Delta\omega' \rangle$$

for all D^∞ vector fields Y along ω. Pick Y to be any D^∞ vector field satisfying $Y(t_j) = \Delta\omega'(t_j)$, $j = 1, \ldots, k-1$. Then $|\Delta\omega'|^2(t_j) = 0$ for each $j = 1, \ldots, k-1$ and $\omega \in C^1$. ■

Theorem II.4.3. (The second variation of arc length) *Let ω and L be as in Theorem* II.4.1, *with $|\omega'| = 1$. Then, for the second derivative of L, we have*

$$(d^2L/d\epsilon^2)(0) = \langle \nabla_\epsilon \partial_\epsilon v \mid_{\epsilon=0}, \omega' \rangle \Big|_\alpha^\beta + \int_\alpha^\beta \{|\nabla_t Y|^2 - \langle R(\omega', Y)\omega', Y \rangle - \langle \omega', \nabla_t Y \rangle^2 - \langle \nabla_t \omega', \nabla_\epsilon \partial_\epsilon v \rangle\}\, dt. \tag{II.4.5}$$

Proof. For the second derivative of L, we start just prior to the integration by parts in the proof of the formula for the first variation, namely,

$$\begin{aligned}\frac{\partial^2 L}{\partial \epsilon^2} &= \partial_\epsilon \int_\alpha^\beta \langle \nabla_t \partial_\epsilon v, \partial_t v/|\partial_t v| \rangle\, dt \\ &= \int_\alpha^\beta \{\langle \nabla_\epsilon \nabla_t \partial_\epsilon v, \partial_t v/|\partial_t v| \rangle + |\nabla_t \partial_\epsilon v|^2/|\partial_t v| \\ &\quad - |\partial_t v|^{-2} \partial_\epsilon(|\partial_t v|) \langle \nabla_t \partial_\epsilon v, \partial_t v \rangle\}\, dt.\end{aligned}$$

If we set $\epsilon = 0$, then we obtain

$$\frac{d^2L}{d\epsilon^2}(0) = \int_\alpha^\beta \{\langle \nabla_\epsilon \nabla_t \partial_\epsilon v, \partial_t v\rangle_{|\epsilon=0} + |\nabla_t Y|^2 - \langle \nabla_t Y, \omega'\rangle^2\}\, dt.$$

But

$$\begin{aligned}\langle \nabla_\epsilon \nabla_t \partial_\epsilon v, \partial_t v\rangle &= \langle \nabla_t \nabla_\epsilon \partial_\epsilon v, \partial_t v\rangle + \langle R(\partial_t v, \partial_\epsilon v)\partial_\epsilon v, \partial_t v\rangle \\ &= \partial_t \langle \nabla_\epsilon \partial_\epsilon v, \partial_t v\rangle - \langle \nabla_\epsilon \partial_\epsilon v, \nabla_t \partial_t v\rangle - \langle R(\partial_t v, \partial_\epsilon v)\partial_t v, \partial_\epsilon v\rangle,\end{aligned}$$

and (II.4.5) follows easily. ■

Theorem II.4.4. *Let $\gamma : [\alpha, \beta] \to M, |\gamma'| = 1$ be a geodesic. For any variation v of γ let*

$$Y_\perp = Y - \langle Y, \gamma'\rangle \gamma'.$$

Then

(II.4.6) $$L'(0) = \langle Y, \gamma'\rangle\Big|_\alpha^\beta,$$

(II.4.7) $$L''(0) = \langle \nabla_Y \partial_\epsilon v, \gamma'\rangle\Big|_\alpha^\beta + \int_\alpha^\beta \{|\nabla_t Y_\perp|^2 - \langle R(\gamma', Y_\perp)\gamma', Y_\perp\rangle\}\, dt.$$

In particular, if v is a homotopy of γ, then

(II.4.8) $$L'(0) = 0,$$

(II.4.9) $$L''(0) = \int_\alpha^\beta \{|\nabla_t Y_\perp|^2 - \langle R(\gamma', Y_\perp)\gamma', Y_\perp\rangle\}\, dt.$$

Proof. One only has to realize that if γ is a geodesic, then

$$\nabla_t Y_\perp = \nabla_t Y - \langle \nabla_t Y, \gamma'\rangle \gamma' = (\nabla_t Y)_\perp,$$

which implies

$$|\nabla_t Y_\perp|^2 = |\nabla_t Y|^2 - \langle \nabla_t Y, \gamma'\rangle^2.$$

■

§II.5. Jacobi's Equation and Criteria

We are given a fixed geodesic $\gamma : [\alpha, \beta] \to M$, $|\gamma'| = 1$, and define Υ_0 to be the vector space of D^1 vector fields X along γ, orthogonal to γ and satisfying $X(\alpha) = X(\beta) = 0$. We may think of Υ_0 as having the inner product

(II.5.1) $$(X, Y) = \int_\alpha^\beta \langle X, Y\rangle\, dt.$$

Definition. On Υ_0, we define the symmetric bilinear form $I(X, Y)$ over $\mathbb{R}$, called the *index form*, by

$$I(X, Y) = \int_\alpha^\beta \{\langle \nabla_t X, \nabla_t Y\rangle - \langle R(\gamma', X)\gamma', Y\rangle\}\, dt.$$

The motivation of considering such a bilinear form is, of course, that if $Y \in D^\infty$ is induced by a homotopy of γ then $L''(0) = I(Y, Y)$. To obtain a self-adjoint (relative to (II.5.1)) operator associated with the index form, we require that $X \in C^2$, in which case integration by parts gives

$$I(X, Y) = -\int_\alpha^\beta \langle {\nabla_t}^2 X + R(\gamma', X)\gamma', Y\rangle\, dt. \tag{II.5.2}$$

So, the operator in question is

$$\mathcal{L}X = -\left\{{\nabla_t}^2 X + R(\gamma', X)\gamma'\right\},$$

and for $X, Y, \in C^2$, we have

$$(\mathcal{L}X, Y) = (X, \mathcal{L}Y) = I(X, Y).$$

The operator is defined, therefore, on C^2 with associated bilinear form defined on at least D^1.

Definition. We define a *Jacobi field along* γ to be a differentiable vector field Y along γ satisfying *Jacobi's equation*

$${\nabla_t}^2 Y + R(\gamma', Y)\gamma' = 0. \tag{II.5.3}$$

Theorem II.5.1. *The set $\mathcal{J}$ of Jacobi fields along γ is a vector space over $\mathbb{R}$ of dimension equal to* 2(dim M). *More particularly, one has: Given any $t_0 \in [\alpha, \beta]$, $\xi, \eta \in M_{\gamma(t_0)}$, there exists a unique $Y \in \mathcal{J}$ satisfying $Y(t_0) = \xi$, $(\nabla_t Y)(t_0) = \eta$.*

Thus, if $Y \in \mathcal{J}$, $Y \neq 0$, then

$$|Y|^2 + |\nabla_t Y|^2 > 0 \tag{II.5.4}$$

on all of γ.

Also, if $Y \in \mathcal{J}$, $Y \neq 0$, and $Y(t_0) = 0$, then there exists an $\epsilon > 0$ such that $Y(t) \neq 0$ for all t satisfying $0 < |t - t_0| < \epsilon$.

Proof. Let $n = \dim M$, $\{e_1, \ldots, e_n\}$ an orthonormal basis of $M_{\gamma(\alpha)}$, and $\{E_1, \ldots, E_n\}$ the parallel vector fields along γ satisfying $E_j(\alpha) = e_j$. Then, $\{E_1(t), \ldots, E_n(t)\}$ is an orthonormal basis of $M_{\gamma(t)}$, for every $t \in [\alpha, \beta]$, and

$Y(t)$ may be written as

$$Y(t) = \sum_{j=1}^{n} Y^j(t)E_j(t).$$

Set

$$\mathsf{R}_j{}^k(t) = \langle R(\gamma', E_j)\gamma', E_k\rangle(t);$$

then, $\mathsf{R}_j{}^k$ is symmetric for every t, and (II.5.3) reads as

(II.5.5) $$(Y^k)'' + \sum_j Y^j \mathsf{R}_j{}^k(t) = 0,$$

where $j, k = 1, \ldots, n$.

The claims of the theorems follow immediately from the theory of linear ordinary differential equations. ∎

Proposition II.5.1. *For $X, Y \in \mathcal{J}$, we have*

(II.5.6) $$\langle \nabla_t X, Y\rangle - \langle X, \nabla_t Y\rangle = \text{const.}$$

Thus, for any $Y \in \mathcal{J}$, we have constants $a, b \in \mathbb{R}$ for which

(II.5.7) $$\langle Y, \gamma'\rangle = at + b.$$

In particular,

$$\mathcal{J}^\perp := \{Y \in \mathcal{J} : \langle Y, \gamma'\rangle = 0 \text{ on } [\alpha, \beta]\}$$

is a subspace of $\mathcal{J}$ with codimension equal to 2.

Proof. Differentiate the left-hand side of (II.5.6). ∎

Definition. Given a real constant κ, we let $\mathbf{S}_\kappa$ denote the solution to the ordinary differential equation

$$\psi'' + \kappa\psi = 0,$$

satisfying the initial conditions

$$\mathbf{S}_\kappa(0) = 0, \qquad \mathbf{S}_\kappa{}'(0) = 1.$$

We also let $\mathbf{C}_\kappa$ denote the solution to the above ordinary differential equation satisfying the initial conditions

$$\mathbf{C}_\kappa(0) = 1, \qquad \mathbf{C}_\kappa{}'(0) = 0.$$

Of course, we have

(II.5.8)
$$\mathbf{S}_\kappa(t) = \begin{cases} (1/\sqrt{\kappa})\sin\sqrt{\kappa}t & \kappa > 0 \\ t & \kappa = 0. \\ (1/\sqrt{-\kappa})\sinh\sqrt{-\kappa}t & \kappa < 0 \end{cases}$$

Also,

(II.5.9)
$$\mathbf{C}_\kappa(t) = \begin{cases} \cos\sqrt{\kappa}t & \kappa > 0 \\ 1 & \kappa = 0. \\ \cosh\sqrt{-\kappa}t & \kappa < 0 \end{cases}$$

Furthermore, we have

$$\mathbf{S}_\kappa{}' = \mathbf{C}_\kappa, \qquad \mathbf{C}_\kappa{}' = -\kappa\mathbf{S}_\kappa, \qquad \mathbf{C}_\kappa{}^2 + \kappa\mathbf{S}_\kappa{}^2 = 1,$$
$$(\mathbf{C}_\kappa/\mathbf{S}_\kappa)' = (\mathbf{S}_\kappa{}'/\mathbf{S}_\kappa)' = -\mathbf{S}_\kappa{}^{-2}.$$

Definition. Given the Riemannian manifold M, we refer to a geodesic γ as a *unit speed geodesic* if γ is parameterized with respect to arc length. When we refer to the *sectional curvature along* γ, we are referring to the sectional curvature of 2–sections determined by γ' and a vector in the tangent space to M at γ.

Given the Riemannian manifold M, γ a unit speed geodesic in M such that the sectional curvature along γ is identically equal to the constant κ. Then, Jacobi's equation (II.5.3) becomes

$$\nabla_t{}^2 Y + \kappa Y = 0 \quad \text{for} \quad Y \in \mathcal{J}^\perp,$$

and (II.5.5) becomes, with $E_n = \gamma'$,

$$(Y^j)'' + \kappa Y^j = 0, \qquad j = 1, \ldots, n-1.$$

For the Jacobi field $Y \in \mathcal{J}^\perp$, we therefore have

(II.5.10)
$$Y(t) = \mathbf{C}_\kappa(t)A(t) + \mathbf{S}_\kappa(t)B(t),$$

where $A(t)$, $B(t)$ are parallel vector fields along γ which are pointwise orthogonal to γ.

Definition. Given the Riemannian manifold M, γ a geodesic in M, a point $\gamma(t_1)$ is said to be *conjugate to* $\gamma(t_0)$ *along* γ if there exists $Y \in \mathcal{J}$, $Y \neq 0$ such that

(II.5.11)
$$Y(t_0) = Y(t_1) = 0.$$

Of course, Y in (II.5.11) must be an element of $\mathcal{J}^\perp$ by (II.5.7). Also, one has immediately

Proposition II.5.2. *If, for a given $t_0 \in (\alpha, \beta]$, $\gamma(t_0)$ is not conjugate to $\gamma(\alpha)$, then, for any $\xi \in \gamma'(t_0)^\perp$, there exists a unique $Y \in \mathcal{J}^\perp$ satisfying*

$$Y(\alpha) = 0, \quad Y(t_0) = \xi.$$

Theorem II.5.2. *If M has constant sectional curvature κ along the unit speed geodesic $\gamma : \mathbb{R} \to M$, then $\gamma(0)$ has a conjugate point along γ if and only if $\kappa > 0$, in which case $\gamma(\ell\pi/\sqrt{\kappa})$ is conjugate to $\gamma(0)$ along γ, for any integer ℓ. Furthermore, these are the only points on γ conjugate to $\gamma(0)$ along γ.*

We recall from the beginning of this section that Υ_0 is the vector space of D^1 vector fields X along γ, orthogonal to γ, and satisfying $X(\alpha) = X(\beta) = 0$.

Theorem II.5.3. *Let $Y_1, \ldots, Y_N \in \mathcal{J}^\perp$ satisfy*

$$\langle Y_j, \nabla_t Y_k\rangle - \langle Y_k, \nabla_t Y_j\rangle = 0 \tag{II.5.12}$$

for $j, k = 1, \ldots, N \leq n-1$, and assume $X \in \Upsilon_0$ has the representation

$$X = \sum_{j=1}^{N} f^j Y_j.$$

Then,

$$I(X, X) = \int_\alpha^\beta \sum_{j=1}^{N} \left| f^{j\prime} Y_j \right|^2 \, dt.$$

Proof. The proof is by direct calculation, namely,

$$\nabla_t X = \sum_j \left\{ f^{j\prime} Y_j + f^j \nabla_t Y_j \right\},$$

$$\langle R(\gamma', X)\gamma', X\rangle = \sum_{j,k} f^j f^k \langle R(\gamma', Y_j)\gamma', Y_k\rangle,$$

(II.5.12) and (II.5.3) combine to imply

$$\begin{aligned}
&|\nabla_t X|^2 - \langle R(\gamma', X)\gamma', X\rangle \\
&\quad = \sum \left\{ f^{j\prime} f^{k\prime} \langle Y_j, Y_k\rangle + f^{j\prime} f^k \langle Y_j, \nabla_t Y_k\rangle + f^j f^{k\prime} \langle \nabla_t Y_j, Y_k\rangle \right. \\
&\qquad \left. + f^j f^k \langle \nabla_t Y_j, \nabla_t Y_k\rangle - f^j f^k \langle R(\gamma', Y_j)\gamma', Y_k\rangle \right\} \\
&\quad = \left|\sum f^{j\prime} Y_j\right|^2 + \sum \{ (f^j f^k)' \langle \nabla_t Y_j, Y_k\rangle + f^j f^k \langle \nabla_t Y_j, \nabla_t Y_k\rangle \\
&\qquad - f^j f^k \langle R(\gamma', Y_j)\gamma', Y_k\rangle \} \\
&\quad = \left|\sum f^{j\prime} Y_j\right|^2 + \left\{ \sum f^j f^k \langle \nabla_t Y_j, Y_k\rangle \right\}'.
\end{aligned}$$

Thus, we have

$$\text{(II.5.13)} \quad |\nabla_t X|^2 - \langle R(\gamma', X)\gamma', X\rangle = \left|\sum f^{j\prime} Y_j\right|^2 + \left\langle \sum f^j \nabla_t Y_j, X\right\rangle',$$

and the claim follows. ∎

Theorem II.5.4. (C. F. Jacobi (1836))[2] *If $\gamma(\alpha)$ has no conjugate points along γ on $(\alpha, \beta]$ then the index form I is positive definite on Υ_0. If, however, we only assume that $\gamma(\alpha)$ has no conjugate points on (α, β), then I is nonnegative on Υ_0; and $I(X, X) = 0$ if and only if $X \in \mathcal{J}^\perp \cap \Upsilon_0$, that is, X is a Jacobi field satisfying $X(\alpha) = X(\beta) = 0$.*

Proof. Suppose we have linearly independent

$$\{Y_j \in \mathcal{J}^\perp : Y_j(\alpha) = 0,\ j = 1, \ldots, n-1\}.$$

Then, for any $X \in \Upsilon_0$, we certainly have the representation

$$X = \sum_{j=1}^{n-1} f^j(t) Y_j(t)$$

for all $t \in (\alpha, \beta)$. (Note the *open* interval (α, β).)

Then, the vector fields $Y_1, \ldots, Y_n$ satisfy (II.5.12), and (II.5.13) implies

$$\begin{aligned} I(X, X) &= \int_\alpha^\beta \{|\nabla_t X|^2 - \langle R(\gamma', X)\gamma', X\rangle\}\, dt \\ &= \lim_{\epsilon \downarrow 0} \int_{\alpha+\epsilon}^{\beta-\epsilon} \{|\nabla_t X|^2 - \langle R(\gamma', X)\gamma', X\rangle\}\, dt \\ &= \lim_{\epsilon \downarrow 0} \left\{ \sum_j f^j \langle \nabla_t Y_j, X\rangle\Big|_{\alpha+\epsilon}^{\beta-\epsilon} + \int_{\alpha+\epsilon}^{\beta-\epsilon} \left|\sum f^{j\prime} Y_j\right|^2 dt \right\}. \end{aligned}$$

Assume, for the moment, that f^j, $j = 1, \ldots, n-1$, are bounded on (α, β). Then,

$$I(X, X) = \lim_{\epsilon \downarrow 0} \int_{\alpha+\epsilon}^{\beta-\epsilon} \left|\sum f^{j\prime} Y_j\right|^2 dt \geq 0.$$

If $I(X, X) = 0$, then $f^j =$ constant for each $j = 1, \ldots, n-1$, that is, $X \in \mathcal{J}^\perp$. We already have $X \in \Upsilon_0$ by assumption, which implies the claim.

[2] See Exercise II.18 in §II.9.

So, we must show that for each j, f^j is bounded near $t = \alpha$ and β. We work with $t = \beta$.

Let $\ell = \dim \Upsilon_0 \cap \mathcal{J}^\perp$, that is, the dimension of the space of Jacobi fields vanishing at $\gamma(\alpha)$, $\gamma(\beta)$; and let $\{Y_1, \ldots, Y_\ell\}$ be a basis of $\Upsilon_0 \cap \mathcal{J}^\perp$. Let $e_\mu = (\nabla_t Y_\mu)(\alpha)$, $\mu = 1, \ldots, \ell$; complete $\{e_\mu : \mu = 1, \ldots, \ell\}$ to a basis $\{e_1, \ldots, e_{n-1}\}$ of $\gamma'(\alpha)^\perp$, and let Y_ν, $\nu = \ell + 1, \ldots, n-1$, be the Jacobi field along γ vanishing at $\gamma(\alpha)$ with $(\nabla_t Y_\nu)(\alpha) = e_\nu$.

Now note that, by assumption, $Y_{\ell+1}(\beta), \ldots, Y_{n-1}(\beta)$ are linearly independent. By (II.5.4), we have $(\nabla_t Y_1)(\beta), \ldots, (\nabla_t Y_\ell)(\beta)$ are linearly independent. From (II.5.12), we have

$$\{(\nabla_t Y_1)(\beta), \ldots, (\nabla_t Y_\ell)(\beta)\} \perp \{Y_{\ell+1}(\beta), \ldots, Y_{n-1}(\beta)\}.$$

Thus,

$$\{(\nabla_t Y_1)(\beta), \ldots, (\nabla_t Y_\ell)(\beta), Y_{\ell+1}(\beta), \ldots, Y_{n-1}(\beta)\}$$

is a basis of $\gamma'(\beta)^\perp$.

Now, Taylor's formula reads as

$$X(t) = \tau_{t_0,t}\{X(t_0) + (t - t_0)(\nabla_t X)(t_0)\} + o(t - t_0),$$

where $\tau_{t_0,t} : M_{\gamma(t_0)} \to M_{\gamma(t)}$ denotes parallel translation.

Next, for $X \in \Upsilon_0$, we have a unique $(\xi^1, \ldots, \xi^{n-1}) \in \mathbb{R}^{n-1}$ such that

$$(\nabla_t X)(\beta) = \sum_{\mu=1}^{\ell} \xi^\mu (\nabla_t Y_\mu)(\beta) + \sum_{\nu=\ell+1}^{n-1} \xi^\nu Y_\nu(\beta),$$

which implies, for sufficiently small $\beta - t > 0$,

$$\begin{aligned} X(t) &= \tau_{\beta,t}\{(t - \beta)(\nabla_t X)(\beta)\} + o(t - \beta) \\ &= \tau_{\beta,t}\left\{\sum_\mu \xi^\mu (t - \beta)(\nabla_t Y_\mu)(\beta) + \sum_\nu (t - \beta)\xi^\nu Y_\nu(\beta)\right\} + o(t - \beta) \\ &= \sum_\mu \xi^\mu Y_\mu(t) + (t - \beta)\sum_\nu \xi^\nu Y_\nu(t) + o(t - \beta), \end{aligned}$$

from which one concludes

$$\lim_{t \uparrow \beta} f^\mu(t) = \xi^\mu, \quad \lim_{t \uparrow \beta} f^\nu(t) = 0.$$

Thus, f^j, $j = 1, \ldots, n-1$, are bounded, which was our claim. ∎

Theorem II.5.5. (C. F. Jacobi (1836)) *Let $\gamma(\alpha)$ have a conjugate point at $t = t_0 \in (\alpha, \beta)$. Then there exists $X \in \Upsilon_0$ such that $I(X, X) < 0$. Thus, a geodesic cannot minimize distance past its first conjugate point.*

Proof. Let $t_0 \in (\alpha, \beta)$, $Y \in \mathcal{J}^\perp$, $Y \neq 0$ such that $Y(\alpha) = Y(t_0) = 0$. Then, for

$$Y_1(t) = \begin{cases} Y(t) & t \in [\alpha, t_0] \\ 0 & t \in [t_0, \beta] \end{cases},$$

one easily has $Y_1 \in \Upsilon_0$, and $I(Y_1, Y_1) = 0$ by (II.5.2). We show how to perturb Y_1 to produce X satisfying $I(X, X) < 0$.

Certainly $(\nabla_t Y)(t_0) \neq 0$. Pick $Z(t)$ to be the parallel field along γ for which $Z(t_0) = -(\nabla_t Y)(t_0)$; let $\varphi : [\alpha, \beta] \to \mathbb{R}$ be differentiable such that $\varphi(\alpha) = \varphi(\beta) = 0$ and $\varphi(t_0) = 1$ and let

$$X_\lambda = Y_1 + \lambda \varphi Z.$$

Then, by explicit calculation and integration by parts, we have

$$\begin{aligned} I(X_\lambda, X_\lambda) &= I(Y_1, Y_1) + 2\lambda I(Y_1, \varphi Z) + O(\lambda^2) \\ &= 2\lambda \int_\alpha^{t_0} \left\{ \langle \nabla_t Y, \nabla_t(\varphi Z)\rangle - \langle R(\gamma', Y)\gamma', \varphi Z\rangle \right\} dt + O(\lambda^2) \\ &= 2\lambda \langle \nabla_t Y, \varphi Z\rangle \big|_\alpha^{t_0} + O(\lambda^2) \\ &= -2\lambda |(\nabla_t Y)(t_0)|^2 + O(\lambda^2), \end{aligned}$$

which is less than 0 for sufficiently small positive λ. The theorem is proven. ∎

§II.6. Elementary Comparison Theorems

We start with the well-known:

Theorem II.6.1. (O. Bonnet (1855), S. B. Myers (1941)) *Let M be a Riemannian manifold, $\gamma : [0, \beta] \to M$ a unit speed geodesic in M such that*

$$\mathrm{Ric}(\gamma', \gamma') \geq (n-1)\kappa > 0$$

on $\gamma([0, \beta])$. If $\beta \geq \pi/\sqrt{\kappa}$, then $\gamma((0, \beta])$ contains a point conjugate to $\gamma(0)$ along γ.

Therefore, if M is a complete Riemannian manifold, of dimension $n \geq 2$, such that there exists a constant $\kappa > 0$ for which

$$\mathrm{Ric}(\xi, \xi) \geq (n-1)\kappa |\xi|^2 \tag{II.6.1}$$

for any $\xi \in TM$, then M is compact with diameter $\leq \pi/\sqrt{\kappa}$.

Proof. Given the unit speed geodesic γ, pick an orthonormal basis $\{e_1, \ldots, e_{n-1}\}$ of $\gamma'(0)^{\perp}$; let E_j be the parallel vector field along γ determined by $E_j(0) = e_j$, $j = 1, \ldots, n-1$; and set, for each $j = 1, \ldots, n-1$,

$$X_j(t) = \sin(\pi t/\beta)E_j(t).$$

Then, $X_j \in \Upsilon_0$, and

$$\begin{aligned}
&\sum_j I(X_j, X_j) \\
&\quad = \int_0^\beta \left\{(n-1)(\pi^2/\beta^2)\cos^2(\pi t/\beta) - \left\{\sum \mathcal{K}(E_j, \gamma')\right\} \sin^2(\pi t/\beta)\right\} dt \\
&\quad = \int_0^\beta \{(n-1)(\pi^2/\beta^2)\cos^2(\pi t/\beta) - \operatorname{Ric}(\gamma', \gamma')\sin^2(\pi t/\beta)\}\, dt \\
&\quad \le \int_0^\beta \{(n-1)(\pi^2/\beta^2)\cos^2(\pi t/\beta) - (n-1)\kappa \sin^2(\pi t/\beta)\}\, dt \\
&\quad = (n-1)(\beta/2)(\pi^2/\beta^2 - \kappa).
\end{aligned}$$

So, if $\beta \ge \pi/\sqrt{\kappa}$, then the index form is no longer positive definite on $\gamma|(0, \beta]$, which implies $\gamma|(0, \beta]$ contains a point conjugate to $\gamma(0)$ along γ. This is the first claim.

For the second claim, we need only note, that given any $p, q \in M$, there exists (by the completeness of M) a unit speed geodesic $\gamma : [0, \beta] \to M$ such that $\gamma(0) = p$, $\gamma(\beta) = q$, and $\beta = \ell(\gamma) = d(p, q)$. Since $\gamma|(0, \beta]$ is a minimizing geodesic, its index form is positive semidefinite, which implies

$$d(p, q) = \beta \le \pi/\sqrt{\kappa}.$$

Thus, M has finite diameter bounded above by $\pi/\sqrt{\kappa}$, which implies M is compact. ■

Theorem II.6.2. (J. Hadamard (1898), E. Cartan (1946)) *Let M be a Riemannian manifold, $\gamma : [0, \beta] \to M$ a unit speed geodesic in M such that*

$$\mathcal{K} \le 0$$

for all sectional curvatures along $\gamma|(0, \beta]$. Then $\gamma((0, \beta])$ contains no point conjugate to $\gamma(0)$ along γ.

Therefore, if M is complete and all its sectional curvatures are nonpositive, then M has no conjugate points.

Proof. Let $\gamma : [\alpha, \beta] \to M$, $|\gamma'| = 1$ be a geodesic, and $X \in \Upsilon_0$. Then,

$$I(X, X) = \int_\alpha^\beta |\nabla_t X|^2 - \mathcal{K}(X, \gamma')|X|^2\, dt \geq \int_\alpha^\beta |\nabla_t X|^2\, dt \geq 0,$$

and the claim follows from Theorem II.5.5. ∎

Theorem II.6.3. (M. Morse (1930), I. J. Schönberg (1932)) *Let M be a Riemannian manifold, $\delta > 0$, $\gamma : [0, \beta] \to M$ a unit speed geodesic in M such that*

$$\mathcal{K} \leq \delta$$

for all sectional curvatures along $\gamma|[0, \beta]$. Then, if $t = \beta$ is a conjugate point of $\gamma(0)$ along γ, we have

$$\beta \geq \pi/\sqrt{\delta}. \tag{II.6.2}$$

Proof. Let $Y \neq 0$, $Y \in \mathcal{J}^\perp \cap \Upsilon_0$. Then, one has

$$\begin{aligned} 0 &= I(Y, Y) \\ &= \int_0^\beta \{|\nabla_t Y|^2 - \mathcal{K}(Y, \gamma')|Y|^2\}\, dt \\ &\geq \int_0^\beta \{|\nabla_t Y|^2 - \delta|Y|^2\}\, dt. \end{aligned}$$

It is an easy consequence of Wirtinger's inequality, for functions vanishing at endpoints of an interval (see Exercise III.42), that

$$\int_0^\beta |\nabla_t Y|^2\, dt \geq (\pi^2/\beta^2) \int_0^\beta |Y|^2\, dt. \tag{II.6.3}$$

Therefore,

$$\pi^2/\beta^2 - \delta \leq 0,$$

which implies the claim. ∎

Theorem II.6.4. (H. E. Rauch (1951)) *Let M be a Riemannian manifold, δ a real constant, $\gamma : [0, \beta] \to M$ a unit speed geodesic in M such that*

$$\mathcal{K} \leq \delta$$

for all sectional curvatures along $\gamma|[0, \beta]$. *If* $Y \in \mathcal{J}^{\perp}$, *then the function* $|Y|$ *along* γ *satisfies the differential inequality*

$$|Y|'' + \delta|Y| \geq 0. \tag{II.6.4}$$

on $[0, \beta)$.

Furthermore, if ψ *denotes the solution on* $[0, \beta]$ *of*

$$\psi'' + \delta\psi = 0, \qquad \psi(0) = |Y|(0), \quad \psi'(0) = |Y|'(0), \tag{II.6.5}$$

and ψ *does not vanish on* $(0, \beta)$, *then*

$$\{|Y|/\psi\}' \geq 0, \tag{II.6.6}$$

$$|Y| \geq \psi, \tag{II.6.7}$$

on $(0, \beta)$.

We have equality in (II.6.6) *at* $t_0 \in (0, \beta)$ *if and only if*

$$\mathcal{K}(Y, \gamma') = \delta$$

on all of $[0, t_0]$, *and there exists a parallel unit vector field E along* γ *for which*

$$Y(t) = \psi(t)E(t)$$

on all of $[0, t_0]$.

Proof. We start with

$$|Y|' = \langle Y, \nabla_t Y\rangle |Y|^{-1},$$

which implies

$$\begin{aligned} |Y|'' &= |Y|^{-1}\{|\nabla_t Y|^2 - \langle Y, R(\gamma', Y)\gamma'\rangle\} - |Y|^{-3}\langle Y, \nabla_t Y\rangle^2 \\ &\geq -\delta|Y| + |Y|^{-3}\{|\nabla_t Y|^2|Y|^2 - \langle Y, \nabla_t Y\rangle^2\} \\ &\geq -\delta|Y| \end{aligned}$$

by the Cauchy–Schwarz inequality, which implies (II.6.4).

For second claim, since

$$\{|Y|/\psi\}' = \{|Y|'\psi - |Y|\psi'\}/\psi^2,$$

we study the function

$$F := |Y|'\psi - |Y|\psi'.$$

Well,

$$F(0) = 0, \quad \text{and} \quad F' = \{|Y|'\psi - |Y|\psi'\}' \geq 0$$

by (II.6.4), (II.6.5). This implies (II.6.6). One immediately has (II.6.7).

If we have equality in (II.6.6) at some $t_0 \in (0, \beta]$, then $F(t_0) = 0$, which implies $F = 0$ on all of $(0, t_0]$, which implies

$$|Y| = \psi$$

on all of $[0, t_0]$. Write

$$Y = \psi E, \qquad |E| = 1$$

along γ. Then,

$$\nabla_t Y = \psi' E + \psi \nabla_t E.$$

Now we have equality in (II.6.4) on $(0, t_0]$, which implies equality in the Cauchy–Schwarz inequality, which implies Y and $\nabla_t Y$ are linearly dependent. Since $\nabla_t Y$ and Y are linearly dependent at every point of $(0, t_0]$, and E is a unit vector field, we have E is parallel along $\gamma|[0, t_0]$. ∎

§II.7. Jacobi Fields and the Exponential Map

Theorem II.7.1. *Let M be a Riemannian manifold of dimension ≥ 2, $\mathcal{T}M$ the domain of the exponential map of the Levi-Civita connection, $p \in M$, $\xi \in \mathcal{T}M \cap M_p$, and $\eta \in M_p$. Then, to calculate $(\exp_p)_{*|\xi} \mathfrak{I}_\xi \eta$ set $\gamma(t) = \exp t\xi$, and let $Y(t)$ be the Jacobi field along γ determined by the initial conditions*

$$Y(0) = 0, \qquad (\nabla_t Y)(0) = \eta. \tag{II.7.1}$$

Then for all t such that $t\xi \in \mathcal{T}M$, we have

$$(\exp_p)_{*|t\xi} \mathfrak{I}_{t\xi} \eta = t^{-1} Y(t). \tag{II.7.2}$$

Proof. (See the argument of Gauss' lemma (Lemma I.6.1).) For a vector space V and a path $\zeta : (-\epsilon_0, \epsilon_0) \to V$, we shall denote the derivative (in contrast to the velocity vector) of ζ by $\dot{\zeta}$. Thus,

$$\zeta'(\epsilon) = \mathfrak{I}_{\zeta(\epsilon)} \dot{\zeta}(\epsilon).$$

Let $V = M_p$ and pick ζ so that

$$\zeta(0) = \xi, \qquad \dot{\zeta}(0) = \eta;$$

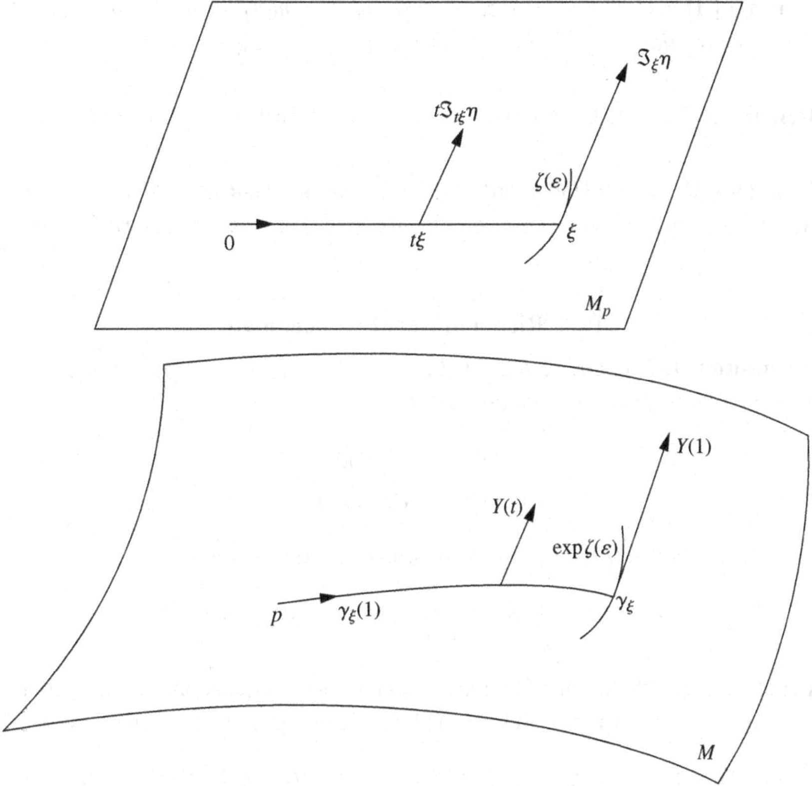

Figure II.3. Differentiating the geodesic variation.

and consider the geodesic variation

$$v(t, \epsilon) = \exp t\zeta(\epsilon).$$

Then, its associated Jacobi field $Y(t) = (\partial_\epsilon v)(t, 0)$ (see Figure II.3) is given by

$$Y(t) = (\exp_p)_{*|t\xi}\, \Im_{t\xi}\, t\dot{\zeta}(0) = t(\exp_p)_{*|t\xi}\, \Im_{t\xi}\, \eta.$$

So, we wish to verify that Y is *the* Jacobi field associated with the initial conditions (II.7.1). Certainly, $Y(0) = 0$. Also, $\nabla_t \partial_\epsilon v = \nabla_\epsilon \partial_t v$, by (I.4.7), implies that

$$(\nabla_t \partial_\epsilon v)(0, \epsilon) = \dot{\zeta}(\epsilon),$$

which implies

$$(\nabla_t Y)(0) = (\nabla_t \partial_\epsilon v)(0, 0) = \eta. \qquad \blacksquare$$

Corollary II.7.1. *The null space of* $(\exp_p)_{*|\xi}$ *is isomorphic to the subspace of Jacobi fields along* $\gamma(t) = \exp t\xi$ *vanishing at* p *and* $\exp \xi$.

Remark II.7.1. See Exercises II.26 and II.27 for further detail on this result.

Corollary II.7.2. (J. Hadamard (1898), E. Cartan (1946)) *If* M *is complete with nonpositive sectional curvature, then* exp *is of maximal rank on all of* TM.

§II.8. Riemann Normal Coordinates

Proposition II.8.1. *Given* $p \in M$, $\xi, \eta, \zeta \in M_p$, $|\xi| = 1$, $\gamma(t) = \exp t\xi$, *and* Y, Z *Jacobi fields along* γ *determined by*

$$Y(0) = 0, \qquad (\nabla_t Y)(0) = \eta,$$
$$Z(0) = 0, \qquad (\nabla_t Z)(0) = \zeta;$$

then the Taylor expansion of $\langle Y, Z\rangle(t)$ *about* $t = 0$ *is given by*

$$\langle Y, Z\rangle(t) = t^2\langle \eta, \zeta\rangle - (t^4/3)\langle R(\xi, \eta)\xi, \zeta\rangle + 0(t^5).$$

Proof. Direct calculation. The idea is that Taylor's expansion, in a neighborhood of $t = 0$, for a vector field $Y(t)$ along the geodesic $\gamma_\xi(t)$ is given by

$$Y(t) = \tau_t \left\{Y(0) + t\nabla_t Y(0) + (t^2/2)\nabla_t{}^2 Y(0) + (t^3/6)\nabla_t{}^3 Y(0)\right\} + O(t^4),$$

where τ_t denotes parallel translation along γ_ξ from p to $\gamma_t(\xi)$. Now one uses the hypotheses of the theorem to calculate the derivatives of $Y(t)$ at $t = 0$. In the inner product, one uses the fact that the parallel translation is an isometry. ■

Now fix $p \in M$ and U an open set about, and starlike with respect to, $0 \in M_p$ for which $\exp|\mathsf{U}$ is a diffeomorphism of U onto its image $U := \exp \mathsf{U}$, an open set in M about p.

Then, every choice of orthonormal basis $\{e_1, \ldots, e_n\}$ of M_p determines a chart $\mathbf{n} : U \to \mathbb{R}^n$, referred to as *Riemann normal coordinates,* given by

$$\mathbf{n}^j(q) = \langle (\exp|\mathsf{U})^{-1}(q), e_j\rangle$$

for $q \in U$, that is, for $v = \sum_j v^j e_j \in \mathsf{U}$, we have

$$\mathbf{n}^j(\exp v) = v^j.$$

In this chart, we have for $\gamma(t) = \exp tv$,

$$\gamma^j(t) := (\mathbf{n}^j \circ \gamma)(t) = tv^j, \qquad \gamma'(t) = \sum_j v^j \partial_{j|\gamma(t)}.$$

If Y_j is the Jacobi field along γ determined by the initial conditions

$$Y_j(0) = 0, \quad (\nabla_t Y_j)(0) = e_j,$$

then (II.7.2) implies

$$\partial_{j|\exp tv} = (\exp_p)_{*|tv} \mathfrak{I}_{tv} e_j = t^{-1} Y_j(t) \tag{II.8.1}$$

for $tv \in \mathsf{U}$.

Theorem II.8.1. *For $v \in \mathsf{U}$,*

$$g_{jk}(\exp v) = \delta_{jk} - (1/3)\langle R(v, e_j)v, e_k\rangle + O(|v|^3)$$

as $v \to 0$.

Proof. Fix $\xi \in \mathsf{S}_p$, and define $\gamma; Y_1, \ldots, Y_n$ as above. Then, (II.7.2) and Proposition (II.8.1) imply

$$\begin{aligned} g_{jk}(\exp t\xi) &= t^{-2}\langle Y_j, Y_k\rangle(t) \\ &= t^{-2}\{t^2\langle e_j, e_k\rangle - (t^4/3)\langle R(\xi, e_j)\xi, e_k\rangle + O(t^5)\} \\ &= \delta_{jk} - (t^2/3)\langle R(\xi, e_j)\xi, e_k\rangle + O(t^3), \end{aligned}$$

which implies the claim. ∎

Corollary II.8.1. *In the above, we also have for $v \in \mathsf{U}$,*

$$\det(g_{jk}(\exp v)) = 1 - (1/3)\mathrm{Ric}\,(v, v) + O(|v|^3)$$

as $v \to 0$.

Proof. This is an immediate consequence of the formula for the derivative of the determinant, namely,

Proposition II.8.2. *Let $A_{jk} : \mathbb{R}^m \to \mathbb{R} \in C^1$, $j, k = 1, \ldots, n$. Then, on the open set for which* $\det(A_{jk}) \neq 0$ *one has, setting $\mathsf{A} = (A_{jk})$,*

$$\frac{\partial}{\partial x^\ell} \ln \det \mathsf{A} = \operatorname{tr} \frac{\partial \mathsf{A}}{\partial x^\ell} \mathsf{A}^{-1} \tag{II.8.2}$$

for $\ell = 1, \ldots, m$.

Proof. It suffices to show that if $A(\epsilon)$ is a differentiable map of the reals to the space of $n \times n$–matrices (with obvious differentiable structure – see Example I.9.4(1)) satisfying $A(0) = I$, the identity matrix, then the Taylor expansion, about $\epsilon = 0$, of the function $\epsilon \mapsto \det A(\epsilon)$ is given by

$$\det A(\epsilon) = 1 + \epsilon \operatorname{tr} A'(0) + O(\epsilon^2).$$

This is a direct consequence of the definition of the determinant. One then has to adjust for $A(0)$ not necessarily equal to the identity matrix I. ∎

Normal Coordinates in Constant Sectional Curvature Spaces

We now assume M has constant sectional curvature κ and consider a unit speed geodesic $\gamma : (\alpha, \beta) \to M$. For convenience, we assume $0 \in (\alpha, \beta)$. By $\mathcal{J}^{\perp}$ we denote, as in §II.5, the Jacobi fields along γ that are orthogonal to γ at every point of γ. Recall that $Y \in \mathcal{J}^{\perp}$ satisfying $Y(0) = 0$ is given by

$$Y = \mathbf{S}_\kappa E, \qquad \nabla_t E = 0.$$

Thus, as proved in Theorem II.5.2, for $\kappa \leq 0$ and M complete, M has no conjugate points (see also Theorem II.6.2). For $\kappa > 0$, and $p \in M$ satisfying $\overline{\mathsf{B}(p; \pi/\sqrt{\kappa})} \subseteq TM$, we conclude that M is compact (hence, complete), and $\exp \mathsf{S}(p; \pi/\sqrt{\kappa})$ consists of precisely one point.

For the general Jacobi field Y along γ, we have $a, b \in \mathbb{R}$ and vector fields E_1, E_2 parallel along γ and orthogonal to γ such that

$$Y(t) = (at + b)\gamma' + \mathbf{C}_\kappa E_1 + \mathbf{S}_\kappa E_2,$$

and for the initial condition $Y(0) = 0$, we have

$$Y(t) = at\gamma' + \mathbf{S}_\kappa E_2.$$

Now, let $p \in M$, and $\{e_1, \ldots, e_n\}$ an orthonormal basis of M_p, thereby determining Riemann normal coordinates on a neighborhood U about p. Let $\xi \in \mathsf{S}_p$, $\gamma(t) = \exp t\xi$ and, as in (II.8.1),

$$t^{-1} Y_j(t) = (\exp_p)_{*|t\xi}\, \mathfrak{I}_{t\xi} e_j = \partial_{j|\exp t\xi}$$

$j = 1, \ldots, n$. Then there exist $a_j \in \mathbb{R}$, and $E_j(t)$ a vector field parallel along γ and orthogonal to γ, such that

$$Y_j(t) = a_j t\gamma' + \mathbf{S}_\kappa E_j.$$

But

$$e_j = (\nabla_t Y_j)(0) = a_j \xi + E_j(0);$$

this implies

$$a_j = \langle \xi, e_j \rangle = \xi^j, \quad E_j(0) = e_j - \xi^j \xi.$$

Direct calculation now yields

$$g_{jk}(\exp t\xi) = \xi^j \xi^k + \frac{\mathbf{S}_\kappa{}^2(t)}{t^2}\{\delta_{jk} - \xi^j \xi^k\}.$$

We therefore have the following:

Theorem II.8.2. *Let M have constant sectional curvature κ, $p \in M$, U a neighborhood about $0 \in M_p$ as above, and $\{e_1, \ldots, e_n\}$ an orthonormal basis of M_p. Then, in the resulting normal coordinate system, we have for $v \in \mathsf{U}$*

$$g_{jk}(\exp v) = \frac{v^j v^k}{|v|^2} + \frac{\mathbf{S}_\kappa{}^2(|v|)}{|v|^2}\left\{\delta_{jk} - \frac{v^j v^k}{|v|^2}\right\}. \tag{II.8.3}$$

In particular, if M, M' are Riemannian manifolds of constant sectional curvature κ, $p \in M$ and $q \in M'$, then there exists $\delta > 0$ such that each choice of orthonormal bases in M_p and M'_q determine an isometry of $B(p;\delta)$ onto $B(q;\delta)$.

Remark II.8.1. We can use (II.8.3) to define, for each $\kappa < 0$, a Riemannian metric on $\mathbb{R}^n$ which is complete, and which has constant sectional curvature κ.

As a prelude to what follows in later work, we give a formal calculation of the Riemannian metric of constant sectional curvature κ in geodesic spherical coordinates, namely, set

$$v = t\xi, \qquad |\xi| = 1.$$

Then

$$dv = t\,d\xi + (dt)\xi,$$

and

$$\langle \xi, t\,d\xi \rangle = 0.$$

Also,

$$\sum_j v^j\,dv^j = t\,dt.$$

Therefore, (II.8.3) simplifies to

(II.8.4) $$ds^2 := \sum_{j,k} g_{jk}(\exp v)\, dv^j dv^k = dt^2 + \mathbf{S}_\kappa{}^2(t)|d\xi|^2,$$

a formula we derived in each of our examples of spaces of constant sectional curvature.

§II.9. Notes and Exercises

Curvature Tensor Estimates

Exercise II.1. (J. P. Bourguignon & H. Karcher (1978)) From Propositions II.1.1 and II.1.2, one concludes that full knowledge of the sectional curvatures at a point determines the curvature tensor itself at that point. We sketch here some practical versions of this fact.

For any $\lambda \in \mathbb{R}$, we let

$$R_\lambda = \lambda R_1,$$

where R_1 is defined in Proposition II.1.2.

Fix a point p in a given Riemannian manifold M. All calculations in this exercise take place in the one tangent space, M_p. We set

$$\kappa = \min \mathcal{K}, \qquad \delta = \max \mathcal{K}.$$

(a) Prove that for any $u_j \in M_p$, $j = 1, \ldots, 4$, we have

$$\begin{aligned} 6\langle R(u_1, u_2)u_3, u_4\rangle = {} & \langle R(u_1, u_2 + u_3)(u_2 + u_3), u_4\rangle \\ & - \langle R(u_1, u_2 - u_3)(u_2 - u_3), u_4\rangle \\ & + \langle R(u_2, u_1 - u_3)(u_1 - u_3), u_4\rangle \\ & - \langle R(u_2, u_1 + u_3)(u_1 + u_3), u_4\rangle. \end{aligned}$$

(b) Also show, for $\xi, \eta, \zeta \in M_p$, that

$$4\langle R(\xi, \eta)\xi, \zeta\rangle = \langle R(\xi, \eta + \zeta)\xi, \eta + \zeta\rangle - \langle R(\xi, \eta - \zeta)\xi, \eta - \zeta\rangle.$$

Thus, (a) and (b) constitute an alternative version of Proposition II.1.1.

(c) Prove, for orthogonal ξ, η, ζ,

$$\langle R(\xi, \eta)\xi, \zeta\rangle \leq \frac{\delta - \kappa}{4}|\xi|^2|\eta + \zeta|^2.$$

(d) Prove, for orthonormal u_j, $j = 1, \ldots, 4$,

$$|\langle R(u_1, u_2)u_3, u_4\rangle| \leq \frac{2}{3}(\delta - \kappa).$$

(e) Prove, for arbitrary ξ, η,

$$|R(\xi, \eta)\xi - R_{(\kappa+\delta)/2}(\xi, \eta)\xi| \leq \frac{\delta - \kappa}{2}|\xi|^2|\eta|.$$

(f) Prove, for arbitrary unit vectors u_j, $j = 1, \ldots, 3$, that

$$|R(u_1, u_2)u_3| \leq \frac{4}{3} \max |\mathcal{K}|.$$

Schur's Theorem

Exercise II.2. Prove (F. Schur (1886))

Theorem. *If M is a Riemannian manifold of* $\dim M \geq 3$*, and there exists a function $\kappa : M \to \mathbb{R}$ such that*

$$R(\xi, \eta)\zeta = \kappa(p)R_1(\xi, \eta)\zeta$$

for all $\xi, \eta, \zeta \in M_p$, $p \in M$, then the function κ must be a constant – so M has constant sectional curvature.

Note II.1. Thus, Schur's theorem provides a striking contrast to Theorem II.1.1. An interesting theorem inspired by Schur's theorem, and the full apparatus of pinching theorems, was proved by E. A. Ruh (1982). It goes as follows:

Definition. The sectional curvature $\mathcal{K}$ of a Riemannian manifold M is said to be *locally δ–pinched* if there exists a positive function $\kappa : M \to \mathbb{R}$ such that

$$\delta\kappa(x) < \mathcal{K} < \kappa(x)$$

at every point $x \in M$.

Theorem. *There exists $\delta = \delta(n)$ with*

$$1/4 < \delta < 1,$$

such that any compact locally δ–pinched Riemannian manifold of dimension n is diffeomorphic to a spherical space form, that is a compact Riemannian manifold of constant sectional curvature equal to 1.

It has been noted in Gribkov (1980) that the compactness is essential, namely, if M is noncompact, then even if δ is arbitrarily close to 1, the variation of sectional curvature over the manifold can still be arbitrarily large.

Ruh's theorem has been refined in Huisken (1985).

Bibliographic Sampler for Curves and Surfaces in Euclidean Space

Note II.2. Our treatment of submanifolds, done at breakneck speed in §II.2, follows that of Hicks (1965), which is highly recommended to the reader.

For traditional treatments, we refer the reader to Struik (1961) and Stoker (1969). More modern treatments of curves and surfaces in space are given in do Carmo (1976), Klingenberg (1976), and O'Neill (1966a). Berger–Gostiaux (1988) starts from manifolds and features a full study of curves in the plane and in space, among other matters. More recent books include Gray (1998), Morgan (1992), and Oprea (1997).

Totally Geodesic Submanifolds

Exercise II.3.

(a) Let M be a submanifold of $\overline{M}$ as described in §II.2. We say that M is *totally geodesic in* $\overline{M}$, if for any geodesic γ in $\overline{M}$, for which there exists t_0 such that $\gamma(t_0) \in M$ and $\gamma'(t_0) \in M_{\gamma(t_0)}$, there exists an $\epsilon > 0$ such that $\gamma|(t_0 - \epsilon, t_0 + \epsilon)$ is completely contained in M. Show that M is totally geodesic if and only if the second fundamental form B vanishes identically on M.

(b) (S. Kobayashi (1958)) Show that if M is a Riemannian manifold possessing an isometry $\phi : M \to M$, then any connected component of the set of all points left fixed by ϕ is totally geodesic.

Two Norms of Linear Transformations

Definition. Let V be a finite-dimensional inner product space, $A : V \to V$ a linear transformation, and $B : V \times B \to \mathbb{R}$ the associated bilinear form given by

$$B(x, y) = \langle Ax, y\rangle.$$

The *Gram–Schmidt norm of* A is given by

$$\|A\|_{\mathrm{gs}} = \mathrm{tr}\,(AA^*),$$

where A^* denotes the adjoint of A.

Note II.3. To give an explicit formula for $\|A\|_{\mathrm{gs}}$, pick an orthonormal basis $\{e_1, \ldots, e_n\}$ of V, and let $\mathfrak{A} = (\mathfrak{a}_{jk})$ denote the matrix of A associated with the basis $\{e_j\}$, that is

$$Ae_j = \sum e_k \mathfrak{a}_{kj}.$$

Then A^* is represented by $\mathfrak{A}^T$, and

$$\operatorname{tr}(\mathfrak{A}\mathfrak{A}^T) = \sum_{j,k} \mathfrak{a}_{jk}{}^2 = \sum_j |Ae_j|^2.$$

Since AA^* is self-adjoint, the arithmetic–geometric mean inequality implies

$$\{\det(AA^*)\}^{1/n} \leq \frac{\operatorname{tr}(AA^*)}{n} \leq \|A\|^2,$$

that is,

$$(\det A)^{1/n} \leq \|A\|.$$

Thus, the Gram–Schmidt norm of A coincides with its norm as a $(1, 1)$–tensor on V. To distinguish it from the usual sup norm of A,

$$\|A\| = \sup\{|Ax| : |x| = 1\},$$

we often use the bilinear form B instead, since the norms of A and B, relative to the inner products associated to the respective tensor spaces generated by V, are equal. Namely,

$$\|B\| = \|A\|_{\mathsf{gs}}.$$

The Second Fundamental Form and Local Convexity

Exercise II.4. Let M be a codimension 1 submanifold of the Riemannian manifold $\overline{M}$, $p \in M$. Let ξ be a unit vector orthogonal to M_p. Let $\tilde{M}_p = \exp M_p$, where exp denotes (for the moment) the exponential map in $\overline{M}$. Show that $\tilde{M}_p$ is a smooth submanifold in some neighborhood of p, and has vanishing second fundamental form at p. (One might refer to $\tilde{M}_p$ as *totally geodesic at* p.) Show that if the second fundamental form of M, with respect to ξ, is positive definite, then p has a neighborhood U in $\overline{M}$ in which

$$M \cap \tilde{M}_p \cap U = \{p\}.$$

Thus, when the second fundamental form is definite, one might say that "M lies, locally, on one side of M_p" – a sort of local convexity.

Exercise II.5. Show that it is impossible to imbed a compact surface, of non-positive Gauss curvature, into $\mathbb{R}^3$.

A much deeper theorem, which goes back to D. Hilbert (1901), states that any complete surface of constant negative curvature cannot be imbedded in

$\mathbb{R}^3$. Later, Efimov (1964) proved a corresponding theorem for variable negative curvature. See T. K. Milnor (1972) for a detailed presentation.

Exercise II.6. Let M^{n-1} be an immersed hypersurface in $\mathbb{R}^n$. Show that its second fundamental form is definite (positive or negative definite, depending on the choice of local unit normal vector field) if and only if all its intrinsic Riemannian sectional curvatures are positive.

Definition. Let $M^{m-1} \subset \overline{M}^m$ be an immersed submanifold. We say a point $p \in M$ is *umbilic* if the second fundamental form of M at p is a scalar multiple of the first fundamental form.

Exercise II.7. Let $\overline{M} = \mathbb{R}^m$. Show that if every point of M is umbilic, then M is a piece of a sphere in $\mathbb{R}^m$. In particular, if M is compact and everywhere umbilic in $\mathbb{R}^m$ then M is a sphere in $\mathbb{R}^m$.

Mean Curvature

$\overline{M}$ is our given m–dimensional Riemannian manifold, and M a connected n–dimensional submanifold of $\overline{M}$, $0 \leq n < m$.

Definition. The *mean curvature vector H of the submanifold M at p* is given by

$$H = \operatorname{tr} \mathfrak{B},$$

where $\mathfrak{B}$ the second fundamental form, and the trace of $\mathfrak{B}$ is taken with respect to the Riemannian metric of $\overline{M}$ restricted to M, that is, the first fundamental form of M in $\overline{M}$.

Thus, if $\{e_1, \ldots, e_n\}$ is an orthonormal basis of M_p, then

$$H = \sum_{j=1}^{n} \mathfrak{B}(e_j, e_j).$$

Exercise II.8. Prove

$$\operatorname{tr} \mathfrak{A}^{\nu} = \langle H, \nu \rangle.$$

for all $\nu \in {M_p}^{\perp}$.

Definition. The manifold M is said to be a *minimal submanifold of $\overline{M}$* if its mean curvature vanishes identically on M.

Note II.4. The literature and ongoing work on minimal submanifolds are enormous. For openers, we refer the reader to Barbosa–Colares (1986), Bombieri (1983), Dierkes–Hildebrandt–Kuester–Wohlrab (1992), Lawson (1980), Nitsche (1989), and Osserman (1986, 1990, 1997). (This list is far from exhaustive, even definitive.)

Hessians, Again

Let $f : M \to \mathbb{R}$ be a differentiable function on the Riemannian manifold M. Define Δf, the *Laplacian of* f, to be the trace of the Hessian (defined just before Exercise I.10) of f relative to the first fundamental form, so

$$\Delta f = \operatorname{tr}(\xi \mapsto \nabla_\xi \operatorname{grad} f).$$

Exercise II.9. Let $p \in M$ such that $f(p) = \alpha$, and grad f does not vanish at p. Then, the level surface of f through p, $f^{-1}[\alpha]$, restricted to a sufficiently small neighborhood of p, is an embedded $(n-1)$–manifold. Show that the Hessian of f at p is given by

$$(\operatorname{Hess} f)_{|(f^{-1}[\alpha])_p} = \mathfrak{B}_{-\operatorname{grad} f_{|p}},$$

the second fundamental form of $f^{-1}[\alpha]$ associated to the normal vector $-$grad f at p.

Now consider M a submanifold of $\overline{M} = \mathbb{R}^m$, $m > n = \dim M$.

Let $A, B, \ldots$ range over $1, \ldots, m$; $j, k, \ldots$ range over $1, \ldots, n$; and $\alpha, \beta, \ldots$ range over $n+1, \ldots, m$. Consider a neighborhood U in $\mathbb{R}^m$, with orthonormal frame field $\{\overline{\mathbf{e}}_1, \ldots, \overline{\mathbf{e}}_m\}$, such that $\{\overline{\mathbf{e}}_1, \ldots, \overline{\mathbf{e}}_n\}$ are tangent to M at all points of $M \cap U$, and $\{\overline{\mathbf{e}}_{n+1}, \ldots, \overline{\mathbf{e}}_m\}$ are normal to M at all points of $M \cap U$. We let $\{\overline{\omega}^A\}$ denote the coframe on U dual to $\{\overline{\mathbf{e}}_A\}$ and $\{\overline{\omega}_B{}^A\}$ the associated connection forms. We also let

$$\mathbf{e}_A = \overline{\mathbf{e}}_A|M, \qquad \omega^A = \overline{\omega}^A|M, \qquad \omega_B{}^A = \overline{\omega}_B{}^A|M.$$

Since $\overline{M} = \mathbb{R}^m$, we have the natural fixed basis $\{\mathfrak{e}_A\}$ of $\mathbb{R}^m$, which we identify with the moving parallel frame

$$E_{A|p} = \mathfrak{I}_p\, \mathfrak{e}_A,$$

via the natural identification $\mathfrak{I}_p$ of $\mathbb{R}^m$ with its tangent space at $p \in \mathbb{R}^m$. Here, we continue the discussion (with the notation) of Exercise I.12.

Henceforth, restrict $\mathbf{x} = \sum_A x^A \mathbf{e}_A$ to M. So, $d\mathbf{x}$ is a form on M with values in a fixed vector space, namely, $\mathbb{R}^m$, and, similarly, for Hess $\mathbf{x}$.

Exercise II.10.

(a) Prove

$$dx^B = \sum_j \omega^j S_j{}^B.$$

Also, prove that

$$\operatorname{grad} x^A = \sum_j T_A{}^j \mathbf{e}_j,$$

from which one has

$$|\operatorname{grad} \mathbf{x}|^2 := \sum_A |\operatorname{grad} x^A|^2 = n,$$

(b) Prove

$$\nabla dx^B = \sum_{j,\alpha} \omega_j{}^\alpha S_\alpha{}^B \otimes \omega^j.$$

So,

$$\operatorname{Hess} \mathbf{x} = \mathfrak{B},$$

the second fundamental form, from which one concludes

$$\Delta \mathbf{x} = H.$$

In particular, the coordinate functions all have vanishing Hessian at a point $p \in M$ if and only if the second fundamental form vanishes at p.

(c) Consider the case of codimension 1, that is, M is a hypersurface in $\mathbb{R}^m$. Then,

$$\mathbf{n} := \mathbf{e}_m = \sum_A S_m{}^A E_A$$

is a unit normal vector field along M. And the mean curvature H is given by

$$H = \mathfrak{h}\mathbf{n},$$

where $\mathfrak{h}$ is the *mean curvature function relative to* $\mathbf{n}$.

Prove

$$d\mathbf{n} = -\sum_{k,j,A} h_{jk} S_j{}^A \omega^k E_A = -\sum_{k,j} h_{jk} \omega^k \mathbf{e}_j,$$

and

$$\begin{aligned}\nabla d\mathbf{n} &= -\sum_{j,k}\left\{dh_{jk} + \sum_{\ell}\{\omega_j{}^{\ell}h_{\ell k} - \omega_k{}^{\ell}h_{j\ell}\}\right\}\otimes\omega^k\,\mathbf{e}_j \\ &\quad -\sum_{j,k,\ell} h_{jk}h_{j\ell}\omega^{\ell}\wedge\omega^k\,\mathbf{n} \\ &:= -\sum_{j,k,\ell} h_{jk\ell}\omega^{\ell}\otimes\omega^k\mathbf{e}_j - \sum_{j,k,\ell} h_{jk}h_{j\ell}\omega^{\ell}\wedge\omega^k\,\mathbf{n}.\end{aligned}$$

Use $0 = d^2\,\mathbf{n}$ to show $h_{jk\ell} = h_{j\ell k}$.

(d) Prove

$$\Delta\,\mathbf{n} = -\mathrm{grad}\,\mathfrak{h} - \|\mathfrak{B}\|^2\mathbf{n}.$$

(e) Prove

$$\Delta\,(\mathbf{x}\cdot\mathbf{n}) = -\mathfrak{h} + \mathbf{x}\cdot\mathrm{grad}\,\mathfrak{h} - \|\mathfrak{B}\|^2(\mathbf{x}\cdot\mathbf{n}).$$

Connections in Normal Bundles of Submanifolds

Let M be a (positive codimension) submanifold of $\overline{M}$, with associated tangent and normal bundles TM and νM, respectively, and Levi-Civita connection $\overline{\nabla}$ on $\overline{M}$. Of course, the Levi-Civita connection in TM is given by

$$\nabla_X Y = (\overline{\nabla}_X Y)^T,$$

where X and Y are sections in TM, and the superscript T denotes the projection of $T\overline{M}$ to TM. A connection $\mathcal{D}$ in the normal bundle is defined by

$$\mathcal{D}_X Z = (\overline{\nabla}_X Z)^N,$$

where X and Z are sections in TM and νM, respectively, and the superscript N denotes the projection of $T\overline{M}$ to νM.

Exercise II.11. Prove that for the curvature tensor $R_{\mathcal{D}}$ of $\mathcal{D}$ (see §1.9) we have, at $p \in M$,

$$\begin{aligned}\langle R_{\mathcal{D}}(\xi,\eta)\sigma,\tau\rangle = \langle\overline{R}(\xi,\eta)\sigma,\tau\rangle + \sum_j &\left\{\langle\mathfrak{B}(e_j,\xi),\sigma\rangle\langle\mathfrak{B}(e_j,\eta),\tau\rangle\right. \\ &\left.-\langle\mathfrak{B}(e_j,\eta),\sigma\rangle\langle\mathfrak{B}(e_j,\xi),\tau\rangle\right\},\end{aligned}$$

where $\xi,\eta \in M_p$ and $\sigma,\tau \in M_p{}^{\perp}$, and $\{e_j\}$ is an orthonormal frame of M_p.

Continue with $\overline{M}$ an m–dimensional Riemannian manifold, and M an n–dimensional submanifold with $n < m$. We consider an orthonormal frame field

$\{\overline{e}_1, \ldots, \overline{e}_m\}$ on some neighborhood U in $\overline{M}$ of some $p \in M$, such that $\{\overline{e}_1, \ldots, \overline{e}_n\}$ are tangent to M at all points of $M \cap U$, and $\{\overline{e}_{n+1}, \ldots, \overline{e}_m\}$ are normal to M at all points of $M \cap U$. Again, let $A, B, \ldots$ range over $1, \ldots, m$; $j, k, \ldots$ range over $1, \ldots, n$; and $\alpha, \beta, \ldots$ range over $n+1, \ldots, m$. We let $\{\overline{\omega}^A\}$ denote the coframe on U dual to $\{\overline{e}_A\}$, and we set

$$e_A = \overline{e}_A|M, \qquad \omega^A = \overline{\omega}^A|M, \qquad {\omega_B}^A = {\overline{\omega}_B}^A|M.$$

Exercise II.12.

(a) Let $\mathcal{D}$ denote the connection in the normal bundle. Show that for any section η in the normal bundle νM, represented by

$$\eta = \sum_\alpha \eta^\alpha e_\alpha,$$

we have

$$\mathcal{D}\eta = \sum_{\alpha,\beta} \left\{ d\eta^\alpha + \sum_\beta \eta^\beta {\omega_\beta}^\alpha \right\} \otimes e_\alpha := \sum_{k,\alpha} {\eta^\alpha}_k \, \omega^k \otimes e_\alpha,$$

and

$$\begin{aligned} \mathcal{D}\mathcal{D}\eta &= \sum_{j,\alpha} \left\{ d{\eta^\alpha}_j - \sum_\ell {\omega_j}^\ell {\eta^\alpha}_\ell + \sum_\beta {\eta^\beta}_j {\omega_\beta}^\alpha \right\} \otimes \omega^j \otimes e_\alpha \\ &:= \sum_{j,k,\alpha} {\eta^\alpha}_{jk} \, \omega^j \otimes \omega^k \otimes e_\alpha. \end{aligned}$$

(b) Show that

$$\begin{aligned} & d\left\{ \sum_{j,\alpha} (-1)^{j-1} \eta^\alpha {\eta^\alpha}_j \, \omega^1 \wedge \cdots \wedge \widehat{\omega^j} \wedge \cdots \wedge \omega^n \right\} \\ & \quad = \sum_{j,\alpha} \left\{ ({\eta^\alpha}_j)^2 + \eta^\alpha {\eta^\alpha}_{jj} \right\} \omega^1 \wedge \cdots \wedge \omega^n. \end{aligned}$$

On Isometries

Exercise II.13. Let M be a complete Riemannian manifold, ϕ, ψ isometries of M. Assume that there exists $p \in M$ such

$$\phi(p) = \psi(p) \qquad \text{and} \qquad \phi_{*|p} = \psi_{*|p}.$$

Show that $\phi = \psi$ on all of M.

Spherical and Hyperbolic Geometry

Exercise II.14.

(a) Use the discussions of Examples I.9.4 and I.9.5 of §I.9 to calculate the curvature of spheres by representing $\mathbb{S}^n$ in $\mathbb{R}^{n+1}$ as the homogeneous space $\mathcal{SO}(n+1)/\mathcal{SO}(n)$, where $\mathcal{SO}$ refers to the *special orthogonal group,* that is, those elements of the orthogonal group with determinant equal to 1. Compare with Exercise I.22.

(b) Of course, $\mathbb{S}^n$ can be represented as $\mathcal{O}(n+1)/\mathcal{O}(n)$. Using induction on n, show that $\mathcal{O}(n)$ is the full group of isometries of $\mathbb{S}^n$ (see Wolf (1967, p. 66)).

Exercise II.15. Consider a connected component M_o of the submanifold M of Minkowski space $\mathbb{R}^{n+1}$, endowed with quadratic form $\mathcal{M}$, as described in Example I.9.1 of §I.9.

(a) Show that M_o is isometric to the hyperbolic space of constant sectional curvature $-1/\rho^2$.

(b) Let $\mathcal{O}(n,1)$ be the *orthogonal group of the quadratic form* $\mathcal{M}$, that is, $\mathcal{O}(n,1)$ is the group of linear transformations of $\mathbb{R}^{n+1}$ that leave $\mathcal{M}$ invariant. Show that every element of $\mathcal{O}(n,1)$, when restricted to M, is an isometry of M to itself. Furthermore, show that, given $p \in M_o$ with orthonormal frame $\{e_{p;1},\ldots,e_{p;n}\}$ of $(M_o)_p$, and $q \in M_o$ with orthonormal frame $\{e_{q;1},\ldots,e_{q;n}\}$ of $(M_o)_q$, then there exists $A \in \mathcal{O}(n,1)$ such that

$$A(p) = q, \qquad A_* e_{p;j} = e_{q;j}.$$

Note II.5. We define a *geodesic triangle* pqr in a Riemannian manifold M to consist of three pairwise distinct points p, q, r in M, and minimizing geodesics $\sigma_{pq}, \sigma_{qr}, \sigma_{rp}$ joining p to q, q to r, and r to p, respectively.

Given a geodesic triangle, whose sides have respective lengths a, b, c and angles at opposite vertices are given, respectively, by α, β, γ, then if M is one of our model spaces of constant sectional curvature κ we have the *Law of Sines*

$$\sin\alpha : \sin\beta : \sin\gamma = \mathbf{S}_\kappa(a) : \mathbf{S}_\kappa(b) : \mathbf{S}_\kappa(c)$$

(where the colon denotes proportion). The *Law of Cosines* reads, when $\kappa \neq 0$, as:

$$\mathbf{C}_\kappa(a) = \mathbf{C}_\kappa(b)\mathbf{C}_\kappa(c) + \kappa\mathbf{S}_\kappa(b)\mathbf{S}_\kappa(c)\cos\alpha.$$

See Berger (1987, Vol. II, pp. 286, 329). Check what happens when $\kappa \to 0$.

For spherical geometry and trigonometry see Berger (1987, Vol. II, Chapter 18); for elementary hyperbolic geometry and trigonometry see Berger (1987, Vol. II, Chapter 19), Fenchel (1989), and Meschkowski (1964). For a view of hyperbolic trigonometry via Exercise II.15, see Buser (1992, Chapter 2). For a comprehensive introduction to the classical geometries, see Ratcliffe (1994).

Note II.6. The Killing–Hopf theorem (Theorem IV.5) states that any complete simply connected Riemannian manifold of constant sectional curvature is isometric to one of the model spaces. For constant positive sectional curvature, a full classification of *all* such spaces has been given by J. A. Wolf. See his presentation in Wolf (1967). He also gives a discussion of manifolds of constant vanishing sectional curvature. For a more extensive treatment, see Charlap (1986).

For the geometry and topology of manifolds of constant sectional negative curvature, start with Beardon (1983), Epstein (1987), Bedford–Keane–Series (1991), and the recent Buser (1992), and progress from there to Thurston (1979). More recently, one has Benedetti–Petronio (1994), Ratcliffe (1994), and Thurston's own Thurston (1997).

A Result of J. L. Synge

Exercise II.16. Consider a geodesic variation in the Riemannian manifold M – one might call it a *ruled surface*. Show, using one of two possible arguments, that the Gauss curvature of the surface is less than or equal to the sectional curvature in M associated to each tangent 2–plane of the surface (Synge (1934)).

On Conjugate Points

Exercise II.17. Assume $\gamma : [0, \beta] \to M, p = \gamma(0), \xi = \gamma'(0)$, is a unit speed geodesic. Show that points along γ, conjugate to p along γ, are isolated.

On Jacobi's Criteria

Theorem II.5.4 states that if $\gamma : [0, \beta] \to M$ is a unit speed geodesic with no points on $(0, \beta]$ conjugate to $\gamma(0)$ along γ, then γ is a "strict local minimum" of the distance function among curves connecting $\gamma(0)$ to $\gamma(\beta)$.

Here, "strict local minimum" is understood in the sense that if $v(t, \epsilon)$ is a homotopy of γ, with length function $L(\epsilon)$, as described in §II.4, where $L(0) = \beta$, then $L'(0) = 0$ and $L''(0) > 0$. In the argument we gave, we followed Ambrose (1961). The original version of the argument in 2 dimensions can be found in Darboux (1894, Vol. III, pp. 95ff).

A different version of "strict local minimum" in Jacobi's theorem goes as follows:

Exercise II.18. Let $p = \gamma(0), \xi = \gamma'(0) \in \mathsf{S}_p$. For ϵ in $(0, 1)$, let $\mathcal{C}_\epsilon(\xi)$ denote the neighborhood of $\xi \in \mathsf{S}_p$ given by

$$\mathcal{C}_\epsilon(\xi) = \{\eta \in \mathsf{S}_p : \langle \xi, \eta \rangle > 1 - \epsilon\};$$

and for any ϵ in $(0, 1)$ and $r > 0$ let

$$\mathcal{C}_{\epsilon,r}(\xi) = \{t\eta \in M_p : t \in [0, r), \eta \in \mathcal{C}_\epsilon(\xi)\},$$

and

$$\mathfrak{C}_{\epsilon,r}(\xi) = \exp \mathcal{C}_{\epsilon,r}(\xi).$$

Prove the following:

Theorem. (C. F. Jacobi (1836)) *Assume $\gamma : [0, \beta] \to M, p = \gamma(0), \xi = \gamma'(0)$, is a unit speed geodesic such that $\gamma|(0, \beta]$ is one-to-one with no points conjugate to p along γ. Then there exist ϵ in $(0, 1), r > \beta$, such that $\mathcal{C}_{\epsilon,r}(\xi) \subseteq \mathcal{T}M$, the domain of the exponential map. Furthermore, there exists sufficiently small $\epsilon > 0$ such that, if ω is a path from p to $\gamma(\beta)$ with image completely contained in $\mathfrak{C}_{\epsilon,r}(\xi)$ then*

$$\ell(\omega) \geq \beta,$$

with equality only if the image of ω is the same as that of γ. (See Darboux (1894, Vol. III, p. 86))

Note II.7. A different proof, for surfaces, of Theorem II.5.5 can be found in Darboux (1894, Vol. III, p. 88).

Geometry of the Index Form

Exercise II.19. Let M be a Riemannian manifold, p a point in M, and r *the distance function on M based at p*, that is, r is given by

$$r(x) = d(p, x).$$

(a) Show, for $r > 0$ sufficiently small, that $r \in C^\infty$ and $|\operatorname{grad} r| = 1$.

(b) Show, with $\beta > 0$ sufficiently small as in (a), $\gamma : [0, \beta] \to M$ a unit speed geodesic emanating from p, that for any Jacobi field Y along γ, vanishing at p and orthogonal to γ along γ, its index form I is given by

$$I(Y, Y) = \langle \nabla_t Y, Y \rangle(\beta) = \mathfrak{B}_{-\operatorname{grad} r_{|\gamma(\beta)}}(Y(\beta), Y(\beta)) = \operatorname{Hess} r(Y(\beta), Y(\beta))$$

where $\mathfrak{B}$ denotes the second fundamental form of the level surface $r^{-1}[\beta]$.

Definition. A function $g : \mathbb{R} \to \mathbb{R}$ is called *convex* if for any $a < b$ and $s \in (0, 1)$ we have

$$g((1-s)a + sb) \leq (1-s)g(a) + sg(b).$$

The function g is called *strictly convex* if the inequality is strict inequality.

A function f on a Riemannian manifold M is (*strictly*) *convex* if for every nontrivial geodesic $\gamma : [0, 1] \to M$ the function $f \circ \gamma$ is (strictly) convex.

A subset A of a Riemannian manifold M is called *convex* if for any $p, q \in A$ there exist a unique unit speed minimizing geodesic γ_{pq} in M connecting p to q, and $\gamma_{pq} \subseteq A$.

Note II.8. For discussion of uniqueness of minimizing geodesics joining points of a Riemannian manifold, see the introductory discussion of §III.2. For other notions of convexity – other than "convex" – see §IX.6.

Exercise II.20.

(a) Assume a function f on the Riemannian manifold M is C^2. Show that f is convex (resp. strictly convex) if Hess f is nonnegative (resp. positive definite). Show that if f is (strictly) convex then Hess f is nonnegative.

(b) Let A be a convex open subset of M, and assume the sectional curvature of M on A is nonpositive. Given $p \in A$, set $r(q) = d(p, q)$. Show that the function $r : A \to \mathbb{R}$ is convex on A.

(c) Show in (b) that the distance function $d : A \times A \to \mathbb{R}$ is convex.

On Manifolds of Positive Curvature

Note II.9. If the sectional curvature of a complete Riemannian manifold is positive everywhere but not bounded away from 0, then one cannot conclude that the manifold is compact – simply consider a paraboloid of revolution. But see Calabi (1967) and Schneider (1972). On the other hand, the condition of strictly positive curvature remains quite restrictive. D. Gromoll and W. Meyer (1969) have shown that noncompact manifolds possessing a complete Riemannian metric of strictly positive sectional curvature are contractible. The topological structure of manifolds possessing a complete Riemannian metric of nonnegative sectional curvature is richer. See the original papers of J. Cheeger and D. Gromoll (1971), (1972), and the updated presentations in Cheeger–Ebin (1975, Chapter 8), Klingenberg (1982, §2.9), and Besse (1987, p. 171ff). See some of our remarks in §IX.9.

On the Morse–Schönberg Theorem

Exercise II.21. The Rauch theorem (Theorem II.6.4) is stronger than the Morse–Schönberg theorem (Theorem II.6.3) and therefore supplies another proof of the Morse–Schönberg theorem. The following theorem is closer to the original argument of Morse and Schönberg.

Prove:

Theorem. *Let M_1, M_2 be Riemannian manifolds of same dimension, with γ_1, γ_2 unit speed geodesics in M_1, M_2, respectively. Let $\mathcal{K}_1$, $\mathcal{K}_2$ denote respective sectional curvatures along γ_1, γ_2, and assume that*

$$\sup \mathcal{K}_{1_{|\gamma(t)}} \leq \min \mathcal{K}_{2_{|\gamma(t)}}$$

for all $t > 0$. Then, the first conjugate point to $\gamma_1(0)$ along γ_1 cannot occur earlier than the first conjugate point to $\gamma_2(0)$ along γ_2.

On the Rauch Theorem

Note II.10. The attribution of Theorem II.6.4 to H. E. Rauch is, actually, off the mark, for the method of the argument certainly goes back to the Sturm's (1836) separation arguments of the nineteenth century, even if the specific Rauch result and its applications (see Chapter IX) are more recent. The theorem more properly belonging to Rauch is his corresponding comparison theorem for the length of Jacobi fields when the sectional curvature is bounded from below. Simply note that, in the argument given in Theorem II.6.4, if the curvature is bounded from below, then the bound on the curvature pushes in the opposite direction from that of the Cauchy–Schwarz inequality. Thus, a genuinely new argument is required. See §IX.2 below.

Exercise II.22.

(a) For the original Sturmian argument, prove the following:

Theorem. *Given continuous functions $K, H \to \mathbb{R}$, functions ϕ, ψ defined on $\mathbb{R}$ satisfying the inequalities*

$$\phi'' + K\phi \geq 0, \qquad \psi'' + H\psi \leq 0,$$

and

$$K \leq H$$

on all of $\mathbb{R}$. *Then, on any domain in* $\mathbb{R}$ *for which* $\phi, \psi > 0$, *we also have*

$$\{\phi'\psi - \phi\psi'\}' \geq (H - K)\phi\psi \geq 0,$$

with equality at any given point if and only if all three given inequalities are actually equalities at that point.

Thus, the additional assumptions

$$\phi(\alpha) \leq \psi(\alpha), \qquad \phi'(\alpha) \geq \psi'(\alpha)$$

imply that

$$\phi'/\phi \geq \psi'/\psi, \qquad (\phi/\psi)' \geq 0$$

on any interval on which $\phi > 0$. *If we also have* $\phi(\alpha) = \psi(\alpha)$, *then we may conclude*

$$\phi/\psi \geq 1$$

on any interval (α, β) *on which* $\phi > 0$.

(b) Characterize the case of equality in any of the above.

(c) Prove the:

Corollary. *Given* $\kappa \in \mathbb{R}$ *and a function* f *satisfying*

$$f'' + \kappa f \leq 0 \qquad f(0) = f'(0) = 0,$$

and assume one has $t > 0$ *for which* $\mathbf{S}_\kappa > 0$ *on all of* $(0, t)$. *Then,* $f' \leq 0$ *on all of* $[0, t]$.

Exercise II.23. Let M_1, M_2 be 2–dimensional Riemannian manifolds, with respective sectional (here, Gauss) curvature functions $\mathcal{K}_1$, $\mathcal{K}_2$. Assume that

$$\sup \mathcal{K}_1 \leq \inf \mathcal{K}_2.$$

Given points $p_1 \in M_1$, $p_2 \in M_2$, and a path $\zeta : [\alpha, \beta] \to \mathbb{R}^2$.

Let $\iota_1 : \mathbb{R}^2 \to (M_1)_{p_1}$ and $\iota_2 : \mathbb{R}^2 \to (M_2)_{p_2}$ be linear isometries (so, $\iota_2 \circ \iota_1{}^{-1}$ is an identification of the two tangent planes),

$$\zeta_1 = \iota_1 \circ \zeta \subseteq \mathcal{T}M_1, \qquad \zeta_2 = \iota_2 \circ \zeta \subseteq \mathcal{T}M_2,$$

and

$$\omega_1 = \exp_{p_1} \circ \zeta_1, \qquad \omega_2 = \exp_{p_2} \circ \zeta_2.$$

Assume that, for every ϵ in $[\alpha, \beta]$, the geodesic segment $\gamma_{\zeta_2(\epsilon)/|\zeta_2(\epsilon)|}|(0, |\zeta_2(\epsilon)|]$ has no points conjugate to p_2.

Then show that

$$\ell(\omega_1) \geq \ell(\omega_2). \tag{II.9.1}$$

Exercise II.24. Assume in the above that the common dimension of M_1, M_2 is $n \geq 2$, and M_2 has constant sectional curvature δ. Then derive (II.9.1).

Note II.11. The above two exercises are, then, the geometric versions of the Rauch theorem (Theorem II.6.4). But, after all, it is only a first step. In practice, one starts with M_1 complete and a path ω_1 in M_1, and wishes to construct a path ω_2 in M_2 with which to compare ω_1. More precisely, suppose M_2 is one of the space forms constructed in §II.3, with constant sectional curvature δ, and we are given $p_1 \in M_1$, $x_1, y_1 \in B(p_1; \pi/\sqrt{\delta})$ (when $\delta \leq 0$, we think of $\pi/\sqrt{\delta}$ as $+\infty$). We wish to estimate $d(x_1, y_1)$ from below by comparison with distances in M_2. The naive way is as follows:

Connect p_1 to x_1, y_1 by minimizing geodesics $\gamma_{\xi_1}, \gamma_{\eta_1}$, where $\xi_1, \eta_1 \in \mathsf{S}_{p_1}$; pick $p_2 \in M_2$; construct a linear isometry

$$\iota : (M_1)_{p_1} \to (M_2)_{p_2},$$

and let

$$\xi_2 = \iota(\xi_1), \qquad \eta_2 = \iota(\eta_1)$$

and

$$x_2 = \gamma_{\xi_2}(d(p_1, x_1)), \qquad y_2 = \gamma_{\eta_2}(d(p_1, y_1)).$$

One would like to show

$$d(x_1, y_1) \geq d(x_2, y_2).$$

The idea would be to let ω_1 be a minimizing geodesic joining x_1 to y_1, lift ω_1 (via the exponential map) to a path ζ_1 in $(M_1)_{p_1}$, proceed as in the previous exercises to construct ω_2, connecting x_2 to y_2, and then argue

$$d(x_1, y_1) = \ell(\omega_1) \geq \ell(\omega_2) \geq d(x_2, y_2).$$

However, the difficulty is in the construction of ζ_1. Rauch's Theorem II.6.4 and Corollary II.7.2 imply that as long as $\omega_1 \subseteq B(p_1; \pi/\sqrt{\delta})$, we can produce a local lift of ω_1 to $(M_1)_{p_1}$ via the inverse of $\exp_{p_1}$. So, to carry out the argument, the first thing we must guarantee is that $\omega_1 \subseteq B(p_1; \pi/\sqrt{\delta})$. (Of course, when $\delta \leq 0$, this is not a problem.) The second difficulty is that – even if we have a lift ζ_1 starting from γ_1 – we have no guarantee that ζ_1 will connect $d(p_1, x_1)\xi_1$ to

$d(p_1, y_1)\eta_1$! Thus, a more penetrating geometric study is required (see Chapter IX).

Riemann Normal Coordinates

Exercise II.25. Given a Riemannian manifold M, $p \in M$. Show that given any $\epsilon > 0$, there exists $\delta > 0$ such that

$$\frac{d(\exp \xi, \exp \eta)}{|\xi - \eta|} = 1 \pm O(\epsilon^2)$$

for all $\xi, \eta \in \mathsf{B}(p; \delta)$.

Exercise II.26. Assume $\dim M = 2$. Let $\gamma : [0, \beta] \to M, p = \gamma(0), \xi = \gamma'(0)$, be a unit speed geodesic with $\gamma(t_1)$ conjugate to p along γ. Show that there exists a neighborhood U of ξ in S_p, and $\epsilon > 0$ such that, for every $\zeta \in \mathsf{U}$, the geodesic γ_ζ intersects γ at some value of t satisfying $|t - t_1| < \epsilon$.

Exercise II.27. Let $\gamma : [0, \beta] \to M$, $p = \gamma(0)$, $\xi = \gamma'(0)$, be a unit speed geodesic with $\gamma(t_1)$ conjugate to p along γ. Prove that $\exp_p$ is not one-to-one on any neighborhood of $t_1\xi$.

III

Riemannian Volume

We begin here our foray into the global theory – where we consider the full Riemannian manifold M. Our very first steps, in this chapter, are devoted to describing the cut locus of a point. In short, for each unit tangent vector ξ at a point p, the cut point of p along the geodesic γ_ξ emanating from p is the point along γ_ξ after which γ_ξ no longer minimizes distance from p. The collection of such cut points of p, the cut locus $C(p)$ of p, determine the topology of M since $M \setminus C(p)$ is diffeomorphic to an n–disk.

In integration theory, the major topic of the chapter, the cut locus $C(p)$ has measure equal to 0, so the topology of M may be effectively disregarded at the early stages of study of the influence the geometry of M has on the volume measure of M. But one cannot be so cavalier. The Gauss–Bonnet theorem (see §V.1) implies that when a connected compact surface has constant Gauss curvature -1, then knowledge of the area (2–dimensional volume) of the surface is equivalent to knowledge of the topology of the surface.

Nevertheless, our study in this chapter does not devote itself to the development of this interplay between volume and topology. Rather, it starts at a more elementary level. It continues the development of the comparison theorems of Chapter II. The basic idea is that when curvature influences the rate at which geodesics emanating from the same point separate, it automatically influences the rate at which the volume grows. Thus, the study of the geodesics is finer than the study of the volume. Nevertheless, the theory of volume comparison theorems is sufficiently elementary and rich to yield, for example, an easy characterization (Theorem 11) of equality in the Bonnet–Myers theorem (Theorem II.6.1).

The study of volume and topology is initiated in the following chapter, in the context of Riemannian coverings and continued in §V.2 with the Gauss–Bonnet theory of surfaces.

The concluding sections of this chapter are for future reference. First we introduce Fermi coordinates based on a submanifold of a given Riemannian manifold and give appropriate generalizations of the comparison theorems discussed heretofore. Then, in the last section, we summarize, for future work, the integration of differential forms on manifolds. We simply recall the necessary definitions and facts through Stokes' theorem, and then do the calculations that pass from the general Stokes' theorem to the Green's formulae in Riemannian manifolds.[1]

In an appendix to this chapter, we apply Green's formulae of §7 to introduce the Laplacian and its associated eigenvalue problems. We then use R. L. Bishop's volume comparison theorems to derive S. Y. Cheng's (1975) lowest-eigenvalue comparison theorems. We wished to include this section here because it contains immediate applications of work carried out in this chapter. However, because its results are not used in the sequel, it seemed best to include it only as an appendix.

§III.1. Geodesic Spherical Coordinates

Given a manifold M, a *coordinate system on* M will be a C^∞ map $\varphi : \mathcal{O} \to M$ from an open set $\mathcal{O}$ in $\mathbb{R}^n$. Most often, the coordinate systems we use are inverse maps of charts on M; however, we prefer not to impose this constraint, so that it might be possible that the map φ might fail to be one-to-one, or of maximal rank.

Recall (from §I.3) that $\mathcal{T}M$ denotes the domain of the exponential map.

Given a Riemannian manifold M, $p \in M$, and a coordinate system $\xi : \mathcal{O} \to \mathsf{S}_p$, a coordinate system v on M is determined by

$$v(t, u) = \exp t\xi(u).$$

The domain of the map v will consist of the collection $\{(t, u)\}$ in $(0, +\infty) \times \mathcal{O}$ for which $t\xi(u) \in \mathcal{T}M$.

In what follows, given $\xi \in \mathsf{S}_p$ we let

$$\tau_{t;\xi} : M_p \to M_{\gamma_\xi(t)}$$

denote parallel translation along γ_ξ, and we write

$$\partial_\alpha \xi := \Im_\xi{}^{-1} \circ \xi_*(\partial/\partial u^\alpha),$$

$$\partial_t v := v_*(\partial/\partial t), \qquad \partial_\alpha v := v_*(\partial/\partial u^\alpha),$$

[1] On first reading, one might pass on these last sections, without disturbing the overall development of ideas and results. One would then return to these sections as needed.

$\alpha = 1, \ldots, n-1$, where $\partial/\partial t$ and $\partial/\partial u^1, \ldots, \partial/\partial u^{n-1}$ are natural coordinate vector fields on $(0, +\infty)$ and $\mathcal{O}$, respectively. Then,

$$(\partial_t v)(t; \xi) = \partial_t v_{|\exp t\xi} = \gamma_\xi{}'(t),$$

and Theorem II.7.1 implies

$$(\partial_\alpha v)(t; \xi) = \partial_\alpha v_{|\exp t\xi} = Y_\alpha(t; \xi),$$

where $Y_\alpha(t; \xi)$ is the Jacobi field along γ_ξ determined by the initial conditions

$$Y_\alpha(0; \xi) = 0, \qquad (\nabla_t Y_\alpha)(0; \xi) = \partial_\alpha \xi.$$

Of course,

$$|\partial_t v| = 1; \tag{III.1.1}$$

and since $\partial_\alpha \xi \perp \xi$, we have by (II.5.7)

$$\langle \partial_t v, \partial_\alpha v \rangle = 0. \tag{III.1.2}$$

(This is the content of the Gauss lemma (Theorem I.6.1).) So, the full knowledge of the Riemannian metric along γ_ξ requires the study of

$$\langle \partial_\alpha v, \partial_\beta v \rangle(\exp t\xi) = \langle Y_\alpha(t; \xi), Y_\beta(t; \xi) \rangle.$$

If M has constant sectional curvature κ along γ_ξ, then

$$Y_\alpha(t; \xi) = \mathbf{S}_\kappa(t) \tau_{t;\xi}(\partial_\alpha \xi);$$

so

$$\langle \partial_\alpha v, \partial_\beta v \rangle(\exp t\xi) = \mathbf{S}_\kappa{}^2(t) \langle \partial_\alpha \xi, \partial_\beta \xi \rangle,$$

which is what we encapsulated in the formal calculations of Chapter II (see (II.8.4)) as

$$ds^2 = dt^2 + \mathbf{S}_\kappa{}^2(t) |d\xi|^2.$$

For the general situation, we proceed as follows: Given $p \in M$, $\xi \in \mathsf{S}_p$, let $\xi^\perp$ denote the orthogonal complement of $\mathbb{R}\xi$ in M_p; and, for each $t > 0$, let

$$\mathsf{R}(t) = R(\gamma_\xi{}'(t), \cdot)\gamma_\xi{}'(t),$$

$$\mathcal{R}(t) = \tau_{t;\xi}{}^{-1} \circ \mathsf{R}(t) \circ \tau_{t;\xi}. \tag{III.1.3}$$

Of course, $\mathcal{R}(t)$ maps $\mathbb{R}\xi$ to 0, so one only considers $\mathcal{R}(t)$ as genuinely acting on $\xi^\perp$.

We set $\mathcal{A}(t;\xi)$ to be the solution of the matrix (more precisely: linear transformation) ordinary differential equation on $\xi^{\perp}$:

(III.1.4)
$$\mathcal{A}'' + \mathcal{R}(t)\mathcal{A} = 0,$$

satisfying the initial conditions

(III.1.5)
$$\mathcal{A}(0;\xi) = 0, \qquad \mathcal{A}'(0;\xi) = I.$$

Then, for each $\eta \in \xi^{\perp}$, the vector field $Y(t)$ along γ_ξ, given by

$$Y(t) = \tau_{t;\xi}\mathcal{A}(t;\xi)\eta,$$

is the Jacobi field along γ_ξ, in $\mathcal{J}^{\perp}$ (see Proposition II.5.1), determined by the initial conditions

$$Y(0) = 0, \qquad (\nabla_t Y)(0) = \eta.$$

We therefore have

$$(\partial_\alpha v)(t;\xi) = Y_\alpha(t;\xi) = \tau_{t;\xi}\mathcal{A}(t;\xi)\partial_\alpha\xi,$$

which implies

(III.1.6)
$$\langle \partial_\alpha v, \partial_\beta v\rangle(\exp t\xi) = \langle \mathcal{A}(t;\xi)\partial_\alpha\xi, \mathcal{A}(t;\xi)\partial_\beta\xi\rangle.$$

In the spirit of our formal expressions above (II.8.4), we write

(III.1.7)
$$ds^2 = dt^2 + |\mathcal{A}(t;\xi)\,d\xi|^2.$$

Of course, for M with constant sectional curvature equal to κ, we have

$$\mathcal{A}(t;\xi) = \mathbf{S}_\kappa(t)I.$$

§III.2. The Conjugate and Cut Loci

M is our given Riemannian manifold.

Definition. Let $p \in M$. The *conjugate locus of p in M_p* (the *tangential conjugate locus*) may be defined in two ways – equivalent to each other by Corollary II.7.1:

(i) It is the subset of $M_p \cap \mathcal{T}M$ consisting of all critical points of $\exp_p$.

(ii) It is the collection of vectors $t\xi \in M_p \cap \mathcal{T}M$, with $t > 0$, $\xi \in \mathsf{S}_p$, for which

(III.2.1)
$$\det \mathcal{A}(t;\xi) = 0,$$

where $\mathcal{A}$ is given by (III.1.4), (III.1.5). (Thus, for a given $\xi \in \mathsf{S}_p$, the nullity of $\exp_p$ at $t_0\xi$ is equal to the order of t_0 as a zero of the function $\mathcal{A}(t;\xi)$ (in t).)

By the *conjugate locus of p in M* (the *conjugate locus*), we mean the image of the tangential conjugate locus under the exponential map $\exp_p$.

Definition. Given $p \in M$, $\xi \in \mathsf{S}_p$, we define $c(\xi)$ the *distance to the cut point of p along* γ_ξ by

$$c(\xi) := \sup\{t > 0 : t\xi \in \mathcal{T}M,\ d(p, \gamma_\xi(t)) = t\}.$$

To appreciate the definition note that:

1. If $d(p, \gamma_\xi(t_1)) = t_1$ for some given $t_1 > 0$, then $d(p, \gamma_\xi(t)) = t$ for all $t \in [0, t_1]$. Indeed, if there exists $T \in [0, t_1]$ such that $d(p, \gamma_\xi(T)) < T$, then the triangle inequality implies

$$\begin{aligned} d(p, \gamma_\xi(t_1)) &\le d(p, \gamma_\xi(T)) + d(\gamma_\xi(T), \gamma_\xi(t_1)) \\ &< T + (t_1 - T) \\ &= t_1. \end{aligned}$$

So, the geodesic minimizes distance between p and $\gamma_\xi(t)$ for *all* $t \in [0, c(\xi))$, and fails to minimize distance for *all* $t > c(\xi)$.

Of course, if $c(\xi)$ is finite, and $c(\xi)\xi \in \mathcal{T}M$, then γ_ξ minimizes distance between ξ and $\gamma_\xi(c(\xi))$, as well.

2. Also, if $t < c(\xi)$, then γ_ξ is the only minimizing geodesic from p to $\gamma_\xi(t)$. If not, then there exists another $\eta \in \mathsf{S}_p$ for which $\gamma_\eta(t) = \gamma_\xi(t)$. But then, for any $T \in (t, c(\xi))$, one could travel along γ_η from p to $\gamma_\xi(t)$ followed by traveling along γ_ξ from $\gamma_\xi(t)$ to $\gamma_\xi(T)$. Then, one would have a minimizing broken geodesic from p to $\gamma_\xi(T)$, a contradiction to Corollary I.6.2.

3. One has a more detailed description of $c(\xi)$. Certainly, if p has a conjugate point $\gamma_\xi(T)$ along γ_ξ, then Jacobi's criterion (Theorem II.5.5) implies that $c(\xi) \le T$. So, one possibility for $c(\xi)$ is that it is the distance along γ_ξ to the first conjugate point of p along γ_ξ. What are the other possibilities? When M is complete, there is only one other possibility, as follows:

Given $\xi \in \mathsf{S}_p$, $c(\xi) < +\infty$, $c(\xi)\xi \in \mathcal{T}M$, consider a strictly decreasing sequence (t_j) with $t_j > c(\xi)$ for all j, and $t_j \to c(\xi)$ as $j \to \infty$. Then there exists (by Theorem I.7.1) sequences of geodesics emanating from p, with initial unit velocity vectors $\eta_j \in \mathsf{S}_p$, and respective lengths $d_j > 0$, such that

$$\begin{aligned} \gamma_\xi(t_j) &= \gamma_{\eta_j}(d_j), \\ d_j &= d(p, \gamma_{\eta_j}(d_j)) < t_j, \end{aligned}$$

that is, γ_{η_j} minimizes distance – strictly less than t_j – from p to $\gamma_\xi(t_j)$ for all j. Of course, it is impossible that infinitely many η_j denote the same element of S_p. Then, (η_j) has a convergent subsequence $(\zeta_k) = (\eta_{j_k})$ with $\zeta_k \to \zeta$ as $k \to \infty$. If $\zeta = \xi$, then $\exp_p$ is not one-to-one in any neighborhood of $c(\xi)\xi$; so $\exp_p$ has a critical point at $c(\xi)\xi$, which implies $\gamma_\xi(c(\xi))$ is conjugate to p along γ_ξ – the first possibility. If $\zeta \neq \xi$ then $d_{j_k} \to c(\xi)$, and $\gamma_\zeta(c(\xi)) = \gamma_\xi(c(\xi))$. So, the second possibility is that there are at least two distinct minimizing geodesics from p to $\gamma_\xi(c(\xi))$.

4. When M is complete then for all unit tangent vectors $\xi \in TM$, for which we have $c(\xi) < +\infty$, we also have

$$c(-\gamma_\xi{}'(c(\xi))) = c(\xi).$$

5. Of course, if M is complete and $\gamma_\xi|[0, t]$ minimizes the distance from p to $\gamma_\xi(t)$ for all $t > 0$, then $c(\xi) = +\infty$.

Notation. In what follows, we let $\mathsf{S}M$ denote the *unit tangent bundle of* M, that is

$$\mathsf{S}M := \{\xi \in TM : |\xi| = 1\},$$

with the natural projection $\pi|\mathsf{S}M$, where π denotes the projection of TM to M. When there is no possibility of confusion, we write π in place of $\pi|\mathsf{S}M$.

We shall now consider the function $c(\xi)$ as defined on $\mathsf{S}M$, and prove

Theorem III.2.1. *The function $c : \mathsf{S}M \to (0, +\infty]$, where c is the distance along γ_ξ from $\pi(\xi)$ to the cut point of $\pi(\xi)$ along γ_ξ, is upper semicontinuous on $\mathsf{S}M$. If M is Riemannian complete, then the function c is continuous on $\mathsf{S}M$.*

Proof. Suppose we are given $\xi \in \mathsf{S}M$, with a sequence (ξ_k) in $\mathsf{S}M$ for which $\xi_k \to \xi$ as $k \to \infty$. Set

$$p = \pi(\xi), \qquad p_k = \pi(\xi_k), \qquad d_k = c(\xi_k).$$

If the sequence (d_k) has an unbounded subsequence $(\delta_j) = (d_{k_j})$ for which $\delta_j \uparrow +\infty$ as $j \to \infty$, then for every $T > 0$ one has, for sufficiently large j, $\delta_j > T$. Then

$$\lim_{j\to\infty} \gamma_{\xi_{k_j}}(T) = \gamma_\xi(T),$$

and

$$d(p, \gamma_\xi(T)) = \lim_{j\to\infty} d(p_{k_j}, \gamma_{\xi_{k_j}}(T)) = T.$$

So, $c(\xi) = +\infty$.

Similarly, if (d_k) has a convergent subsequence

$$(\delta_j) = (d_{k_j}) \to \delta$$

as $j \to \infty$, then again one has for all positive $\epsilon < \delta$

$$\begin{aligned} d(p, \gamma_\xi(\delta - \epsilon)) &= \lim_{j\to\infty} d(p_{k_j}, \gamma_{\xi_{k_j}}(\delta_j - \epsilon)) \\ &= \lim_{j\to\infty} \delta_j - \epsilon \\ &= \delta - \epsilon. \end{aligned}$$

So, $c(\xi) \geq \delta$. In sum, we have

$$\limsup_{k\to\infty} c(\xi_k) \leq c(\xi).$$

It therefore remains to show that if M is complete, then

$$\text{(III.2.2)} \qquad \liminf_{k\to\infty} c(\xi_k) \geq c(\xi).$$

It suffices to assume that the sequence $(c(\xi_k))$ converges to $\delta < +\infty$ as $k \to \infty$. So, we wish to show that γ_ξ cannot minimize past $\gamma_\xi(\delta)$. By passing to a subsequence if necessary, we may assume that either (i) $\gamma_{\xi_k}(c(\xi_k))$ is conjugate to p_k along γ_{ξ_k} for all k, or (ii) to each k one has $\eta_k \in \mathsf{S}M$, $\eta_k \neq \xi_k$ for all k, for which

$$\pi(\eta_k) = \pi(\xi_k) = p_k \qquad \text{and} \qquad \gamma_{\eta_k}(c(\xi_k)) = \gamma_{\xi_k}(c(\xi_k))$$

for all k.

In case (i), $\gamma_\xi(\delta)$ is certainly conjugate to p along γ_ξ; so, $c(\xi) \leq \delta$. In case (ii), by passing to a subsequence if necessary, we may assume the existence of $\eta \in \mathsf{S}M$ for which $\eta_k \to \eta$ as $k \to \infty$. But then, $\pi(\eta) = \pi(\xi) = p$ and $\gamma_\eta(\delta) = \gamma_\xi(\delta)$. If $\eta \neq \xi$, then certainly $c(\xi) \leq \delta$. If $\eta = \xi$, then the map $\pi \times \exp$ is not a diffeomorphism on a neighborhood of $(\delta\xi, \delta\xi)$ in $TM \times TM$. This implies $\gamma_\xi(\delta)$ is conjugate to p along γ_ξ. Again, we have $c(\xi) \leq \delta$. This concludes the proof of (III.2.2), and, with it, the proof of the theorem. ■

Definition. For every $p \in M$, we define the *cut locus of p in M_p* (the *tangential cut locus*), $\mathsf{C}(p)$, by

$$\mathsf{C}(p) := \{c(\xi)\xi : c(\xi) < +\infty,\ \xi \in \mathsf{S}_p\} \cap TM,$$

and the *cut locus of p in M* (the *cut locus*), $C(p)$, by

$$C(p) := \exp \mathsf{C}(p).$$

Also, we set

$$\mathsf{D}_p := \{t\xi : 0 \le t < c(\xi), \xi \in \mathsf{S}_p\},$$

$$D_p := \exp \mathsf{D}_p.$$

One immediately has

Theorem III.2.2. *The domain* D_p *is the largest domain, starlike with respect to the origin of* M_p*, for which* $\exp_p$ *restricted to that domain is a diffeomorphism. Furthermore,*

$$D_p = M \setminus C(p).$$

Note that item **4** above implies that, for $p, q \in M$, we have $q \in C(p)$ if and only if $p \in C(q)$.

Also note that when M is complete, one always has, for all $p \in M, \delta > 0$,

$$B(p;\delta) = \exp \mathsf{B}(p;\delta), \tag{III.2.3}$$

which we already know (see Corollary I.6.1), and

$$S(p;\delta) \cap D_p = \exp \mathsf{S}(p;\delta) \cap \mathsf{D}_p. \tag{III.2.4}$$

Definition. Given any $p \in M$, we define the *injectivity radius of* p, inj p, by

$$\operatorname{inj} p := \inf\{c(\xi) : \xi \in \mathsf{S}_p\};$$

the *injectivity radius of* M, inj M, will be defined by

$$\operatorname{inj} M := \inf\{\operatorname{inj} p : p \in M\}.$$

It is already an immediate consequence of Corollary I.6.3 that inj p is positive for every $p \in M$, and that inj M is positive whenever M is compact. Of course, these facts follow from the more detailed result Theorem III.2.1. We leave it to the reader to prove

Theorem III.2.3. *The function* $\operatorname{inj} : M \to (0, +\infty]$ *is continuous.*

Theorem III.2.4. (Klingenberg's lemma (1959)) *Let* M *be a complete Riemannian manifold,* $p \in M$*, and* $q \in C(p)$ *such that*

$$d(p,q) = d(p, C(p)),$$

that is, q is the point in $C(p)$ closest to p. If q is not conjugate to p along a minimizing geodesic connecting p to q, then q is the midpoint of a geodesic loop, starting and ending at p.

In particular, if M is compact and the sectional curvatures of M satisfy

$$\mathcal{K} \leq \delta,$$

then

$$\operatorname{inj} M \geq \min\{\pi/\sqrt{\delta}, \ell(M)/2\},$$

where $\ell(M)$ is the length of the shortest simple closed geodesic in M.

Proof. Given p and q as described previously, if q is not conjugate to p along any minimizing geodesic connecting p to q, then there exist two distinct unit speed minimizing geodesic segments γ_1 and γ_2 from p to q. Neither contain any points conjugate to p. Let L denote the common length of γ_1 and γ_2,

$$\gamma_1(0) = \gamma_2(0) = p.$$

Then, one has two hypersurfaces given by

$$\{\gamma_\xi(L)\} \qquad \text{and} \qquad \{\gamma_\eta(L)\},$$

where ξ varies over a neighborhood of $\gamma_1{}'(0)$ in S_p, and η varies over a neighborhood of $\gamma_2{}'(0)$ in S_p. If $\gamma_1{}'(L) \neq -\gamma_2{}'(L)$, then the two hypersurfaces intersect transversally at q. This implies that, for varying ξ and η,

$$\{\gamma_\xi(L-\epsilon)\} \cap \{\gamma_\eta(L-\epsilon)\} \neq \emptyset,$$

(see Fig. III.1) for sufficiently small $\epsilon > 0$, which contradicts the assumption that q is the point in $C(p)$ closest to p.

The second claim follows easily from the first claim and the Morse–Schönberg theorem (Theorem II.6.3). ∎

§III.3. Riemannian Measure

We start with the formula for change of variables of integral calculus: Let D, Ω, be domains in $\mathbb{R}^n$, $n \geq 1$, let

$$\varphi : D \to \Omega$$

be a C^1 diffeomorphism, and let $J_\varphi(x)$ denote the Jacobian matrix associated to φ at x. Then, for any L^1 function f on Ω, we have

$$\text{(III.3.1)} \qquad \int_D (f \circ \varphi)|\det J_\varphi|\, dV = \int_\Omega f\, dV.$$

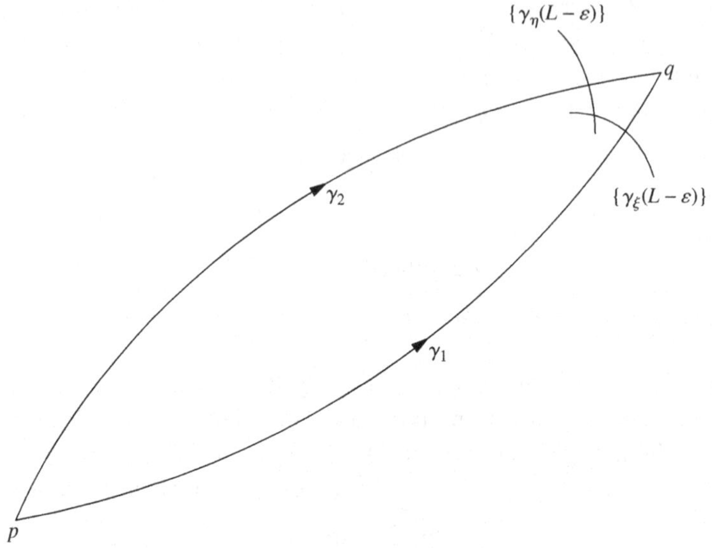

Figure III.1. For Klingenberg's lemma.

Now, let M be a Riemannian manifold, and let $x : U \to \mathbb{R}^n$ be a chart on M. Then, for each $p \in U$, we let (as in §I.5) $G^x(p)$ denote the matrix given by

$$G^x(p) = (g^x_{ij}(p)), \qquad g^x_{ij}(p) = \left\langle \frac{\partial}{\partial x^i}|_p, \frac{\partial}{\partial x^j}|_p \right\rangle,$$

and we set

$$g^x := \det G^x > 0.$$

What if we are given a different chart $y : U \to \mathbb{R}^n$ on the same U in M? Then, we relate the formulae as follows: Set J to be the Jacobian matrix

$$J_{rj} = \frac{\partial (y \circ x^{-1})^r}{\partial x^j};$$

then we have

$$\frac{\partial}{\partial x^j} = \sum_r \frac{\partial}{\partial y^r} J_{rj},$$

which implies

$$G^x = J^T G^y J,$$

where J^T denotes the transpose of J, which implies

$$\sqrt{g^x} = \sqrt{g^y} |\det J|.$$

Thus, we have the local densities

$$\text{(III.3.2)} \qquad \sqrt{g^x}\, dx^1 \cdots dx^n = \sqrt{g^y}\, dy^1 \cdots dy^n,$$

by which we mean that the integral

$$I(f;U) = \int_{x(U)} (f\sqrt{g^x})\circ x^{-1}\, dx^1 \cdots dx^n$$

depends only on f and U – not on the particular choice of chart x.

We now turn the local Riemannian measure to a *global Riemannian measure* on all of M: First, pick an atlas

$$\{x_\alpha : U_\alpha \to \mathbb{R}^n : \alpha \in A\},$$

and subordinate partition of unity

$$\{\phi_\alpha : \alpha \in A\}.$$

Then define the global Riemannian measure dV by

$$dV := \sum_{\alpha\in A} \phi_\alpha \sqrt{g^{x_\alpha}}\, dx_\alpha^1 \cdots dx_\alpha^n,$$

or, equivalently,

$$\int_M f\, dV = \sum_{\alpha\in A} I(\phi_\alpha f; U_\alpha).$$

One easily checks that the measure is well-defined, that is, it is independent of both the particular choices of atlas and subordinate partition of unity.

One easily checks that a function f is measurable with respect to dV if and only if $f\circ x^{-1}$ is measurable on $x(U)$ for any chart $x : U \to \mathbb{R}^n$.

In all that follows, we work with this measure.

Definition. For any measurable B in M, we let $V(B)$ denote the measure of B and refer to $V(B)$ as the *volume of* B. If Γ is an $(n-1)$–dimensional submanifold of M, then we usually denote its Riemannian measure by dA; and for any measurable Λ in Γ, we denote its measure by $A(\Lambda)$, and refer to $A(\Lambda)$ as the *area of* Λ.

The Effective Calculation of Integrals

If the manifold M is diffeomorphic to $\mathbb{R}^n$, then one has, possibly, a convenient way to literally calculate an integral, by referring the calculation to one coordinate system. However, as soon as one cannot cover the manifold with one "naturally" chosen chart, one would then be forced to literally pick an atlas and subordinate partition of unity! This would not go well, at all.

The simplest overarching approach is to use the geometry of the Riemannian manifold to indicate a judicious choice of a set of measure 0 to delete, which will thereby leave an open set that is the domain of a chart on M. The quickest example that comes to mind is the stereographic projection of the sphere $\mathbb{S}^n$ to $\mathbb{R}^n$ (see §II.3), in which the domain of the chart covers all of $\mathbb{S}^n$ less the pole of the projection. So any integral on the sphere may be referred to this one chart.

Before proceeding, we note that (III.3.2) implies that the notion of a set of measure 0 depends only on the differentiable structure of the manifold. It makes no difference whether we are referring to a local measure on M induced by Lebesgue measure on the image of a chart on M, or whether we are referring to Riemannian measure.

Continuing, we work, in our setting, with spherical coordinates as follows: For convenience, we assume that M is complete. For any point $p \in M$, introduce geodesic spherical coordinates about p, as described in §III.1. It is important to remember that, there, the spherical coordinates actually describe (locally) a differentiable map of $(0, +\infty) \times \mathsf{S}_p$ into $M \setminus \{p\}$, given by

$$(t, \xi) \mapsto \exp t\xi.$$

The map may fail to be the inverse map of a chart on $M \setminus \{p\}$ since the map may fail to be a diffeomorphism; also, since S_p is not diffeomorphic to a subset of $\mathbb{R}^{n-1}$, one cannot use ξ, literally, as an $(n-1)$–dimensional coordinate. The second difficulty is simply addressed by picking a chart on S_p. It need never be explicit, since the final formulae never require it (unless, possibly, in some unusual circumstances – our goal is never to have to contend with them). The first difficulty must be dealt with by restricting the geodesic spherical coordinates to $\mathsf{D}_p \setminus \{p\}$.

Thus, a chart on $M \setminus C(p) = D_p$ is given by

$$(\exp_p |\mathsf{D}_p \setminus \{p\})^{-1} : D_p \setminus \{p\} \to \mathsf{D}_p \setminus \{p\};$$

and the Riemannian measure is given on D_p by

$$dV(\exp t\xi) = \sqrt{\mathbf{g}}(t; \xi)\, dt\, d\mu_p(\xi),$$

for some function $\sqrt{\mathbf{g}}$ on D_p, where $d\mu_p(\xi)$ denotes the Riemannian measure on S_p induced by the Euclidean Lebesgue measure on M_p.

Of course, the set $\{p\}$ has measure 0; so, we never have to explicitly include it in, or exclude it from, our discussion of integrals. More significantly, $C(p)$ has measure 0. Indeed, the function $c(\xi)$ is continuous on all of $\mathsf{S}M$, so its restriction to S_p is certainly continuous. Thus, the tangential cut locus of p is the image of the continuous map $\xi \mapsto c(\xi)\xi$ from S_p to M_p, and therefore has

Lebesgue measure equal to 0. The image of the tangential cut locus of p under the differentiable exponential map is the cut locus of p in M, $C(p)$. Therefore,

Proposition III.3.1. *For any $p \in M$, the cut locus $C(p)$ of p is a set of measure* 0.

Thus, for any $p \in M$, and integrable function f on M, we have

$$\text{(III.3.3)} \qquad \int_M f\, dV = \int_{\mathsf{D}_p} f(\exp t\xi)\sqrt{\mathbf{g}}(t;\xi)\, dt\, d\mu_p(\xi)$$

$$\text{(III.3.4)} \qquad = \int_{\mathsf{S}_p} d\mu_p(\xi) \int_0^{c(\xi)} f(\exp t\xi)\sqrt{\mathbf{g}}(t;\xi)\, dt$$

$$\text{(III.3.5)} \qquad = \int_0^{+\infty} dt \int_{t^{-1}\mathsf{S}(p;t)\cap \mathsf{D}_p} f(\exp t\xi)\sqrt{\mathbf{g}}(t;\xi)\, d\mu_p(\xi),$$

where $t^{-1}\mathsf{S}(p;t) \cap \mathsf{D}_p$ is the subset of S_p obtained by dividing each of the elements of $\mathsf{S}(p;t) \cap \mathsf{D}_p$ by t.

It remains to calculate $\sqrt{\mathbf{g}}(t;\xi)$.

Theorem III.3.1. *We have*

$$\sqrt{\mathbf{g}}(t;\xi) = \det \mathcal{A}(t;\xi),$$

where $\mathcal{A}(t;\xi)$ is given by (III.1.4) *and* (III.1.5).

Proof. Let u be a chart on S_p, $\xi = u^{-1}$, and let x be a chart on $D_p \setminus \{p\}$ given by

$$x = (u \circ \{(\exp|D_p)^{-1}/|(\exp|D_p)^{-1}|\}, |(\exp|D_p)^{-1}|).$$

Then, what in §III.1 was called $\partial_t v$ is here equal to $\partial/\partial x^n$; and what was in §1 referred to as $\partial_\alpha v$ is here equal to $\partial/\partial x^\alpha$, $\alpha = 1, \dots, n-1$.

We let G denote the matrix of the Riemannian metric on M associated to the chart x, and we let H denote the matrix of the Riemannian metric on S_p associated to the chart u. Then, equation (III.1.6) translates to our language here as

$$g_{\alpha\beta} = \sum_{\gamma,\delta} \mathcal{A}^*{}_{\alpha\gamma} h_{\gamma\delta} \mathcal{A}_{\delta\beta}, \qquad \alpha, \beta, \gamma, \delta = 1, \dots, n-1,$$

and (III.1.1) and (III.1.2) translate to

$$g_{nn} = 1, \qquad g_{\alpha n} = g_{n\alpha} = 0, \qquad \alpha = 1, \dots, n-1.$$

We conclude that

$$\sqrt{g} = \sqrt{h}\det \mathcal{A},$$

which implies the claim. ■

For Euclidean space $\mathbb{R}^n$, with respect to spherical coordinates, we have the previously known

$$dV(t\xi) = t^{n-1}\,dt\,d\mu_{n-1}(\xi), \tag{III.3.6}$$

where $d\mu_{n-1}$ denotes the Riemannian measure of $\mathbb{S}^{n-1}$.

For a specific calculation of the volume of $\mathbb{S}^{n-1}$, that is

$$\mathbf{c_{n-1}} := \int_{\mathbb{S}^{n-1}} d\mu_{n-1},$$

we introduce the classical Gamma function $\Gamma(x)$, given by

$$\Gamma(x) := \int_0^{+\infty} e^{-t}t^{x-1}\,dt.$$

Then the integral converges for $x > 0$, and one verifies that

$$\Gamma(1) = 1, \tag{III.3.7}$$

and, by integration by parts, that

$$\Gamma(x+1) = x\Gamma(x). \tag{III.3.8}$$

Also, one has

$$\begin{aligned}\left\{\int_{\mathbb{R}} e^{-t^2}\,dt\right\}^n &= \int_{\mathbb{R}^n} e^{-|x|^2}\,dV(x)\\ &= \int_{\mathbb{S}^{n-1}} d\mu_{n-1}(\xi)\int_0^{+\infty} t^{n-1}e^{-t^2}\,dt\\ &= \mathbf{c_{n-1}}\int_0^{+\infty} t^{n-1}e^{-t^2}\,dt\\ &= \mathbf{c_{n-1}}\Gamma(n/2)/2,\end{aligned}$$

that is,

$$\left\{\int_{\mathbb{R}} e^{-t^2}\,dt\right\}^n = \frac{\mathbf{c_{n-1}}}{2}\Gamma(n/2). \tag{III.3.9}$$

Therefore, to evaluate $\mathbf{c_{n-1}}$, we must calculate the classical integral $\int e^{-t^2}\,dt$ over $\mathbb{R}$. In our context, we argue as follows: One easily has, by deleting one point, that $\mathbf{c_1} = 2\pi$. Therefore, by (III.3.9),

$$\left\{\int_{\mathbb{R}} e^{-t^2}\,dt\right\}^2 = 2\pi\Gamma(1)/2 = \pi,$$

which implies

$$\int_{\mathbb{R}} e^{-t^2}\, dt = \sqrt{\pi}.$$

We therefore conclude

$$\mathbf{c_{n-1}} = \frac{2\pi^{n/2}}{\Gamma(n/2)}. \tag{III.3.10}$$

To evaluate $\Gamma(n/2)$, we have, for even n, $n/2$ an integer; one then uses (III.3.7) and (III.3.8). To evaluate $\Gamma(n/2)$, we have, for odd n, $n/2$ a half-integer; one then uses (III.3.8), and the fact that (III.3.9) for $n = 1$ reads as

$$\Gamma(1/2) = \sqrt{\pi}.$$

If M has constant curvature κ, then in spherical geodesic coordinates about any point $p \in M$ we have

$$\sqrt{\mathbf{g}}(t;\xi) = \mathbf{S}_\kappa{}^{n-1}(t). \tag{III.3.11}$$

In particular, if $\kappa = 1$, we conclude immediately that

$$\mathbf{c_n} = \mathbf{c_{n-1}} \int_0^\pi \sin^{n-1} t\, dt. \tag{III.3.12}$$

If one assumes the integral as known, then one uses (III.3.12) to calculate the volume of $\mathbb{S}^n$ in a new way. If one assumes the knowledge of the volume of spheres, then one has a new calculation of the integral

$$\int_0^\pi \sin^{n-1} t\, dt.$$

Finally, we note that (III.3.6) and (III.3.10) imply that for $\boldsymbol{\omega_n}$, *the volume of the unit disk* $\mathbb{B}^n$ *in* $\mathbb{R}^n$, we have

$$\boldsymbol{\omega_n} = \frac{\mathbf{c_{n-1}}}{n} = \frac{\pi^{n/2}}{\Gamma(n/2+1)}. \tag{III.3.13}$$

Volume of Metric Disks

We now apply this approach to calculating volumes in the general Riemannian setting.

Notation. Given $x \in M$, we let $V(x;r)$ denote the volume of $B(x;r)$, that is,

$$V(x;r) = \int_{B(x;r)} dV.$$

Definition. For any $x \in M$, $r > 0$, we define $\mathsf{D}_x(r)$ to be the subset of S_x consisting of those elements ξ for which $r\xi \in \mathsf{D}_x$, which we write as

$$r\mathsf{D}_x(r) = \mathsf{S}(x;r) \cap \mathsf{D}_x.$$

Given $x \in M$, we let $\mathfrak{A}(x;r)$ denote the *lower area of* $S(x;r)$; that is,

$$\mathfrak{A}(x;r) := \int_{\mathsf{D}_x(r)} \det \mathcal{A}(r;\xi)\, d\mu_x(\xi).$$

Remark III.3.1. We refer to $\mathfrak{A}$ as the "lower area" because (i) $\mathfrak{A}(x;r)$ is guaranteed to be the genuine $(n-1)$–dimensional area of $S(x;r)$ only if r is less than $\operatorname{inj} x$, and (ii) for any reasonable definition of area of $S(x;r)$, when $r > \operatorname{inj} x$, one will have $\mathfrak{A}(x;r)$ less than or equal to the area of $S(x;r)$. An example of the difficulties when r is allowed to be greater than or equal to $\operatorname{inj} x$ would be $\mathbb{P}^n$, the real n–dimensional projective space of constant curvature equal to 1 (see §IV.2). Then, the distance sphere of x, $S(x;\pi/2)$, with radius equal to $\pi/2$ (the hyperplane at infinity) is, in fact, an $(n-1)$–dimensional manifold with area equal to $\mathbf{c_{n-1}}/2$. But the definition determines $\mathfrak{A}(x;\pi/2)$ to be equal to 0, since it excludes the intersection of $\mathsf{S}(x;\pi/2)$ with the tangential cut locus. On the other hand, if we were to include the intersection of $\mathsf{S}(x;\pi/2)$ with the tangential cut locus, then the integral would be taken over *all* of $\mathsf{S}(x;\pi/2)$, in which case each point in $S(x;\pi/2)$ in the manifold would be counted twice, with the result that $\mathfrak{A}(x;\pi/2)$ would then be equal to $\mathbf{c_{n-1}}$.

We discuss the definition of the area of $S(x;r)$, for arbitrary r, in §III.5.

For the above definition of $\mathfrak{A}(x;r)$, we have, using (III.3.4), (III.3.5),

$$\begin{aligned} V(x;r) &= \iint_{\mathsf{D}_x \cap \mathsf{B}(x;r)} \det \mathcal{A}(t;\xi)\, dt\, d\mu_x(\xi) \\ &= \int_0^r dt \int_{\mathsf{D}_x(t)} \det \mathcal{A}(t;\xi)\, d\mu_x(\xi) \\ &= \int_0^r \mathfrak{A}(x;t)\, dt. \end{aligned}$$

We immediately have the first claims of:

Proposition III.3.2. *The lower area function* $\mathfrak{A}(x;r)$ *is integrable with respect to* r, $V(x;r)$ *is continuous for all* $r > 0$, *and differentiable for almost all* r *– in which case its derivative is given by* $\mathfrak{A}(x;r)$.

Even if $V(x;r)$ is not differentiable for all r, one has the inequality

$$\limsup_{\epsilon\downarrow 0} \frac{V(x;r+\epsilon)-V(x;r)}{\epsilon} \leq \mathfrak{A}(x;r)$$

for all $r > 0$.

Proof. We first note that $r < R$ implies

$$\mathsf{D}_x(R) \subseteq \mathsf{D}_x(r).$$

Then

$$\begin{aligned}
\frac{V(x;r+\epsilon)-V(x;r)}{\epsilon} &= \frac{1}{\epsilon}\int_r^{r+\epsilon} \mathfrak{A}(x;s)\,ds \\
&= \frac{1}{\epsilon}\int_r^{r+\epsilon} ds \int_{\mathsf{D}_x(s)} \sqrt{\mathbf{g}}(s;\xi)\,d\mu_x(\xi) \\
&\leq \frac{1}{\epsilon}\int_r^{r+\epsilon} ds \int_{\mathsf{D}_x(r)} \sqrt{\mathbf{g}}(s;\xi)\,d\mu_x(\xi) \\
&= \int_{\mathsf{D}_x(r)} d\mu_x(\xi)\frac{1}{\epsilon}\int_r^{r+\epsilon} \sqrt{\mathbf{g}}(s;\xi)\,ds.
\end{aligned}$$

Let $\epsilon \downarrow 0$. Then, Lebesgue's dominated convergence theorem implies the claim. ∎

Also note that, even when $V(x;r)$ might not be differentiable with respect to r at some r_0, we still have that $V(x;r)$ is locally uniformly Lipschitz with respect to r. Indeed, given $R > 0$, let κ_R denote the infimum of the Ricci curvature on $B(x;R)$. Then, for $s, r \in (0, R)$, $s < r$, we have, by (III.4.13),

$$\begin{aligned}
\frac{V(x;r)-V(x;s)}{r-s} &= \frac{1}{r-s}\int_{\mathsf{D}_x(s)} d\mu_x(\xi) \int_s^{\min\{r,c(\xi)\}} \det \mathcal{A}(t;\xi)\,dt, \\
&\leq \mathbf{c_{n-1}} \max_{[0,R]} \mathbf{S}_\kappa{}^{n-1},
\end{aligned}$$

which implies the claim.

§III.4. Volume Comparison Theorems

We start with some preliminaries.

Definition. We let $\mathbb{M}_\delta$ denote the *space form of constant sectional curvature* δ, namely, (i) the n–sphere of constant sectional curvature δ, when $\delta > 0$, (ii) $\mathbb{R}^n$, when $\delta = 0$, and (iii) the hyperbolic space of constant sectional curvature δ, when $\delta < 0$.

Notation. Here, and in *all* that follows, we let $\pi/\sqrt{\delta} := +\infty$ when $\delta \leq 0$.

Notation. We denote the volume of the disk of radius r in $\mathbb{M}_\delta$ by

$$V_\delta(r) = \mathbf{c_{n-1}} \int_0^r \mathbf{S}_\delta{}^{n-1}(t)\,dt. \tag{III.4.1}$$

We denote the area of the disk of radius r in $\mathbb{M}_\delta$ by

$$A_\delta(r) = \mathbf{c_{n-1}} \mathbf{S}_\delta{}^{n-1}(r). \tag{III.4.2}$$

Next we restate Rauch's comparison theorem (Theorem II.6.4) in the form we require for the current discussion. Let M be a Riemannian manifold; for convenience, we assume that M is complete.

Let $p \in M$, $\xi \in \mathsf{S}_p$, such that all the sectional curvatures along γ_ξ are less than or equal to some constant δ. Then, for any Jacobi field Y along γ_ξ, pointwise orthogonal to γ_ξ, and vanishing at $p = \gamma_\xi(0)$, we have

$$\frac{|Y|'}{|Y|} \geq \frac{\mathbf{S}_\delta{}'}{\mathbf{S}_\delta}, \tag{III.4.3}$$

$$|Y| \geq |\nabla_t Y|(0)\mathbf{S}_\delta, \tag{III.4.4}$$

for all $t < \pi/\sqrt{\delta}$.

We have equality in (III.4.3) at $t = t_0 \in (0, \pi/\sqrt{\delta}]$ if and only if there exists a parallel vector field E along γ_ξ such that

$$Y(t) = \mathbf{S}_\delta(t) E(t), \qquad \mathcal{R}(t)\,E(t) = \delta E(t) \tag{III.4.5}$$

for all $t \in (0, t_0]$ ($\mathcal{R}(t)$ is defined by (III.1.3)).

In particular, we have

$$(\mathcal{A}^*\mathcal{A})(t;\xi) \geq \mathbf{S}_\delta^2(t) I, \tag{III.4.6}$$

where $\mathcal{A}^*$ denotes the adjoint of the linear transformation $\mathcal{A}$ for all $t \in (0, \pi/\sqrt{\delta}]$, with equality in (III.4.6) at a $t_0 \in (0, \pi/\sqrt{\delta}]$ if and only if

$$\mathcal{A}(t;\xi) = \mathbf{S}_\delta(t) I, \qquad \mathcal{R}(t) = \delta I \tag{III.4.7}$$

for all $t \in (0, t_0]$.

We now consider the corresponding comparison theorem for $\det \mathcal{A}(t;\xi)$. In that which follows, we shall write $\mathcal{A}(t)$ for $\mathcal{A}(t;\xi)$, since the geodesic γ_ξ is fixed in the discussion.

Theorem III.4.1. (P. Günther (1960), R. L. Bishop (1964)) *Assume we have the geodesic γ_ξ as described above, with all sectional curvatures along γ_ξ less* *than or equal to* δ. *Then*

$$\frac{(\det \mathcal{A})'}{\det \mathcal{A}} \geq (n-1)\frac{\mathbf{S}_\delta{}'}{\mathbf{S}_\delta}, \tag{III.4.8}$$

on $(0, \pi/\sqrt{\delta})$, *and*

$$\det \mathcal{A} \geq \mathbf{S}_\delta{}^{n-1} \tag{III.4.9}$$

on $(0, \pi/\sqrt{\delta}]$.

We have equality in (III.4.8) *at a* $t_0 \in (0, \pi/\sqrt{\delta}]$ *if and only if* (III.4.7) *is valid on all of* $[0, t_0]$.

Proof. Instead of working with $\mathcal{A}$, we work with

$$\mathcal{B} := \mathcal{A}^*\mathcal{A},$$

which is self-adjoint. Of course,

$$\frac{(\det \mathcal{A})'}{\det \mathcal{A}} = \frac{1}{2}\frac{(\det \mathcal{B})'}{\det \mathcal{B}}.$$

Given $\tau \in (0, \pi/\sqrt{\delta})$, let $\{e_1, \ldots, e_{n-1}\}$ be an orthonormal basis of $\xi^\perp$ consisting of eigenvectors of $\mathcal{B}(\tau)$, and consider the solutions $\{\eta_1(t), \ldots, \eta_{n-1}(t)\}$ to the vector Jacobi equation in $\xi^\perp$:

$$\eta'' + \mathcal{R}(t)\eta = 0,$$

given by

$$\eta_\alpha(t) = \mathcal{A}(t)e_\alpha, \qquad \alpha = 1, \ldots, n-1.$$

Then, by Proposition II.8.2 and (III.4.3),

$$\frac{1}{2}\frac{(\det \mathcal{B})'}{\det \mathcal{B}}(\tau) = \frac{1}{2}\operatorname{tr} \mathcal{B}'\mathcal{B}^{-1}(\tau) = \sum_{\alpha=1}^{n-1} \frac{\langle \eta_\alpha{}', \eta_\alpha\rangle}{\langle \eta_\alpha, \eta_\alpha\rangle}(\tau) \geq (n-1)\frac{\mathbf{S}_\delta{}'}{\mathbf{S}_\delta}(\tau),$$

which implies (III.4.8), and, from it, (III.4.9).

The case of equality in (III.4.8) is easy, and is left to the reader. ∎

Theorem III.4.2. (P. Günther (1960), R. L. Bishop (1964)) *Assume that the sectional curvatures of* M *are all less than or equal to* δ. *Then, for every* $x \in M$, *we have*

$$V(x;r) \geq V_\delta(r) \tag{III.4.10}$$

for all $r \leq \min\{\operatorname{inj} x, \pi/\sqrt{\delta}\}$, *with equality for some fixed* r *if and only if* $B(x;r)$ *is isometric to the disk of radius* r *in the constant curvature space form* $\mathbb{M}_\delta$.

Proof. The result is an immediate consequence of (III.3.5), Theorems III.3.1 and III.4.1, and the derivation of Theorem II.8.2. ■

We now turn to volume comparison theorems when the curvature is bounded from below. Here, the good news is that one uses the Ricci curvature instead of the sectional curvature (in the spirit of the hypothesis of the Bonnet–Myers theorem (Theorem II.6.1)), and the lower bound on the Ricci curvature yields an upper bound on volume growth – valid beyond the injectivity radius! The first step, however, is to give the upper estimate on the logarithmic derivative of det $\mathcal{A}$, valid up to the first conjugate point along each geodesic.

Definition. Given a geodesic $\gamma_\xi : [0, \beta) \to M$ in the Riemannian manifold M, $p = \gamma_\xi(0)$, $\xi = \gamma_\xi{}'(0)$, the *first conjugate point of p along γ_ξ, $\gamma(t_o)$*, is the point for which t_o is the infimum of all t' for which $\gamma_\xi(t')$ is conjugate to p along γ_ξ. We denote t_o by conj ξ.

The function $(t, \xi) \mapsto \mathcal{A}(t; \xi)$ is continuous on $[0, \infty) \times \mathsf{S}M$, and $\mathcal{A}(0; \xi) = I$ for all ξ. The zeroes of $t \mapsto \mathcal{A}(t; \xi)$, for each fixed ξ, characterize the conjugate points of $p = \pi(\xi)$ along γ_ξ. So, for each ξ, conj ξ is indeed a minumum (not just an infimum) and conj $\xi > 0$.

Theorem III.4.3. (R. L. Bishop (1964)) *Assume we are given a real constant κ and the fixed geodesic γ_ξ, with the Ricci curvature along γ_ξ greater than or equal to $(n-1)\kappa$, that is,*

(III.4.11) $$\mathrm{Ric}\,(\gamma_\xi{}'(t), \gamma_\xi{}'(t)) = \mathrm{tr}\,\mathcal{R}(t) \geq (n-1)\kappa$$

for all $t \in (0, \mathrm{conj}\,\xi]$. Then

(III.4.12) $$\frac{(\det \mathcal{A})'}{\det \mathcal{A}} \leq (n-1)\frac{\mathbf{S}_\kappa{}'}{\mathbf{S}_\kappa},$$

on $(0, \mathrm{conj}\,\xi)$, and

(III.4.13) $$\det \mathcal{A} \leq \mathbf{S}_\kappa{}^{n-1}$$

on $(0, \mathrm{conj}\,\xi]$.

We have equality in (III.4.12) *at $t = t_0 \in (0, \mathrm{conj}\,\xi)$ if and only if*

(III.4.14) $$\mathcal{A}(t) = \mathbf{S}_\kappa(t) I, \qquad \mathcal{R}(t) = \kappa I$$

for all $t \in (0, t_0]$.

Remark III.4.1. Note that (III.4.13) implies the Bonnet–Myers theorem, in that it implies that det $\mathcal{A}(t)$ must have a zero not later than the first zero of $\mathbf{S}_\kappa(t)$, that is, when $\kappa > 0$, not later than $\pi/\sqrt{\kappa}$.

Proof of Theorem III.4.3. Again, note that by Proposition II.8.2 we have

$$\frac{(\det \mathcal{A})'}{\det \mathcal{A}} = \operatorname{tr} \mathcal{A}'\mathcal{A}^{-1}.$$

Next, set

$$\mathbf{Ct}_\kappa(t) := \mathbf{S}_\kappa{}'(t)/\mathbf{S}_\kappa(t), \tag{III.4.15}$$

$\mathbf{arcCt}_\kappa$ the inverse function of $\mathbf{Ct}_\kappa$ and consider

$$\psi := (n-1)\mathbf{Ct}_\kappa.$$

Then, ψ satisfies the scalar *Riccati equation*

$$\psi' + \frac{\psi^2}{n-1} + (n-1)\kappa = 0.$$

Also, $\psi(t)$ is strictly decreasing with respect to t, for all t; and, when $\kappa \le 0$, has limiting value, as $t \uparrow +\infty$, equal to $(n-1)\sqrt{-\kappa}$.

Given linear transformations $\mathsf{A}(t), \mathsf{B}(t) : V \to V$, depending differentiably on t (where V denotes some finite-dimensional vector space), we associate their Wronskian $\mathsf{W}(t)$ defined by

$$\mathsf{W}(\mathsf{A}, \mathsf{B}) := \mathsf{A}'^*\mathsf{B} - \mathsf{A}^*\mathsf{B}'.$$

One verifies that, for our $\mathcal{A}(t)$, we have $\mathsf{W}(\mathcal{A}, \mathcal{A}) = 0$. Set

$$\mathcal{U} := \mathcal{A}'\mathcal{A}^{-1}.$$

Then

$$\mathcal{U}^* - \mathcal{U} = (\mathcal{A}^{-1})^*\mathsf{W}(\mathcal{A}, \mathcal{A})\mathcal{A}^{-1} = 0$$

so $\mathcal{U}$ is self-adjoint. Also, $\mathcal{U}$ satisfies the *matrix* (more precisely: linear transformation) *Riccati equation*

$$\mathcal{U}' + \mathcal{U}^2 + \mathcal{R} = 0, \tag{III.4.16}$$

which implies

$$(\operatorname{tr}\mathcal{U})' + \operatorname{tr}\mathcal{U}^2 + \operatorname{tr}\mathcal{R} = 0.$$

Now the Cauchy–Schwarz inequality implies

$$\operatorname{tr}\mathcal{U}^2 \ge \frac{(\operatorname{tr}\mathcal{U})^2}{n-1}, \tag{III.4.17}$$

which implies, for

$$\phi := \operatorname{tr}\mathcal{U} = \operatorname{tr}\mathcal{A}'\mathcal{A}^{-1} = \frac{(\det \mathcal{A})'}{\det \mathcal{A}},$$

the differential inequality

$$\phi' + \frac{\phi^2}{n-1} + (n-1)\kappa \leq 0. \tag{III.4.18}$$

We, therefore, wish to compare ϕ with ψ.

As mentioned, we have

$$\Psi := \frac{\psi^2}{n-1} + (n-1)\kappa > 0$$

on all of $(0, \pi/\sqrt{\kappa})$. Next, note that

$$\phi \sim \frac{n-1}{t}$$

as $t \downarrow 0$. So, there exists $\epsilon_0 > 0$ such that

$$\Phi := \frac{\phi^2}{n-1} + (n-1)\kappa > 0$$

on $(0, \epsilon_0)$.

Assume that $\Phi > 0$ on all of $(0, t)$, $t \in (0, \operatorname{conj} \xi)$. Then, the inequality (III.4.18) implies

$$\frac{-\phi'}{\frac{\phi^2}{n-1} + (n-1)\kappa} \geq 1, \tag{III.4.19}$$

which implies

$$\int_0^s \frac{-\phi'}{\frac{\phi^2}{n-1} + (n-1)\kappa}(\tau)\, d\tau \geq s \quad \forall\ s \in (0, t]. \tag{III.4.20}$$

That is,

$$\mathbf{arcCt}_\kappa \frac{\phi(s)}{(n-1)} \geq s \quad \forall\ s \in (0, t],$$

which implies

$$\phi \leq \psi \qquad \text{on } [(0, t],$$

which is (III.4.12). Of course, (III.4.13) follows easily.

If we have equality in (III.4.12) at some $t_0 \in (0, t]$, then the equality in (III.4.20) at $s = t_0$ implies we have equality in (III.4.17) and (III.4.18) on all of $(0, t_0]$. This, in turn, implies

$$\phi = \psi, \qquad \operatorname{tr} \mathcal{R} = (n-1)\kappa,$$

and $\mathcal{U}$ is a scalar multiple of the identity for each $s \in (0, t_0]$. Since $\mathcal{U}$ is a scalar multiple of the identity at each t, the Riccati equation (III.4.16) implies that is

$\mathcal{R}$ is a scalar multiple of the identity for each s. This implies $\mathcal{R}(s) = \kappa I$ for all $s \in (0, t_0]$. Finally, since $\mathcal{U}$ is a scalar multiple of the identity and its trace is identically equal to $(n-1)\mathbf{S}_\kappa{}'(s)/\mathbf{S}_\kappa(s)$, we have

$$\mathcal{A}'\mathcal{A}^{-1}(s) = \frac{\mathbf{S}_\kappa{}'(s)}{\mathbf{S}_\kappa(s)} I$$

for all $s \in (0, t_0]$. But this then implies $\mathcal{A}(s) = \mathbf{S}_\kappa(s)I$ for all $s \in (0, t_0]$, which is (III.4.14).

Now, let t be arbitrary in $(0, \operatorname{conj} \xi]$, and assume we do not have $\phi \leq \psi$ on all of $(0, t)$. Then there exists a maximal $t_1 \in (0, t)$ such that $\phi \leq \psi$ on $(0, t_1)$. In particular, $\phi = \psi$ at t_1. Then, $\Phi(t_1) > 0$ and there exists $\epsilon_1 > 0$ such that $\Phi|[t_1, t_1 + \epsilon_1) > 0$, which implies (III.4.19) is valid from t_1 to any $s \in (t_1, t_1 + \epsilon_1)$, which implies $\phi \leq \psi$ on $(t_1, t_1 + \epsilon_1)$ – a contradiction to the maximality of t_1. So, we have (III.4.12) on all of $(0, t]$.

To consider the case of equality, it suffices to consider the case where there exists $t_2 \in [0, t]$ such that $\phi < \psi$ on $(0, t_2)$ and $\phi(t_2) = \psi(t_2)$. But then, $\Phi(t_2) = \Psi(t_2) > 0$, which implies there exists $\epsilon > 0$ such that (III.4.19) is valid on $(t_2 - \epsilon, t_2]$. For any $t \in (t_2 - \epsilon, t_2]$, integrate (III.4.19) from t to t_2. One obtains $\phi(t) \geq \psi(t)$ – a contradiction. ∎

Theorem III.4.4. (R. L. Bishop (1964)) *Assume that the Ricci curvatures of M are all greater than or equal to $(n-1)\kappa$. Then for every $x \in M$ and every $r > 0$ we have*

$$V(x; r) \leq V_\kappa(r), \tag{III.4.21}$$

with equality for some fixed r if and only if $B(x; r)$ is isometric to the disk of radius r in the constant curvature space form $\mathbb{M}_\kappa$.

Proof. For any $r > 0$, we have

$$\begin{aligned} V(x; r) &= \int_{\mathbf{S}_x} d\mu_x(\xi) \int_0^{\min\{c(\xi), r\}} \det \mathcal{A}(t; \xi)\, dt \\ &\leq \int_{\mathbf{S}_x} d\mu_x(\xi) \int_0^{\min\{c(\xi), r\}} \mathbf{S}_\kappa{}^{n-1}(t)\, dt \\ &\leq \int_{\mathbf{S}_x} d\mu_x(\xi) \int_0^{r} \mathbf{S}_\kappa{}^{n-1}(t)\, dt \\ &= V_\kappa(r). \end{aligned}$$

The case of equality is easy. ∎

Proposition III.4.1. *Assume that the Ricci curvatures of M are all greater than or equal to $(n-1)\kappa$. Then for every $x \in M$, we have*

$$\frac{\mathfrak{A}(x;r)}{A_\kappa(r)} \tag{III.4.22}$$

is decreasing with respect to r.

Proof. We recall that $r < R$ implies

$$\mathsf{D}_x(R) \subseteq \mathsf{D}_x(r).$$

Now (III.4.12) is equivalent to saying that

$$\frac{\det \mathcal{A}(t;\xi)}{\mathbf{S}_\kappa{}^{n-1}(t)}$$

is decreasing with respect to t, for each $\xi \in \mathsf{S}_x$. Therefore,

$$\begin{aligned}
\frac{\mathfrak{A}(x;r)}{A_\kappa(r)} &= \mathbf{c_{n-1}}^{-1} \int_{\mathsf{D}_x(r)} \frac{\det \mathcal{A}(r;\xi)}{\mathbf{S}_\kappa{}^{n-1}(r)}\, d\mu_x(\xi) \\
&\geq \mathbf{c_{n-1}}^{-1} \int_{\mathsf{D}_x(R)} \frac{\det \mathcal{A}(r;\xi)}{\mathbf{S}_\kappa{}^{n-1}(r)}\, d\mu_x(\xi) \\
&\geq \mathbf{c_{n-1}}^{-1} \int_{\mathsf{D}_x(R)} \frac{\det \mathcal{A}(R;\xi)}{\mathbf{S}_\kappa{}^{n-1}(R)}\, d\mu_x(\xi) \\
&= \frac{\mathfrak{A}(x;R)}{A_\kappa(R)},
\end{aligned}$$

which implies the claim. ∎

Lemma III.4.1. (M. Gromov (1982, 1986)) *Suppose f and g are positive integrable functions, of a real variable r, for which*

$$f/g$$

is decreasing with respect to r. Then, the function

$$\int_0^r f \Big/ \int_0^r g$$

is also decreasing with respect to r.

Proof. Consider $r < R$. Then,

$$\int_0^r f \int_0^R g = \int_0^r f \int_0^r g + \int_0^r f \int_r^R g$$

and

$$\int_0^R f \int_0^r g = \int_0^r f \int_0^r g + \int_r^R f \int_0^r g.$$

Now we wish to show that

$$\int_0^r f \int_0^R g \geq \int_0^R f \int_0^r g,$$

which is therefore equivalent to showing

$$\int_0^r f \int_r^R g \geq \int_r^R f \int_0^r g.$$

Set $f = gh$. Then, by hypothesis, h is decreasing. This implies

$$\int_0^r f \int_r^R g = \int_0^r gh \int_r^R g \geq h(r) \int_0^r g \int_r^R g \geq \int_0^r g \int_r^R hg = \int_0^r g \int_r^R f,$$

which is the claim. ■

Theorem III.4.5. (M. Gromov (1982, 1986)) *Assume that the Ricci curvatures of M are all greater than or equal to $(n-1)\kappa$. Then, for every $x \in M$, we have*

$$\frac{V(x;r)}{V_\kappa(r)} \tag{III.4.23}$$

is decreasing with respect to r.

Proof. The theorem is an immediate consequence of Proposition III.4.1 and Lemma III.4.1. ■

Recall that during this whole discussion we have assumed that M is complete. Now suppose that our constant κ is positive, and that all Ricci curvatures are bounded from below by $(n-1)\kappa$. Then, the Bonnet–Myers theorem states that M is compact, with diameter less than or equal to $\pi/\sqrt{\kappa}$. The Bishop theorem then implies that $V(M)$ is less than or equal to $V(\mathbb{M}_\kappa)$, with equality if and only if M is isometric to $\mathbb{M}_\kappa$.

We now ask: what if the diameter of M is equal to $\pi/\sqrt{\kappa}$?

Theorem III.4.6. (V. A. Toponogov (1959), S. Y. Cheng (1975)) *Given M Riemannian complete, with all Ricci curvatures bounded from below by $(n-1)\kappa$, $\kappa > 0$. If the diameter of M is equal to $\pi/\sqrt{\kappa}$, then M is isometric to the standard sphere of constant sectional curvature equal to κ.*

Proof. (Shiohama (1983)) Pick points $x, y \in M$ so that $d(x, y) = \pi/\sqrt{\kappa}$. Then, the previous theorem implies

$$\frac{V(x; \pi/2\sqrt{\kappa})}{V(\mathbb{M}_\kappa)/2} \geq \frac{V(x; \pi/\sqrt{\kappa})}{V(\mathbb{M}_\kappa)} = \frac{V(M)}{V(\mathbb{M}_\kappa)}.$$

Therefore,

$$V(x; \pi/2\sqrt{\kappa}) \geq V(M)/2.$$

Similarly,

$$V(y; \pi/2\sqrt{\kappa}) \geq V(M)/2.$$

But

$$B(x; \pi/2\sqrt{\kappa}) \cap B(y; \pi/2\sqrt{\kappa}) = \emptyset.$$

Therefore,

$$V(x; \pi/2\sqrt{\kappa}) = V(y; \pi/2\sqrt{\kappa}) = V(M)/2.$$

Thus, $V(M) = V(\mathbb{M}_\kappa)$, and M is isometric to $\mathbb{M}_\kappa$. ∎

§III.5. The Area of Spheres

As usual, M is our Riemannian manifold. For convenience, we assume here that M is complete.

We now consider the area of metric spheres $S(x; r)$ for arbitrary r – even when $r \geq \operatorname{inj} x$. For now, $S(x; r)$ is no longer guaranteed to be a smooth imbedded $(n-1)$–dimensional submanifold of M. However, we shall be able to extend the notion of $(n-1)$–dimensional measure in such a fashion that, except for a set of $(n-1)$–dimensional measure (in this new sense) equal to 0, $S(x; r)$ is a Borel subset (in the relative topology) of an immersed $(n-1)$–dimensional submanifold (not necessarily connected) and therefore possesses a well-defined area. This more general measure is Hausdorff measure. In what follows, we only summarize the basic facts that we require from geometric measure theory. The reader might start with F. Morgan's guide (1988) and the references to other introductions therein, before approaching the classic Federer (1969). We also considered Hausdorff measure in Chapter IV of Chavel (2001).

Definition. Given any metric space X, a subset S in X, we define, as usual, its *diameter* by

$$\operatorname{diam} S = \sup \{d(x, y) : \; x, y \in S\}.$$

For any integer $k \geq 0$, we define the *δ–approximate k–dimensional Hausdorff measure of S*, $\mathcal{H}^k_\delta(S)$, by

$$\mathcal{H}^k_\delta(S) = \inf \sum_j \omega_{\mathbf{k}} \left\{ \frac{\operatorname{diam} S_j}{2} \right\}^k,$$

where the infimum is taken over all countable covers $\{S_j\}$ of S for which $\operatorname{diam} S_j \leq \delta$ for all j; we define the *k–dimensional Hausdorff measure of S*, $\mathcal{H}^k(S)$, by

$$\mathcal{H}^k(S) = \lim_{\delta \downarrow 0} \inf \mathcal{H}^k_\delta(S).$$

A subset E is called *$\mathcal{H}^k$–measurable* if

$$\mathcal{H}^k(E \cap S) + \mathcal{H}^k(E \cap (X \setminus S)) = \mathcal{H}^k(E)$$

for *all* subsets S of X.

Remark III.5.1. One may extend the definition of Hausdorff measure to non-integral dimension, by replacing $\omega_{\mathbf{k}}$ by its corresponding expression in terms of gamma functions (III.3.13).

The collection of $\mathcal{H}^k$–measurable subsets of X form a σ–algebra, and this is the one with which we work.

For Euclidean space $\mathbb{R}^n$, $n \geq 1$, with its usual metric, one always has

$$d\mathcal{H}^n = dV, \tag{III.5.1}$$

where dV is Lebesgue measure on $\mathbb{R}^n$. Furthermore, one has:

Theorem III.5.1. (The area formula) *Let $\phi : \mathbb{R}^k \to \mathbb{R}^n$, $k \leq n$, be a Lipschitz function on $\mathbb{R}^k$. Then*:

(i) *For any Lebesgue measurable subset E of $\mathbb{R}^k$, we have*

$$\int_E |\det J_\phi|\, dV = \int_{\mathbb{R}^n} \operatorname{card} E \cap \phi^{-1}[x]\, d\mathcal{H}^k(x),$$

where card *denotes cardinality.*

(ii) *If f is any L^1 Lebesgue integrable function on $\mathbb{R}^k$, then*

$$\int_{\mathbb{R}^k} f|\det J_\phi|\, dV = \int_{\mathbb{R}^n} \sum_{y \in \phi^{-1}[x]} f(y)\, d\mathcal{H}^k(x).$$

Thus, the area formula generalizes (III.3.1).

Theorem III.5.2. (The coarea formula) *Let $f : \mathbb{R}^n \to \mathbb{R}$ be a Lipschitz function on $\mathbb{R}^n$, with gradient vector field* grad f *(defined almost everywhere). Then, for any measurable subset E of $\mathbb{R}^n$, we have*

$$\int_E |\operatorname{grad} f|\, dV = \int_{\mathbb{R}} \mathcal{H}^{n-1}(E \cap f^{-1}[t])\, dt.$$

Given M an n–dimensional Riemannian manifold, $n \geq 1$, one uses the distance function determined by the Riemannian metric on M to determine the collection of Hausdorff measures on M. Again, one has (III.5.1).

Definition. We define the *area of* $S(x;r)$, $A(x;r)$, by

$$A(x;r) = \mathcal{H}^{n-1}(S(x;r)).$$

Proposition III.5.1. *For any $x \in M$, we have*

$$A(x;r) = \mathfrak{A}(x;r)$$

for almost all $r \in \mathbb{R}$.

Proof. Clearly,

$$A(x;r) - \mathfrak{A}(x;r) = \mathcal{H}^{n-1}(C(x) \cap S(x;r)).$$

In the coarea formula, set $E = C(x)$ and consider the distance function $f(y) = d(x, y)$ on M. Then, f is Lipschitz, with gradient of length equal to 1 almost everywhere on M (actually, everywhere except at x and $C(x)$). Then, Proposition III.3.1 and the coarea formula yield

$$0 = \int_{\mathbb{R}} \mathcal{H}^{n-1}(C(x) \cap f^{-1}[r])\, dr = \int_{\mathbb{R}} \mathcal{H}^{n-1}(C(x) \cap S(x;r))\, dr,$$

which implies the claim. ∎

§III.6. Fermi Coordinates

In this section we take note that one may consider the distance function based on a submanifold, namely, let M be our given n–dimensional Riemannian manifold, let $\mathfrak{M}$ be a connected k–dimensional submanifold of M, $0 \leq k < n$, and consider the distance function r to be the function on M given by

$$r(q) = d(q, \mathfrak{M}).$$

Here, one has a corresponding apparatus of first and second variations of arc length, elementary comparison theorems, Fermi coordinates based on $\mathfrak{M}$

(in place of geodesic spherical coordinates based on a point), focal cut and conjugate loci, and volume comparison theorems. In what follows, we merely give the definitions and results, leaving the extension of earlier arguments to the reader. In places where more detail is warranted, we try to supply it.

We first note that the gradient vector field of r, when it is well-defined, has unit length; thus, its integral curves are geodesics emanating from $\mathfrak{M}$ (see Exercise I.3).

Next, we note that if $\mathfrak{M}$ is compact, then there exists an $\epsilon > 0$ such that $\operatorname{grad} r$ is defined and smooth on all of $r^{-1}[(0, \epsilon)]$. Also, when $\mathfrak{M}$ has dimension greater than 0, all the integral curves of $\operatorname{grad} r$ on $r^{-1}[[0, \epsilon)]$ intersect $\mathfrak{M}$ orthogonal to $\mathfrak{M}$ at the point of intersection. Indeed, it suffices to check the following: Given any $p \in \mathfrak{M}$, there exists a neighborhood U of p in $\mathfrak{M}$, and $\epsilon_p > 0$, such that, if π_ν denotes the projection of the normal bundle $\nu\mathfrak{M}$ to $\mathfrak{M}$, then $\exp|{\pi_\nu}^{-1}[U] \cap \mathsf{B}(U; \epsilon_p)$ is a diffeomorphism of ${\pi_\nu}^{-1}[U] \cap \mathsf{B}(U; \epsilon_p)$ onto its image in M. (For the notation $\mathsf{B}(U; \epsilon)$, see §I.6.)

We now state the appropriate generalizations of Theorem I.6.2 and the arguments used to derive it. First, set

$$\nu\mathsf{S}_p = \mathsf{S}_p \cap {\mathfrak{M}_p}^\perp,$$

the fiber over p in the *unit* normal bundle of $\mathfrak{M}$.

Assume $\mathfrak{M}$ is compact. Let $\epsilon > 0$ be given as above, $p \in \mathfrak{M}, \xi \in \nu\mathsf{S}_p, \gamma(t) = \gamma_\xi(t)$, $t_0 \in (0, \epsilon)$, $q = \gamma(t_0)$. Then, for any path ω starting in $\mathfrak{M}$ and ending at q, that is, $\omega : [0, 1] \to M$ with $\omega(0) \in \mathfrak{M}$, $\omega(1) = q$, we have

$$\ell(\omega) \geq \ell(\gamma) = t_0 = r(q),$$

with equality only if the image of ω is the same as the image of γ. Conversely, start with any $q \in M$. Then, for any unit speed path ω connecting a point p in $\mathfrak{M}$ to q, we have

$$\ell(\omega) = d(q, \mathfrak{M})$$

only if ω is a geodesic, and $\omega'_{|p} \in {\mathfrak{M}_p}^\perp$.

Before proceeding, first recall, from §II.2, that if $\mathfrak{M}$ is a k–dimensional submanifold of the n–dimensional Riemannian manifold M, then the second fundamental form of $\mathfrak{M}$ in M is, at each point $p \in \mathfrak{M}$, a vector-valued symmetric bilinear form $\mathfrak{B} : \mathfrak{M}_p \times \mathfrak{M}_p \to {\mathfrak{M}_p}^\perp$, given by

$$\mathfrak{B}(\xi, \eta) = (\nabla_\xi Y)^N,$$

where Y is any extension of η to a tangent vector field on $\mathfrak{M}$, ∇ denotes the Levi-Civita connection of the Riemannian metric on M, and the superscript N

denotes projection onto $\mathfrak{M}_p{}^\perp$. To every vector $\nu \in \mathfrak{M}_p{}^\perp$ one has the real-valued bilinear form

$$\mathfrak{b}_\nu(\xi, \eta) = \langle \mathfrak{B}(\xi, \eta), \nu \rangle,$$

and Weingarten map $\mathfrak{A}^\nu : \mathfrak{M}_p \to \mathfrak{M}_p$ given by

$$\langle \mathfrak{A}^\nu \xi, \eta \rangle = \mathfrak{b}_\nu(\xi, \eta)$$

for $\xi, \eta \in \mathfrak{M}_p$ – so

$$\mathfrak{A}^\nu \xi = -(\nabla_\xi V)^T,$$

where V is an extension of ν to a normal vector field on $\mathfrak{M}$, and the superscript T denotes projection onto $\mathfrak{M}_p$.

We now present the formulation of results associated with the first and second variations of arc length from the submanifold $\mathfrak{M}$ to a fixed point in M. Let $\gamma : [0, \beta] \to M$ be a unit speed geodesic, such that

$$p = \gamma(0) \in \mathfrak{M}, \quad \xi = \gamma'(0) \in \nu\mathsf{S}_p, \qquad \text{and} \qquad q = \gamma(\beta).$$

Consider a variation $v : [0, \beta] \times (-\epsilon_0, \epsilon_0) \to M$, where

$$v(0, \epsilon) \in \mathfrak{M}, \quad v(\beta, \epsilon) = q \qquad \text{for all } \epsilon, \text{ and} \qquad v(t, 0) = \gamma(t).$$

Set

$$Y(t) = (\partial_\epsilon v)(t, 0), \qquad \eta = Y(0),$$

and

$$L(\epsilon) = \int_0^\beta |\partial_t v|(t, \epsilon)\, dt.$$

Then

$$L'(0) = 0,$$

and

$$L''(0) = -\mathfrak{b}_\xi(\eta, \eta) + \int_0^\beta \left\{ |\nabla_t Y_\perp|^2 - \langle R(\gamma', Y_\perp)\gamma', Y_\perp \rangle \right\} dt.$$

One also has

$$\langle (\nabla_t Y_\perp)(0), Y_\perp(0) \rangle = -\mathfrak{b}_\xi(\eta, \eta).$$

Integration by parts then implies

$$L''(0) = -\int_0^\beta \langle \nabla_t{}^2 Y_\perp + R(\gamma', Y_\perp)\gamma', Y_\perp \rangle\, dt.$$

Therefore, it is natural[2] to consider the collection $\mathfrak{T}$ of *transverse vector fields X along γ*, that is, those vector fields X along γ for which X is pointwise orthogonal to γ, with initial data

$$X(0) \in \mathfrak{M}_p, \qquad (\nabla_t X)(0) + \mathfrak{A}^\xi X(0) \perp \mathfrak{M}_p.$$

Let $\mathfrak{T}_0$ be the subcollection of $\mathfrak{T}$ for which we also have

$$X(\beta) = 0.$$

Define the *index form I on $\mathfrak{T}$* by

$$I(X_1, X_2) = -\mathfrak{b}_\xi(X_1(0), X_2(0)) + \int_0^\beta \langle \nabla_t X_1, \nabla_t X_2 \rangle - \langle R(\gamma', X_1)\gamma', X_2 \rangle \, dt;$$

then when the index form I and the linear operator

$$\mathcal{L} = -\{\nabla_t{}^2 X + R(\gamma', X)\gamma'\}$$

are restricted to $\mathfrak{T}_0$, we have I is the symmetric bilinear form of $\mathcal{L}$.

We now consider transverse Jacobi fields. For any such transverse Jacobi field $Y \in \mathfrak{T}$ along γ, the index form I is given by

$$I(Y, Y) = \langle \nabla_t Y, Y \rangle(\beta). \tag{III.6.1}$$

Note that the collection of transverse Jacobi fields $\mathfrak{T}$ along γ is an $(n-1)$–dimensional vector space. Indeed, it is rather easy to show that the collection is a vector space. To calculate the dimension, first note that the *full* collection of Jacobi fields Y with $Y(0) \in \mathfrak{M}_p$, pointwise orthogonal to γ, is $(n+k-1)$–dimensional. Then, map this $(n+k-1)$–dimensional vector space to M_p by

$$Y \mapsto (\nabla_t Y)(0) + \mathfrak{A}^\xi Y(0),$$

and show that $\mathfrak{M}_p$ is in the range of this map.

Now assume the sectional curvatures along γ are all equal to κ, and the Weingarten map of ξ, $\mathfrak{A}^\xi$, is given by

$$\mathfrak{A}^\xi = \lambda I. \tag{III.6.2}$$

Then, the collection of transverse Jacobi fields along γ, pointwise orthogonal to γ, are given as sums of the vectors fields:

$$Y(t) = \mathbf{S}_\kappa(t)\tau_t\eta, \qquad Z(t) = (\mathbf{C}_\kappa - \lambda\mathbf{S}_\kappa)(t)\tau_t\zeta,$$

where τ_t denotes parallel translation along γ from p to $\gamma(t)$, $\eta \in \xi^\perp \cap \mathfrak{M}_p^\perp$, and $\zeta \in \mathfrak{M}_p$.

[2] Here, "natural" is a bit too naive.

Definition. Given M, $\mathfrak{M}$, $p \in \mathfrak{M}$, $\xi \in \nu\mathsf{S}_p$ as above, $\gamma = \gamma_\xi$, a point $\gamma(t)$ is said to be *focal to* $\mathfrak{M}$ *along* γ if there exists a nontrivial transverse Jacobi field Y such that $Y(t) = 0$.

Assume the sectional curvatures along γ are all equal to the constant κ, and the Weingarten map $\mathfrak{A}^\xi$ satisfies (III.6.2). One can easily determine conditions on κ and λ which characterize the existence of focal points.

The Jacobi criteria (Theorems II.5.4 and II.5.5), on the positivity of the index form and the nonexistence of focal points, remain valid in this setting with an important added comment. Continue with $M, \mathfrak{M}, p \in \mathfrak{M}, \xi \in \nu\mathsf{S}_p$. Expand the domain of the index form I to Υ, the collection of *all* vector fields X along γ, pointwise orthogonal to γ, for which $X(0) \in \mathfrak{M}_p$, and let Υ_0 consist of those elements of Υ that vanish at $t = \beta$. On Υ, define the index form as above. One can now verify, that the argument of Theorem II.5.4 remains valid for the index form on Υ_0, not just on $\mathfrak{T}_0$. This will ease the proof of some of the comparison theorems in this setting. One can also check for the corresponding version of Theorem II.5.5.

We now sketch appropriate versions of the volume results of this chapter – some aspects of which lead to deep results later on. Early arguments were given in Grossman (1967), and then extended and deepened in Heintze–Karcher (1978). To start:

Let M be a complete n–dimensional Riemannian manifold, and $\mathfrak{M}$ a k–dimensional submanifold (we still include the case of dimension of $\mathfrak{M}$ equal to 0). Let

$$\mathrm{Exp} = \exp|\nu\mathfrak{M},$$

where $\nu\mathfrak{M}$ denotes the normal bundle of $\mathfrak{M}$ in M, with natural projection π_ν; also let

$$\nu\mathsf{S}\mathfrak{M} = \nu\mathfrak{M} \cap \mathsf{S}M$$

denote the unit normal bundle of $\mathfrak{M}$. Map $E : [0, +\infty) \times \nu\mathsf{S}\mathfrak{M} \to M$ by

$$E(t, \xi) = \mathrm{Exp}\, t\xi,$$

so E determines radial coordinates on M, also known as *Fermi coordinates*, associated wih the distance function $r : M \to \mathbb{R}$ (defined previously).

For $p \in \mathfrak{M}$, $\xi \in \nu\mathsf{S}_p$, one calculates the Riemannian metric along the geodesic

$$\gamma(t) = \mathrm{Exp}\, t\xi$$

as follows (Figure III.2):

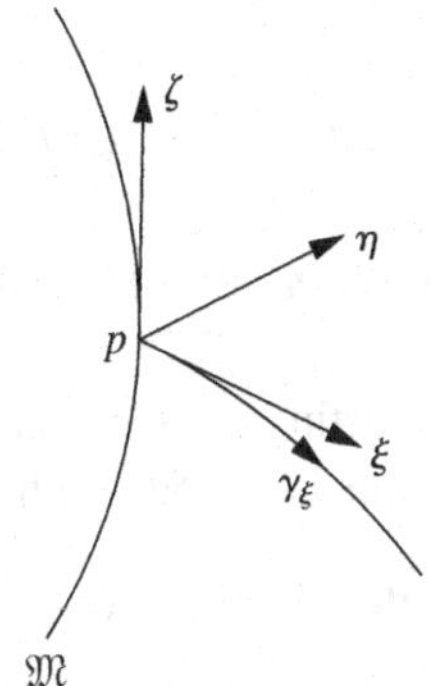

Figure III.2. 1–dimensional $\mathfrak{M}$ in the 3–dimensional M.

(a) Let $\zeta \in \mathfrak{M}_p$. Then, $E_{*|(t,\xi)}\mathfrak{I}_{t\xi}\zeta$, for fixed ξ and varying t, is the Jacobi field Z_ζ along γ, pointwise orthogonal to γ, determined by the initial conditions

$$Z_\zeta(0) = \zeta, \qquad (\nabla_t Z_\zeta)(0) = -\mathfrak{A}^\xi \zeta.$$

Indeed, for any path $\omega(\epsilon)$ in $\mathfrak{M}$ satisfying $\omega(0) = p, \omega'(0) = \zeta$, one first shows that ξ can be extended to a normal unit vector field X along ω such that $\nabla_\epsilon X \in T\mathfrak{M}$ for all ϵ, that is, ξ may be extended to a local vector field along $\omega(\epsilon)$ that is *parallel in the connection of the normal bundle* $\nu\mathfrak{M}$ (see Exercise II.12). Now one can easily imitate the argument of Theorem II.7.1.

(b) Let $\eta \in \mathfrak{M}_p{}^\perp \cap \xi^\perp = \mathfrak{I}_\xi{}^{-1}((\nu\mathsf{S}_p)_\xi)$ – that is, consider η, orthogonal to $\mathfrak{M}_p$ and ξ, as a tangent vector at ξ to the unit normal ($(n-k-1)$–dimensional) sphere at p. Then, $E_{*|(t,\xi)}t\mathfrak{I}_{t\xi}\eta$ is the Jacobi field Y_η along γ, pointwise orthogonal to γ, determined by the initial conditions

$$Y_\eta(0) = 0, \qquad (\nabla_t Y_\eta)(0) = \eta,$$

for fixed ξ and varying t.

We therefore let $\mathcal{A}(t;\xi)$ denote the matrix solution to Jacobi's equation along γ, pulled back to $\xi^\perp$, as in §III.1:

$$\mathcal{A}'' + \mathcal{R}\mathcal{A} = 0,$$

subject to the initial conditions

$$\mathcal{A}(0;\xi)|\mathfrak{M}_p = I, \qquad \mathcal{A}'(0;\xi)|\mathfrak{M}_p = -\mathfrak{A}^\xi,$$

and

$$\mathcal{A}(0;\xi)|\mathfrak{M}_p{}^\perp \cap \xi^\perp = 0, \qquad \mathcal{A}'(0;\xi)|\mathfrak{M}_p{}^\perp \cap \xi^\perp = I;$$

and we write

$$ds^2 = dt^2 + |\mathcal{A}(t;\xi)\,d\xi|^2.$$

Definition. The *focal locus of* $\mathfrak{M}$ *in* $\nu\mathfrak{M}$ (the *tangential focal locus*) may be defined in two ways:

(i) It is the subset of $\nu\mathfrak{M}$ consisting of all critical points of Exp.

(ii) It is the collection of vectors $t\xi \in \nu\mathfrak{M}$, with $t > 0$, $\xi \in \nu\mathsf{S}\mathfrak{M}$, for which

$$\det\, \mathcal{A}(t;\xi) = 0,$$

where $\mathcal{A}$ is given above. (Thus, for a given $\xi \in \nu\mathsf{S}\mathfrak{M}$ the nullity of Exp at $t_0\xi$ is equal to the order of t_0 as a zero of the function (in t) $\mathcal{A}(t;\xi)$.)

By the *focal locus of* $\mathfrak{M}$ *in* M (the *focal locus*), we mean the image of the tangential focal locus of $\mathfrak{M}$ under the exponential map Exp.

Definition. Given $\xi \in \nu\mathsf{S}\mathfrak{M}$, we define $c_\nu(\xi)$ the *distance to the focal cut point of* $\mathfrak{M}$ *along* γ_ξ by

$$c_\nu(\xi) := \sup\,\{t > 0 :\; d(\mathfrak{M}, \gamma_\xi(t)) = t\}.$$

We define the *focal cut locus of* $\mathfrak{M}$ *in* $\nu\mathfrak{M}$ (the *tangential focal cut locus*), $\nu\mathsf{C}(\mathfrak{M})$, by

$$\nu\mathsf{C}(\mathfrak{M}) := \{c_\nu(\xi)\xi : c_\nu(\xi) < +\infty,\ \xi \in \nu\mathsf{S}\mathfrak{M}\},$$

and the *focal cut locus of* $\mathfrak{M}$ *in* M (the *focal cut locus*), $\nu C(\mathfrak{M})$, by

$$\nu C(\mathfrak{M}) := \mathrm{Exp}\ \nu\mathsf{C}(\mathfrak{M}).$$

Also, we set

$$\nu\mathsf{D}_{\mathfrak{M}} := \{t\xi : 0 \le t < c_\nu(\xi), \xi \in \nu\mathsf{S}\mathfrak{M}\},$$

$$\nu D_{\mathfrak{M}} := \mathrm{Exp}\ \nu\mathsf{D}_{\mathfrak{M}}.$$

One now has corresponding versions of item **1** through item **5**, and Theorems 1, 2, and 4 of §III.2.

We may now consider integration. For, $f \in L^1(M)$, we have the analogue of (III.6.3)

$$\int_M f\,dV = \int_{\mathfrak{M}} dV_k(p) \int_{\nu\mathsf{S}_p} d\mu_{n-k-1,p}(\xi) \int_0^{c_\nu(\xi)} f(\mathrm{Exp}\,t\xi)\sqrt{\mathbf{g}}(t;\xi)\,dt$$

with

$$\sqrt{\mathbf{g}}(t;\xi) = \det \mathcal{A}(t;\xi),$$

where dV_k the (k–dimensional) Riemannian measure of $\mathfrak{M}$, and $d\mu_{n-k-1,p}$ the standard $(n-k-1)$–dimensional measure on $\nu\mathsf{S}_p$.

To generalize the Bishop theorem (Theorem III.4.3) in the spirit of our previous argument, one must restrict oneself to the case

$$k = n - 1.$$

(Otherwise, the argument, as presented, does not go through. Why?) The result goes as follows:

Theorem III.6.1. *Fix $\xi \in \nu\mathsf{S}\mathfrak{M}$, $\gamma(t) = \exp t\xi$. Assume that all Ricci curvatures along γ are bounded below by $(n-1)\kappa$, and assume that*

$$\operatorname{tr}\mathfrak{A}^{\xi} \geq (n-1)\lambda.$$

Let $\beta_{\kappa,\lambda} \in (0, +\infty]$ denote the first positive zero of $(\mathbf{C}_\kappa - \lambda\mathbf{S}_\kappa)(t)$, should such a zero exist; otherwise, set $\beta_{\kappa,\lambda} = +\infty$. Then, $\mathfrak{M}$ has a focal point along γ at distance $t_0 \leq \beta_{\kappa,\lambda}$, and

$$\det \mathcal{A} \leq (\mathbf{C}_\kappa - \lambda\mathbf{S}_\kappa)^{n-1} \tag{III.6.4}$$

on all of $[0, t_0]$. One has equality in (III.6.4) at $\tau \in (0, t_0]$ if and only if $\mathcal{A} = (\mathbf{C}_\kappa - \lambda\mathbf{S}_\kappa)I$ on all of $t \in [0, \tau]$, in which case we have $\mathfrak{A}^{\xi} = \lambda I$, and $\mathcal{R} = \kappa I$ on all of $t \in [0, \tau]$.

Remark III.6.1. When $k < n-1$ one may still obtain results using the arguments of Exercise II.21, namely, suppose we are given, in addition to M_1 and M_2, submanifolds of the same dimension $\mathfrak{M}_j$, $p_j \in \mathfrak{M}_j$, $\xi_j \in \nu\mathsf{S}_{p_j}$, with attendant second fundamental forms $\mathfrak{b}_{\xi_j}$, $j = 1, 2$. Assume

$$\mathfrak{b}_{\xi_1} \leq \mathfrak{b}_{\xi_2}$$

in the sense that the highest eigenvalue of $\mathfrak{b}_{\xi_1}$ is less than or equal to the lowest eigenvalue of $\mathfrak{b}_{\xi_2}$, and also assume the sectional curvature condition given in Exercise II.21,

$$\sup \mathcal{K}_{1_{|\gamma_{\xi_1}(t)}} \leq \min \mathcal{K}_{2_{|\gamma_{\xi_2}(t)}}$$

for all $t > 0$. Then, one can use the argument of Exercise II.21 to obtain the analogue of the Morse–Schönberg theorem, that the first focal point to $\gamma_{\xi_1}(0)$ along γ_{ξ_1} cannot occur earlier than the first focal point to $\gamma_{\xi_2}(0)$ along γ_{ξ_2}. Note that this method of argument gives a Bonnet–Myers theorem with the assumption of a positive lower bound on the sectional curvature – both in the conjugate and focal point versions.

Also, one has corresponding volume comparison theorems for sectional curvature bounded from below. See our presentation in §IX.3.

§III.7. Integration of Differential Forms

Let E be a real vector space, E^* its dual, and $\Lambda^k E^*$ the space of alternating k–covectors. Recall that, for any given $\xi \in E$, one defines $\mathrm{i}(\xi) : \Lambda^k E^* \to \Lambda^{k-1} E^*$ by

$$(\mathrm{i}(\xi)\alpha)(\xi_1, \ldots, \xi_{k-1}) = \alpha(\xi, \xi_1, \ldots, \xi_{k-1})$$

when $k \geq 1$, and $\mathrm{i} = 0$ when $k = 0$.

Let M be an n–dimensional differentiable manifold, and $\Lambda^k T M^*$ the alternating k–cotangent bundle with its associated natural differentiable structure. The differentiable differential k–forms on M are, then, differentiable sections of M in $\Lambda^k T M^*$.

Notation. We denote the collection of differentiable differential k–forms on M by $\mathfrak{S}^k(M)$. And, we let $\mathfrak{S}^k_c(M)$ denote the compactly supported differentiable differential k–forms on M.

Let X be a differentiable vector field on M, with associated flow $\phi_t : M \to M$. Then, for any $\omega \in \mathfrak{S}^k$, the *Lie derivative of* ω *with respect to* X, $\mathcal{L}_X\omega$, is defined by

$$\mathcal{L}_X\omega = ({\phi_t}^*\omega)'(0),$$

where ${\phi_t}^*\omega$ denotes the pull-back of ω by ϕ_t, and the prime denotes differentiation with respect to t. Then, it is well-known that $\mathcal{L}_X$ is given by

$$\mathcal{L}_X = d \circ \mathrm{i}(X) + \mathrm{i}(X) \circ d, \tag{III.7.1}$$

where d denotes exterior differentiation.

As usual, our manifold M is connected.

Now, for any $p \in M$, the dimension of $\Lambda^n(M_p)^*$ is, of course, equal to 1. Therefore, if $\mathfrak{o}$ denotes the zero section of $M \to \Lambda^n T M^*$, then $\Lambda^n T M^* \setminus \mathfrak{o}(M)$ has *at most* 2 components. M is *orientable* if $\Lambda^n T M^* \setminus \mathfrak{o}(M)$ has, in fact, 2 components; and an *orientation* of M is a choice of one of those components.

It is well-known that M is orientable if and only if there exists a cover of M by charts $\{x_\alpha : U_\alpha \to \mathbb{R}^n : \alpha \in \mathcal{I}\}$ for which

$$\det\left(\frac{\partial(x_\beta \circ x_\alpha{}^{-1})^j}{\partial x_\alpha{}^k}\right) > 0$$

on $x_\alpha(U_\alpha \cap U_\beta)$, for all $\alpha,\ \beta \in \mathcal{I}$. Also, M is orientable if and only if M possesses a nowhere vanishing differentiable differential n–form.

In establishing the above, one uses a useful calculation, namely, if E is an n–dimensional real vector space, and $\beta_1, \ldots, \beta_n \in E^*$, and the forms $\alpha^j \in E^*$, $j = 1, \ldots, n$, are given by

$$\alpha^j = \sum_k \beta^k a_k{}^j,$$

then

$$\alpha^1 \wedge \cdots \wedge \alpha^n = (\det(a_k{}^j))\beta^1 \wedge \cdots \wedge \beta^n. \tag{III.7.2}$$

In particular, for charts $x : U \to \mathbb{R}^n$, $y : U \to \mathbb{R}^n$ on M, we have

$$dy^j = \sum_k \frac{\partial(y \circ x^{-1})^j}{\partial x^k}\, dx^k,$$

and

$$dy^1 \wedge \cdots \wedge dy^n = \det\left(\frac{\partial(y \circ x^{-1})^j}{\partial x^k}\right) dx^1 \wedge \cdots \wedge dx^n.$$

When M is orientable with fixed orientation Λ, and $\{e_1, \ldots, e_n\}$ is a basis of some tangent space to M, then $\{e_1, \ldots, e_n\}$ is *positively oriented* if for its dual basis $\{\omega^1, \ldots, \omega^n\}$ we have $\omega^1 \wedge \cdots \wedge \omega^n \in \Lambda$.

Let M be (henceforth) orientable, with given fixed orientation Λ. Then, a chart $x : U \to \mathbb{R}^n$ on M is *positively oriented* if $dx^1 \wedge \cdots \wedge dx^n \in \Lambda$ on all of U. Otherwise, the chart x is *negatively oriented.*

Given a differentiable n–form ω on M, then for any chart $x : U \to \mathbb{R}^n$ on M, one has a function $f : U \to \mathbb{R}^n \in C^\infty$ such that

$$\omega | U = f\, dx^1 \wedge \cdots \wedge dx^n.$$

One then defines, for $\omega \in \mathfrak{S}^n_c(U)$,

$$I(\omega; U) = (\operatorname{sign} x) \int_{x(U)} f\, dx^1 \cdots dx^n,$$

where $\operatorname{sign} x$ is $+1$ or -1 depending on whether x is positively oriented or negatively oriented. One easily checks that I depends only on ω and U, and not on the choice of x. Therefore, for any $\omega \in \mathfrak{S}^n(M)$, and for any cover of M by charts $x_\alpha : U_\alpha \to \mathbb{R}^n$ (where α belongs to some index set $\mathcal{I}$) with subordinate partition of unity $\{\phi_\alpha\}$, the integral

$$\int_U \omega := \sum_{\alpha \in \mathcal{I}} I(\phi_\alpha \omega; U_\alpha \cap U),$$

for any open U in M, is well-defined and depends only on ω and U. The definition of $\int_\cdot \omega$ is consistent with the definition of $I(\omega; \cdot)$, in the sense that if U is the domain of any chart, then

$$\int_U \omega = I(\omega; U).$$

In particular, for any U, we have

$$\int_U \omega = \sum_{\alpha \in \mathcal{I}} \int_U \phi_\alpha \omega.$$

Let M_1, M_2 be orientable, with respective orientations Λ_1, Λ_2, and let $\Phi : M_1 \to M_2$ be a diffeomorphism. Then, Φ is orientation preserving if, for any $\omega \in \Lambda_2$, we have $\Phi^* \omega \in \Lambda_1$. Otherwise, Φ is orientation reversing. For any integrable nowhere vanishing $\omega \in \mathfrak{S}^n(M_2)$, we then have

$$\int_{M_2} \omega = (\operatorname{sign} \Phi) \int_{M_1} \Phi^* \omega.$$

One immediately has, for any vector field X on M and $\omega \in \mathfrak{S}_c^n(M)$,

$$\int_M \mathcal{L}_X \omega = 0.$$

When M is Riemannian orientable, one can write the Riemannian measure as a global n–form σ, the *volume form of the Riemannian metric*, on M. In any chart $x : U \to \mathbb{R}^n$ on M, we have

$$\sigma = (\operatorname{sign} x)\sqrt{g^x}\, dx^1 \wedge \cdots \wedge dx^n.$$

Therefore, for integrable f on M, and any open U in M, we have

$$\int_U f\, dV = \int_U f\sigma.$$

Theorem III.7.1. (Stokes' theorem I) *If M is oriented and $\omega \in \mathfrak{S}_c^{n-1}(M)$, then*

$$\int_M d\omega = 0.$$

If M is oriented n–dimensional, Ω an open subset of M with smooth boundary $\partial\Omega$, $p \in \partial\Omega$, and $\xi \in M_p, \xi \neq 0$, we say that ξ is an *outward vector* if $\xi \notin (\partial\Omega)_p$ and there exists $\epsilon > 0$, $\gamma : (-\epsilon, \epsilon) \to M \in C^1$ such that $\gamma(0) = p$, $\gamma'(0) = \xi$, $\gamma(t) \in \Omega$ for $t < 0$ (in particular, $\gamma(t) \notin \Omega$ for sufficiently small $t > 0$). A basis $\{\xi_2, \ldots, \xi_n\}$ of $(\partial\Omega)_p$ is *positively oriented*, if $\{\xi, \xi_2, \ldots, \xi_n\}$ is a positively oriented basis of M_p for an outward vector ξ. The definition is independent of the particular choice of outward vector ξ. Thus, in this manner, the orientation of M determines an orientation on $(\partial\Omega)_p$. Note that, here, we do not require that Ω and $\partial\Omega$ be connected.

Theorem III.7.2. (Stokes' theorem II) *Let M, Ω, and $\partial\Omega$ be as previously described. Then, for any $\omega \in \mathfrak{S}_c^{n-1}(M)$, we have*

$$\iint_{\Omega} d\omega = \int_{\partial\Omega} \omega.$$

Now, let $\omega \in \mathfrak{S}^n(M)$ be nowhere vanishing. Then, for any C^1 vector field X on M, we define the *divergence of X with respect to ω*, $\operatorname{Div}_\omega X$, by

$$(\operatorname{Div}_\omega X)\,\omega = \mathcal{L}_X\omega.$$

Theorem III.7.3. (Divergence theorem I) *For any compactly supported C^1 vector field X on M, and nowhere vanishing $\omega \in \mathfrak{S}^n(M)$, we have*

$$\int_M (\operatorname{Div}_\omega X)\,\omega = 0; \tag{III.7.3}$$

and

$$\iint_{\Omega} (\operatorname{Div}_\omega X)\,\omega = \int_{\partial\Omega} \mathrm{i}(X)\omega, \tag{III.7.4}$$

for any Ω with smooth boundary.

Green's Formulae in Riemannian Manifolds

Let M be an n–dimensional Riemannian manifold. Then, the Riemannian metric induces a natural bundle isomorphism $\theta : TM \to TM^*$ given by

$$\theta(\xi)(\eta) = \langle \xi, \eta \rangle,$$

for all $p \in M$ and $\xi,\ \eta \in M_p$.

Definition. For any C^1 function f on M, the *gradient vector field of f on M*, $\operatorname{grad} f$, is defined by

$$\operatorname{grad} f = \theta^{-1}(df).$$

That is, for any $\xi \in TM$, we have

$$\langle \operatorname{grad} f, \xi \rangle = df(\xi) = \xi f.$$

For C^1 functions $f,\ h$ on M, we have

$$\operatorname{grad}(f + h) = \operatorname{grad} f + \operatorname{grad} h,$$

and

$$\operatorname{grad} fh = f \operatorname{grad} h + h \operatorname{grad} f.$$

If $x : U \to \mathbb{R}^n$ is a chart on M then

$$\operatorname{grad} f = \sum_{j,k} \frac{\partial(f \circ x^{-1})}{\partial x^j} g^{jk} \frac{\partial}{\partial x^k}.$$

Let ∇ denote the Levi-Civita connection of the Riemannian metric.

Definition. For any C^1 vector field X on M, we define the *divergence of X with respect to the Riemannian metric*, div X, by

$$\operatorname{div} X = \operatorname{tr}(\xi \mapsto \nabla_\xi X).$$

For the C^1 function f and vector fields X, Y on M, we have

$$\operatorname{div}(X + Y) = \operatorname{div} X + \operatorname{div} Y,$$

and

$$\operatorname{div} fX = \langle \operatorname{grad} f, X\rangle + f \operatorname{div} X.$$

Proposition III.7.1. *Assume also that M is oriented, σ the volume form on M. Then,*

$$\operatorname{div} X = \operatorname{Div}_\sigma X$$

for all C^1 vector fields X on M.

Proof. Let $x : U \to \mathbb{R}^n$ be a positively oriented chart on M, and

$$X|U = \sum_j \xi^j \frac{\partial}{\partial x^j}.$$

Then,

$$\begin{aligned}
(\operatorname{Div}_\sigma X)\sigma &= \mathcal{L}_X \sigma \\
&= (d\mathrm{i}(X))\sigma \\
&= d(\mathrm{i}(X)\sqrt{g}\, dx^1 \wedge \cdots \wedge dx^n) \\
&= d\left(\sum_{j=1}^n (-1)^{j-1} \sqrt{g}\xi^j\, dx^1 \wedge \cdots \wedge \widehat{dx^j} \wedge \cdots \wedge dx^n\right) \\
&= \sum_{j=1}^n (-1)^{j-1} d(\sqrt{g}\xi^j) \wedge dx^1 \wedge \cdots \wedge \widehat{dx^j} \wedge \cdots \wedge dx^n \\
&= \sum_{j=1}^n \frac{\partial(\sqrt{g}\xi^j)}{\partial x^j} dx^1 \wedge \cdots \wedge dx^n \\
&= \left\{ \frac{1}{\sqrt{g}} \sum_{j=1}^n \frac{\partial(\sqrt{g}\xi^j)}{\partial x^j} \right\} \sigma,
\end{aligned}$$

that is,

$$\mathrm{Div}_\sigma X = \frac{1}{\sqrt{g}} \sum_{j=1}^{n} \frac{\partial(\sqrt{g}\xi^j)}{\partial x^j}.$$

On the other hand,

$$\nabla_{\partial/\partial x^j} X = \sum_\ell \left\{ \frac{\partial \xi^\ell}{\partial x^j} + \sum_k \Gamma_{kj}{}^\ell \xi^k \right\} \frac{\partial}{\partial x^\ell},$$

which implies

$$\mathrm{div}\, X = \sum_\ell \left\{ \frac{\partial \xi^\ell}{\partial x^\ell} + \sum_k \Gamma_{k\ell}{}^\ell \xi^k \right\},$$

and

$$\begin{aligned} \Gamma_{k\ell}{}^\ell &= \frac{1}{2} \sum_{r,\ell} g^{\ell r} \{\partial_k g_{r\ell} + \partial_\ell g_{kr} - \partial_r g_{k\ell}\} \\ &= \frac{1}{2} \sum_{r,\ell} g^{\ell r} \partial_k g_{r\ell} \\ &= \frac{1}{2} \frac{\partial_k g}{g} \\ &= \frac{\partial_k \sqrt{g}}{\sqrt{g}} \end{aligned}$$

(the third equality follows from Proposition II.8.2). So,

$$\mathrm{div}\, X = \sum_\ell \left\{ \frac{\partial \xi^\ell}{\partial x^\ell} + \frac{\xi^\ell}{\sqrt{g}} \frac{\partial \sqrt{g}}{\partial x^\ell} \right\} = \frac{1}{\sqrt{g}} \sum_{j=1}^{n} \frac{\partial(\sqrt{g}\xi^j)}{\partial x^j},$$

which implies the claim. ■

One verifies that if X has compact support on M then, without requiring any orientability of M, we have the *Riemannian divergence theorem*:

(III.7.5) $$\int_M \mathrm{div}\, X \, dV = 0.$$

Definition. Let f be a C^2 function on M. Then, we define the *Laplacian of f*, Δf, by

$$\Delta f = \mathrm{div}\,\mathrm{grad}\, f.$$

The function f is said to be *harmonic* if its Laplacian vanishes identically on M.

Thus, in a chart $x : U \to \mathbb{R}^n$, we have

$$\Delta f = \frac{1}{\sqrt{g}} \sum_{j,k=1}^{n} \frac{\partial}{\partial x^j} \left\{ \sqrt{g} g^{jk} \frac{\partial (f \circ x^{-1})}{\partial x^k} \right\}.$$

Furthermore, for C^2 functions f and h on M, we have

$$\Delta(f + h) = \Delta f + \Delta h,$$

and

$$\operatorname{div} f \operatorname{grad} h = f \Delta h + \langle \operatorname{grad} f, \operatorname{grad} h \rangle,$$

which implies

$$\Delta\, fh = f \Delta h + 2\langle \operatorname{grad} f, \operatorname{grad} h \rangle + h \Delta f.$$

Theorem III.7.4. (Green's formulae I) *Let $f : M \to \mathbb{R} \in C^2(M)$, $h : M \to \mathbb{R} \in C^1(M)$, with at least one of them compactly supported. Then,*

(III.7.6)
$$\int_M \{h \Delta f + \langle \operatorname{grad} h, \operatorname{grad} f \rangle\}\, dV = 0.$$

If both f and h are C^2, then

(III.7.7)
$$\int_M \{h \Delta f - f \Delta h\}\, dV = 0.$$

Corollary III.7.1. *The only compactly supported harmonic functions are the constant functions.*

Theorem III.7.5. (Divergence theorem II) *Let M be oriented, Ω a domain in M with smooth boundary $\partial\Omega$, ν the outward unit vector field along $\partial\Omega$ which is pointwise orthogonal to $\partial\Omega$ (there is only one such vector field). Then, for any compactly supported C^1 vector field X on M we have*

(III.7.8)
$$\iint_\Omega \operatorname{div} X\, dV = \int_{\partial\Omega} \langle X, \nu \rangle\, dA.$$

Proof. It suffices to show that if σ is the volume form of dV, and τ that of dA, then

(III.7.9)
$$\mathrm{i}(X)\sigma = \langle X, \nu \rangle \tau$$

on all of $\partial\Omega$.

Lemma III.7.2. *Let M be oriented, σ the volume form of dV. If $p \in M$, and $\{e_1, \ldots, e_n\}$ is a positively oriented orthonormal basis of M_p with dual basis $\{\omega^1, \ldots, \omega^n\}$, then*

$$\sigma = \omega^1 \wedge \cdots \wedge \omega^n.$$

Proof. Let $x : U \to \mathbb{R}^n$ be a positively oriented chart about p. Then there exist unique matrices A and B satisfying

$$e_j = \sum_k A_j{}^k \frac{\partial}{\partial x^k}, \qquad \omega^r = \sum_s dx^s \, B_s{}^r.$$

Of course, $B = A^{-1}$. But then

$$I = AGA^T,$$

which implies

$$\sqrt{g} = \det B.$$

Therefore, by (III.7.2),

$$\begin{aligned} \omega^1 \wedge \cdots \wedge \omega^n &= (\det B)\, dx^1 \wedge \cdots \wedge dx^n \\ &= \sqrt{g}\, dx^1 \wedge \cdots \wedge dx^n \\ &= \sigma. \end{aligned}$$

■

Corollary III.7.2. *For any $\xi \in M_p$, we have*

$$\mathrm{i}(\xi)\sigma = \sum_j (-1)^{j-1} \langle \xi, e_j \rangle \omega^1 \wedge \cdots \wedge \widehat{\omega^j} \wedge \cdots \wedge \omega^n.$$

The proof of (III.7.9) is now immediate. ■

Theorem III.7.6. (Green's formulae II) *Given M, Ω, and ν as in the divergence theorem* (II), *and given $f \in C^2(M)$, $h \in C^1(M)$, at least one of them compactly supported. Then,*

$$\text{(III.7.10)} \qquad \iint_\Omega \{h\Delta f + \langle \operatorname{grad} f, \operatorname{grad} h \rangle\}\, dV = \int_{\partial\Omega} h \langle \nu, \operatorname{grad} f \rangle \, dA.$$

If both f and h are C^2, then

$$\text{(III.7.11)} \qquad \iint_\Omega \{h\Delta f - f\Delta h\}\, dV = \int_{\partial\Omega} \{h\langle \nu, \operatorname{grad} f \rangle - f \langle \nu, \operatorname{grad} h \rangle\}\, dA.$$

§III.8. Notes and Exercises

Riemannian Symmetric Spaces

Exercise III.1. Use geodesic spherical coordinates to prove the:

(Local) Cartan–Ambrose–Hicks Theorem. *Suppose we are given Riemannian manifolds M_1, M_2 of the same dimension, $p_j \in M_j$, and $\delta \in (0, \min\{\text{inj } p_1, \text{inj } p_2\})$. Let*

$$\iota : (M_1)_{p_1} \to (M_2)_{p_2}$$

denote some fixed linear isometry, and map

$$\phi : B(p_1 : \delta) \to B(p_2; \delta)$$

by

$$\phi(q) = \exp \circ \iota \circ (\exp |\mathsf{B}(p_1; \delta))^{-1}(q). \tag{III.8.1}$$

For $\xi_j \in \mathsf{S}_{p_j}$, $j = 1, 2$, let $\mathcal{R}_{\xi_j}(t)$ denote the respective curvature map of M_{p_j} to itself as given by (III.1.3), *and assume that for all $\xi \in \mathsf{S}_{p_1}$, $t \in (0, \delta)$, we have*

$$\mathcal{R}_{\iota(\xi)}(t) \circ \iota = \iota \circ \mathcal{R}_\xi(t).$$

Then, ϕ is an isometry of $B(p_1; \delta)$ onto $B(p_2; \delta)$.

(See Cartan (1946), Ambrose (1956), Hicks (1965).) Note that we used precisely this argument to prove Theorem II.8.2 and employed it in the case of equality in the Bishop and Gromov theorems.

Definition. We say that a Riemannian manifold M is *locally symmetric* if $\nabla R = 0$ on all of M.

Exercise III.2.

(a) Show that the Riemannian manifold M is locally symmetric if and only if for every geodesic $\gamma(t)$, and parallel vector fields $X_1, \ldots, X_4$ along γ, one has

$$\langle R(X_1, X_2)X_3, X_4\rangle(t) = \text{const.}$$

(b) Let M be Riemannian locally symmetric. Show that for any unit speed geodesic γ there exist constants $\kappa_1, \ldots \kappa_{n-1}$, and parallel pointwise orthonormal vector fields $E_1, \ldots E_{n-1}$ along γ, such that the space of Jacobi fields along γ, pointwise orthogonal to γ, is spanned by Jacobi fields of the form

$$Y_j(t) = \{\alpha_j \mathbf{C}_{\kappa_j}(t) + \beta_j \mathbf{S}_{\kappa_j}(t)\}E_j(t), \qquad j = 1, \ldots, n-1.$$

(c) Let M_1, M_2 be Riemannian locally symmetric of the same dimension, $p_j \in M_j$, and $\delta \in (0, \min\{\operatorname{inj} p_1, \operatorname{inj} p_2\})$. Suppose we have a linear isometry

$$\iota : (M_1)_{p_1} \to (M_2)_{p_2}$$

such that

$$\iota \circ R(\xi, \eta)\zeta = R(\iota(\xi), \iota(\eta))\iota(\zeta),$$

for all $\xi, \eta, \zeta \in M_{p_1}$. Show that ϕ defined by (III.8.1) is an isometry of $B(p_1; \delta)$ onto $B(p_2; \delta)$.

Definition. Given a Riemannian manifold M, $p \in M$. We define the *local geodesic symmetry through* p to be the map on $B(p; \operatorname{inj} p)$ given by

$$\mathfrak{s}_p = \exp \circ -\operatorname{id}_{M_p} \circ (\exp |\mathsf{B}(p; \operatorname{inj} p))^{-1}.$$

Exercise III.3. Prove that M is locally symmetric if and only if the local geodesic symmetry $\mathfrak{s}_p$ is an isometry of $B(p; \operatorname{inj} p)$ onto itself, for every $p \in M$.

Definition. Given a Riemannian manifold M. We say that M is *Riemannian symmetric* if, for each $p \in M$, the local geodesic symmetry $\mathfrak{s}_p$ through p can be extended to a global isometry of M.

Exercise III.4.

(a) Prove that given a Riemannian manifold with involutive isometry ϕ to itself (i.e., $\phi^2 = \operatorname{id}_M$) for which p is an *isolated* fixed point, then ϕ is the geodesic symmetry through p.

(b) Given M Riemannian symmetric. Show that if $\gamma : \mathbb{R} \to M$ is a geodesic, $\gamma(0) = p$, $\gamma(\alpha) = q$, then

$$(\mathfrak{s}_q \circ \mathfrak{s}_p)(\gamma(t)) = \gamma(t + 2\alpha)$$

for all t. Also show that if X is a parallel vector field along γ, then

$$(\mathfrak{s}_q \circ \mathfrak{s}_p)_* X(t) = X(t + 2\alpha)$$

for all t. Furthermore, M is complete, and, therefore, Riemannian homogeneous, that is, to each p and q in M, there exists an isometry ϕ of M onto itself such that $\phi(p) = q$. Finally, show that if $\gamma : \mathbb{R} \to M$ is a geodesic for which there exists $L \in \mathbb{R}$ such that $\gamma(L) = \gamma(0)$, then γ is L–periodic.

Exercise III.5. Given M a symmetric space.

(a) Let $p \in M$, ξ a unit tangent vector in M_p, $\gamma(t) = \exp t\xi$. Show that the family of isometries

$$\Gamma_\xi = \{\phi_t = \mathfrak{s}_{\gamma(t/2)} \circ \mathfrak{s}_p : t \in \mathbb{R}\}$$

is a 1–parameter group of isometries of M.

(b) Assume M has a periodic geodesic. Show that sectional curvatures along the geodesic are nonnegative.

(c) Similarly, show that if M is compact then all sectional curvatures of M are nonnegative.

Note III.1. The study of symmetric spaces is an immense subject, initiated and developed by É. Cartan. Standard works include Helgason (1962), Kobayashi–Nomizu (1969, Vol. II); for symmetric spaces of strictly positive curvature from a Riemannian perspective, see Chavel (1972, Chapters III and IV), Cheeger–Ebin (1975, Chapter III), Besse (1978, Chapter III); and for a more extended Riemannian approach, see Klingenberg (1982, pp. 141–158).

To solve Jacobi's equations (Exercise III.3) on the more general naturally reductive Riemannian homogeneous spaces (Example I.9.5), see Chavel (1967; 1972, Chapter III).

The Cut Locus

Note III.2. The first explicit discussion of the cut locus seems to be in Poincaré (1905). For introductory work on the cut locus, see Myers (1935), Weinstein (1968), and Bishop (1977). For the conjugate locus, see Warner (1965).

Two-Point Homogeneous Spaces

Recall that we say that a connected Riemann manifold M is *homogeneous,* or *the collection of isometries of M acts transitively on M,* if to each $p, q \in M$ there exists an isometry ϕ of M, such that $\phi(p) = q$.

Of course, the collection $I(M)$ of isometries of *any* Riemannian manifold M is a group; less trivial, $I(M)$ is a Lie group (Myers–Steenrod (1939)). Now, for any $\phi \in I(M)$, one has the associated action of ϕ_* on TM. Should ϕ leave a point p invariant, then $\phi_{*|M_p}$ is an orthogonal transformation of M_p to itself.

Exercise III.6.

(a) Let H_p denote the *isotropy group of* p, that is, the collection of isometries that leave p fixed. Prove that if M is complete, then H_p is isomorphic to a subgroup of $\mathcal{O}(n)$ ($n = \dim M$).

(b) Assume $G = I(M)$ acts transitively on M. Prove that M is complete.

(c) As in (b), assume $I(M)$ acts transitively on M. Show that M may be represented as the left cosets G/H_p. Also show that, for any $p, q \in M$, one has H_p, H_q *conjugate in* G, that is, there exists $g \in G$ such that $H_q = gH_pg^{-1}$.

More is known, namely, that when G acts transitively on M then the space M is diffeomorphic to G/H_p, for any $p \in M$ (Warner (1971, p. 123)).

Once one is considering the Riemannian homogeneous space $M = G/H$, one naturally inquires as to transitivity of the associated action of G on TM, given by

$$(\tau_g)_* : TM \to TM, \qquad g \in G,$$

where τ_g denotes the left translation of G/H associated to g (as described in Example I.9.5). Since τ_g is an isometry, the action cannot be transitive since lengths of tangent vectors are preserved by $(\tau_g)_*$; rather, one should really consider the associated action of G on $\mathsf{S}M$. Certainly, if the action of G on $\mathsf{S}M$ is transitive on the collection of orthonormal *pairs* of vectors in $\mathsf{S}M$, then M has constant sectional curvature. The weaker hypothesis is therefore that one only knows that the associated action of G is transitive (not knowing to what extent) on $\mathsf{S}M$.

Definition. A connected Riemannian manifold M is said to be *two-point homogeneous* if the associated action of $I(M)$ on $\mathsf{S}M$ is transitive.

Exercise III.7. Show that the following are equivalent:

(a) M is two-point homogeneous.

(b) Given any p_1, p_2 and q_1, q_2 in M such that $d(p_1, p_2) = d(q_1, q_2)$, there exist an isometry $g \in I(M)$ such that $g \cdot p_j = q_j$, $j = 1, 2$.

(c) For any $p \in M$, the isotropy group H_p acts transitively on $S(p; r)$, for all $r > 0$.

Note III.3. H. C. Wang (1952) and J. Tits (1955) proved, using classification arguments, that two-point homogeneous spaces are Riemannian symmetric. Direct proofs of this theorem were given for the noncompact case in Helgason (1959) and Nagano (1959). A direct proof for the compact case was given recently, in Szabo (1990, 1991).

On the Riemannian Measure

Exercise III.8. Given p in the Riemannian manifold M, $\sqrt{g_{\mathrm{rnc}}}$ the volume density of the Riemannian measure in Riemann normal coordinates on the geodesic disk $B(p; \mathrm{inj}\, p)$. Show that

$$\sqrt{g_{\mathrm{rnc}}}(\exp t\xi) = t^{1-n}\sqrt{\mathbf{g}}(t;\xi)$$

for all $t \in (0, c(\xi))$, $\xi \in \mathsf{S}_p$.

Exercise III.9. (Folk result) Given a Riemannian manifold M, $p \in M$, $\xi \in \mathsf{S}_p$, $q = \exp \rho\xi$, with $\rho < c(\xi)$. Show that

$$\det \mathcal{A}(\rho;\xi) = \det \mathcal{A}(\rho; -\gamma_\xi{}'(\rho)).$$

Note III.4. One could define Riemannian measure abstractly (Gromov (1999)), namely, the functional *n–dimensional volume* of an n–dimensional Riemannian manifold

(i) endows the unit n–cube with volume equal to 1, and

(ii) has the *monotonicity property* that if M_1 admits a one-to-one onto map to M_2, which does not increase distances, then the volume of M_1 is not less than the volume of M_2.

The main result is that Riemannian volume is characterized by these two properties – see Burago–Burago–Ivanov (2001, pp. 193–195).

Burago–Burago–Ivanov (2001, pp. 201–205) also has a proof of *Besikovitch's inequality,* namely, let $I = [0, 1]$, $I^n = [0, 1]^n \subset \mathbb{R}^n$ denote the unit n–cube, endowed with some Riemannian metric g. For each $i = 1, \ldots, n$, let d_i denote the distance in this metric g between the faces F_i^0 and F_i^1, where F_i^0 (resp. F_i^1) denotes the set of points in I^n whose i–th coordinate is 0 (resp. 1). Then, the g–volume $V_g(I^n)$ of I^n satisfies

$$V_g(I^n) \geq \prod_{i=1}^{n} d_i.$$

Definition. A Riemannian manifold is said to be *harmonic* if to each $p \in M$ there exist $\epsilon_p \in (0, +\infty]$ and a function $\sigma_p : [0, \epsilon_p) \to [0, +\infty)$ such that for every $\xi \in \mathsf{S}M$, we have

$$|\det \mathcal{A}(t;\xi)| = \sigma_{\pi(\xi)}(t).$$

The manifold is referred to as *globally harmonic* if ϵ_p is always equal to $+\infty$. Otherwise, M is called *locally harmonic*.

Exercise III.10.

(a) Show that a two-point homogeneous space is globally harmonic.

(b) Show that σ_p is independent of p, more precisely, show that if M is harmonic there exists a function $\sigma : [0, \sup_p \epsilon_p) \to [0, +\infty)$ such that

$$\sigma_p = \sigma | [0, \epsilon_p).$$

Note III.5. The *Lichnerowicz conjecture* states that a harmonic space is two-point homogeneous. One does not distinguish between local and global harmonic in the conjecture, since the two definitions coincide if the Riemannian manifold is real analytic. But one knows that local harmonic implies real analytic; therefore, the two are equivalent. Most recently, Z. I. Szabo verified the conjecture for compact harmonic spaces with finite fundamental group. See Ruse–Walker–Willmore (1961) for the early classical discussion, Besse (1978, Chapter VI) for a modern introduction, and Szabo (1990) for the solution. Also, counterexamples to the conjecture in the noncompact case were announced in Damek–Ricci (1992).

The Area Formula

Some discussion of Hausdorff measure – including a proof that in the top dimension it is equal to Riemannian measure, and a proof of the area formula – is presented in Chapter IV of Chavel (2001).

The Smooth Coarea Formula

Let M^m, N^n be C^r Riemannian manifolds, $m \geq n$, and $\Phi : M \to N$ a C^1 map from M to N. We want to give an effective calculation of the volume disortion of the map, namely, of

$$\mathcal{J}_\Phi(x) = \sqrt{\det\ \Phi_* \circ (\Phi_*)^{\mathrm{adj}}}.$$

Exercise III.11. Prove

$$\mathcal{J}_\Phi(x) = \begin{cases} 0 & \text{rank } \Phi_* < n \\ \left|\det\,(\Phi_* | (\ker \Phi_*)^\perp)\right| & \text{rank } \Phi_* = n \end{cases}.$$

Let $\{e_A\}$, $A = 1, \ldots, m$, an orthonormal moving frame on M with dual coframe $\{\omega^A\}$, and $\{E_j\}$, $j = 1, \ldots, n$, an orthonormal moving frame on N with dual coframe $\{\theta^j\}$. Then, the local volume forms on M and N are given by $\omega^1 \wedge \cdots \wedge \omega^m$ and $\theta^1 \wedge \cdots \wedge \theta^n$, respectively.

There exist functions $\sigma^j{}_A$ on M such that

$$\Phi^*\theta^j = \sum_A \sigma^j_A \omega^A.$$

Exercise III.12. Prove:

(a) If at some $x \in M$, we have $\dim \ker \Phi_{*|x} > m - n$, then

$$\Phi^*(\theta^1 \wedge \cdots \wedge \theta^n) = 0 \quad \text{at } x.$$

(b) If at some $x \in M$, we have $\dim \ker \Phi_{*|x} = m - n$, then

$$\mathcal{J}_\Phi\, \omega^1 \wedge \cdots \wedge \omega^m = \Phi^*(\theta^1 \wedge \cdots \wedge \theta^n) \wedge \omega^{n+1} \wedge \cdots \wedge \omega^m.$$

(c) Let M, N be C^r Riemannian manifolds, with $m = \dim M \geq \dim N = n$, $r > m - n$, and let $\Phi : M \to N \in C^r$. Then, for any measurable function $f : M \to \mathbb{R}$, which is everywhere nonnegative or is in $L^1(M)$, one has

$$\int_M f \mathcal{J}_\Phi \, dV_m = \int_N dV_n(y) \int_{\Phi^{-1}[y]} (f | \Phi^{-1}[y])\, dV_{m-n},$$

where, for any k, dV_k denotes k–dimensional volume.

(d) Let M^m be a C^m Riemannian manifold, and let $\Phi : M \to \mathbb{R} \in C^m$. Then, for any measurable function $f : M \to \mathbb{R}$, which is everywhere nonnegative or is in $L^1(M)$, one has

$$\int_M f |\text{grad}\, \Phi|\, dV = \int_{\mathbb{R}} dy \int_{\Phi^{-1}[y]} (f | \Phi^{-1}[y])\, dA.$$

(e) Let M^{k-1} be a hypersurface in $\mathbb{R}^k$, given by the graph of a C^1 function $\phi : G \to \mathbb{R}$, where G is open in $\mathbb{R}^{k-1}$; so M is given by

$$x^k = \phi(x^1, \ldots, x^{k-1}), \qquad (x^1, \ldots, x^{k-1}) \in G.$$

Then, the surface area element on M, dA, is given by

$$dA = \sqrt{1 + |\text{grad}_{k-1} \phi|^2}\, d\mathbf{v}_{k-1},$$

where $d\mathbf{v}_{k-1}$ denotes Lebesgue measure on $\mathbb{R}^{k-1}$, and grad_{k-1} denotes the gradient of functions on $\mathbb{R}^{k-1}$.

(f) If $\Omega \subset\subset \mathbb{R}^n$ is a domain with C^1 boundary, ν the exterior unit normal vector field along $\partial\Omega$, and, for a given $\xi \in \mathbb{S}^{n-1}$, $p_\xi : \partial\Omega \to \xi^\perp$ is the projection, then

$$\int_{\partial\Omega} |\nu_w \cdot \xi|\, dA(w) = \int_{\xi^\perp} \text{card}\,(\partial\Omega \cap p_\xi{}^{-1}[y])\, d\mathbf{v}_{n-1}(y).$$

(g) If $\Omega \subset\subset \mathbb{R}^n$ is a domain with C^1 boundary, and for every $\xi \in \mathbb{S}^{n-1}$, $p_\xi : \partial\Omega \to \xi^\perp$ denotes the projection, then

$$A(\partial\Omega) = \frac{1}{2\boldsymbol{\omega}_{\mathbf{n-1}}} \int_{\mathbb{S}^{n-1}} d\mu_{n-1}(\xi) \int_{\xi^\perp} \operatorname{card}(\partial\Omega \cap p_\xi{}^{-1}[y])\, d\mathbf{v}_{n-1}(y).$$

(h) If Ω is convex, $\partial\Omega \in C^1$, then

$$A(\partial\Omega) = \frac{1}{\boldsymbol{\omega}_{\mathbf{n-1}}} \int_{\mathbb{S}^{n-1}} \mathbf{v}_{n-1}(p_\xi(\Omega))\, d\mu_{n-1}(\xi).$$

(i) If Ω is convex, $\partial\Omega \in C^1$, and Ω_o is open containing Ω, then

$$A(\partial\Omega) \le A(\partial\Omega_o).$$

The First Variation of Area

We continue as in §III.6, where $\mathfrak{M}$ is a k–dimensional submanifold of the n–dimensional Riemannian manifold M, $0 \le k < n$. Let $\phi : \mathfrak{M} \to M$ denote the inclusion of $\mathfrak{M}$ in M, and let

$$\Phi : \mathfrak{M} \times I \to U$$

be a *variation of* ϕ, that is, Φ is a differentiable map such that: I is an open interval about 0 in $\mathbb{R}$; for each $\epsilon \in I$, $\Phi|\mathfrak{M} \times \{\epsilon\}$ is an imbedding; and $\Phi|\mathfrak{M} \times \{0\} = \phi$.

For each fixed $\epsilon \in I$, we define the imbedding $\phi_\epsilon : \mathfrak{M} \to M$ by

$$\phi_\epsilon(x) = \Phi(x, \epsilon).$$

So,

$$\Phi(\mathfrak{M} \times \{\epsilon\}) = \phi_\epsilon(\mathfrak{M}).$$

Let

$$X := \Phi_* \partial_\epsilon = \xi + \eta,$$

where ∂_ϵ is the natural coordinate vector field on I, and ξ is tangent to $\phi_\epsilon(\mathfrak{M})$, η normal to $\phi_\epsilon(\mathfrak{M})$, for all ϵ. Then, for each fixed ϵ, X can be viewed as a vector field on M along the mapping ϕ_ϵ.

Exercise III.13. Verify that if Σ is a k–form on M, then

$$\frac{d}{d\epsilon} \int_{\phi_\epsilon(\mathfrak{M})} \Sigma = \int_{\mathfrak{M}} \frac{d}{d\epsilon} \phi_\epsilon{}^* \Sigma = \int_{\mathfrak{M}} \phi_\epsilon{}^* \{(d \circ \mathrm{i}(X) + \mathrm{i}(X) \circ d)\, \Sigma\}.$$

Let $p \in \mathfrak{M}$, U a neighborhood of p in M, and

$$e_A : U \times I \to TM, \qquad A = 1, \ldots, n,$$

be vector fields on M such that: for each $(q, \epsilon) \in U \times I$ the collection $\{e_{A|(q,\epsilon)} : A = 1, \ldots, n\}$ is an orthonormal basis of M_q; and such that for each ϵ in I, $e_r|\mathfrak{M} \times \{\epsilon\}$ is tangent to $\phi_\epsilon(\mathfrak{M})$, $r = 1, \ldots, k$, and $e_\alpha|\mathfrak{M} \times \{\epsilon\}$ is normal to $\phi_\epsilon(\mathfrak{M})$, $\alpha = k+1, \ldots, n$. Finally, let $\{\omega^1, \ldots, \omega^n, d\epsilon\}$ be the coframe dual to $\{e_1, \ldots, e_n, \partial_\epsilon\}$ on $U \times I$.

Then for each $\epsilon \in I$,

$$A_k(\epsilon) = \int_{\phi_\epsilon(\mathfrak{M})} \omega^1 \wedge \cdots \wedge \omega^k = \int_{\mathfrak{M}} {\phi_\epsilon}^*(\omega^1 \wedge \cdots \wedge \omega^k)$$

is the k–dimensional area of $\phi_\epsilon(M)$.

Exercise III.14. Prove that if X is compactly supported, then

$$A_k{}'(\epsilon) = -\int_{\phi_\epsilon \mathfrak{M})} \langle \eta, H \rangle\, \omega^1 \wedge \cdots \wedge \omega^k = -\int_{\phi_\epsilon(\mathfrak{M})} \langle \eta, H \rangle\, dV_k,$$

where H denotes the mean curvature vector field of the imbedding of $\mathfrak{M}$ in M (see §II.9), and dV_k denotes k–dimensional Riemannian measure.

In particular, $\mathfrak{M}$ is a stationary point of the k–dimensional area functional if and only if $\mathfrak{M}$ *is minimal in* M, that is, if and only if H vanishes identically on all of $\mathfrak{M}$. Compare the result with Theorems II.4.1 and II.4.2.

For the second variation, see Exercise III.32.

Note III.6. Note that the calculations and final result are valid if ϕ_ϵ is a 1–parameter family of *immersions* of $\mathfrak{M}$ into M.

Exercise III.15. Recall (Exercise II.5) that one cannot imbed a compact orientable surface M of nonpositive Gauss curvature $\mathcal{K}$ isometrically in $\mathbb{R}^3$. Show that, in particular, there are no compact orientable minimal surfaces imbedded in $\mathbb{R}^3$.

Note III.7. Now that we know that there are no compact orientable minimal surfaces in $\mathbb{R}^3$ (we shall show in Exercise III.29 that there are no *immersed* compact minimal surfaces in $\mathbb{R}^3$), one might ask whether there are any complete noncompact examples. For a long time, the only available examples of such minimal surfaces in $\mathbb{R}^3$ with finite topological type were the plane, the catenoid, and the helicoid. However, C. J. Costa (1984) constructed a new example of a complete noncompact immersed minimal surface in $\mathbb{R}^3$, and D. Hoffman and

W. Meeks (1985) subsequently showed this example to be imbedded. For recent papers, to work one's way back, see Hoffman–Karcher (1997).

Hypersurfaces of Constant Mean Curvature

Let M be our Riemannian manifold, Ω a relatively compact domain in M with smooth boundary $\mathfrak{M}$, and exterior unit normal vector field ν along $\mathfrak{M}$. Then, we may write

$$H = \mathfrak{h}\nu,$$

where $\mathfrak{h}$ is now the *mean curvature function* of the compact hypersurface $\mathfrak{M}$. Furthermore, the vector field η on $\mathfrak{M}$ of any normal variation of $\mathfrak{M}$ is given by

$$\eta = \varphi\nu.$$

So, the first variation formula (Exercise III.14) reads as

$$A'(0) = -\int_{\mathfrak{M}} \varphi\mathfrak{h}\, dA.$$

Exercise III.16. Consider the variation $\Psi : \overline{\Omega} \times I \to M$ (where I is an open interval in $\mathbb{R}$ containing the origin), $\Psi_\epsilon(x) = \Psi(x, \epsilon)$, of the closed domain $\overline{\Omega}$, that is, $\Psi|\overline{\Omega} \times \{0\} = \mathrm{id}_{\overline{\Omega}}$; and set

$$\varphi = \langle (\partial\Psi/\partial\epsilon)_{|\mathfrak{M}\times\{0\}}, \nu\rangle.$$

Denote the variation of volume of Ω by

$$V(\epsilon) = \int_{\Psi_\epsilon(\Omega)} dV,$$

and prove

$$V'(0) = \int_{\mathfrak{M}} \varphi\, dA.$$

Exercise III.17. So, the previous exercise says that if the variation Ψ preserves volume, then

$$\int_{\mathfrak{M}} \varphi\, dA = 0.$$

Prove the converse, namely, prove that given any function $f : \mathfrak{M} \to \mathbb{R}$ satisfying

$$\int_{\mathfrak{M}} f\, dA = 0,$$

there exists a vector field Z on M whose flow Σ_ϵ on M preserves the volume of Ω, that is, $V(\Sigma_\epsilon(\Omega)) = V(\Omega)$, for sufficiently small ϵ, and satisfies $Z|\mathfrak{M} = f\nu$.

Exercise III.18. Prove that if $\mathfrak{M}$ has minimum $(n-1)$–area among all boundaries of domains with volume equal to that of Ω, then $\mathfrak{M}$ has constant mean curvature, that is, $\mathfrak{h}$ is identically constant on $\mathfrak{M}$.

The Gauss Map

Exercise III.19. Let M be an oriented surface in $\mathbb{R}^3$, that is, M is an imbedded 2–dimensional submanifold of $\mathbb{R}^3$ with inherited Riemannian metric, and smooth unit normal vector field $\mathbf{n}$ on M. Consider $\mathbf{n}$ as a smooth map of M to $\mathbb{S}^2$. Let $\sigma_{\mathbb{S}^2}$ denote the area 2–form on $\mathbb{S}^2$, σ the area 2–form on M, and $\mathcal{K}$ the Gauss curvature function on M. Prove that

$$\mathbf{n}^*(\sigma_{\mathbb{S}^2}) = \mathcal{K}\,\sigma.$$

Note III.8. For discussion of the Gauss map in greater generality, start with Hoffman–Osserman (1980) and Osserman (1980).

On the Günther–Bishop Theorems

Note III.9. Note that, for the volume comparison theorems, we only cited Günther (1960) when discussing the lower bound of $V(x;r)$ under the hypothesis of an upper bound on sectional curvature, and the restriction $r < \operatorname{inj} x$ (Theorems III.4.1 and III.4.2). Günther also considered comparison theorems for upper bounds on $V(x;r)$, but, he assumed that the *sectional* curvature – not just the Ricci curvature – was bounded below, and he only considered $r < \operatorname{inj} x$. Thus, for upper bounds, the Bishop theorems were a significant improvement.

Note III.10. P. Kröger (2004) has given a proof of the Günther theorem – that is, when all sectional curvatures are bounded from above by a constant δ, and the disks under consideration have radius less than the injectivity radius of their common center – in the spirit of the Günther–Bishop theorem with the Ricci curvature bounded from below.

Namely, along every geodesic $\gamma(t) = \exp t\xi$, he proves the inequality

$$\frac{d^2}{dt^2}\left(\frac{\mathcal{A}(t;\xi)}{\mathbf{S}_\delta(t)^{n-2}}\right) + \kappa\frac{\mathcal{A}(t;\xi)}{\mathbf{S}_\delta(t)^{n-2}} \geq 0,$$

note the exponent of $\mathbf{S}_\delta(t)$! Once one has the inequality, then for

$$\phi(t) := \frac{\mathcal{A}(t;\xi)}{\mathbf{S}_\delta(t)^{n-2}}$$

we have

$$\phi'' + \kappa\phi \geq 0, \qquad \phi(0) = 0,$$

which implies $\phi'\mathbf{S}_\delta - \phi\mathbf{S}'_\delta \geq 0$, which then implies the result.

Note III.11. Assume M is complete, with Ricci curvature bounded from below by $(n-1)\kappa$. Fix $x \in M$, $V(r) := V(x;r)$. Then, the Gromov theorem implies that $V'(r)/\mathbf{S}_\kappa^{n-1}(r)$ is a decreasing function of r. In particular, $V'(r)$ has right- and left-handed limits at every value of r. Then, R. Grimaldi and P. Pansu (1994) have shown that

$$\lim_{s\downarrow r} V'(s) - \lim_{s\uparrow r} V'(s) = -2\mathcal{H}^{n-1}(S(x;r) \cap C(x)),$$

where $\mathcal{H}^{n-1}$ is $(n-1)$–dimensional Hausdorff measure, and $C(x)$ is the cut locus of x. (See the proof of Proposition III.5.1.)

In particular, if, in addition, M is noncompact and real analytic, then $V(r) \in C^1$ for all $r > 0$.

On Volumes of Disks

Exercise III.20. (E. Calabi, S. T. Yau) Show that if M is complete noncompact, with nonnegative Ricci curvature, then for any $x \in M$ we have

$$V(x; R) \geq \text{const.}_x\, R$$

as $R \uparrow +\infty$. Note that the result is sharp, in the sense that for M consisting of the flat cylinder $M = \mathbb{S}^{n-1} \times \mathbb{R}$, we have

$$V(x; R) \sim 2\mathbf{c}_{\mathbf{n-1}}R$$

as $R \uparrow +\infty$.

Note III.12. Other results and examples can be found in Croke–Karcher (1988).

Fermi Coordinates

Exercise III.21. (Remark in a discussion with J. Velling) Show that if $\mathfrak{M}$ is compact, M has nonpositive sectional curvature, and r the distance function on M based on $\mathfrak{M}$, as described in §III.6, then there exists $\rho_0 > 0$ such that for all $\rho > \rho_0$, the level surface $r^{-1}[\rho]$ is an immersed differentiable $(n-1)$–manifold.

Exercise III.22. (See Velling (1999)) Use Theorem II.2.1 and (III.6.1) to show that if M is a 3–dimensional space form $\mathbb{M}_\kappa$, and $\mathfrak{M}$ is a geodesic in $\mathbb{M}_\kappa$, then the cylinder $S(\mathfrak{M}; r)$ always has Gauss curvature equal to 0.

Exercise III.23. Prove the following result of T. Frankel (1961):

Theorem. *Let M be a complete connected Riemannian manifold of positive sectional curvature, and let $\mathfrak{M}_1$ and $\mathfrak{M}_2$ be two compact totally geodesic submanifolds of M. If*

$$\dim \mathfrak{M}_1 + \dim \mathfrak{M}_2 \geq \dim M,$$

then $\mathfrak{M}_1$ and $\mathfrak{M}_2$ have nonempty intersection.

H. Weyl's Formula for the Volume of Tubes

Note III.13. We refer the reader to a result of H. Weyl (1939) which states that, for any k–dimensional submanifold $\mathfrak{M}$ of $\mathbb{R}^n$ with compact closure, $k < n$, the volume of the tubular neighborhood $B(\mathfrak{M}; r)$ of $\mathfrak{M}$ of radius r is given by a polynomial in r with coefficients depending only on the *intrinsic* geometry of $\mathfrak{M}$, not on the imbedding. For a full treatment of the result, with further developments, see the survey Gray (1990).

Geometric Interpretation of the Riccati Equation

The following view has been effectively emphasized in Eschenburg (1987), Grove (1987), and Karcher (1989).

Exercise III.24. Given $\mathfrak{M}$ a connected k–dimensional submanifold of M, then for any $\xi \in \nu\mathsf{S}\mathfrak{M}$, $t > 0$, $\mathsf{U}^{-\gamma_\xi{}'(t)}$ will denote the Weingarten map (here we change the notation for the Weingarten map – the reason will immediately become obvious) of $S(\mathfrak{M}; t)_{\gamma_\xi(t)}$ at a regular point $\gamma_\xi(t)$ of $S(\mathfrak{M}; t)$.

(a) Show that if Y is any transverse Jacobi field along γ_ξ, then

$$(\nabla_t Y)(t) = \mathsf{U}^{-\gamma_\xi{}'(t)} Y(t).$$

(b) Show that if

$$\mathcal{U} = \mathcal{A}'\mathcal{A}^{-1},$$

where $\mathcal{A}$ is given as in §III.6, then

$$\mathcal{U} = \tau_t{}^{-1} \circ \mathsf{U}^{-\gamma_\xi{}'(t)} \circ \tau_t,$$

where τ_t denotes parallel translation along γ_ξ from $M_{\gamma_\xi(0)}$ to $M_{\gamma_\xi(t)}$, and $\mathcal{U}$ satisfies the Riccati equation (III.4.16)

$$\mathcal{U}' + \mathcal{U}^2 + \mathcal{R} = 0.$$

Note that the self-adjointness of $\mathcal{U}$ is now immediate.

Exercise III.25. (See Velling (1999)) Show that if M is n–dimensional hyperbolic space, and $\mathfrak{M}$ a k–dimensional submanifold of M, then for any $\xi \in \nu S\mathfrak{M}$, the Weingarten map $\mathsf{U}^{-\gamma_\xi{}'(t)}$ of $S(\mathfrak{M};t)_{\gamma_\xi(t)}$ is asymptotic to the identity map I of $S(\mathfrak{M};t)_{\gamma_\xi(t)}$, as $t \uparrow +\infty$.

Manifolds With No Conjugate Points

Exercise III.26. Let M be a complete Riemannian manifold with no conjugate points. For any $x, y \in M$, let $N_T(x, y)$ denote the number of geodesic segments of length $\leq T$ joining x to y.

(a) Show that $N_T(x, y)$ is finite for all x, y, T.

(b) Prove

$$\int_M N_T(x, y)\, dV(y) = \int_0^T dt \int_{\mathsf{S}_x} \sqrt{\mathbf{g}}(t;\xi)\, d\mu_x(\xi).$$

The Laplacian

Exercise III.27. Note that we have given two definitions of the Laplacian of a function: one here just prior to Theorem III.7.4, and the other just prior to Exercise II.9. Prove that the two coincide, that is, prove that, for a C^2 function on a Riemannian manifold M, we have

$$\operatorname{div}\operatorname{grad} f = \operatorname{tr}\operatorname{Hess} f.$$

Exercise III.28. Let M be a complete Riemannian manifold, $o \in M$. Then, for functions f defined on M, we define the *averaging operator* av_o *based at* o by

$$(\mathsf{av}_o f)(x) = A(o;r)^{-1} \int_{S(o;r)} f\, dA, \qquad r = d(o,x).$$

Show that M is harmonic if and only if the Laplacian commutes with the averaging operator at o, that is,

$$\Delta \circ \mathsf{av}_o = \mathsf{av}_o \circ \Delta,$$

for all $o \in M$.

Exercise III.29.

(a) Prove that an immersed submanifold M of the Euclidean space $\mathbb{R}^m$ is minimal in $\mathbb{R}^m$ if and only if the restriction of the natural coordinate functions of $\mathbb{R}^m$ to M are all harmonic on M (i.e., their Laplacian – relative to the metric on M – vanishes identically on M).

(b) Prove that there are no compact minimal immersed submanifolds of $\mathbb{R}^m$. (See Exercise III.15)

Exercise III.30. In these problems, we think of $\mathbf{x}$ as a tangent vector at the position x.

(a) Prove *Minkowski's formula*: If $\Omega \subset\subset \mathbb{R}^n$ is a domain with differentiable boundary, then

$$V(\Omega) = \frac{1}{n} \int_{\partial\Omega} \mathbf{x} \cdot \mathbf{n}\, dA,$$

where $\mathbf{n}$ is the unit normal exterior vector field of Ω in M along $\partial\Omega$.

(b) Prove *Jellet's formula* (J. H. Jellet (1853)): If Ω is a domain in M, an n–dimensional submanifold of $\mathbb{R}^m$, with Ω having compact closure and differentiable boundary Γ in M , then

$$\iint_{\Omega} \{n + \langle \mathbf{x}, H\rangle\}\, dV = \int_{\Gamma} \langle \mathbf{x}, \mathbf{n}\rangle\, dA,$$

where H is the mean curvature vector field of M in $\mathbb{R}^m$.

(c) Prove: For *any* closed hypersurface $\mathcal{S}$ in $\mathbb{R}^n$, we have

$$A(S) = \frac{1}{n} \int_S \mathfrak{h} \langle \mathbf{x}, \mathbf{n}\rangle\, dA,$$

where $\mathbf{n}$ is a unit normal vector field of $\mathcal{S}$ in $\mathbb{R}^n$, and $\mathfrak{h}$ is the mean curvature function of $\mathcal{S}$ relative to $\mathbf{n}$.

(d) Prove: If $\mathcal{S}$ is a closed hypersurface in $\mathbb{R}^n$ with nonzero constant mean curvature $\mathfrak{h}$, then

$$A(\mathcal{S}) = -\frac{1}{\mathfrak{h}} \int_S \|\mathfrak{B}\|^2 \langle \mathbf{x}, \mathbf{n}\rangle\, dA.$$

(e) Also, to complete the apparatus of formuale, we will include Exercise II.10(d)–(e): Prove

$$\Delta\, \mathbf{n} = -\operatorname{grad} \mathfrak{h} - \|\mathfrak{B}\|^2 \mathbf{n},$$

and

$$\Delta\, (\mathbf{x} \cdot \mathbf{n}) = -\mathfrak{h} + \mathbf{x} \cdot \operatorname{grad} \mathfrak{h} - \|\mathfrak{B}\|^2 (\mathbf{x} \cdot \mathbf{n}).$$

The Second Variation of Area

Consider the k–dimensional submanifold $\mathfrak{M}$ of the n–dimensional Riemannian manifold M, with inclusion $\phi : \mathfrak{M} \to M$, and variation $\Phi : \mathfrak{M} \times I \to M$ of ϕ, as described prior to Exercise III.14 (the first variation of area). Again, we write

$$\Phi_*\partial_\epsilon := \xi + \eta,$$

where ξ is tangent, and η is normal, to $\Phi(\mathfrak{M} \times \{\epsilon\})$, for all ϵ. For $p \in \mathfrak{M}$, U a neighborhood of p in M, we consider the vector fields

$$e_A : U \times I \to TM, \qquad A = 1, \ldots, n,$$

on M such that: for each $(q, \epsilon) \in U \times I$ the collection $\{e_{A|(q,\epsilon)} : A = 1, \ldots, n\}$ is an orthonormal basis of M_q; and such that for each ϵ in I, $e_r|\mathfrak{M} \times \{\epsilon\}$ is tangent to $\phi_\epsilon(\mathfrak{M})$, $r = 1, \ldots, k$, and $e_\alpha|\mathfrak{M} \times \{\epsilon\}$ is normal to $\phi_\epsilon(\mathfrak{M})$, $\alpha = k + 1, \ldots, n$. Finally, we let $\{\omega^1, \ldots, \omega^n, d\epsilon\}$ be the coframe dual to $\{e_1, \ldots, e_n, \partial_\epsilon\}$ on $U \times I$.

To pose the problem of calculating the second variation of area $A_k''(0)$, we require a few preliminaries.

Definition. Consider the k–dimensional submanifold $\mathfrak{M}$ of the n–dimensional Riemannian manifold M, with $\mathcal{D}$ the connection in the normal bundle (see §II.9, especially Exercise II.12). For a normal vector field η on $\mathfrak{M}$, define the *Laplacian of* η, $\Delta\eta$, by

$$\Delta\eta := \sum_r (\mathcal{D}\mathcal{D}\eta)(e_r, e_r) = \sum_r \{\mathcal{D}_r{}^2 - \mathcal{D}_{\nabla_r e_r}\}\eta.$$

We define the *normal Ricci transformation* $\mathfrak{Ric} : \nu\mathfrak{M} \to \nu\mathfrak{M}$ to be the bundle map given by

$$\mathfrak{Ric}\,\eta = \sum_r \{R(e_r, \eta)e_r\}^\perp.$$

In what follows, we always let $\mathfrak{B}$ denote the second fundamental form of $\mathfrak{M}$ in M relative to η with Weingarten map $\mathfrak{A}^\eta$ at each point of $\mathfrak{M}$.

Exercise III.31. Show that, for compactly supported η, we have

$$-\int_{\mathfrak{M}} \langle \Delta\eta, \eta\rangle \, dA_k = \int_{\mathfrak{M}} |\mathcal{D}\eta|^2 \, dA_k,$$

where dA_k denotes the Riemannian measure on $\mathfrak{M}$.

Exercise III.32. Prove the second variation formula:

$$A_k''(0) = \int_{\mathfrak{M}} \{|\mathcal{D}\eta|^2 + \langle H, \eta\rangle^2 - \|\mathfrak{B}\|^2 - \langle \nabla_\eta \eta, H\rangle - \langle \mathfrak{Ric}\,\eta, \eta\rangle\}\, dA_k.$$

Exercise III.33. Conclude that if $k = n - 1$, that is, $\mathfrak{M}$ is a hypersurface in M, then any normal vector field η along $\mathfrak{M}$ is given by

$$\eta = \varphi\nu, \qquad \varphi = \varphi(x, \epsilon),$$

where ν is a unit normal vector field along $\phi_\epsilon(\mathfrak{M})$; $H = \mathfrak{h}\nu$, that is, the mean curvature is essentially scalar, in which case the variation formulae read as:

$$\frac{dA}{d\epsilon}(0) = -\int_{\mathfrak{M}} \varphi\mathfrak{h}\, dA,$$

and

$$\frac{d^2A}{d\epsilon^2}(0) = \int_{\mathfrak{M}} \left\{|d\varphi|^2 - \langle \nabla_\eta \eta, \mathfrak{h}\nu\rangle + \varphi^2\{\mathfrak{h}^2 - \|\mathfrak{B}\|^2 - \langle \mathfrak{Ric}\,\nu, \nu\rangle\}\right\} dA,$$

where $\mathfrak{B}$ is the second fundamental form relative to ν.

Note III.14. The approach in Exercise III.14 and in Exercise III.33 via moving frames (as described in our sketches for solutions to the exercises) is highly influenced by S. S. Chern's (1968).

Note III.15. Recall that the hypersurface $\mathfrak{M}$ is minimal in M when $\mathfrak{H}$ vanishes on all of $\mathfrak{M}$. So, one can now consider the stability of such submanifolds, namely, under what circumstances the second variation of area is nonnegative, that is,

$$\int_{\mathfrak{M}} \left\{|d\varphi|^2 - \varphi^2\left(\|\mathfrak{B}\|^2 + \langle \mathfrak{Ric}\,\nu, \nu\rangle\right)\right\} dA \geq 0,$$

for all compactly supported functions φ on $\mathfrak{M}$. For two basic studies, see Barbosa–do Carmo (1976) and Fischer–Colberie–Schoen (1980). See also Osserman (1980).

Exercise III.34. In a similar vein, show that if D is a relatively compact domain in M with smooth boundary $\mathfrak{M} = \partial D$ having constant mean curvature, and then the second variation of area of $\mathfrak{M}$, subject to deformations of Ω that preserve the volume of D is

$$\int_{\partial D} \left\{|d\varphi|^2 - \varphi^2\left(\|\mathfrak{B}\|^2 + \langle \mathfrak{Ric}\,\nu, \nu\rangle\right)\right\} dA,$$

for all functions φ on ∂D satisfying $\int_{\partial D} \varphi\, dA = 0$.

Definition. We say that an immersed closed hypersurface $\Psi : \mathfrak{M} \to M$ of constant mean curvature is *stable* if its "second variation" is nonnegative, that is,

$$\int_{\mathfrak{M}} \left\{|d\varphi|^2 - \varphi^2 \left(\|\mathfrak{B}\|^2 + \langle \mathfrak{Ric}\, \nu, \nu\rangle\right)\right\} dA \geq 0,$$

for all functions φ on $\mathfrak{M}$ satisfying $\int_{\mathfrak{M}} \varphi\, dA = 0$.

Exercise III.35. Prove the following:

Theorem. (Barbosa–do Carmo (1984)) *If* $\mathbf{x} : M^{n-1} \to \mathbb{R}^n$, *$M$ compact, is an immersion with nonzero constant mean curvature, and the immersion is stable, then* $\mathbf{x}(M)$ *is a sphere.*

One can find a recent survey of the case of nonzero constant mean curvature of a hypersurface, with discussion of and references to stability questions, in do Carmo (1989).

§III.9. Appendix: Eigenvalue Comparison Theorems

In this section, we introduce a study of functions on Riemannian manifolds that highlights the interplay of the geometry–topology of the manifold with the analytic properties of the functions under consideration. We give very few details of the general analytic theory. We only highlight those aspects required for application of the volume comparison theorems to the study of analysis. The subject has recently become quite vast, and we recommend the student to Courant–Hilbert (1967) for the early most basic background, to Gilbarg–Trudinger (1977) for the fundamental results on elliptic equations, and to Berger–Gauduchon–Mazet (1974) and Bérard (1986) for the explicitly Riemannian character of the theory. See also Jost (1995). For recent developments, see Buser (1992). Much of what we say here was presented in Chavel (1984), and the reader is referred there for more results and details. More recently, one has Davies–Safarov (1999).

To start, we let $L^2(M)$ denote the space of measurable functions f on M for which

$$\int_M |f|^2\, dV < +\infty.$$

On $L^2(M)$, we consider the usual inner product, and induced norm, given by

$$(f, h) = \int_M f h\, dV, \qquad \|f\|^2 = (f, f),$$

for f, h in $L^2(M)$. With this inner product $L^2(M)$ is a Hilbert space.

One considers, among others, the following two eigenvalue problems.

Closed eigenvalue problem. Let M be compact, connected. Find all real numbers λ for which there exists a nontrivial solution $\phi \in C^2(M)$ to the equation

$$\Delta\phi + \lambda\phi = 0. \tag{III.9.1}$$

Dirichlet eigenvalue problem. For M connected with compact closure and smooth boundary, find all real numbers λ for which there exists a nontrivial solution $\phi \in C^2(M) \cap C^0(\overline{M})$ to (III.9.1) satisfying the boundary condition

$$\phi \mid \partial M = 0. \tag{III.9.2}$$

The desired numbers λ are referred to as *eigenvalues* of Δ, and the vector space of solutions of the eigenvalue problem with given λ – it is a linear problem in both of these instances – its *eigenspace*. The elements of the eigenspace are called *eigenfunctions*. The basic result is:

Theorem III.9.1. *For each of the above eigenvalue problems, the set of eigenvalues consists of a sequence*

$$0 \leq \overline{\lambda}_1 < \overline{\lambda}_2 < \ldots \uparrow +\infty,$$

and each associated eigenspace is finite dimensional. Eigenspaces belonging to distinct eigenvalues are orthogonal in $L^2(M)$, and $L^2(M)$ is the direct sum of all the eigenspaces. Furthermore, each eigenfunction is in $C^\infty(\overline{M})$.

We give a few comments on the theorem. First, as soon as one knows that the eigenfunction $\phi \in C^2(M) \cap C^1(\overline{M})$, one also knows that its eigenvalue must be nonnegative. Indeed, one sets $f = h = \phi$ and applies the appropriate Green formula ((III.7.6) or (III.7.10)) to obtain

$$\lambda = \|\phi\|^{-2} \int_M |\operatorname{grad} \phi|^2 \, dV \geq 0. \tag{III.9.3}$$

From (III.9.3), one has: $\lambda = 0$ implies that ϕ is a constant function. Therefore, in the closed eigenvalue problem, we have $\overline{\lambda}_1 = 0$, and in the Dirichlet eigenvalue problem, we have $\overline{\lambda}_1 > 0$.

Also note that the orthogonality of distinct eigenspaces is a direct consequence of the Green formulae (III.7.7) and (III.7.11). Indeed, let ϕ, ψ be eigenfunctions of the respective eigenvalues λ, τ. Then,

$$0 = \int_M \{\phi\Delta\psi - \psi\Delta\phi\} \, dV = (\lambda - \tau) \int_M \phi\psi \, dV,$$

and the remark follows.

The dimension of each eigenspace is referred to as the *multiplicity of the eigenvalue*. It will be convenient to (henceforth) list the eigenvalues as

$$0 \leq \lambda_1 \leq \lambda_2 \leq \cdots \uparrow +\infty,$$

with each eigenvalue repeated according to its multiplicity.

If $\{\phi_1, \phi_2, \ldots\}$ is an orthonormal sequence in $L^2(M)$ of eigenfunctions so that ϕ_j is an eigenfunction of λ_j for each $j = 1, 2, \ldots$, then $\{\phi_1, \phi_2, \ldots\}$ is a complete orthonormal sequence of $L^2(M)$. In particular, for $f \in L^2(M)$, we have

$$f = \sum_{j=1}^{\infty} (f, \phi_j)\phi_j \text{ in } L^2(M), \qquad \text{and} \qquad \|f\|^2 = \sum_{j=1}^{\infty} (f, \phi_j)^2.$$

These last two formulae are referred to as the *Parseval identities*.

If we think of the Laplacian as an operator on C^2 functions, then for M compact the Laplacian is symmetric in the sense that

$$(\text{III.9.4}) \qquad (\Delta\phi, \psi) = (\phi, \Delta\psi) = -\int_M \langle \operatorname{grad} \phi, \operatorname{grad} \psi \rangle \, dV;$$

and a similar comment holds when M has compact closure and smooth boundary for functions in $C^2(M) \cap C^1(\overline{M})$ that vanish on ∂M. So, the bilinear form representing $-\Delta$ is given by the *Dirichlet* or *energy integral*

$$\mathcal{D}[f, h] = \int_M \langle \operatorname{grad} f, \operatorname{grad} h \rangle \, dV.$$

Note, however, that the bilinear form $\mathcal{D}$ only involves the first derivatives of f and h, and we may therefore extend its definition to $C^1(\overline{M})$, with vanishing boundary data (III.9.2) when there is nonempty boundary.

Moreover, when $\phi \in C^2(M) \cap C^1(\overline{M})$ satisfies (III.9.2), one has

$$(\text{III.9.5}) \qquad (\Delta\phi, \psi) = -\int_M \langle \operatorname{grad} \phi, \operatorname{grad} \psi \rangle \, dV,$$

even if we relax the differentiability on ψ – that we only require $\psi \in C^1(\overline{M})$. Therefore, in what follows, we let f range over $C^1(M)$ when considering the closed eigenvalue problem, and over functions in $C^1(\overline{M})$ satisfying (III.9.2) when considering the Dirichlet eigenvalue problem.

Theorem III.9.2. (Lord Rayleigh) *For all $f \neq 0$, we have*

$$(\text{III.9.6}) \qquad \lambda_1 \leq \frac{\mathcal{D}[f, f]}{\|f\|^2}$$

with equality if and only if f is an eigenfunction of λ_1.

If $\{\phi_1, \phi_2, \ldots\}$ *is a complete orthonormal basis of* $L^2(M)$ *such that* ϕ_j *is an eigenfunction of* λ_j *for each* $j = 1, 2, \ldots$, *then for* $f \neq 0$, *satisfying*

$$(f, \phi_1) = \cdots = (f, \phi_{k-1}) = 0, \tag{III.9.7}$$

we have the inequality

$$\lambda_k \leq \frac{\mathcal{D}[f, f]}{\|f\|^2}, \tag{III.9.8}$$

with equality if and only if f *is an eigenfunction of* λ_k.

Proof. The argument is based on (III.9.5). Given $f \neq 0$, set

$$\alpha_j = (f, \phi_j).$$

Then, (III.9.7) is equivalent to saying that

$$\alpha_1 = \cdots = \alpha_{k-1} = 0.$$

So, for all $k = 1, 2, \ldots$, and $r = k, k+1, \ldots$, we have

$$\begin{aligned} 0 &\leq \mathcal{D}\left[f - \sum_{j=k}^{r} \alpha_j \phi_j, f - \sum_{j=k}^{r} \alpha_j \phi_j\right] \\ &= \mathcal{D}[f, f] - 2\sum_{j=k}^{r} \alpha_j \mathcal{D}[f, \phi_j] + \sum_{j,l=k}^{r} \alpha_j \alpha_l \mathcal{D}[\phi_j, \phi_l] \\ &= \mathcal{D}[f, f] + 2\sum_{j=k}^{r} \alpha_j (f, \Delta\phi_j) - \sum_{j,l=k}^{r} \alpha_j \alpha_l (\phi_j, \Delta\phi_l) \\ &= \mathcal{D}[f, f] - \sum_{j=k}^{r} \lambda_j {\alpha_j}^2. \end{aligned}$$

We conclude that

$$\sum_{j=k}^{\infty} \lambda_j {\alpha_j}^2 < +\infty,$$

and

$$\mathcal{D}[f, f] \geq \sum_{j=k}^{\infty} \lambda_j {\alpha_j}^2 \geq \lambda_k \sum_{j=k}^{\infty} {\alpha_j}^2 = \lambda_k \|f\|^2,$$

by the Parseval identities. The case of equality follows easily. ■

Thus the eigenvalues of the Laplacian may be realized via a variational problem on C^1 functions. Moreover, the variational problem actually lives on a larger space. Namely, one might consider all those functions that are limits of C^1 functions relative to the Dirichlet and L^2 integrals. Then, the eigenvalues would satisfy (III.9.6) and (III.9.8) for this larger class of functions. More precisely, we endow $C^1(M)$ with the metric

$$(f, h)_H := (f, h) + \mathcal{D}[f, h], \tag{III.9.9}$$

and let $H(M)$ denote the completion of $C^1(M)$ with respect to this metric. It is known that $C^\infty(\overline{M})$ is dense in $H(M)$. Also, it is known that uniform Lipschitz functions may be realized as elements of $H(M)$, with their gradient – defined almost everywhere in M – in $L^2(M)$. (One uses a standard regularization argument; see Adams (1975, pp. 29ff).)

Notation. For the closed eigenvalue problem, we let $\mathfrak{H}(M)$ denote $H(M)$. For the Dirichlet eigenvalue problem, we let $\mathfrak{H}(M)$ denote the closure in $H(M)$ of those functions in $C^1(\overline{M})$ vanishing on ∂M. It is known that, in this latter case, $C_c^\infty(M)$ is dense in $\mathfrak{H}(M)$.

Thus, (III.9.6) and (III.9.8) are valid – by definition (for $\mathcal{D}$ is defined on all of $\mathfrak{H}(M)$ by taking limits) – if we allow f to range over all of $\mathfrak{H}(M)$. The characterization of equality is a more delicate matter. However, the result remains valid that, if for some $f \in \mathfrak{H}(M)$, we have equality in (III.9.6) or (III.9.8), then f is an eigenfunction of the eigenvalue in question.

We now sketch the argument that ϕ_1 *the eigenfunction of* λ_1 *never vanishes on* M. (Courant–Hilbert (1967, Vol. I, pp. 451ff). The close details were first worked out in Bérard–Meyer (1982).) Certainly, in the closed eigenvalue problem, we have ϕ_1 equal to a constant and there is very little to discuss. So, we wish to consider the Dirichlet eigenvalue problem.

Before we begin this argument, we note that one may be in situations in which M has compact closure, but nothing is known about the smoothness of the boundary. So, the existence theory of eigenvalues–eigenfunctions is far more delicate. It is nevertheless possible, at the most elementary level, to simply consider minimizing the functional

$$f \mapsto \frac{\mathcal{D}[f, f]}{\|f\|^2},$$

where f varies over $\mathfrak{H}(M)$, the completion of $C_c^\infty(M)$ relative to the metric (III.9.9). We then define the *fundamental tone of* M, $\lambda^*(M)$ by

$$\lambda^*(M) = \inf_{f \in \mathfrak{H}(M);\, f \neq 0} \frac{\mathcal{D}[f, f]}{\|f\|^2}.$$

Clearly, if

$$M_1 \subseteq M_2, \qquad \dim M_1 = \dim M_2,$$

then

$$\lambda^*(M_1) \geq \lambda^*(M_2).$$

We now proceed to show that, when M has compact closure and smooth boundary, the eigenfunction ϕ_1 for the Dirichlet eigenvalue problem does not vanish. The first step is to use the strong maximum principle Courant–Hilbert (1967, Vol. I, p. 326ff), or the unique continuation principle Aronszajn (1957), to show that if ϕ ever vanishes in M, then it must change sign. Next, we argue here that it is impossible that ϕ_1 ever change sign in M – therefore, ϕ_1 never vanishes in M.

This, itself, consists of two steps. First, we show that if D is any domain in M for which

$$\phi_1 \mid D > 0, \qquad \phi_1 \mid \partial D = 0, \tag{III.9.10}$$

then

$$\int_D |\text{grad}\, \phi_1|^2\, dV = \lambda_1 \int_D {\phi_1}^2\, dV. \tag{III.9.11}$$

Indeed, if one knew that ∂D was C^1, then the result would follow from the Green formula (III.7.10). When we know nothing of the smoothness of ∂D, we argue as follows: Let ϵ be any regular value of $\phi_1 \mid D$, and

$$D_\epsilon = \{x \in D : \phi_1(x) > \epsilon\}.$$

Then

$$\begin{aligned}
\iint_{D_\epsilon} |\text{grad}\, \phi_1|^2\, dV &= \lambda_1 \iint_{D_\epsilon} {\phi_1}^2\, dV + \int_{\partial D_\epsilon} \phi_1 \frac{\partial \phi_1}{\partial \nu}\, dA \\
&= \lambda_1 \iint_{D_\epsilon} {\phi_1}^2\, dV + \epsilon \int_{\partial D_\epsilon} \frac{\partial \phi_1}{\partial \nu}\, dA \\
&= \lambda_1 \iint_{D_\epsilon} {\phi_1}^2\, dV + \epsilon \iint_{D_\epsilon} \Delta \phi_1\, dV \\
&= \lambda_1 \iint_{D_\epsilon} \{{\phi_1}^2 - \epsilon \phi_1\}\, dV \\
&\leq \lambda_1 \iint_D {\phi_1}^2\, dV.
\end{aligned}$$

If we let $\epsilon \downarrow 0$, we obtain

$$\lambda_1 \geq \iint_D |\text{grad}\, \phi_1|^2 / \iint_D {\phi_1}^2 \geq \lambda^*(D) \geq \lambda^*(M) = \lambda_1$$

by Rayleigh's theorem, which implies (III.9.11). Furthermore, it also implies

$$\lambda^*(D) = \lambda_1. \tag{III.9.12}$$

We now conclude the proof that ϕ_1 cannot change sign. Since M has smooth boundary, the strong maximum principle implies that the normal derivative of ϕ_1 along ∂M never vanishes. So, ϕ_1 cannot change sign in a neighborhood of the boundary. Assume that $\phi_1 < 0$ in this neighborhood. If ϕ_1 changes sign in M, then one has a domain $D \subset\subset M$ on which ϕ_1 satisfies (III.9.10). One then also has a nonnegative function $\psi \in C_c^\infty(M)$ such that

$$\psi \mid D = 1.$$

Set $\Omega = M \setminus \overline{D}$ and consider the functions ϕ and ψ_ϵ in $\mathfrak{H}(M)$ defined by

$$\phi \mid D = \phi_1 \mid D, \qquad \phi \mid \Omega = 0,$$

and

$$\psi_\epsilon = \phi + \epsilon\psi.$$

Then

$$\iint_M \psi_\epsilon{}^2\, dV = \iint_D \{\phi_1{}^2 + 2\epsilon\phi_1\}\, dV + \epsilon^2 \iint_M \psi^2\, dV,$$

$$\iint_M |\mathrm{grad}\, \psi_\epsilon|^2\, dV = \lambda_1 \iint_D \phi_1{}^2\, dV + \epsilon^2 \iint_\Omega |\mathrm{grad}\, \psi|^2\, dV.$$

Now, one easily shows that for sufficiently small $\epsilon > 0$

$$\frac{\mathcal{D}[\psi_\epsilon, \psi_\epsilon]}{\|\psi_\epsilon\|^2} < \lambda_1,$$

which contradicts Rayleigh's theorem.

An immediate consequence of the positivity of ϕ_1 on all of M is that the multiplicity of the eigenvalue λ_1 is precisely 1. Indeed, if the multiplicity is greater than 1, then λ_1 has an eigenfunction ψ L^2–orthogonal to ϕ_1, which implies ψ cannot be everywhere nonzero – a contradiction.

We now wish to consider comparison theorems for λ^*. In the spirit of the earlier comparison theorems, we wish to compare what happens in our Riemannian manifold with the corresponding phenomena in a space form of constant sectional curvature. Recall from §III.4, for each fixed $\kappa \in \mathbb{R}$, we let $\mathbb{M}_\kappa$ denote space form of constant sectional curvature κ as described in §II.3 (we assume the dimension is some fixed $n > 1$).

We let $\mathcal{L}$ denote the Laplacian on $\mathbb{S}^{n-1}$. We also fix a point o in $\mathbb{M}_\kappa$ and consider spherical geodesic coordinates in $\mathbb{M}_\kappa$ centered at o. Then, for $r > 0$

and $\xi \in \mathsf{S}_o$, and a function f on $\mathbb{M}_\kappa$, one easily has

(III.9.13) $$(\Delta f)(\exp r\xi) = \mathbf{S}_\kappa{}^{1-n}\partial_r(\mathbf{S}_\kappa{}^{n-1}\partial_r f) + \mathbf{S}_\kappa{}^{-2}\mathcal{L}_\xi f,$$

where, when writing $\mathcal{L}_\xi f$ we mean that $f|\mathsf{S}(o;r)$ is to be considered as a function on S_o with associated Laplacian $\mathcal{L}$.

We now consider the Dirichlet eigenvalue problem on

$$M := \mathbb{B}_\kappa(o;R) \subset \mathbb{M}_\kappa.$$

We let $\mathcal{E}_o$ denote the radial functions with respect to o (i.e., functions that depend only on distance from o) in $L^2(M)$. The orthogonal complement $\mathcal{E}_o{}^\perp$ of $\mathcal{E}_o$ is seen to be given by

$$\mathcal{E}_o{}^\perp = \{G : M \to \mathbb{R} \in L^2 : \int_{\mathbb{S}_\kappa(o;r)} G\, dA = 0\ \forall r > 0\},$$

where dA denotes the $(n-1)$–measure on $\mathbb{S}_\kappa(o;r)$.

Now Δ certainly maps $\mathcal{E}_o \to \mathcal{E}_o$. Let f be an eigenfunction of

$$\lambda_\kappa(R) := \lambda_1(M),$$

and write f as

$$f = F + G,$$

where $F \in \mathcal{E}_o$ and $G \in \mathcal{E}_o{}^\perp$. Then,

(III.9.14) $$F(\exp r\xi) = \frac{1}{\mathbf{c_{n-1}}}\int_{\mathsf{S}_o} f(\exp rv)\, d\mu_o(v),$$

which implies by (III.9.13)

$$\Delta F + \lambda_\kappa(R)F = 0,$$

which therefore yields

$$\Delta G + \lambda_\kappa(R)G = 0.$$

Since $f|\mathbb{S}_\kappa(o;R) = 0$, we have, by (III.9.14), $F|\mathbb{S}_\kappa(o;R) = 0$, which implies, $G|\mathbb{S}_\kappa(o;R) = 0$. Thus, F and G themselves are eigenfunctions of $\lambda_\kappa(R)$, so either $F = 0$ or $G = 0$. Since the Dirichlet eigenfunction of $\lambda_\kappa(R)$ never vanishes, but $\int G\, dA = 0$ over every $\mathbb{S}_\kappa(o;r)$, we have $G = 0$ on all of $\mathbb{B}_\kappa(o;R)$. Said differently, the eigenfunction of the lowest Dirichlet eigenvalue of $\mathbb{B}_\kappa(o;R) \subset \mathbb{M}_\kappa$ is radial with respect to o.

We are now given a Riemannian manifold M. For convenience, assume that M is complete. For any x in M and $\rho > 0$, let $\lambda(x;\rho)$ denote the lowest Dirichlet eigenvalue of $B(x;\rho)$, when the boundary $S(x;\rho)$ is smooth – for example, when

$\rho < \operatorname{inj} x$. When the boundary is not smooth – for example, when $\rho \geq \operatorname{inj} x$ – let $\lambda(x;\rho)$ denote the fundamental tone $\lambda^*(B(x;\rho))$ of $B(x;\rho)$.

Theorem III.9.3. (S. Y. Cheng (1975)) *Assume that the sectional curvatures of M are all less than or equal to δ. Then, for every $x \in M$, we have*

$$\lambda(x;\rho) \geq \lambda_\delta(\rho)$$

for all $\rho \leq \min\{\operatorname{inj} x, \pi/\sqrt{\delta}\}$, with equality for some fixed ρ if and only if $B(x;\rho)$ is isometric to the disk of radius ρ in the constant curvature space form $\mathbb{M}_\delta$.

Lemma III.9.1. (J. Barta (1937)) *For any function $f \in C^2(B(x;\rho)) \cap C^1(\overline{B(x;\rho)})$ with*

$$f|B(x;\rho) > 0, \qquad f|S(x;\rho) = 0,$$

we have

$$\inf_{B(x;\rho)} \frac{\Delta f}{f} \leq -\lambda(x;\rho) \leq \sup_{B(x;\rho)} \frac{\Delta f}{f}.$$

Proof. Let ϕ be an eigenfunction $\lambda(x;\rho)$ with

$$\phi|B(x;\rho) > 0, \qquad \phi|S(x;\rho) = 0;$$

and set

$$h = \phi - f.$$

Then

$$-\lambda(x;\rho) = \frac{\Delta\phi}{\phi} = \frac{\Delta f}{f} + \frac{f\Delta h - h\Delta f}{f(f+h)}.$$

Since $f(f+h)|B(x;\rho) > 0$, and

$$\int_{B(x;\rho)} \{f\Delta h - h\Delta f\}\,dV = 0,$$

the claim follows. ∎

Proof of Theorem III.9.3. We let ϕ denote an eigenfunction of $\lambda_\delta(\rho)$ with $\phi|\mathbb{B}_\delta(o;\rho) > 0$. Then, one has

$$\phi(\exp r\xi) = \Phi(r),$$

where exp denotes the exponential map in $\mathbb{M}_\delta$, and Φ satisfies

$$\partial_r{}^2\Phi + (n-1)\frac{\mathbf{C}_\delta}{\mathbf{S}_\delta}\partial_r\Phi + \lambda_\delta(\rho)\Phi = 0, \tag{III.9.15}$$

with boundary conditions

$$(\partial_r\Phi)(0) = \Phi(\rho) = 0.$$

One may write (III.9.15) as

$$\mathbf{S}_\delta{}^{1-n}\partial_r(\mathbf{S}_\delta{}^{n-1}\partial_r\Phi) + \lambda_\delta(\rho)\Phi = 0,$$

which implies

$$(\mathbf{S}_\delta{}^{n-1}\partial_r\Phi)(r) = -\lambda_\delta(\rho)\int_0^r \mathbf{S}_\delta{}^{n-1}\Phi < 0$$

on all of $(0, \rho)$ – so Φ is strictly decreasing with respect to r.

Now consider the function $F : B(x;\rho) \to \mathbb{R}$ given by

$$F(\exp r\xi) = \Phi(r),$$

where exp denotes here the exponential map from M_x to M. Then,

$$\begin{aligned}
\frac{\Delta F}{F}(\exp r\xi) &= \frac{\partial_r\{\sqrt{\mathbf{g}}(r;\xi)\partial_r F\}}{\sqrt{\mathbf{g}}(r;\xi)F} \\
&= \frac{1}{F}\left\{\partial_r{}^2 F + \frac{\partial_r\sqrt{\mathbf{g}}(r;\xi)}{\sqrt{\mathbf{g}}(r;\xi)}\partial_r F\right\} \\
&\le \frac{1}{F}\left\{\partial_r{}^2 F + (n-1)\frac{\mathbf{C}_\delta}{\mathbf{S}_\delta}\partial_r F\right\} \\
&= -\lambda_\delta(\rho)
\end{aligned}$$

by Theorem III.4.1 and the negativity of $\partial_r F$; so

$$-\lambda(x;\rho) \le \sup\frac{\Delta F}{F} \le -\lambda_\delta(\rho).$$

The case of equality is handled easily. ∎

Theorem III.9.4. (S. Y. Cheng (1975)) *Assume that the Ricci curvatures of M are all greater than or equal to $(n-1)\kappa$. Then, for every $x \in M$, $\rho > 0$, we have*

$$\lambda(x;\rho) \le \lambda_\kappa(\rho),$$

with equality for some fixed ρ if and only if $B(x;\rho)$ is isometric to the disk of radius ρ in the constant curvature space form $\mathbb{M}_\kappa$.

Proof. Here, we have no guarantee that $S(x;\rho)$, the boundary of $B(x;\rho)$, is smooth, so we may only think of $\lambda(x;\rho)$ as the fundamental tone of $B(x;\rho)$.

Thus, we wish to show that given any $\epsilon > 0$ there exists a function F on $B(x;\rho)$, approximated relative to the metric (III.9.9) by functions in $C_c^\infty(B(x;\rho))$, such that

$$\mathcal{D}[F,F] \le \{\lambda_\kappa(\rho)+\epsilon\}\|F\|^2.$$

The function F is constructed as above, namely, we let ϕ denote an eigenfunction of $\lambda_\kappa(\rho)$ with $\phi|\mathbb{B}_\kappa(o;\rho) > 0$, where $\mathbb{B}_\kappa(o;\rho)$ denotes the disk in $\mathbb{M}_\kappa$. Then, one has

$$\phi(\exp r\xi) = \Phi(r),$$

where exp denotes the exponential map in $\mathbb{M}_\kappa$, and Φ satisfies

$$\partial_r{}^2\Phi + (n-1)\frac{\mathbf{C}_\kappa}{\mathbf{S}_\kappa}\partial_r\Phi + \lambda_\kappa(\rho)\Phi = 0, \tag{III.9.16}$$

with boundary conditions

$$(\partial_r\Phi)(0) = \Phi(\rho) = 0.$$

Again,

$$(\partial_r\Phi)(r) < 0$$

on all of $(0,\rho)$ – so Φ is strictly decreasing with respect to r. Then, we define $F : B(x;\rho) \to \mathbb{R}$ by

$$F(\exp r\xi) = \Phi(r),$$

where the exp denotes the exponential map from M_x to M. We note that the function F is defined for $r\xi \in \overline{\mathsf{B}(x;\rho)} \cap \mathsf{D}_x$, that is, for those $r\xi$ inside the tangent cut locus of x.

The function F is well defined on *all* of $\overline{B(x;\rho)}$, since if two minimizing geodesics, emanating from x, intersect at y, then both geodesics have the same length. The function F is continuous, since the function $c(\xi)$, $\xi \in \mathsf{S}_x$, the distance along the geodesic γ_ξ to its cut point, is continuous. Also, for $r\xi \in B(x;\rho) \cap \mathsf{D}_x$, we have

$$|(\operatorname{grad} F)(\exp r\xi)| = |\partial_r\Phi(r)|;$$

so grad F has bounded length on $B(x;\rho)$. Since grad F is continuous everywhere except, possibly, on $C(x) \cap \overline{B(x;\rho)}$ – a set of Riemannian n–measure equal to 0 – we conclude that $F \in H(B(x;\rho))$.

We now wish to show that $F \in \mathfrak{H}(B(x;\rho))$, that is, that F is approximated in $H(B(x;\rho))$ by a function $G \in C_c{}^\infty(B(x;\rho))$.

Let $L : [0, \infty) \to \mathbb{R} \in C^\infty$ with $L'(0) = 0$, and $\operatorname{supp} L \subseteq [0, \rho_1]$ for some $\rho_1 < \rho$; and let $G : B(x;\rho) \to \mathbb{R}$ be defined by

$$G(\exp r\xi) = L(r)$$

for all $r\xi \in \overline{\mathsf{B}(x;\rho)} \cap \overline{\mathsf{D}_x}$. Then, $G \in H(B(x;\rho))$ (as F is) and has compact support. Also, we have

$$\begin{aligned}
\|F - G\|^2 &= \int_{\mathsf{S}_x} d\mu_x(\xi) \int_0^{\min\{c(\xi),\rho\}} (\Phi - L)^2(r)\sqrt{\mathbf{g}}(r;\xi)\,dr \\
&\le \int_{\mathsf{S}_x} d\mu_x(\xi) \int_0^{\min\{c(\xi),\rho\}} (\Phi - L)^2(r)\mathbf{S}_\kappa{}^{n-1}(r)\,dr \\
&\le \int_{\mathsf{S}_x} d\mu_x(\xi) \int_0^{\rho} (\Phi - L)^2(r)\mathbf{S}_\kappa{}^{n-1}(r)\,dr \\
&= \mathbf{c_{n-1}} \int_0^{\rho} (\Phi - L)^2(r)\mathbf{S}_\kappa{}^{n-1}(r)\,dr,
\end{aligned}$$

which is the L^2 distance of functions on $\mathbb{B}_\kappa(o;\rho)$ determined by the functions Φ and L. A similar estimate holds for $\mathcal{D}[F - G, F - G]$. So, any degree of approximation achieved in $H(\mathbb{B}_\kappa(o;\rho))$ is achieved automatically in $H(B(x;\rho))$. Thus, F is an admissible function, that is, an element of $\mathfrak{H}(B(x;\rho))$.

Let

$$b(\xi) = \min\{c(\xi), \rho\}.$$

Then, it suffices to establish

$$\int_0^{b(\xi)} (\partial_r\Phi)^2\sqrt{\mathbf{g}}(r;\xi)\,dr \le \lambda_\kappa(\rho) \int_0^{b(\xi)} \Phi^2\sqrt{\mathbf{g}}(r;\xi)\,dr$$

for every $\xi \in \mathsf{S}_x$. Well,

$$\begin{aligned}
\int_0^{b(\xi)} (\partial_r\Phi)^2\sqrt{\mathbf{g}}(r;\xi)\,dr &= \Phi(\partial_r\Phi)\sqrt{\mathbf{g}}(r;\xi)\Big|_0^{b(\xi)} - \int_0^{b(\xi)} \Phi\partial_r\{(\partial_r\Phi)\sqrt{\mathbf{g}}(r;\xi)\}\,dr \\
&= (\Phi\partial_r\Phi)(b(\xi))\sqrt{\mathbf{g}}(b(\xi);\xi) \\
&\quad - \int_0^{b(\xi)} \Phi\partial_r\{(\partial_r\Phi)\sqrt{\mathbf{g}}(r;\xi)\}\,dr \\
&\le - \int_0^{b(\xi)} \Phi\left\{\partial_r{}^2\Phi + (\partial_r\Phi)\frac{\partial_r\sqrt{\mathbf{g}}(r;\xi)}{\sqrt{\mathbf{g}}(r;\xi)}\right\}\sqrt{\mathbf{g}}(r;\xi)\,dr \\
&\le - \int_0^{b(\xi)} \Phi\left\{\partial_r{}^2\Phi + (n-1)\frac{\mathbf{C}_\kappa}{\mathbf{S}_\kappa}\partial_r\Phi\right\}\sqrt{\mathbf{g}}(r;\xi)\,dr \\
&= \lambda_\kappa(\rho)\int_0^{b(\xi)} \Phi^2\sqrt{\mathbf{g}}(r;\xi)\,dr,
\end{aligned}$$

which is our claim. Note that we have used (i) $\Phi|[0, \rho) > 0$, (ii) $\partial_r \Phi|(0, \rho] < 0$ and (iii) Theorem III.4.3.

The case of equality is easily handled. ∎

Notes and Exercises

Max–Min Methods

Exercise III.36. Prove the following important generalization of Rayleigh's theorem:

Max–Min Theorem. *Consider the Dirichlet or closed eigenvalue problem. Given any* $v_1, \ldots, v_{k-1}$ *in* $L^2(M)$*, let*

$$\mu = \inf \mathcal{D}[f, f]/\|f\|^2,$$

where f *varies over the subspace (less the origin) of functions in* $\mathfrak{H}(M)$ *orthogonal to* $v_1, \ldots, v_{k-1}$ *in* $L^2(M)$*. Then, for the eigenvalue* λ_k *(the counting is with multiplicity), we have*

$$\mu \leq \lambda_k.$$

Of course, if $v_1, \ldots, v_{k-1}$ *are orthonormal, with each* v_l *an eigenfunction of* λ_ℓ*,* $\ell = 1, \ldots, k-1$*, then* $\mu = \lambda_k$*.*

Note III.14. Important consequences may be obtained from the arguments of the previous exercise. They include theorems on (i) domain monotonicity of eigenvalues, and (ii) the number of nodal domains of an eigenfunction (i.e., the number of connected components where the eigenfunction does not vanish) – Courant's nodal domain theorem.

Similarly, one can replace vanishing Dirichlet boundary data with vanishing Neumann boundary data, with a similar (although more restrictive) max–min theorem.

For these and other matters, see the classic Courant–Hilbert (1967) and the more recent Chavel (1984).

Exercise III.37. Let $j_{n,k}$ denote the k–th Dirichlet eigenvalue of $\mathbb{B}^n$ (with eigenvalues repeated according to multiplicity).

(a) Show that $\lambda_k(\mathbb{B}^n(\epsilon))$, the k–th Dirichlet eigenvalue of $\mathbb{B}^n(\epsilon)$, is given by

$$\lambda_k(\mathbb{B}^n(\epsilon)) = \frac{j_{n,k}}{\epsilon^2}.$$

(b) Let M be an n–dimensional Riemannian manifold, $x \in M$. Use Riemann normal coordinates (§II.8) and the max–min theorem to prove

$$\lambda_k(B(x;\epsilon)) \sim \frac{j_{n,k}}{\epsilon^2}$$

as $\epsilon \downarrow 0$.

Exercise III.38. We use the following variant of the argument of the max–min theorem. Consider the case of M compact Riemannian, with closed eigenvalue problem. List the eigenvalues as

$$\{\lambda_1 = 0 < \lambda_2 \leq \lambda_3 \leq \ldots \uparrow +\infty\},$$

with eigenvalues repeated according to multiplicity, and with corresponding $L^2(M)$–orthonormal eigenfunctions

$$\{\phi_1, \phi_2, \phi_3, \ldots\}.$$

Consider k pairwise disjoint domains $\Omega_1, \ldots, \Omega_k$, each with compact closure and smooth boundary, and let $\lambda(\Omega_j)$ be the lowest Dirichlet eigenvalue of Ω_j. Prove

$$\lambda_k(M) \leq \sup_{j=1,\ldots,k} \lambda(\Omega_j).$$

Weyl's Asymptotic Formula

Note III.15. The celebrated Weyl formula (1911, 1912) states that

$$N(\lambda) \sim \frac{\boldsymbol{\omega}_{\mathbf{n}} V(M)}{(2\pi)^n} \lambda^{n/2},$$

as $\lambda \uparrow +\infty$, where $N(\lambda)$ denotes the number of eigenvalues, counted with multiplicity, $\leq \lambda$. Similarly,

$$\lambda_\ell \sim \frac{(2\pi)^2}{{\boldsymbol{\omega}_{\mathbf{n}}}^{2/n}} \left\{ \frac{\ell}{V(M)} \right\}^{2/n}$$

as $\ell \uparrow +\infty$. See Courant–Hilbert (1967, Vol. I, Chapter VI) and Chavel (1984) for discussions of the result.

The Weyl formula allows one to determine $V(M)$ once one knows the spectrum of the Laplacian of M. It was then asked whether the knowledge of the spectrum determines the Riemannian metric itself (see Kac (1966)). The answer was known quite early in the game that counterexamples exist (Milnor (1964)), and a vigorous search for very simple and elementary examples has ensued. See Buser (1992) for extended discussions, and Gordon–Webb–Wolpert (1992a,

1992b), Buser–Conway–Doyle–Semmler (1994), Gordon–Webb (1996), and Gordon (2000) for the most recent progress, with references.

Exercise III.39. Solve the closed eigenvalue problem on the circle and the Dirichlet eigenvalue on an interval. Verify Weyl's asymptotic formula.

Exercise III.40. Let M be compact n–dimensional with Ricci curvature bounded below by $(n-1)\kappa$. A subset $\mathcal{G}$ of M is said to be ϵ*–separated* if any two distinct points of $\mathcal{G}$ have distance at least ϵ. So, if $\mathcal{G}$ is maximal ϵ–separated, then all disks of radius $\epsilon/2$ centered at points of $\mathcal{G}$ are pairwise disjoint, and

$$M = \bigcup_{\xi \in \mathcal{G}} B(\xi; \epsilon).$$

Show that, for such a $\mathcal{G}$, we have

$$\operatorname{card} \mathcal{G} \geq \frac{V(M)}{V_\kappa(\epsilon)}.$$

(The exercise is Lemma IV.4.1 in Chapter IV. It is given here for use in the following exercise.)

Exercise III.41. Let M be compact n–dimensional with Ricci curvature bounded below by $(n-1)\kappa$. Show that there exists a constant depending only on n and κ such that

$$\lambda_\ell(M) \leq c(n, \kappa) \left\{ \frac{\ell}{V(M)} \right\}^{2/n}.$$

Eigenvalues and Wirtinger's Inequality

Exercise III.42. (Wirtinger's inequality) Prove that for $f : \mathbb{R} \to \mathbb{R} \in D^1$ which is L–periodic, and that satisfies

$$\int_0^L f = 0,$$

one has

$$\int_0^L (f')^2 \geq \frac{4\pi^2}{L^2} \int_0^L f^2,$$

with equality if and only if

$$f(t) = \alpha \cos \frac{2\pi t}{L} + \beta \sin \frac{2\pi t}{L}.$$

The "fixed-endpoint" version goes as follows: Prove that for $f : [0, L] \to \mathbb{R} \in D^1$, which satisfies

$$f(0) = f(L) = 0,$$

one has

$$\int_0^L (f')^2 \geq \frac{\pi^2}{L^2} \int_0^L f^2,$$

with equality if and only if

$$f(t) = \beta \sin \frac{2\pi t}{L}.$$

Hint: Usually, it is an exercise in Fourier series. In the context here, it is Rayleigh's theorem applied to Exercise III.39.

When the Fundamental Tone Is Bounded Away From 0

Exercise III.43.

(a) Let M^2 be a 2–dimensional Riemannian manifold minimally imbedded in $\mathbb{R}^3$. Show that, for every relatively compact domain Ω in $M^2 \cap \mathbb{B}^3(R)$, one has

$$\lambda^*(\Omega) > \frac{1}{4R^2}.$$

In particular, if M^2 is complete, and contained in $\mathbb{B}^3(R)$ for some fixed $R > 0$, then $\lambda^*(\Omega) >$ const. > 0 for all relatively compact Ω.

(b) Also show, if M is any complete Riemannian manifold with $\lambda^*(\Omega)$ uniformly bounded away from 0 for all relatively compact Ω, then for any $o \in M$ and fixed $k > 0$, $\limsup V(o; r)/r^k = +\infty$, as $r \uparrow +\infty$.

Lichnerowicz's Formula

Exercise III.44. Prove the Lichnerowicz formula (1958)

$$\frac{1}{2}\Delta(|dF|^2) = |\text{Hess}\, F|^2 + \langle \text{grad}\, \Delta F, \text{grad}\, F\rangle + \text{Ric}\,(\text{grad}\, F, \text{grad}\, F)$$

for all smooth functions on the Riemannian manifold M.

Exercise III.45. Use the Lichnerowicz formula to prove that if M is compact n–dimensional, with Ricci curvature bounded below by $(n-1)\kappa$, then the first nonzero eigenvalue $\lambda_2(M)$ (for the closed eigenvalue problem the lowest eigenvalue is equal to 0) satisfies

$$\lambda_2(M) \geq n\kappa.$$

We note that a result of M. Obata (1962) states that one has equality if and only if M is an n–sphere of constant sectional curvature κ. We suggest a slick argument in Exercise VI.6. (See the hint there.) Also see the discussion (based on Cheng (1975)) in Chavel (1984, pp. 82–84).

Exercise III.46. Given $\mathbb{M}_\kappa$, $\kappa > 0$. Show that, for any fixed $o \in \mathbb{M}_\kappa$, $\phi(x) = \cos\sqrt{\kappa}d(o, x)$ is an eigenfunction of $\lambda_2(\mathbb{M}_\kappa) = n\kappa$.

IV

Riemannian Coverings

In this chapter, we continue the development of the global theory, wherein we emphasize volume and integration. The major theme is the study of the volume growth of Riemannian manifolds, and the fundamental approach is to reduce the study of the volume growth to a corresponding discrete problem.

Such a study presupposes that the Riemannian manifold has sufficient local uniformity to allow us to disregard local fluctuations of the geometric data. The primary example is that of a noncompact manifold covering a compact Riemannian manifold, where the Riemannian metric on the cover is the lift of the Riemannian metric on the compact via the covering (this example was first considered in this context by V. Efremovič (1953), A. S. Svarc (1955), and J. Milnor (1968)). The covering determines a discrete group of isometries of the cover, which, in turn, induces a tiling of the cover by relatively compact fundamental domains – each isometric to the other. Thus, the estimate of the volume of a metric disk in the cover is reduced to counting the number of fundamental domains contained in the disk, and the smallest number of fundamental domains containing the disk. Since the action of the group is free, counting fundamental domains is the same as counting elements of the group. The quantitative estimates used here are the Bishop theorems of §III.4.

More generally, one may relax the degree of local uniformity and nevertheless obtain similar discretizations of Riemannian manifolds. Here, for the calibration of discrete to the continuous, one must use Gromov's refinement of the Bishop theorems (see §III.4). Our treatment, in §IV.4, follows that of M. Kanai (1985).

We consider a number of other matters along the way. First, we give a skeleton summary of basic background on coverings and fundamental groups. We finish a matter first discussed in §II.8, namely, the determination of the Riemannian metric of constant sectional curvature. Here, we derive the theorem of W. Killing (1891, 1893), and H. Hopf (1925), which states that any complete

Riemannian manifold of constant sectional curvature is covered by one of the models presented in §II.3, namely, the spheres, Euclidean space, hyperbolic space; and that any simply connected complete Riemannian manifold of constant sectional curvature is isometric to one of these models. At the end of the chapter, we also give the proof that any nontrivial free homotopy class of a compact Riemannian manifold possesses a closed geodesic.

§IV.1. Riemannian Coverings

For the necessary background from topology, and for details that we do not discuss here, we refer the reader to Massey (1967).

Definition. If $\widetilde{M}, M$ are connected topological manifolds, we say that a map $\psi : \widetilde{M} \to M$ is a *covering* if every $p \in M$ has a connected open neighborhood U such that ψ maps each component of $\psi^{-1}[U]$ homeomorphically onto U.

It is standard that given the above covering, a point $p \in M$, and a path $\omega : [0, \beta] \to M \in C^0$ such that $\omega(0) = p$, then to each $\widetilde{p} \in \psi^{-1}[p]$ there exists a unique *lift* $\widetilde{\omega} : [0, \beta] \to \widetilde{M} \in C^0$ satisfying $\widetilde{\omega}(0) = \widetilde{p}$ and $\psi \circ \widetilde{\omega} = \omega$. This is referred to as the *unique lifting lemma*.

Definition. If $\widetilde{M}, M$ are connected differentiable manifolds, then $\psi : \widetilde{M} \to M$ is a *differentiable covering* if ψ is a covering, and ψ is differentiable of maximum rank on all of $\widetilde{M}$.

If $\widetilde{M}, M$ are connected Riemannian manifolds, then $\psi : \widetilde{M} \to M$ is a *Riemannian covering* if ψ is a differentiable covering which is a local isometry of $\widetilde{M}$ onto M.

Proposition IV.1.1. *If $\psi : \widetilde{M} \to M$ is a Riemannian covering, then M is complete if and only if $\widetilde{M}$ is complete.*

Proof. One uses the fact that

$$\exp \psi_* \widetilde{\xi} = \psi(\exp \widetilde{\xi}) \qquad \text{(IV.1.1)}$$

for all $\widetilde{\xi} \in T\widetilde{M}$ (see the generalities on isometries in §II.3) as follows:

If $\widetilde{M}$ is complete, let $\gamma : I \to M$ be a maximal geodesic with interval I containing the origin of $\mathbb{R}$. Pick $\widetilde{p}, \widetilde{\xi} \in \widetilde{M}_{\widetilde{p}}$ so that $\phi(\widetilde{p}) = \gamma(0)$, $\phi_{*|\widetilde{p}}\widetilde{\xi} = \gamma'(0)$. Then $\phi(\widetilde{\gamma}_{\widetilde{\xi}}(t)) = \gamma(t)$ for all $t \in I$. But, $\phi(\widetilde{\gamma}_{\widetilde{\xi}}(t))$ is defined for all $t \in \mathbb{R}$. This implies (since I is maximal) that $I = \mathbb{R}$. So, every geodesic in M is infinitely extendable in both directions. Therefore, M is complete.

If M is given to be complete, and $\widetilde{\gamma}(t)$, $t \in \widetilde{I}$ is a maximal geodesic in $\widetilde{M}$, then $\gamma(t) := \phi(\widetilde{\gamma}(t))$ is a geodesic in M. But γ can be defined on all of $\mathbb{R}$. The unique lifting lemma will then imply that $\widetilde{\gamma}$ can be defined on all of $\mathbb{R}$. ■

Theorem IV.1.1. (S. B. Myers (1941)) *If M satisfies the hypotheses of the Bonnet–Myers Theorem* (Theorem II.12)*, that is, if M is complete and the Ricci curvature of M is bounded from below by a positive constant, then not only is M compact, but also any cover of M, $\widetilde{M}$, is compact.*

Proof. Exercise for the reader. ■

Proposition IV.1.2. *Let X, Y be Riemannian manifolds, $\phi : X \to Y$ a local isometry. Then, for any $p \in X$, there exists an $\epsilon > 0$ such that*

$$\phi|B(p;\epsilon) : B(p;\epsilon) \to B(\phi(p);\epsilon)$$

is an isometry.

Proof. Let ϵ_p and $\epsilon_{\phi(p)}$ satisfy

$$\exp|\mathsf{B}(p;\epsilon_p) : \mathsf{B}(p;\epsilon_p) \to B(p;\epsilon_p)$$
$$\exp|\mathsf{B}(\phi(p);\epsilon_{\phi(p)}) : \mathsf{B}(\phi(p);\epsilon_{\phi(p)}) \to B(\phi(p);\epsilon_{\phi(p)})$$

be diffeomorphisms, and pick $\epsilon = \min\{\epsilon_p, \epsilon_{\phi(p)}\}$. The (Euclidean) disks $\mathsf{B}(p;\epsilon)$ and $\mathsf{B}(\phi(p);\epsilon)$ are isometric under $\phi_{*|p}$, and

$$\phi|B(p;\epsilon) = \{\exp|\mathsf{B}(\phi(p);\epsilon)\}\circ\phi_{*|p}\circ\{\exp|\mathsf{B}(p;\epsilon)\}^{-1}$$

is, therefore, a diffeomorphism, which implies the proposition. ■

Theorem IV.1.2. *Let $\widetilde{M}$ be connected and complete, and let $\psi : \widetilde{M} \to M$ be a local isometry of $\widetilde{M}$ onto M. Then, ψ is a covering.*

Proof. M is certainly connected since ψ is continuous.

Let $\gamma : [0, T_o] \to M$ be any geodesic segment in M, $\widetilde{p} \in \widetilde{M}$ such that $\psi(\widetilde{p}) = \gamma(0)$. Then, γ has a unique lift in $\widetilde{M}$ starting at $\widetilde{p}$. Indeed, consider $\widetilde{\xi} \in T\widetilde{M}$ such that $\psi_{*|\widetilde{p}}\widetilde{\xi} = \gamma'(0)$. Then (by Proposition IV.1.2), there exists an $\epsilon > 0$ such that $\widetilde{\gamma}_{\widetilde{\xi}}(t)$ is defined for $t \in [0, \epsilon)$. Therefore, if we set

$$T = \sup\{\tau : \gamma|[0,\tau] \text{ has lift starting at } \widetilde{p}\},$$

then $T > 0$ and $\psi(\gamma_{\widetilde{\xi}}(t)) = \gamma(t)$ for all $t \in [0, T)$; also

$$\lim_{t \uparrow T} \gamma(t) = \psi(\widetilde{\gamma}_{\widetilde{\xi}}(T)).$$

If $T < T_o$, the lift can be defined at T and beyond, which implies $T = T_o$.

Next, suppose we are given $p \in M$; we wish to construct a connected open neighborhood U of p such that ψ maps each component of $\psi^{-1}[U]$ homeomorphically onto U. To this end, fix $\epsilon > 0$ so that $\exp|\mathsf{B}(p;\epsilon)$ is a diffeomorphism of $\mathsf{B}(p;\epsilon)$ onto $B(p;\epsilon)$.

Now one shows

$$\psi^{-1}[B(p;\epsilon)] = \bigcup_{\widetilde{p} \in \psi^{-1}[p]} B(\widetilde{p};\epsilon).$$

Indeed, if

$$\widetilde{q} \in \bigcup_{\widetilde{p} \in \psi^{-1}[p]} B(\widetilde{p};\epsilon),$$

then there exists a path $\widetilde{\omega}$ joining some $\widetilde{p} \in \psi^{-1}[p]$ to $\widetilde{q}$ having length less than ϵ. This implies that $\omega := \psi(\widetilde{\omega})$ joins p to $\psi(\widetilde{q})$ and has length less than ϵ. So, $\psi(\widetilde{q}) \in B(p;\epsilon)$, that is, $\widetilde{q} \in \psi^{-1}[B(p;\epsilon)]$.

On the other hand, if $\widetilde{q} \in \psi^{-1}[B(p;\epsilon)]$, then $q := \psi(\widetilde{q}) \in B(p;\epsilon)$, which implies there exists a geodesic $\gamma_{q,p} : [0, T] \to M$ from q to p, with length less than ϵ. Then, $\gamma_{q,p}$ has a lift $\widetilde{\gamma} : [0, T] \to \widetilde{M}$ starting at $\widetilde{q}$; so $\psi(\widetilde{\gamma}(T)) = p$, and $d(\widetilde{q}, \widetilde{\gamma}(T)) < \epsilon$, which implies

$$\widetilde{q} \in \psi^{-1}[B(p;\epsilon)],$$

which implies the claim.

Finally, by the triangle inequality, given $\widetilde{p}_1, \widetilde{p}_2 \in \psi^{-1}[p]$, $\widetilde{p}_1 \neq \widetilde{p}_2$, $\epsilon < d(\widetilde{p}_1, \widetilde{p}_2)$, one has

$$B(\widetilde{p}_1;\epsilon/3) \cap B(\widetilde{p}_2;\epsilon/3) = \emptyset.$$

But the claims of the preceding paragraph are also valid for when ϵ is replaced by $\epsilon/3$. Thus, the desired neighborhood U about p is $B(p;\epsilon/3)$. ■

Corollary IV.1.1. *The map* $\mathcal{E}^i : \mathbb{R} \to \mathbb{S}^1$ *given by*

$$\theta \mapsto (\cos\theta, \sin\theta) := \mathcal{E}^{i\theta}$$

is a covering.

Theorem IV.1.3. (J. Hadamard (1898), E. Cartan (1946))[1] *If M is complete, and all of its sectional curvatures are nonpositive, then for any $p \in M$, $\exp_p : M_p \to M$ is a covering.*

Proof. Let g denote the Riemannian metric on M and consider the Riemannian metric $(\exp_p)^* g$ on M_p. Then, straight lines emanating from the origin of M_p are geodesics in the Riemannian metric $(\exp_p)^* g$.

By Corollary I.7.2, $(\exp_p)^* g$ is a complete Riemannian metric on M_p. The theorem now follows from Corollary II.7.2 and Theorem IV.1.2. ■

Theorem IV.1.4. *Let M be a complete Riemannian manifold of constant sectional curvature $\kappa > 0$, $\dim M = n \geq 2$. Then there exists a Riemannian covering $\psi : \mathbb{S}^n(1/\sqrt{\kappa}) \to M$.*

We first require two lemmata.

Lemma IV.1.1. (S. B. Myers & N. Steenrod (1939)) *Let M be a Riemannian manifold, and $\varphi : M \to M$ an onto map (not assumed to be continuous) such that $d(\varphi(p), \varphi(q)) = d(p, q)$ for all $p, q \in M$. Then, φ is an isometry, that is, φ is a diffeomorphism preserving the Riemannian metric.*

Proof. Besides the original proof of Myers and Steenrod, one can also refer to the proof in Kobayashi–Nomizu (1969, Vol. I, p. 169), based on Palais (1957). See Exercise IV.3. ■

Lemma IV.1.2. (W. Blaschke (1967), L. W. Green (1963)) *Given $\kappa > 0$, and M a complete Riemannian manifold such that for every $p \in M$,*

$$\exp |\mathsf{B}(p; \pi/\sqrt{\kappa}) \text{ has maximal rank,}$$

and

$$\exp_* |T(\mathsf{S}(p; \pi/\sqrt{\kappa})) = 0.$$

Then (i) *for every $p \in M$, the image* $\exp(\mathsf{S}(p; \pi/\sqrt{\kappa}))$ *in M consists of precisely one point. Thus, the map $Q : M \to M$ given by*

$$Q(p) = \exp(\mathsf{S}(p; \pi/\sqrt{\kappa})) \tag{IV.1.2}$$

[1] The 2–dimensional version, of any two points being joined by a unique geodesic, goes back to H. von Mangolt (1881).

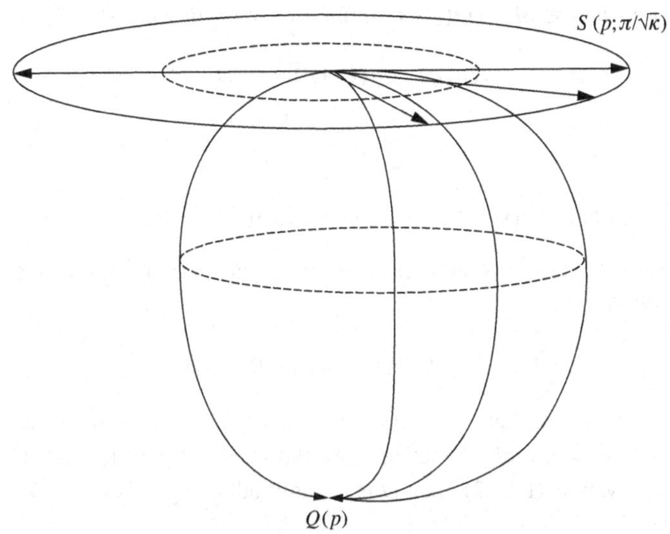

Figure IV.1. A *wiedersehnspunkte.*

is well defined. Moreover, (ii) *one has*

$$Q^2 = \mathrm{id}_M, \tag{IV.1.3}$$

(iii) *Q is an isometry of M, and* (iv) *every unit speed geodesic γ on M is periodic, with period equal to $2\pi/\sqrt{\kappa}$.*

Finally, (v) *M is diffeomorphically covered by the sphere.*

Proof. (i) is straightforward.

To prove (ii), one simply notes that, if $p \in M$, $\xi \in \mathsf{S}_p$, and $\gamma_\xi(t) = \exp t\xi$ then,

$$p = \exp(-(\pi/\sqrt{\kappa})\gamma_\xi'(\pi/\sqrt{\kappa})).$$

To prove (iii) that Q is an isometry, we show that Q satisfies the hypothesis of Lemma 1, that is, Q preserves distances. First note that

$$Q(\gamma_\xi(t)) = \gamma_\xi(t + \pi/\sqrt{\kappa}). \tag{IV.1.4}$$

Next, note that if $p, q \in M$, then there exists $\xi \in \mathsf{S}_p$ such that $q = \gamma_\xi(d(p, q))$. Then, by (IV.1.4)

$$\begin{aligned} d(Q(p), Q(q)) &= d(\gamma_\xi(\pi/\sqrt{\kappa}), \gamma_\xi(d(p, q) + \pi/\sqrt{\kappa})) \\ &\leq d(p, q), \end{aligned}$$

that is, for all $p, q \in M$, $d(Q(p), Q(q)) \leq d(p, q)$. But with (IV.1.3), we have

$$\begin{aligned} d(p,q) &= d(Q^2(p), Q^2(q)) \\ &\leq d(Q(p), Q(q)) \\ &\leq d(p,q), \end{aligned}$$

which implies Q preserves the distance metric $d(\ ,\ )$.

To prove (iv) that every unit speed geodesic has period $2\pi/\sqrt{\kappa}$, one need only argue via (IV.1.3), (IV.1.4), that

$$\gamma_\xi(t) = Q^2(\gamma_\xi(t)) = Q(\gamma_\xi(t + \pi/\sqrt{\kappa})) = \gamma_\xi(t + 2\pi/\sqrt{\kappa}).$$

We now prove (v), that M is diffeomorphically covered by the sphere. Let $\widetilde{M} = \mathbb{S}^n(1/\sqrt{\kappa})$. Certainly, $\widetilde{M}$ satisfies the hypothesis of the lemma. If $\widetilde{Q}$ is the map for $\widetilde{M}$ given by (IV.1.2), then $\widetilde{Q}$ is the antipodal map of $\mathbb{S}^n(1/\sqrt{\kappa})$, that is, it is the restriction of $-\mathrm{id}_{\mathbb{R}^{n+1}}$ to $\mathbb{S}^n(1/\sqrt{\kappa})$.

The idea in the explicit construction of the covering is to fix two points, one in M, the other in $\widetilde{M}$, fix an orthonormal frame in each respective tangent space, determine associated Riemann normal coordinates in each of the two manifolds, and then match the points in the manifold by their coordinates. One must then realize that, even though this mapping has maximal rank on the open disks of radius $\pi/\sqrt{\kappa}$, in each manifold, there is still one point left over, and one must guarantee that extending the map to this extra point remains differentiable of maximal rank. The construction of the map itself in more direct coordinate-free language goes as follows:

Pick $p \in M$, $\widetilde{p} \in \widetilde{M}$, a linear isometry $\varphi : \widetilde{M}_{\widetilde{p}} \to M_p$, and define ψ by:

(IV.1.5)
$$\begin{aligned} \psi | B(\widetilde{p}; \pi/\sqrt{\kappa}) &= \exp \circ \varphi \circ (\exp | \mathsf{B}(\widetilde{p}; \pi/\sqrt{\kappa}))^{-1}, \\ \psi(\widetilde{Q}(\widetilde{p})) &= Q(p). \end{aligned}$$

Then, (IV.1.5) automatically guarantees that $\psi | B(\widetilde{p}; \pi/\sqrt{\kappa})$ is a local diffeomorphism onto its image $B(p; \pi/\sqrt{\kappa})$. Also, note that

(IV.1.6)
$$\psi_{*|\widetilde{p}} = \varphi.$$

Now, we consider the map ψ at $Q(p)$. To do so, we claim

(IV.1.7)
$$\psi | B(\widetilde{Q}(\widetilde{p}); \pi/\sqrt{\kappa}) = \exp \circ Q_* \circ \varphi \circ \widetilde{Q}_* \circ (\exp | \mathsf{B}(\widetilde{Q}(\widetilde{p}); \pi/\sqrt{\kappa}))^{-1}.$$

Indeed, if $q \in M$, $\eta \in \mathsf{S}_q$, $t \in [0, \pi/\sqrt{\kappa}]$, then we have

$$\exp t\eta = \exp(-(\pi/\sqrt{\kappa} - t)\gamma_\eta{}'(\pi/\sqrt{\kappa})) = \exp(-(\pi/\sqrt{\kappa} - t)Q_*(\eta));$$

and a similar statement certainly holds in $\widetilde{M}$. Therefore, if $\widetilde{q} = \widetilde{Q}(\widetilde{p})$, $\widetilde{\eta} \in \mathsf{S}_{\widetilde{q}}$, we have

$$\begin{aligned}\psi(\exp t\widetilde{\eta}) &= \psi(\exp(-(\pi/\sqrt{\kappa} - t)\widetilde{Q}_*(\widetilde{\eta})))\\ &= \exp \circ \varphi \circ (-(\pi/\sqrt{\kappa} - t)\widetilde{Q}_*(\widetilde{\eta}))\\ &= \exp(-(\pi/\sqrt{\kappa} - t))\varphi \circ \widetilde{Q}_*(\widetilde{\eta}))\\ &= \exp Q_* \circ \varphi \circ \widetilde{Q}_*(\exp t\eta),\end{aligned}$$

that is, (IV.1.7).

Thus, ψ is differentiable of maximal rank. Since M is compact, ψ is a covering by Theorem IV.1.3. ∎

Proof of Theorem IV.1.4. One immediately has via Theorem II.8.2 that ψ constructed above is a local isometry. ∎

§IV.2. The Fundamental Group

In this section, we summarize the background material in a fashion suitable for our subsequent use. Again, the reader can find the necessary background in Massey (1967).

1. Let M be a topological manifold, Γ a group of homeomorphisms of M. We say that Γ *acts properly discontinuously on* M if to each $p \in M$ there is a neighborhood U of p such that the collection of open sets $\{\varphi(U) : \varphi \in \Gamma\}$ are pairwise disjoint.

In particular, $\varphi \neq \mathrm{id}_M$ implies $\varphi(p) \neq p$ for all $p \in M$. Thus, the action of ϕ is *free* on M.

Let $\psi : \widetilde{M} \to M$ be a covering. We say that a homeomorphism $\varphi : \widetilde{M} \to \widetilde{M}$ is a *deck transformation of the covering* ψ if $\psi \circ \varphi = \psi$. The collection of deck transformations form a group under composition. One checks that the deck transformation group acts properly discontinuously on $\widetilde{M}$.

Conversely, given M, and Γ acting properly discontinuously on M, let M/Γ be the orbit space of Γ, that is, for $p \in M$, let $[p] =: \{\varphi(p) : \varphi \in \Gamma\}$, $\pi : M \to M/\Gamma$ the projection $\pi(p) = [p]$, and endow M/Γ with the quotient topology. One sees that π is a covering with deck transformation group Γ.

Let Γ act properly discontinuously on M, a differentiable manifold. If each $\varphi \in \Gamma$ is a diffeomorphism, then M/Γ has the natural structure of a differentiable manifold and $\pi : M \to M/\Gamma$ is a differentiable covering. Furthermore, the group $\Gamma_* = \{\varphi_* : TM \to TM : \varphi \in \Gamma\}$ acts properly discontinuously on TM, and TM/Γ_* is naturally diffeomorphic to $T(M/\Gamma)$. Therefore, if, in addition, M is Riemannian and Γ consists of isometries of M, then M/Γ has a natural Riemannian metric for which π is a Riemannian covering.

2. Let $M = \mathbb{S}^n(1/\sqrt{\kappa})$, $\kappa > 0$, and let

$$\Gamma = \{\mathrm{id}_{\mathbb{R}^{n+1}}|\mathbb{S}^n(1/\sqrt{\kappa}), -\mathrm{id}_{\mathbb{R}^{n+1}}|\mathbb{S}^n(1/\sqrt{\kappa})\}.$$

Then the quotient Riemannian manifold M/Γ is called $\mathbb{P}^n(1/\sqrt{\kappa})$, *n–dimensional real projective space* (and when $n \geq 2$) *of constant sectional curvature* κ. One easily checks that $\mathbb{P}^n(1/\sqrt{\kappa})$ has diameter equal to $\pi/2\sqrt{\kappa}$, that all geodesics are simply closed of length $\pi/\sqrt{\kappa}$, that the cut locus of any point consists of the metric sphere centered at that point with radius $\pi/2\sqrt{\kappa}$, that the volume of $\mathbb{P}^n(1/\sqrt{\kappa})$ is half that of $\mathbb{S}^n(1/\sqrt{\kappa})$, and that the cut locus of any point is isometric to the projective space $\mathbb{P}^{n-1}(1/\sqrt{\kappa})$ of dimension $n-1$.

Let $\mathbb{Z}^n$ be the n–fold Cartesian product of the integers and V an n–dimensional real vector space. Fix a basis $\{e_1, \ldots, e_n\}$ of V. Then to each $(\alpha^1, \ldots, \alpha^n) \in \mathbb{Z}^n$ is associated a transformation of V, α, given by

$$\alpha(\xi) = \xi + \sum_{j=1}^{n} \alpha^j e_j.$$

The resulting group of transformations Γ is isomorphic to $\mathbb{Z}^n$ and acts properly discontinuously on $M := V$. The quotient manifold thus obtained is diffeomorphic to the torus

$$\mathbb{T}^n := (\mathbb{S}^1)^n$$

(see Corollary IV.1.1).

If V is an inner product space with its induced standard Riemannian metric, then M/Γ is *flat*, that is, all its sectional curvatures are equal to 0. Note, however, that a change of basis of V changes Γ, and there is no a priori expectation that the new quotient space (which is also flat) is isometric to the old one. (See Exercises IV.4–IV.8)

3. Again, start with M a fixed topological manifold. Two continuous paths $\omega_0 : [\alpha, \beta] \to M$, $\omega_1 : [\alpha, \beta] \to M$ with the same endpoints, that is,

$$\omega_0(\alpha) = \omega_1(\alpha), \quad \omega_0(\beta) = \omega_1(\beta),$$

are said to be *homotopic* if there exists a continuous map $\nu : [\alpha, \beta] \times [0, 1] \to M$ satisfying

$$\nu(t, 0) = \omega_0(t), \quad \nu(t, 1) = \omega_1(t), \qquad \text{for all } t \in [\alpha, \beta],$$

$$\nu(\alpha, \epsilon) = \omega_0(\alpha) = \omega_1(\alpha), \quad \nu(\beta, \epsilon) = \omega_0(\beta) = \omega_1(\beta), \qquad \text{for all } \epsilon \in [0, 1].$$

When ω_0, ω_1 are homotopic to each other, we write $\omega_0 \sim \omega_1$. The relation "homotopic" is an equivalence relation.

Given continuous paths $\omega : [0, 1] \to M$, $\gamma : [0, 1] \to M$ satisfying $\omega(1) = \gamma(0)$, one *composes* ω, γ to define $\omega \cdot \gamma$ by

$$(\omega \cdot \gamma)(t) = \begin{cases} \omega(2t) & 0 \leq t \leq 1/2 \\ \gamma(2t-1) & 1/2 \leq t \leq 1. \end{cases}$$

One easily proves that if $\omega_0 \sim \omega_1, \gamma_0 \sim \gamma_1$ and $\omega_0(1) = \omega_1(1) = \gamma_0(0) = \gamma_1(0)$, then $\omega_0 \cdot \gamma_0 \sim \omega_1 \cdot \gamma_1$. So, the multiplication may be defined on the homotopy classes.

Given $p \in M$, let $\pi(M, p)$ be the homotopy classes of *loops* $\omega : [0, 1] \to M \in C^0$ satisfying $\omega(0) = \omega(1) = p$. Then, $\pi(M, p)$ has the structure of a group and is called the *fundamental group of M based at p*.

If $p_0, p_1 \in M$ then any path $\gamma : [0, 1] \to M \in C^0$ satisfying $\gamma(0) = p_0$, $\gamma(1) = p_1$ determines an isomorphism $\pi(M, p_0) \to \pi(M, p_1)$ via

$$\omega \mapsto \gamma^{-1} \cdot \omega \cdot \gamma.$$

M is called *simply connected* if $\pi(M, p)$ consists of the unit element alone, that is, all loops based at p may be deformed to p. One can show that M is simply connected if and only if any two paths with same endpoints are homotopic.

Of course, $\mathbb{R}^n$ is simply connected for all $n \geq 1$.

Let $\psi : \widetilde{M} \to M$ be a covering. Recall that if ω is a path in M with initial point p, then to each $\widetilde{p} \in \psi^{-1}[p]$ there exists a unique lift of ω in $\widetilde{M}$ with initial point $\widetilde{p}$. Actually, a stronger statement, known as *homotopy lifting lemma* is true: Given a homotopy $v : [0, 1] \times [0, 1] \to M$ with $p = v(0, \epsilon)$ for all ϵ in $[0, 1]$, and $\widetilde{p} \in \psi^{-1}[p]$, there exists a unique $\widetilde{v} : [0, 1] \times [0, 1] \to \widetilde{M}$ such that $\widetilde{v}(0, \epsilon) = \widetilde{p}$ for all ϵ in $[0, 1]$, and $\psi \circ \widetilde{v} = v$. In particular, $\widetilde{v}(1, 0) = \widetilde{v}(1, \epsilon)$ for all ϵ in $[0, 1]$.

4. We now describe the universal covering of M. Fix $p \in M$ and let Ω_p consist of all continuous paths in M starting at p, that is,

$$\Omega_p = \{\omega : [0, 1] \to M \in C^0 : \omega(0) = p\}.$$

On Ω_p introduce the equivalence relation given by homotopy of two paths starting at p, and let $M_0 = \Omega_p / \sim$, the space of homotopy classes of paths starting at p. Since homotopic paths have the same endpoints one obtains a natural projection $\Psi : M_0 \to M$. Then, M_0 can be endowed with a topology for which it is simply connected, and Ψ is a covering called the *universal covering* of M.

Furthermore (and this is the "universal" property of Ψ), given any covering $\psi : \widetilde{M} \to M$ there exists a covering $\psi_0 : M_0 \to \widetilde{M}$ such that $\Psi = \psi \circ \psi_0$. The map ψ_0 is defined as follows: Fix $\widetilde{p} \in \psi^{-1}[p]$. For $\omega \in \Omega_p$, map $\omega \to \widetilde{\omega}(1) \in \widetilde{M}$, where $\widetilde{\omega}$ is the unique lift of ω through $\widetilde{p}$. Actually, by the homotopy lifting lemma, the map is defined on the equivalence class of ω – so it is a map from M_0 to $\widetilde{M}$.

Finally, if $\widetilde{M}$ itself is simply connected then ψ_0 is a homeomorphism.

5. An immediate consequence of the universal property and Theorem 4 is:

Theorem IV.2.1. (W. Killing (1891, 1893), H. Hopf (1925)) *$\mathbb{S}^n$ is simply connected for $n \geq 2$.*

Furthermore, a complete simply connected Riemannian manifold of constant sectional curvature κ is uniquely determined up to isometry. In particular such a space is isometric to the appropriate model among those discussed in §II.3.

6. Let $\psi : \widetilde{M} \to M$ be the universal covering of M, and fix $p \in M, \widetilde{p} \in \psi^{-1}[p]$. Let Γ be the deck transformation group of the covering. Then, Γ is isomorphic to $\pi(M, p)$.

The map is given as follows: given $\gamma \in \Gamma$, all paths joining $\widetilde{p}$ to $\gamma(\widetilde{p})$ are homotopic (since $\pi_1(\widetilde{M})$ is trivial), and therefore project to a well-defined element of $\pi(M, p)$. The map is clearly a homomorphism. By the homotopy lifting lemma, the map is one-to-one into. Should one start with a homotopy class in $\alpha \in \pi(M, p)$, one proceeds to obtain the element $\varphi_\alpha \in \Gamma$ as follows: Let $\widetilde{q} \in \widetilde{M}$. Then, $\widetilde{q}$ is determined by some homotopy class ξ of paths in M joining p to $q := \psi(\widetilde{q})$. Then define $\varphi_\alpha(\widetilde{q})$ to be the point in $\widetilde{M}$ corresponding to the homotopy class $\omega \cdot \xi$ of paths joining p to q, where the loop ω is a representative of the class α. Certainly φ_α commutes with ψ, and is one-to-one.

Also, one has that Γ acts transitively on $\psi^{-1}[p]$ for each $p \in M$.

Theorem IV.2.2. (S. B. Myers (1941)) *Let M be a complete Riemannian manifold, of dimension $n \geq 2$, such that there exists a constant $\kappa > 0$ for which*

$$\operatorname{Ric}(\xi, \xi) \geq (n-1)\kappa|\xi|^2 \tag{IV.2.1}$$

for any $\xi \in TM$, then $\pi_1(M)$ is finite.

Proof. The universal cover of M is complete, with the same estimate on the Ricci curvature. Hence, by the original Bonnet–Myers theorem (Theorem II.6.1), the universal cover of M is compact, and the the number of preimages of any p in M must be finite. ■

§IV.3. Volume Growth of Riemannian Coverings

We start with an elementary result on the growth of groups.

Definition. Let Γ be a finitely generated group, with $\{\gamma_1, \ldots, \gamma_k\}$ a given set of generators. To every $\gamma \in \Gamma$, we associate the *norm of* γ, $|\gamma|$, defined to be the minimum length of γ as a word in the given set of generators $\{\gamma_1, \ldots, \gamma_k\}$. We also define the *counting function* $n(\lambda)$ by

$$n(\lambda) = \operatorname{card}\{\gamma : |\gamma| \le \lambda\}.$$

Note that

$$n(\lambda) \le (2k)(2k-1)^{\lambda-1}.$$

Proposition IV.3.1. *The limit of* $n(\lambda)^{1/\lambda}$*, as* $\lambda \uparrow +\infty$*, exists.*

Proof. Certainly,

$$n(\lambda + \tau) \le n(\lambda)n(\tau), \tag{IV.3.1}$$

which implies

$$n(\ell\lambda) \le n(\lambda)^{\ell}.$$

Given any $\lambda, t > 1$ then pick

$$\ell := [\lambda/t] + 1 \ge \lambda/t,$$

where $[x]$ denotes the largest integer $< x$. Then $n(\lambda) \le n(\ell t) \le n(t)^{\ell} \le n(t)^{\lambda/t+1}$, from which one concludes

$$n(\lambda)^{1/\lambda} \le n(t)^{1/t+1/\lambda}.$$

Thus, one has

$$\limsup_{\lambda\uparrow+\infty} n(\lambda)^{1/\lambda} \le n(t)^{1/t}$$

for all $t > 1$. Now, let $t \uparrow +\infty$. Then,

$$\limsup_{\lambda\uparrow+\infty} n(\lambda)^{1/\lambda} \le \liminf_{\lambda\uparrow+\infty} n(\lambda)^{1/\lambda},$$

which implies the claim. ∎

Now we consider what happens if we use a different set of generators $\{\gamma_1^*, \ldots, \gamma_\ell^*\}$, with counting function $n^*(\lambda)$. Let

$$N = \max\{|\gamma_r^*|_{\{\gamma_1,\ldots,\gamma_k\}} : r = 1, \ldots, \ell\}.$$

Then

$$n^*(\lambda) \leq n(N\lambda),$$

which implies

$$\frac{\ln n^*(\lambda)}{\lambda} \leq N \frac{\ln n(N\lambda)}{N\lambda}.$$

Therefore, the change of generators of Γ leaves invariant the vanishing or nonvanishing of the $\lim \{\ln n(\lambda)\}/\lambda$, as $\lambda \uparrow +\infty$.

Definition. We say that Γ has *exponential growth* if

$$\limsup_{\lambda \uparrow +\infty} \frac{\ln n(\lambda)}{\lambda} > 0.$$

Otherwise, we refer to Γ as having *subexponential growth.*

Definition. Given a Riemannian manifold M, we say that M has *exponential volume growth* if

$$\limsup_{r \uparrow +\infty} \frac{\ln V(x;r)}{r} > 0$$

for some (therefore, for all) x in M. Otherwise, we refer to M as having *subexponential volume growth.*

Proposition IV.3.2. *For any Riemannian manifold M, $x \in M$,*

$$\limsup_{r \uparrow +\infty} \frac{\ln V(x;r)}{r}$$

is independent of the choice of x.

Proof. Given x and y in M, we have

$$B(x;r) \subseteq B(y;r + d(x, y)),$$

which implies

$$\limsup_{r \uparrow +\infty} \frac{\ln V(x;r)}{r} \leq \limsup_{r \uparrow +\infty} \frac{\ln V(y;r + d(x, y))}{r} = \limsup_{r \uparrow +\infty} \frac{\ln V(y;r)}{r},$$

and the same inequality with the roles of x and y interchanged. Therefore, the two limsups are equal. ∎

As the above proof indicates, there can be no a priori expectation that the limsup is uniform with respect to x. However, when M is a noncompact cover of a compact, then we prove below that the limsup is uniform; furthermore, the limsup is actually a limit.

Definition. Let $\psi : \widetilde{M} \to M$ be a covering, with deck transformation group Γ. We say that a domain $\Omega \subseteq \widetilde{M}$ is a *fundamental domain of the covering* if

$$\gamma(\Omega) \cap \Omega = \emptyset \quad \text{for all } \gamma \in \Gamma, \quad \text{and} \quad \psi(\overline{\Omega}) = M.$$

When Γ acts transitively on the fibers $\psi^{-1}[p]$ for each $p \in M$, then $\psi(\overline{\Omega}) = M$ is equivalent to saying that

$$\bigcup_{\gamma \in \Gamma} \gamma(\overline{\Omega}) = \widetilde{M}. \tag{IV.3.2}$$

When considering a Riemannian covering one may construct a fundamental domain using cut loci. Namely, assume $\widetilde{M}$ is complete, $\widetilde{p} \in \widetilde{M}$, $p = \psi(\widetilde{p}) \in M$. Then $\psi_{*|\widetilde{p}} : \widetilde{M}_{\widetilde{p}} \to M_p$ is a linear isometry, and, of course, one has (IV.1.1) for all $\widetilde{\xi} \in \widetilde{M}_{\widetilde{p}}$. Therefore to each $p \in M$, we have

$$p \mapsto \mathsf{D}_p \subseteq M_p \mapsto {\psi_{*|\widetilde{p}}}^{-1}(\mathsf{D}_p) \subseteq \widetilde{M}_{\widetilde{p}} \mapsto \exp_{\widetilde{p}}({\psi_{*|p}}^{-1}(\mathsf{D}_p)),$$

which is a fundamental domain in $\widetilde{M}$.

In what follows here, we shall change notation slightly, in that we have M the cover of M_o.

Theorem IV.3.1. (A. Manning (1979)) *Let M be the universal cover of M_o, M_o compact. Then, for $x \in M$ the limit*

$$\mu := \lim_{r \uparrow +\infty} \frac{\ln V(x;r)}{r}$$

exists, the value is independent of x, and the convergence is uniform with respect to x.

Proof. We already know that if the limit exists, then its value is independent of x.

Furthermore, if we let F denote a fundamental domain of M_o in M, and d its diameter, then for all x, y in F and $r > d$ we have

$$B(x;r-d) \subseteq B(y;r) \subseteq B(x;r+d),$$

which implies

$$V(x;r-d) \leq V(y;r) \leq V(x;r+d) \tag{IV.3.3}$$

for all x, y in M. Therefore, since any $p \in M$ can be translated to F by the deck transformation group of the covering, any convergence will be uniform with respect to the choice of origin x.

To show that the limit itself exists, one is guided by the fact that the local behavior of the Riemannian metric is uniform to a very high degree. Indeed, all data – the metric, the density of the Riemannian measure, the various curvatures, etc. – vary over some fundamental domain, and are then translated by the deck transformation group over all of M, as mentioned in the previous paragraph. Therefore, one must only consider how to piece together this locally uniform behavior to obtain the global result. The key element of this uniformity, in our situation here, is that, for any fixed $r > 0$, the constant

$$c_r := \inf\{V(z;r/2) : z \in M\} > 0.$$

(Why is the constant positive?) With this in hand, we now prove the existence of the limit μ. In analogy with the argument of the previous proposition, we must find an analogue of (IV.3.1).

Fix x; for any $r, s > 0$, we have

$$B(x;r+s) = \bigcup_{y \in B(x;r)} B(y;s).$$

Fix some $b > 0$. If Y is any subset of $B(x;r)$ whose points are, pairwise, at least distance b apart from each other, then

$$\bigcup_{y \in Y} B(y;b/2) \subseteq B(x;r+b/2),$$

where the left-hand side is a disjoint union, which implies

$$V(x;r+b/2) \geq \sum_{y \in Y} V(y;b/2) \geq c_b \operatorname{card} Y;$$

so

$$\operatorname{card} Y \leq {c_b}^{-1} V(x;r+b/2). \tag{IV.3.4}$$

Now choose $Y \subseteq B(x;r)$ to be maximal with respect to the property that all its points, pairwise, have distance at least b from each other. Since Y is maximal in this sense, then every point of $B(x;r)$ is within distance b of Y, which implies

$$B(x;r+s) \subseteq \bigcup_{y \in Y} B(y;s+b)$$

for all $s > 0$. Therefore,

$$\begin{aligned} V(x;r+s) &\leq (\operatorname{card} Y)\max_{y\in Y} V(y;s+b) \\ &\leq {c_b}^{-1}V(x;r+b/2)V(x;s+b+d), \end{aligned}$$

by (IV.3.3), (IV.3.4), for all $r, s, b > 0$, which may be rewritten as

$$V(x;r+b/2+s-b/2) \leq {c_b}^{-1}V(x;r+b/2)V(x;s-b/2+3b/2+d).$$

By changing $r+b/2$ to r, and $s-b/2$ to s, one obtains

$$V(x;r+s) \leq {c_b}^{-1}V(x;r)V(x;s+3b/2+d) \tag{IV.3.5}$$

for all $r > b/2$, $s > -b/2$, $b > 0$. This is the desired analogue of (IV.3.1).

The end of the proof follows the same lines: Set

$$\alpha = {c_b}^{-1}, \qquad A = 3b/2+d.$$

Then

$$V(x;r+s) \leq \alpha V(x;r)V(x;s+A)$$

for r and s as above. We conclude

$$V(x;(k+1)r) \leq \alpha^k V(x;r+A)^{k+1}$$

for all $r > b/2$, $k = 1, 2, \ldots$. Therefore, given any $r > b/2$, $\delta \in (0, r)$, and $k = 1, 2, \ldots$, we have

$$V(x;kr+\delta) \leq V(x;(k+1)r) \leq \alpha^k V(x;r+A)^{k+1},$$

which implies

$$\frac{\ln V(x;kr+\delta)}{kr+\delta} \leq \frac{k\ln\alpha}{kr+\delta} + \frac{k+1}{kr+\delta}\ln V(x;r+A),$$

which implies, by fixing r and letting $k \uparrow +\infty$,

$$\limsup_{s\uparrow+\infty}\frac{\ln V(x;s)}{s} \leq \frac{\ln\alpha}{r} + \frac{\ln V(x;r+A)}{r}$$

for all $r > b/2$, which implies

$$\limsup_{s\uparrow+\infty}\frac{\ln V(x;s)}{s} \leq \liminf_{s\uparrow+\infty}\frac{\ln V(x;s)}{s},$$

which implies the theorem. ∎

We now pursue this passage from the locally uniform to the global in a more detailed manner. Now, our consideration is: Since every fundamental domain has the same volume, then the growth of $V(x;r)$, with respect to r, is essentially equivalent to the growth of the number of images of x, under the action of the deck transformation group, in $B(x;r)$, for large r. Since the action of the deck transformation group on the covering space is free, then the counting of images of x, under this action, may be calibrated to the counting of elements of the deck transformation group. The details of the calibration – we follow Milnor (1968) – (most explicit in (IV.3.6), (IV.3.10) below) go as follows:

Lemma IV.3.1. *Let M be a complete Riemannian manifold, let Γ be any finitely generated subgroup of isometries M acting properly discontinuously on M. Then, for any $x \in M$ and set of generators of Γ, there exist positive numbers μ and ϵ, depending only on x and the choice of generators, such that*

(IV.3.6)
$$n(\lambda) \leq \frac{V(x;\lambda\mu + \epsilon)}{V(x;\epsilon)}$$

for all $\lambda > 0$.

If M/Γ is compact then μ and ϵ may be chosen independently of x.

Proof. Given a collection of generators $\{\gamma_1, \ldots, \gamma_k\}$ of Γ, set

$$\mu = \max_{j=1,\ldots,k} d(x, \gamma_j(x)).$$

Then, for any $\gamma \in \Gamma$, we have, by the triangle inequality,

(IV.3.7)
$$d(x, \gamma(x)) \leq |\gamma| \max_{j=1,\ldots,k} d(x, \gamma_j(x)) = |\gamma|\mu,$$

which implies that, for any $\lambda > 1$, $B(x;\lambda\mu)$ contains at least $n(\lambda)$ *distinct* images of x under the action of Γ.

Now there exists $\epsilon > 0$ such that

$$B(x;\epsilon) \cap \gamma(B(x;\epsilon)) = \emptyset$$

for all $\gamma \in \Gamma$, $x \in M$. One immediately has (IV.3.6). ∎

One immediately concludes from Bishop's comparison theorem (Theorem IIII.4.4):

Theorem IV.3.2. (J. Milnor (1968)) *Let M be a complete Riemannian manifold with nonnegative Ricci curvature, and let Γ be any finitely generated subgroup of isometries M acting properly discontinuously on M. Then, for any* *set of generators of* Γ, *we have*

$$\lim_{\lambda\uparrow+\infty} \frac{\ln n(\lambda)}{\lambda} = 0.$$

Remark IV.3.1. Of course, the more precise version of the theorem is that since M has polynomial growth, that is,

$$V(x;r) \le \text{const.}r^n$$

for all x and r, we also have (from (IV.3.6)) that Γ has polynomial growth, that is,

$$n(\lambda) \le \text{const.}\lambda^n$$

for all λ.

Lemma IV.3.2. *Let M be a complete Riemannian manifold, Γ a finitely generated subgroup of isometries M acting properly discontinuously on M, such that M/Γ is compact; and let E be a compact neighborhood for which*

$$\bigcup_{\gamma\in\Gamma} \gamma(E) = M.$$

(i) *Then $\{\gamma(E) : \gamma \in \Gamma\}$ is a locally finite cover of M by compact neighborhoods.*

(ii) *Let*

$$\Gamma_E = \{\gamma \in \Gamma : \gamma(E) \cap E \neq \emptyset\};$$

then, Γ_E generates E. Furthermore, if we set

$$\nu = \inf_{\gamma\notin\Gamma_E} d(\gamma(E), E),$$

then we have, for any given $\gamma \in \Gamma$,

$$|\gamma| \le \left[\frac{d(y, \gamma(x))}{\nu}\right] + 1, \tag{IV.3.8}$$

for all $x, y \in E$, where the length of γ is measured relative to the elements in Γ_E.

Proof. (i) If the cover is not locally finite, there would exist $y \in M$ and $r > 0$ for which $\overline{B(y;r)}$ contains points from infinitely many distinct $\gamma(E)$. If d denotes the diameter of E, then $\overline{B(y;r+d)}$ contains infinitely many distinct $\gamma(E)$, which contradicts the proper discontinuity of the action of Γ on M.

(ii) We prove both of the claims of (ii) together.

For $|\gamma| = 1$ the estimate (IV.3.8) is obviously true.

So let $|\gamma| > 1$. Given any $x, y \in E$, connect y to $\gamma(x)$ by a minimizing unit speed geodesic segment ω. Let k denote any integer for which

$$\frac{d(y, \gamma(x))}{k} < \nu, \tag{IV.3.9}$$

and set

$$z_j = \omega(jd(y, \gamma(x))/k),$$

$j = 0, \ldots, k$. Then, of course,

$$d(z_{j-1}, z_j) < \nu.$$

So, if we let γ_0 denote the identity transformation of M, and pick $\gamma_j \in \Gamma$ so that $z_j \in \gamma_j(E)$ for $j = 1, \ldots, k$, then $\gamma_j {\gamma_{j-1}}^{-1} \in \Gamma_E$ for all $j = 1, \ldots, k$. Furthermore,

$$\gamma = (\gamma_k {\gamma_{k-1}}^{-1}) \cdots (\gamma_1 {\gamma_0}^{-1}),$$

which implies γ is generated by Γ_E, and $|\gamma| \leq k$. But the most efficient choice of k for which (IV.3.9) is valid is

$$k = \left[\frac{d(y, \gamma(x))}{\nu}\right] + 1,$$

which implies the lemma. ∎

Lemma IV.3.3. *Let M be a complete Riemannian manifold, and let Γ be any finitely generated subgroup of isometries M acting properly discontinuously on M, such that M/Γ is compact. Then there exist positive numbers ν and δ, such that*

$$n(\lambda) \geq \frac{V(x; \lambda\nu - (\nu + 2\delta))}{V(x; \delta)} \tag{IV.3.10}$$

for all $\lambda \geq 1 + 3\delta/\nu$, $x \in M$.

Proof. Let Ω be a fundamental domain of Γ, and δ equal to the diameter of Ω. Then, $B(x; \delta)$ contains $\overline{\Omega}$ for all $x \in \overline{\Omega}$. Apply the Lemma IV.3.2 to $E = B(x; \delta)$; denote the ν corresponding to x, defined in the statement of the lemma, by ν_x, and let

$$\nu = \inf_{x \in M} \nu_x.$$

For any $\ell > 0$, set

$$\Gamma_\ell = \{\gamma \in \Gamma : \gamma(B(x; \delta)) \cap B(x; \ell + \delta) \neq \emptyset\}.$$

Then

$$B(x;\ell+\delta) \subseteq \Gamma_\ell(B(x;\delta)),$$

which implies

$$V(x;\ell+\delta) \le (\mathrm{card}\,\Gamma_\ell)V(x;\delta).$$

To estimate card Γ_ℓ from above, we note that the argument of Lemma IV.3.2 implies

$$|\gamma| \le \frac{\ell+3\delta}{\nu}+1 \qquad \text{for all } \gamma \in \Gamma_\ell,$$

which implies

$$\text{(IV.3.11)} \qquad n\left(\frac{\ell+3\delta}{\nu}+1\right) \ge \mathrm{card}\,\Gamma_\ell \ge \frac{V(x;\ell+\delta)}{V(x;\delta)},$$

which implies the claim. ■

The Günther–Bishop comparison theorem (Theorem III.4.2) then implies:

Theorem IV.3.3. (J. Milnor (1968)) *Let M_o be a compact Riemannian manifold of strictly negative curvature. Then, the fundamental group of M_o, $\pi_1(M_o, x_o)$ for any x_o, has exponential growth.*

§IV.4. Discretization of Riemannian Manifolds

We now view the deck transformation group of a covering as a discretization of the manifold. The highlight of the above section was the ability to employ the strong local uniformity of the Riemannian geometry of the covering to calibrate growth of volume by growth of number of fundamental domains, that is, by growth of the deck transformation group. In this section, we digress from the specific geometry of coverings to show how to preserve this local uniformity in more general settings. But, first, some preliminaries.

Definition. Let X and Y be metric spaces with map $\phi : X \to Y$. We say that ϕ is a *quasi-isometry* if there exists a constant $c \ge 1$ such that

$$c^{-1}d(x_1,x_2) \le d(\phi(x_1),\phi(x_2)) \le cd(x_1,x_2)$$

for all x_1, x_2 in X.

There is no claim that ϕ is onto. So a quasi-isometry is not necessarily a Lipschitz homeomorphism. See the discussion in the example that follows. Of course, $\phi : X \to \phi(X)$ is a Lipschitz homeomorphism.

Example IV.4.1. We recast the Milnor theorems in our current setting.

Let Γ be a finitely generated group, with $\mathcal{A} = \{\gamma_1, \ldots, \gamma_k\}$ a given set of generators. Recall that with every $\gamma \in \Gamma$ is associated the *norm of* γ, $|\gamma|_{\mathcal{A}}$, defined to be the minimum length of γ as a word in the given set of generators $\mathcal{A}$. Note that

$$|\gamma|_{\mathcal{A}} \geq 0, \quad \text{with} \quad |\gamma|_{\mathcal{A}} = 0 \;\Leftrightarrow\; \gamma = \mathrm{id},$$
$$|\beta\gamma|_{\mathcal{A}} \leq |\beta|_{\mathcal{A}} + |\gamma|_{\mathcal{A}}, \quad \text{and} \quad |\gamma^{-1}|_{\mathcal{A}} = |\gamma|_{\mathcal{A}},$$

for all β, γ in Γ. The *word metric on* Γ is then given by

$$\delta_{\mathcal{A}}(\beta, \gamma) = |\beta^{-1}\gamma|_{\mathcal{A}}.$$

If we use a different set of generators $\mathcal{B} := \{\gamma_1^*, \ldots, \gamma_\ell^*\}$ then, as described in §IV.3, we have the metrics induced by $\mathcal{A}$ and $\mathcal{B}$ quasi-isometric to each other, namely, let

$$N = \max\{|\gamma_r^*|_{\mathcal{A}} : r = 1, \ldots, \ell\};$$

then

$$N^{-1}|\gamma|_{\mathcal{A}} \leq |\gamma|_{\mathcal{B}} \leq N|\gamma|_{\mathcal{A}}$$

for all $\gamma \in \Gamma$.

Now, let M be a complete Riemannian manifold, and Γ a finitely generated subgroup of isometries M acting freely and properly discontinuously on M, such that M/Γ is compact. For each $x \in M$, let $\|\cdot\|_x$ denote the *displacement norm on* Γ, given by

$$\|\gamma\|_x = d(x, \gamma\cdot x)$$

for all $\gamma \in \Gamma$, where d denotes distance in M. Then, again, we have

$$\|\gamma\|_x \geq 0, \quad \text{with} \quad \|\gamma\|_x = 0 \;\Leftrightarrow\; \gamma = \mathrm{id},$$
$$\|\beta\gamma\|_x \leq \|\beta\|_x + \|\gamma\|_x, \quad \text{and} \quad \|\gamma^{-1}\|_x = \|\gamma\|_x,$$

for all β, γ in Γ. Then, the results of Milnor (IV.3.7) and (IV.3.8) imply the existence of a constant $a \geq 1$ such that

$$a^{-1}|\gamma|_{\mathcal{A}} \leq \|\gamma\|_x \leq a|\gamma|_{\mathcal{A}}$$

for all $\gamma \in \Gamma$, which implies that the induced metrics are quasi isometric.

Furthermore, the map $\phi : \Gamma \to M$, given by

$$\phi(\gamma) = \gamma \cdot x,$$

satisfies

$$a^{-1}\delta_{\mathcal{A}}(\beta, \gamma) \leq d(\phi(\beta), \phi(\gamma)) \leq a\delta_{\mathcal{A}}(\beta, \gamma)$$

for all β, γ in Γ – so ϕ is a quasi-isometry.

Finally, (IV.3.6) and (IV.3.10) imply the existence of constants $\mathsf{a} \geq 1, \mathsf{b} \geq 0$, and $\mathsf{c} \geq 1$ such that

$$\text{(IV.4.1)} \qquad \mathsf{c}^{-1} n_{\mathcal{A}}(\mathsf{a}^{-1}\lambda - \mathsf{b}) \leq V(x; \lambda) \leq \mathsf{c} n_{\mathcal{A}}(\mathsf{a}\lambda + \mathsf{b}),$$

where $n_{\mathcal{A}}(\lambda)$ denotes the counting function of $\mathcal{A}$, given by

$$n_{\mathcal{A}}(\lambda) = \text{card}\,\{\gamma : |\gamma|_{\mathcal{A}} \leq \lambda\}.$$

Definition. Let X and Y be metric spaces with map $\phi : X \to Y$. We say that ϕ is a *rough isometry* if there exist constants $a \geq 1, b > 0$, and $\epsilon > 0$ such that

$$a^{-1}d(x_1, x_2) - b \leq d(\phi(x_1), \phi(x_2)) \leq ad(x_1, x_2) + b$$

for all x_1, x_2 in X, and ϕ is *ϵ–full*, that is,

$$\bigcup_{x \in X} B(\phi(x); \epsilon) = Y.$$

Note that the definition of rough isometry does not require that the map ϕ be continuous.

Also, the following proposition shows that "X is roughly isometric to Y" is an equivalence relation.

Proposition IV.4.1. *If $\phi : X_1 \to X_2$ and $\psi : X_2 \to X_3$ are rough isometries, then so is $\psi \circ \phi$.*

If $\phi : X \to Y$ is a rough isometry, then there exists $\phi^- : Y \to X$ a rough isometry, for which both $d(\phi^- \circ \phi(x), x)$ and $d(\phi \circ \phi^-(y), y)$ are uniformly bounded on X and Y, respectively.

Any two spaces of finite diameter are roughly isometric.

If X and Y are roughly isometric, then X and $Y \times K$ are roughly isometric for any compact metric space K.

Proof. We only comment on the second claim. By definition, $\phi(X)$ is ϵ–full in Y for some $\epsilon > 0$. Then, given any $y \in Y$, there exists an $x \in X$ for which $d(\phi(x), y) < \epsilon$. Then, define $\phi^-(y) := x$. One checks that ϕ^- is a mapping from Y to X satisfying the claim of the proposition. ■

We are now given a countable set $\mathcal{G}$, such that for each $\xi \in \mathcal{G}$, there is a finite nonempty subset $\mathsf{N}(\xi) \subseteq \mathcal{G} \setminus \{\xi\}$, of cardinality $m(\xi)$, each element of which is referred to as a *neighbor of* ξ. Furthermore, we require that $\eta \in \mathsf{N}(\xi)$ if and only if $\xi \in \mathsf{N}(\eta)$. Then, one determines a graph structure $\mathbf{G}$ by postulating the existence of precisely one oriented edge from any ξ to each of its neighbors, that is, the elements of $\mathsf{N}(\xi)$. We refer to $m(\xi)$ as the *valence of* $\mathbf{G}$ *at* ξ.

Definition. We say that the graph $\mathbf{G}$ has *bounded geometry* if the valence function $m(\xi)$ is bounded uniformly from above on all of $\mathcal{G}$.

A sequence of points $(\xi_0, \ldots, \xi_k)$ is a *combinatorial path of length k* if $\xi_j \in \mathsf{N}(\xi_{j-1})$ for all $j = 1, \ldots, k$. The graph $\mathbf{G}$ is called *connected* if any two points are connected by a path. Note that $m(\xi) \geq 1$ for all ξ if $\mathbf{G}$ is connected.

For any two vertices ξ and η in the connected graph $\mathbf{G}$, one defines their *distance* $\mathsf{d}(\xi, \eta)$ to be the infimum of the length of all paths connecting ξ to η. We also refer to d as the *combinatorial metric*. We set the notations for the respective metric "disks" and their "bounding spheres":

$$\beta(\xi; k) = \{\eta \in \mathcal{G} : \mathsf{d}(\eta, \xi) \leq k\} \quad \text{and} \quad \sigma(\xi; k) = \{\eta \in \mathcal{G} : \mathsf{d}(\eta, \xi) = k\},$$

for any $\xi \in \mathcal{G}$.

Proposition IV.4.2. *Assume* $\mathbf{G}$ *is connected, with more than one edge and with bounded geometry. Set*

$$\mathbf{m} = \max_{\xi \in \mathcal{G}} m(\xi).$$

Then for any finite $\mathcal{K} \subseteq \mathcal{G}$, *and* $k > 0$, *we have*

$$\operatorname{card}\{\eta \in \mathcal{G} : \mathsf{d}(\eta, \mathcal{K}) = k\} \leq \mathbf{m}^k \operatorname{card} \mathcal{K};$$

from which one also has

$$\operatorname{card}\{\eta \in \mathcal{G} : \mathsf{d}(\eta, \mathcal{K}) < k\} \leq \mathbf{m}^k \operatorname{card} \mathcal{K}.$$

Also, if $\mathbf{G}, \mathbf{F}$ *are connected graphs,* $\mathbf{G}$ *with bounded geometry, and*

$$\phi : \mathbf{G} \to \mathbf{F}$$

is a rough isometry, then there exists $\mu \geq 1$ *for which*

$$\operatorname{card} \mathcal{K} \leq \mu \operatorname{card} \phi(\mathcal{K})$$

for all finite subsets $\mathcal{K}$ *of* $\mathcal{G}$.

Remark IV.4.1. Of course, one *always* has: $\operatorname{card} \phi(\mathcal{K}) \leq \operatorname{card} \mathcal{K}$.

Proof. The first claim is obvious. For the second claim, we wish to show that there exists $\mu \geq 1$ such that, given any $\eta \in \mathcal{F}$, we have

$$\operatorname{card} \phi^{-1}[\eta] \leq \mu.$$

Well, the rough isometry property implies that, for all ξ, ξ' in $\phi^{-1}[\eta]$, we have

$$a^{-1}\mathsf{d}(\xi, \xi') - b \leq 0,$$

that is, $\mathsf{d}(\xi, \xi') \leq ab$, which implies $\phi^{-1}[\eta] \in \beta(\xi; ab)$ for any $\xi \in \phi^{-1}[\eta]$. Therefore,

$$\operatorname{card} \phi^{-1}[\eta] \leq \mathbf{m}^{ab} + 1 := \mu. \qquad \blacksquare$$

Example IV.4.2. Let Γ be a finitely generated group, with generator set $\mathcal{A}$, as in Example IV.4.1. Given any $\gamma \in \Gamma$, we let

$$\mathsf{N}(\gamma) = \gamma(\mathcal{A} \cup \mathcal{A}^{-1})$$

be the neighbors of γ. Then, the combinatorial metric of the graph structure coincides with the word metric. It is common to refer to this graph as the *Cayley graph of* Γ.

Definition. Let M be a complete Riemannian manifold. A *graph* $\mathbf{G}$ *in* M is a discrete subset $\mathcal{G}$ of M, for which there exists $R > 0$ such that

$$M = \bigcup_{\xi \in \mathcal{G}} B(\xi; R), \tag{IV.4.2}$$

with the graph structure $\mathbf{G}$ determined by the collection of neighbors of ξ,

$$\mathsf{N}(\xi) := \{\mathcal{G} \cap B(\xi; 3R)\} \setminus \{\xi\},$$

for each $\xi \in \mathcal{G}$.

We refer to R as the *covering radius of* the graph $\mathbf{G}$.

Theorem IV.4.1. *Let M be Riemannian complete, and $\mathbf{G}$ a graph in M, covering radius R. Then there exists a constant $a \geq 1$ such that*

$$\frac{1}{3R} d(\xi_1, \xi_2) \leq \mathsf{d}(\xi_1, \xi_2) \leq \frac{1}{R} d(\xi_1, \xi_2) + 1 \tag{IV.4.3}$$

for all ξ_1, ξ_2 in $\mathcal{G}$. Thus, the inclusion map of $\mathbf{G}$ into M is a rough isometry.

Proof. Given a combinatorial path $\xi_1 = \eta_0, \ldots, \eta_\ell = \xi_2$ in $\mathcal{G}$, connecting ξ_1 to ξ_2 of length ℓ. Then,

$$d(\xi_1, \xi_2) \leq \sum_{j=1}^{\ell} d(\eta_{j-1}, \eta_j) \leq 3R\ell,$$

so

$$d(\xi_1, \xi_2) \leq 3R\mathsf{d}(\xi_1, \xi_2), \tag{IV.4.4}$$

which is the lower bound.

Given ξ_1, ξ_2 in $\mathcal{G}$ joined by a minimizing geodesic γ. Let ℓ be the integer for which

$$(\ell - 1)R \leq d(\xi_1, \xi_2) < \ell R,$$

and $\xi_1 = \eta_0, \ldots, \eta_\ell = \xi_2$ evenly spaced points on γ. So,

$$d(\eta_{j-1}, \eta_j) = \frac{d(\xi_1, \xi_2)}{\ell} < R.$$

To each η_α, $\alpha = 1, \ldots, \ell - 1$, there exists $\zeta_\alpha \in \mathcal{G}$ such that $d(\zeta_\alpha, \eta_\alpha) < R$, which implies, by the triangle inequality, $d(\zeta_{j-1}, \zeta_j) < 3R$ for $j = 1, \ldots, \ell$ (we are setting $\xi_1 = \zeta_0$, $\xi_2 = \zeta_\ell$), which implies $\mathsf{d}(\zeta_{j-1}, \zeta_j) < 1$ for $j = 1, \ldots, \ell$, which implies

$$\mathsf{d}(\xi_1, \xi_2) \leq \ell \leq \frac{1}{R} d(\xi_1, \xi_2) + 1,$$

which implies the claim. ∎

Definition. Let M be a Riemannian manifold. A subset $\mathcal{G}$ of M is said to be ϵ–*separated*, $\epsilon > 0$, if the distance between any two distinct points of $\mathcal{G}$ is greater than or equal to ϵ.

Lemma IV.4.1. *Let M be complete, with*

$$\mathrm{Ric} \geq (n-1)\kappa, \qquad \kappa \leq 0 \tag{IV.4.5}$$

on all of M, and $\mathcal{G}$ an ϵ–separated subset of M. Then,

$$\mathrm{card}\,\{\mathcal{G} \cap B(x;r)\} \leq \frac{V_\kappa(2r + \epsilon/2)}{V_\kappa(\epsilon/2)}$$

for all $x \in M$ and $r > 0$.

Proof. We first comment that, without the lower bound on the Ricci curvature, one knows that there are only a finite number of elements of $\mathcal{G}$ in $B(x;r)$.

Indeed,

$$B(x;r+\epsilon/2) \supseteq \bigcup_{\xi \in \mathcal{G}\cap B(x;r)} B(\xi;\epsilon/2),$$

where the union on the right-hand side is disjoint union. Therefore,

$$\begin{aligned} V(x;r+\epsilon/2) &\geq \sum_{\xi \in \mathcal{G}\cap B(x;r)} V(\xi;\epsilon/2) \\ &\geq \operatorname{card}\{\mathcal{G}\cap B(x;r)\} \inf_{\eta \in \mathcal{G}\cap B(x;r)} V(\eta;\epsilon/2). \end{aligned}$$

But Corollary II.8.1 implies $\inf_{\eta\in\mathcal{G}\cap B(x;r)} V(\eta;\epsilon/2) > 0$. So, the real question is to obtain an upper bound for card $\mathcal{G}\cap B(x;r)$ that depends only on r and ϵ.

Since card $\mathcal{G}\cap B(x;r)$ is finite, there exists $\xi \in \mathcal{G}\cap B(x;r)$ such that

$$V(\xi;\epsilon/2) = \inf_{\eta\in\mathcal{G}\cap B(x;r)} V(\eta;\epsilon/2).$$

Therefore, we have, using Theorem III.4.5,

$$\operatorname{card}\{\mathcal{G}\cap B(x;r)\} \leq \frac{V(x;r+\epsilon/2)}{V(\xi;\epsilon/2)} \leq \frac{V(\xi;2r+\epsilon/2)}{V(\xi;\epsilon/2)} \leq \frac{V_\kappa(2r+\epsilon/2)}{V_\kappa(\epsilon/2)}. \quad ■$$

Definition. Let M be a complete Riemannian manifold. A *discretization of M* is a graph **G** determined by an ϵ–separated subset $\mathcal{G}$ of M, for which there exists $R > 0$ such that

$$M = \bigcup_{\xi\in\mathcal{G}} B(\xi;R). \tag{IV.4.6}$$

Then, ϵ is called the *separation radius*, and R the *covering radius of the discretization*. As before, the graph structure **G** is determined by the collection of neighbors of ξ,

$$\mathsf{N}(\xi) := \{\mathcal{G}\cap B(\xi;3R)\}\setminus\{\xi\},$$

for each $\xi \in \mathcal{G}$.

Remark IV.4.2. Note card $\mathsf{N}(\xi) \geq 1$ for all ξ.

Remark IV.4.3. To achieve the local uniformity of the geometry required to calibrate volumes in the manifold by those in in discretizations, we will require Gromov's improvement (Theorem III.10) of Bishop's theorem (Theorem III.9). In the case of strong unformity of coverings of compact manifolds, we only required Bishop's theorem.

Note that when the Ricci curvature is bounded from below as in (IV.4.5), then for the graph $\mathbf{G}$ we have

$$1 + m(\xi) \leq \frac{V_\kappa(4R + \epsilon/2)}{V_\kappa(\epsilon/2)} := \mathbf{M}_{\epsilon,2R}$$

for all $\xi \in \mathbf{G}$ – so $\mathbf{G}$ has bounded geometry.

To conveniently formulate for future reference,

Corollary IV.4.1. *If M is complete, then any two discretizations are roughly isometric.*

On the collection of vertices $\mathcal{G}$, we have two natural measures. The first is simply the *counting measure* $d\iota$; thus, for any subset $\mathcal{K}$ of $\mathcal{G}$ we have

$$\iota(\mathcal{K}) = \operatorname{card} \mathcal{K}.$$

The second is what we call the *volume measure* $d\mathsf{V}$ *on* $\mathcal{G}$, defined by

$$d\mathsf{V}(\xi) = m(\xi)\, d\iota(\xi).$$

Of course, when $\mathbf{G}$ has bounded geometry, the two measures are commensurate in the sense that the Radon–Nikodym derivative of $d\mathsf{V}$ with respect to $d\iota$ is uniformly bounded away from 0 and $+\infty$. Since in what follows we generally discuss graphs of bounded geometry, and we are only interested in qualitative estimates on volumes, we shall work with the counting measure $d\iota$ – even when we announce the results in terms of the volume measure $d\mathsf{V}$.

Definition. We define

$$\mathsf{V}(\xi; r) = \mathsf{V}(\beta(\xi; r))$$

for any $\xi \in \mathcal{G}, r > 0$.

We say that $\mathbf{G}$ *has exponential volume growth* if

$$\limsup_{r \uparrow +\infty} \frac{\ln \mathsf{V}(\xi; r)}{r} > 0;$$

otherwise, we say that $\mathbf{G}$ *has subexponential volume growth*. Also, we say that $\mathbf{G}$ has *polynomial volume growth* if there exists $k > 0$ such that

$$\mathsf{V}(\xi; r) \leq \text{const.} r^k$$

for sufficiently large $r > 0$.

Theorem IV.4.2. (M. Kanai (1985)) *Let* $\mathbf{G}, \mathbf{F}$ *be connected, roughly isometric graphs, both with bounded geometry. Then,* $\mathbf{G}$ *has polynomial (resp., exponential) volume growth if and only if* $\mathbf{F}$ *has polynomial (resp., exponential) volume growth.*

Proof. If $\phi : \mathbf{G} \to \mathbf{F}$ is a rough isometry, then

$$a^{-1}\mathsf{d}(\xi_1, \xi_2) - b \leq \mathsf{d}(\phi(\xi_1), \phi(\xi_2)) \leq a\mathsf{d}(\xi_1, \xi_2) + b$$

for all ξ_1, ξ_2 in $\mathbf{G}$, which implies (by Proposition IV.4.2)

$$\operatorname{card} \beta(\xi; r) \leq \mu \operatorname{card} \phi(\beta(\xi; r)) \leq \mu \operatorname{card} \beta(\phi(\xi); ar + b)$$

which implies the claim. ∎

Lemma IV.4.2. *Let* M *be a complete Riemannian manifold, with Ricci curvature bounded from below as in* (IV.4.5), *and assume there exist positive constants* r_0 *and* V_0 *such that*

$$V(x; r_0) \geq V_0$$

for all $x \in M$. *Then, for any* $r > 0$, *one has a positive constant* const._r *such that*

$$V(x; r) \geq \text{const.}_r$$

for all $x \in M$.

Proof. If $r > r_0$, then simply use V_0. If $r < r_0$, then simply note that the Bishop–Gromov theorem implies

$$V(x; r) \geq \frac{V_\kappa(r)}{V_\kappa(r_0)} V(x; r_0) \geq \frac{V_\kappa(r)}{V_\kappa(r_0)} V_0,$$

which implies the claim. ∎

Theorem IV.4.3. (M. Kanai (1985)) *Let* M *be a complete Riemannian manifold, with Ricci curvature bounded from below as in* (IV.4.5). *Then, for any discretization* $\mathbf{G}$ *of* M, $\mathbf{G}$ *has polynomial (resp., exponential) volume growth only if (resp., if)* M *has polynomial (resp., exponential) volume growth.*

If, on the other hand, there exist positive constants r_0 *and* V_0 *such that*

$$V(x; r_0) \geq V_0$$

for all $x \in M$, then for any discretization **G** *of* M, **G** *has polynomial (resp., exponential) volume growth if (resp., only if)* M *has polynomial (resp., exponential) volume growth.*

Proof. If $\xi \in \mathbf{G}$, $y \in B(\xi;r)$, then there exists $\eta \in \mathcal{G} \cap B(y;R)$, which implies

$$d(\xi,\eta) < r + R, \quad \Rightarrow \quad B(\xi;r) \subseteq \bigcup_{\eta\in\mathcal{G}\cap B(\xi;r+R)} B(\eta;R),$$

which implies

$$V(\xi;r) \leq V_\kappa(R)\,\mathrm{card}\,\mathcal{G} \cap B(\xi;r+R).$$

But

$$\mathsf{d}(\xi_1,\xi_2) \leq A\,d(\xi_1,\xi_2) + B$$

by (IV.4.3), which implies

$$V(\xi;r) \leq \text{const.}\,\mathrm{card}\,\beta(\xi;A(r+R)+B).$$

Therefore, **G** (resp., M) has polynomial (resp., exponential) volume growth only if the same holds for M (resp., **G**).

For the second claim, we have, by the previous lemma, for every $\xi \in \mathcal{G}$, $\rho > 0$,

$$\text{const.}_{\epsilon/2}\,\mathrm{card}\,\mathcal{G} \cap B(\xi;\rho) \leq \sum_{\eta\in\mathcal{G}\cap B(\xi;\rho)} V(\eta;\epsilon/2) \leq V(\xi;\epsilon/2+\rho)$$

(where ϵ is the separation of the discretization), that is,

$$\text{const.}_{\epsilon/2}\,\mathrm{card}\,\mathcal{G} \cap B(\xi;\rho) \leq V(\xi;\epsilon/2+\rho).$$

Now (IV.4.4) implies

$$\beta(\xi;\rho) \subseteq B(\xi;2R\rho),$$

which implies

$$\text{const.}_{\epsilon/2}\,\mathrm{card}\,\beta(\xi;\rho) \leq V(\xi;\epsilon/2+2R\rho).$$

Therefore, if M (resp., **G**) has polynomial (resp., exponential) volume growth, then so does **G** (resp., M). ∎

Corollary IV.4.2. *Suppose both M_1, M_2 are complete Riemannian manifolds with Ricci curvature bounded from below, for which there exist $r_j, V_j > 0$ such that*

$$V(x_j;r_j) \geq V_j$$

for all $x_j \in M_j$, $j = 1, 2$. If M_1 and M_2 are roughly isometric, then they both have the same type of volume growth.

Proof. The corollary follows directly from the above two theorems. ■

§IV.5. The Free Homotopy Classes

M is our given Riemannian manifold.

Definition. A *loop in M* is a map $\Gamma : \mathbb{S}^1 \to M \in C^0$.

Of course, any loop Γ in M is equivalent to a 2π–periodic map $\gamma : \mathbb{R} \to M$ determined by

$$\gamma(\theta) = \Gamma(\mathcal{E}^{i\theta}). \tag{IV.5.1}$$

(See Corollary IV.1.1) As a map of manifolds, for any $k \geq 1$, $\Gamma \in C^k$ if and only if $\gamma \in C^k$. Similarly, we shall say $\Gamma \in D^k$ if and only if for any bounded interval $[\alpha, \beta] \subseteq \mathbb{R}$, $\gamma|[\alpha, \beta] \in D^k$ (see §I.6).

Definition. For any $\Gamma \in C^1$, and $p \in \mathbb{S}^1$, we define the *velocity vector of* Γ *at* $\Gamma(p)$, $\Gamma'(p)$, by

$$\Gamma'(p) = \gamma'(\theta_0), \quad p = \mathcal{E}^{i\theta_0}, \tag{IV.5.2}$$

where γ is given by (IV.5.1). For $\Gamma \in D^1$, we may define the *length of* Γ, $\ell(\Gamma)$, by

$$\ell(\Gamma) = \int_0^{2\pi} |\gamma'|. \tag{IV.5.3}$$

Definition. A *closed geodesic in M* is a differentiable loop $\Gamma : \mathbb{S}^1 \to M$ such that γ given by (IV.5.1) is a geodesic.

Definition. Two loops, Γ_0 and Γ_1, in M are *freely homotopic* if there exists $\Omega : \mathbb{S}^1 \times [0, 1] \to M \in C^0$ such that

$$\Omega|\mathbb{S}^1 \times \{0\} = \Gamma_0, \quad \Omega|\mathbb{S}^1 \times \{1\} = \Gamma_1. \tag{IV.5.4}$$

"Free homotopy" determines an equivalence relation on the class of loops in M, but has no obvious group structure.

Definition. A *trivial free homotopy class* will be one which contains a constant map.

Theorem IV.5.1. *Let Λ be a nontrivial free homotopy class of loops in a differentiable manifold M. Then, Λ contains loops that are D^1. If M is compact Riemannian, then*

$$\lambda := \inf\{\ell(\Gamma) : \Gamma \in \Lambda \cap D^1\} > 0. \tag{IV.5.5}$$

In this case, there actually exists $\Gamma_0 \in \Lambda \cap D^1$ with length equal to λ. Finally, Γ_0 is a closed geodesic.

Proof. The first claim is easy and will be left to the reader.

To show that λ is strictly positive when M is compact Riemannian, we note that the compactness of M implies (Theorem I.3.2) that $\epsilon := \operatorname{inj} M$ is positive. Thus, if $\Gamma \in \Lambda \cap D^1$ and $\ell(\Gamma) < \epsilon$, then $\Gamma(\mathbb{S}^1) \subseteq B(p;\epsilon)$ for any $p \in \Gamma(\mathbb{S}^1)$, which would imply that Γ is homotopic to a constant map – a contradiction. So, λ is positive.

If there exists $\Gamma_0 \in \Lambda \cap D^1$ satisfying $\ell(\Gamma_0) = \lambda$, then Γ_0 is a geodesic, by the first variation formula (Theorem II.4.1) (there are no boundary terms by the periodicity of γ associated with Γ) and the argument of Theorem II.4.2.

It remains to show that if M is compact Riemannian, then there exists $\Gamma_0 \in \Lambda \cap D^1$ for which $\ell(\Gamma_0) = \lambda$. Well, assume we are given a sequence $\Gamma_j : \mathbb{S}^1 \to M \in \Lambda \cap D^1$ for which $\ell(\Gamma_j) \downarrow \lambda$ as $j \to +\infty$. We leave it to the reader to verify that we may assume $|\Gamma_j{}'|$ is constant on $\mathbb{S}^1$ for each j – the constant will be equal to $\ell(\Gamma_j)/2\pi$. Let $\alpha = \sup \ell(\Gamma_j) < +\infty$. Consider $\mathbb{S}^1$ as a compact Riemannian manifold with standard metric. Then, for $q, q^* \in \mathbb{S}^1$ we have

$$d(\Gamma_j(q), \Gamma_j(q^*)) \leq (\alpha/2\pi) d(q, q^*). \tag{IV.5.6}$$

Let ϵ be as above, that is, for any $p \in M$, $\exp|\mathsf{B}(p;\epsilon)$ is a diffeomorphism and fix an integer $N > \max\{\alpha/\epsilon, 2\}$. Pick $q_k = e^{i2\pi k/N}$, $k = 0, 1, \ldots, N$ (of course, $q_0 = q_N$), and let ω_k be the closed segment on $\mathbb{S}^1$ of length $2\pi/N$ from q_{k-1} to q_k. Let $p_{k;j} = \Gamma_j(q_k)$, let $\gamma_{k;j} : \omega_k \to M$ denote the minimizing geodesic joining $p_{k-1;j}$ to $p_{k;j}$, and let $\overline{\Gamma}_j$ denote the piecewise-geodesic loop given by

$$\overline{\Gamma}_j|\omega_k = \gamma_{k;j}, \qquad k = 1, \ldots, N.$$

Then, $\Gamma_j|\omega_k$ and $\gamma_{k;j}$ are contained in $B(p_{k-1;j};\epsilon)$, which implies that $\overline{\Gamma}_j$ is freely homotopic to Γ_j, which implies

$$\lambda \leq \ell(\overline{\Gamma}_j) \leq \ell(\Gamma_j)$$

for all j, which implies $\ell(\overline{\Gamma}_j) \to \lambda$ as $j \to 0$. So, we only have to show that there exists a subsequence of $\overline{\Gamma}_j$ which converges to a D^1 loop.

Set $\xi_{k;j} = \overline{\gamma}_j{}'(q_{k-1}+)$, that is, $\xi_{k,j}$ is the initial unit velocity vector of the geodesic segment $\gamma_{k;j}$. Since M is compact, so is its unit tangent bundle. We therefore have a subsequence $\{j_r\}$, and unit tangent vectors ξ_k, $k = 1, \ldots, N$ such that

$$\xi_{k;j_r} \to \xi_k, \qquad k = 1, \ldots, N.$$

The continuity of the exponential map easily implies the existence of a limit piecewise-geodesic loop of some subsequence of $\overline{\Gamma}_j$. ∎

Remark IV.5.1. One can carry out the limit argument using the Arzela–Ascoli theorem. Namely, (IV.5.6) implies that the sequence of mappings (Γ_j) is equicontinuous. Since M is compact, the Arzela–Ascoli Theorem implies that (Γ_j) converges uniformly to a loop $\overline{\Gamma} : \mathbb{S}^1 \to M \in C^0$. Fix an integer $N > \max\{\alpha/\epsilon, 2\}$, and set $q_k = e^{i2\pi k/N}$, $k = 0, 1, \ldots, N$, $p_\ell = \overline{\Gamma}(q_\ell)$.

Note that $\overline{\Gamma}$ also satisfies (IV.5.6). Therefore, $d(p_{k-1}, p_k) < \epsilon$ for all $k = 1, \ldots, N$. Let $\gamma_k : \omega_k \to M$ be the unit speed geodesic of length $d(p_{k-1}, p_k)$ joining p_{k-1} to p_k, and define Γ_0 by $\Gamma_0|\omega_k = \gamma_k$, $k = 1, \ldots, N$. Since $\gamma_k(\omega_k) \subseteq B(p_{k-1}; \epsilon)$ for each k one easily has $\Gamma_0 \in \Lambda$. Of course, $\Gamma_0 \in D^1$. To evaluate $\ell(\Gamma_0)$, one has by definition $\ell(\Gamma_0) \geq \lambda$; but, on the other hand,

$$\begin{aligned}\ell(\Gamma_0) &= \sum_{k=1}^{N} \ell(\gamma_k) = \lim_{j\to\infty} \sum_{k=1}^{N} d(\Gamma_j(q_{k-1}), \Gamma_j(q_k)) \\ &\leq \lim_{j\to\infty} \sum_{k=1}^{N} \ell(\Gamma_j|\omega_k) = \lim_{j\to\infty} \ell(\Gamma_j) = \lambda,\end{aligned}$$

which implies $\ell(\Gamma_0) = \lambda$.

For an argument in a similar spirit, see Exercise I.7.

§IV.6. Notes and Exercises

Parallel Translation and Curvature

Exercise IV.1. Let M be a Riemannian manifold, $\mathcal{T}^k M$ the alternating k–vector bundle over M, with naturally induced metrics on the fibers. Thus, if $p \in M$, $\{e_1, \ldots, e_n\}$ an orthonormal basis of M_p, then an orthonormal basis of the fiber over p is given by the collection of k–vectors

$$\{e_{j_1} \wedge \cdots \wedge e_{j_k} : 1 \leq j_1 < \ldots < j_k \leq n\}.$$

Show that if $k \leq n$, $\phi : G \to M$ is an imbedding of an open subset G of $\mathbb{R}^k$ into M (thus, ϕ^{-1} is a chart on the k–dimensional submanifold $\phi(G)$), and we set

$$\partial_j \phi = \phi_* \frac{\partial}{\partial u^j},$$

$j = 1, \ldots, k$, where $\partial/\partial u^j$ are natural coordinate vector fields on $\mathbb{R}^k$, then

$$dV_{\phi(G)} = |\partial_1\phi \wedge \cdots \wedge \partial_k\phi|\, du^1 \cdots du^k,$$

relative to the chart ϕ^{-1} on $\phi(G)$.

Exercise IV.2. (a) Prove that for a C^1 path ω in a Riemannian manifold M, and a C^1 vector field X along ω, one has

$$|X|' \leq |\nabla_t X|.$$

(b) Let $v : [0, 1] \times [0, 1] \to M \in D^1$ be a homotopy with fixed endpoints

$$p = v(0, s), \qquad q = v(1, s).$$

Let $X = X(t, s)$ be a vector field along v such that

$$X(0, s) = x_0 \in M_p, \qquad \nabla_t X = 0.$$

We want to estimate, quantitatively, the difference of parallel translation along $t \mapsto v(t, 0)$ from $t \mapsto v(t, 1)$, or, equivalently, $|X(1, 1) - X(1, 0)|$. Prove

$$|X(1, 1) - X(1, 0)| \leq \frac{4}{3}\{\sup |X|\}\Lambda \int_0^1 ds \int_0^1 |\partial_t v \wedge \partial_s v|\, dt,$$

where $\Lambda = \sup |\mathcal{K}|$.

The Myers–Steenrod Theorem

Exercise IV.3. Prove Lemma IV.1.1 in the following steps.

(a) Show that φ is a homeomorphism.

(b) Fix p and $\varphi(p)$ in M. Let δ_1 denote the injectivity radius of M at p. Show that we have a well-defined map $F : \mathsf{B}(p; \delta_1) \to \mathsf{B}(\varphi(p); \delta_1)$ defined by

$$F(\xi) = (\exp |\mathsf{B}(\varphi(p); \delta_1))^{-1} \circ \varphi \circ \exp \xi.$$

(c) Show that for $\xi \in \mathsf{B}(p; \delta_1)$, $s \in [0, 1]$ one has

$$F(s\xi) = sF(\xi). \tag{IV.6.1}$$

Then show that F may be extended to all of M_p so that it satisfies (IV.6.1) and

$$|F(\xi)| = |\xi| \tag{IV.6.2}$$

for all $\xi \in M_p$, $s \geq 0$.

(d) Next, use Exercise II.25 to show that, given any $\epsilon > 0$, there exists sufficiently small $\delta > 0$ so that

$$|F(\xi) - F(\eta)| = |\xi - \eta|\{1 \pm O(\epsilon^2)\}$$

for all $\xi, \eta \in \mathsf{B}(p; \delta)$.

(e) Next, show $|F(\xi) - F(\eta)| = |\xi - \eta|$ for all $\xi, \eta \in M_p$.

(f) Let $|\xi| = |\eta| = 1$. Use the formula

$$|\xi - \eta| = 2 \sin \frac{1}{2} \measuredangle(\xi, \eta)$$

and (IV.6.2) to show that

$$\sin \frac{1}{2} \measuredangle(F(\xi), F(\eta)) = \sin \frac{1}{2} \measuredangle(\xi, \eta),$$

which therefore implies

$$\cos \measuredangle(F(\xi), F(\eta)) = \cos \measuredangle(\xi, \eta),$$

which implies F preserves the inner product.

(g) Use the expansion of vectors with respect to an orthonormal basis of an inner product space to show that F is additive, and therefore, linear. Then show that $F = \varphi_{*|p}$, which implies the lemma.

Note IV.1. The argument presented from (c) onward also proves that any metric preserving transformation Φ of $\mathbb{R}^n$ is a Euclidean transformation, that is, it is given by

$$\Phi(x) = Ax + a,$$

where a is a vector in $\mathbb{R}^n$, and A is an element of the orthogonal group $\mathcal{O}(n)$ of $\mathbb{R}^n$.

Deck Transformation Groups, Discrete Groups, and Tori

Exercise IV.4. Let $\pi_1 : M_o \to M_1$ and $\pi_2 : M_o \to M_2$ denote two Riemannian universal coverings (so M_o is simply connected) with respective deck transformation groups Γ_1 and Γ_2. Show that M_1 and M_2 are isometric if and only if Γ_1 and Γ_2 are conjugate subgroups of the full group of isometries M_o.

Exercise IV.5. Let $\mathcal{O}(n)$ denote the orthogonal group of $\mathbb{R}^n$, and $\mathcal{E}(n)$ the Euclidean transformation group of $\mathbb{R}^n$ – so $T \in \mathcal{E}(n)$ if

$$Tx = Ax + a, \qquad A \in \mathcal{O}(n), \;\; a \in \mathbb{R}^n.$$

(a) Show that a subgroup of $\mathcal{O}(n)$ is discrete if and only if it is finite.

(b) Show that any discrete subgroup Γ of translations of $\mathbb{R}^n$ must be of the form

$$\Gamma = \left\{ x \mapsto x + a : a = \sum_{j=1}^{k} n_j \mathbf{v}_j \right\},$$

where n_j, $j = 1, \ldots, k$, vary over the integers, and $\mathbf{v}_1, \ldots, \mathbf{v}_k$ are k fixed linearly independent vectors in $\mathbb{R}^n$.

(c) Let $Tx = Ax + a$ be a Euclidean transformation of $\mathbb{R}^n$ with no fixed points. Show that there exists a line along which T is a translation.

Exercise IV.6. Let $\pi : \mathbb{R}^n \to M_o$ be a Riemannian covering, M_o compact, Γ the deck transformation group of the covering. Show that Γ is a discrete subgroup of $\mathcal{E}(n)$.

Exercise IV.7. Let $n = 2$. Categorize the discrete subgroups of $\mathcal{E}(2)$ and thereby characterize the 2–dimensional compact flat Riemannian manifolds.

Exercise IV.8. A slightly different problem is, for example, to determine when two different parallelograms in $\mathbb{R}^2$ determine the same Riemannian torus.

Global Cartan–Ambrose–Hicks Theorem

Exercise IV.9. Let M be a connected Riemannian manifold, ϕ and ψ two isometries of M onto itself. Suppose there exists a point $p \in M$ for which $\phi(p) = \psi(p)$ and $\phi_{*|p} = \psi_{*|p}$. Show that $\phi = \psi$.

Definition. Let M, N be Riemannian manifolds of the same dimension, U_j, $j = 1, 2$, domains in M with nonempty intersection, and ϕ_j, $j = 1, 2$ isometries of the domains U_j into N, such that $\phi_1|U_1 \cap U_2 = \phi_2|U_1 \cap U_2$. Then, we refer to ϕ_1 and ϕ_2 as *immediate continuations, one of the other.*

Let ϕ be an isometry of a domain $U \subseteq M$ onto a domain in N. Let $\omega(t)$, $0 \le t \le 1$, be a continuous curve in M such that $\omega(0) \in U$. The isometry ϕ is said to be *extendable along* ω if for each $t \in [0, 1]$ there exists an isometry ϕ_t of a domain U_t containing $\omega(t)$ onto an open subset of N such that $\phi_0 = \phi$, and such that ϕ_t, ϕ_s are immediate continuations whenever $|t - s|$ is sufficiently small. The family $\{\phi_t : t \in [0, 1]\}$ is called a *continuation of* ϕ *along* ω.

Exercise IV.10.

(a) Let M, N be complete real analytic Riemannian manifolds, and ϕ an isometry of a domain $U \subseteq M$ onto a domain in N. Let $\omega(t)$, $t \in [0, 1]$, be a continuous curve in M such that $\omega(0) \in U$. Prove that ϕ is extendable along ω.

(b) Suppose, in addition, that the path $\sigma(t)$, $t \in [0, 1]$, is continuous, and homotopic (fixed endpoints) to ω. Let $\{\phi_t\}$, $\{\psi_t\}$ be continuations of ϕ along ω, σ, respectively. Show that $\phi_1 = \psi_1$ on some neighborhood of $\omega(1) = \sigma(1)$.

Note IV.2. One can find details for the two previous exercises in Helgason (1962, pp. 62–64).

Exercise IV.11. (See Exercises III.1–III.5.)

(a) Show that if M, N are locally symmetric then we do not require the hypothesis of real analyticity in Exercise IV.10(b) above (even though it is a general theorem that a locally symmetric space is real analytic (Helgason (1962, p. 187))).

(b) Show that a simply connected complete locally symmetric Riemannian manifold is Riemannian symmetric.

On the Myers Comparison Theorem

Note IV.3. One has the following generalization of Theorem IV.2.2. Assume M is complete, with nonnegative Ricci curvature. Fix $x \in M$. Then, the Gromov theorem (Theorem III.4.5) implies that $V(x;r)/\omega_{\mathbf{n}}r^n$ is a decreasing function of r. Set

$$\alpha_M = \lim_{r \uparrow +\infty} \frac{V(x;r)}{\omega_{\mathbf{n}}r^n}.$$

Then, α_M is independent of x – of course $\alpha_M \leq 1$. M. T. Anderson (1990b) has proved that if $\alpha_M > 0$, then the order of $\pi_1(M)$ is bounded above by $1/\alpha_M$.

Manifolds of Nonpositive Curvature

It is an immediate consequence of the Hadamard–Cartan theorem (Theorem IV.1.3) that a complete simply connected manifold of nonpositive curvature is diffeomorphic to Euclidean of the same dimension.

Exercise IV.12. (A. Preissmann (1943)) Prove that, if M is a complete simply connected Riemannian manifold of nonpositive curvature, then

(a) every two points of M are connected by precisely one geodesic; the geodesic is minimizing, and it varies differentiably with respect to its endpoints;

(b) if given a geodesic triangle with sides a, b, c and angle θ at the vertex opposite the side of length c, then

$$c^2 \geq a^2 + b^2 - 2ab\cos\theta;$$

what if all the sectional curvatures are bounded above by the constant $\delta < 0$?;

(c) the sum of the angles of a geodesic triangle is less than or equal to π, with equality if and only if the geodesics span a totally geodesic surface isometric to a Euclidean triangle;

(d) the sum of the angles of a geodesic quadrilateral is less than or equal to 2π, with equality if and only if the geodesics span a totally geodesic surface isometric to a Euclidean quadrilateral.

Exercise IV.13. Let M be as in the previous exercise, and let $\gamma : M \to M$ be an isometry. Define

$$\delta_\gamma = \inf_x d(x, \gamma \cdot x),$$

and assume $\delta_\gamma > 0$ (of course, if γ has a fixed point, then $\delta_\gamma = 0$). An *axis of* γ is a unit speed D^1 path $\omega : \mathbb{R} \to M$ such that

$$\gamma \cdot \omega(t) = \omega(t + \delta_\gamma).$$

(a) Prove that an axis is a geodesic.

(b) Prove that if $\delta_\gamma > 0$ and there exists an $x \in M$ such that

$$\delta_\gamma = d(x, \gamma \cdot x),$$

then the geodesic containing the minimizing geodesic segment joining x to $\gamma \cdot x$ is an axis.

(c) Prove that, except for shift and reorientation of the parameter t, the isometry γ possesses more than one axis only if any two such axes bound a totally geodesic surface isometric to a flat infinite strip $(\alpha, \beta) \times \mathbb{R}$ with canonical Euclidean metric.

(d) Now assume that the curvature of M is strictly negative. Show that any isometry has at most one axis.

(e) Continue with the assumption of strictly negative curvature. Let γ_1, γ_2 be isometries of M which commute (in their action on M). Assume γ_1 possesses an axis ω_1. Prove that $\gamma_2|\omega_1$ maps ω_1 to itself. Show that if γ_2 also has an axis, then the axis must be ω_1.

Exercise IV.14. Prove:

Preissmann's Theorem (1943). *Let $\pi : M \to M_o$ be a covering by simply connected M with strictly negative curvature and Γ_0 an abelian subgroup of Γ, the deck transformation group of the covering. Then, M_o compact implies that Γ_0 is cyclic.*

Note IV.4. See Eberlein–O'Neill (1973) for extensive discussion of complete Riemannian manifolds of negative curvature, as generalizations of hyperbolic geometry. More recent, and still fuller, discussion is to be found in Ballman–Gromov–Schroeder (1985).

Fundamental Domains

Exercise IV.15. Here is a different construction of a fundamental domain of a covering $\pi : M \to M_o$, with deck transformation group Γ. Fix $x \in M$ and define *the Dirichlet fundamental domain based at* x, Dir_x, by

$$\mathrm{Dir}_x = \{y \in M : d(x, y) < d(x, \gamma\cdot y)\ \forall\gamma \in \Gamma, \gamma \neq \mathrm{id}_M\}.$$

Prove:

(a) Dir_x is, in fact, a fundamental domain of the covering;

(b) when M_o is compact, with diameter $d(M_o)$, then $\mathrm{Dir}_x \subseteq B(x; d(M_o))$.

(c) Let $\Gamma_x = \{\gamma \neq \mathrm{id}_M : \overline{\gamma(\mathrm{Dir}_x)} \cap \overline{\mathrm{Dir}_x} \neq \emptyset\}$; and show that

$$\mathrm{Dir}_x = \{y \in M : d(x, y) < d(x, \gamma\cdot y)\ \forall\gamma \in \Gamma_x, \gamma \neq \mathrm{id}_M\}.$$

Exercise IV.16. Let M be a noncompact covering a compact, with deck transformation group Γ and fundamental domain F. Show that there exists a positive constant so that for every $x \in M$, we have

$$A(S(x; r) \cap F) \leq \text{const.}$$

for almost all $r > 0$.

Coverings by Compacta

Exercise IV.17. Let $\pi : M \to M_o$ be a a nonsingular differentiable mapping of M onto M_o. Assume that M is compact. Show that π is a covering.

Exercise IV.18. Let $\pi : M \to M_o$ be a Riemannian covering with deck transformation group Γ. Assume that M is compact. Show that

$$V(M) = V(M_o)\mathrm{card}\,\Gamma.$$

(We casually assumed this result, when stating in §III.3 that the volume of real projective space $\mathbb{P}^n$ is $1/2$ that of the sphere $\mathbb{S}^n$.)

Exercise IV.19. Let $\mathbf{x} : M \to \mathbb{R}^n$ be a Riemannian immersion of the compact $(n-1)$–manifold M into $\mathbb{R}^n$, where all of the Riemannian sectional curvatures of M are positive. Show that the associated Gauss map $\mathbf{n} : M \to \mathbb{S}^2$ is a diffeomorphism.

On Theorem IV.4.1

Note IV.5. The theorem was originally presented in our first edition following the formulation and proof of M. Kanai (1985). For a discrete subset $\mathcal{G}$ whose disks of radius $R > 0$ cover the Riemannian manifold M, he defined the associated graph by $\mathsf{N}(\xi) = \{\eta \in \mathcal{G} \setminus \{\xi\} : d(\eta, \xi) < 2R\}$, for every $\xi \in \mathcal{G}$. He had to postulate that the Ricci curvature was bounded uniformly from below. I. Holopainen (1994) realized that if one defined $\mathsf{N}(\xi)$ by

$$\mathsf{N}(\xi) = \{\eta \in M \setminus \{\xi\} : d(\eta, \xi) < 3R\},$$

then one could drop the hypothesis of Ricci curvature from below!

Homotopy Considerations

Exercise IV.20. Prove:

J. L. Synge's Lemma (1936). *Let M be a compact, even-dimensional orientable Riemannian manifold with strictly positive curvature. Then, M is simply connected.*

Exercise IV.21. Use the argument of Theorem IV.5.1 to show that if M is compact, then given any real $\rho > 0$, there are at most a finite number of free homotopy classes with minimizing geodesic having length less than or equal to ρ.

Exercise IV.22. Formulate and prove the corresponding version of Theorem IV.5.1 for homotopy classes of closed paths with a fixed base point. Note that here the manifold need not be compact – only complete.

Exercise IV.23. (A. Preissmann (1943, pp. 191ff)) The proof given in Theorem IV.5.1 and the above two exercises exist "downstairs"– in the Riemannian manifold itself. Another approach is to go "upstairs" – to use the universal covering. It goes as follows: Given a Riemannian manifold M_o, with universal cover $\pi : M \to M_o$ and associated deck transformation group Γ. Prove:

(a) Given $x_o \in M_o$, $x \in \pi^{-1}[x_o]$, and $\gamma \in \Gamma$. Then, $d(x, \gamma \cdot x)$ is the minimum length of all D^1 paths in the homotopy class in $\pi_1(M_o, x_o)$ determined by γ. Furthermore, the minimum length is realized by projecting, under π, a minimizing geodesic segment joining x to $\gamma \cdot x$ to a geodesic loop in M based at x.

(b) The collection of free homotopy classes of M_o are in one-to-one correspondence with the elements of Γ.

(c) If M_o is compact, then any fundamental domain F of M_o in M has compact closure. Now use the argument of (a) to derive a second proof of Theorem IV.5.1.

(d) If M_o is compact with nonpositive sectional curvature, show that, for any nontrivial free homotopy class, the minimizing closed geodesic is covered by an axis of the element in Γ associated to the class.

(e) If M_o is compact, with strictly negative curvature, then any nontrivial free homotopy class has precisely one minimizing geodesic.

Also, if ω_o is the minimizer in M_o of the free homotopy class associated with $\gamma \in \Gamma$, then the minimizer of the free homotopy class determined by γ^k, $k \in \mathbb{Z}$, is the geodesic ω_o covered $|k|$ times, in the appropriate direction.

The Results in Length Spaces

See the discussion of length spaces in §I.9.

Exercise IV.24. (See Remark IV.5.1.) Let X be a compact length space.

(a) Show that every nontrivial free homotopy class in X has a minimizing geodesic. (See Remark IV.5.1.)

(b) Also show that given any real $\rho > 0$, there are at most a finite number of free homotopy classes with minimizing geodesic having length less than or equal to ρ.

Exercise IV.25. Formulate and prove corresponding versions of Theorem IV.5.1 and the previous exercise for homotopy classes of closed paths with a fixed base point.

On the Displacement Norm

Given a Riemannian manifold M with a group Γ of isometries acting freely and properly discontinuously on M, then to each $x \in M$, we associate the norm (see §IV.4) on Γ defined by $\|\gamma\|_x = d(x, \gamma \cdot x)$.

Exercise IV.26. Prove the following:

Theorem. (M. Gromov (1981, p. 43)) *Given a compact Riemannian manifold M_o, with universal cover $\pi : M \to M_o$ and associated deck transformation group Γ. Then, the fundamental group Γ is generated by those elements γ for which*

$$\|\gamma\|_x \leq 2d(M), \tag{IV.6.3}$$

where $d(M)$ denotes the diameter of M.

Exercise IV.27. Continue as in the previous exercise. Fix $x \in M$. Show that given any $\epsilon > 0$ there exists a positive constant σ, such that any $\gamma \in \Gamma$ can be written as a word

$$\gamma = \prod_j \gamma_j,$$

where

$$\|\gamma_j\|_x \le \sigma \quad \forall\, j, \qquad \sum_j \|\gamma_j\|_x \le (1+\epsilon)\|\gamma\|_x.$$

Existence of Closed Geodesics

Note IV.6. If a compact Riemnannian manifold is simply connected, then one cannot use Theorem IV.5.1 to guarantee the existence of simple closed geodesics on M.

An interesting approach, for Riemannian metrics on $\mathbb{S}^2$, was first posed by Poincaré (1905). The idea is to consider the variational problem of minimizing the length of those smooth simple closed curves on $\mathbb{S}^2$ that divide $\mathbb{S}^2$ into two domains of equal total Gauss curvature (i.e., $\int K\, dA = 2\pi$ for both domains). One can easily check (do it!) that should a smooth simply closed curve achieve the minimum length in this class, then it must be a geodesic. The existence of the minimum, and the positivity of its length, were carefully worked out in Croke (1982).

The existence of more than one geodesic on spheres, and other simply connected compact Riemannian manifolds, is the subject of much research. See Klingenberg (1982, §3.6) and his detailed monograph (1978).

V
Surfaces

In this chapter, we present, before resuming the general theory in all dimensions, a variety of results for oriented 2–dimensional Riemannian manifolds – surfaces (the phrase "Riemann surface" reserved for when the surface is orientable with constant curvature equal to -1). So, in all that follows,

Definition. A *surface* will be an oriented 2–dimensional Riemannian manifold.

We start with a topic motivated by the concluding one of Chapter IV. Namely, once one knows that in a nontrivial free homotopy class of a compact Riemannian manifold M there is a minimizing closed geodesic, one may ask for geometric estimates on its length, for example, to estimate its length against the volume of the manifold. Or, one may ask such a question for any homology class. Here, for surfaces, we estimate the length of the shortest homotopically nontrivial closed geodesic (among *all* homotopically nontrivial closed curves) against the area of M. This study was initiated by C. Loewner and P. Pu in the 1950s, almost completely dormant for 30 years, and resuscitated in the 1980s by M. Gromov. Here, we only introduce the subject.

Then, we turn (§V.2) to the celebrated Gauss–Bonnet theorem and formula, followed by (§V.3) B. Randol's collar theorem for compact Riemann surfaces, that is, surfaces of constant curvature -1. The result quite fundamental in the geometry of Riemann surfaces and in analysis on them and the proof is quite beautiful in its own right.

In §V.4, we begin discussion of one of the major themes of the rest of the book, the isoperimetric problem in Riemannian manifolds. The problem has its roots in classical antiquity, features a rich history of results and methods, and is still a subject of current research. In this chapter, we concentrate on two versions of the problem on surfaces: (i) for surfaces with curvature bounded from above (starting in the 1930s and 1940s, but updated in the 1980s) and

(ii) the isoperimetric problem for paraboloids of revolution (1980s and 1990s). In subsequent chapters, we consider different aspects of the problem in higher dimensions.

§V.1. Systolic Inequalities

Henceforth, M is compact.

Definition. The *systolic length of* M, $\ell(M)$, is the shortest homotopically nontrivial closed geodesic in M. The geodesic itself is referred to as the *systole.*

Theorem V.1.1. (C. Loewner) (P. M. Pu (1962)) *Let $\mathcal{G}$ denote a Riemannian metric on the 2–dimesional torus $\mathbb{T}^2$, with total area $A_{\mathcal{G}}$ and systole $\ell_{\mathcal{G}}$. Then,*

$$\frac{{\ell_{\mathcal{G}}}^2}{A_{\mathcal{G}}} \leq \frac{2}{\sqrt{3}}, \tag{V.1.1}$$

with equality if and only if $\mathcal{G}$ is a flat metric on $\mathbb{T}^2$ generated by the equilateral triangle.

Proof. We refer the reader to Exercises IV.6–IV.8 for background on tori covered by $\mathbb{R}^n$.

The uniformization theorem (Farkas–Kra (1980, Chapter IV)) implies that, given the Riemannian metric $\mathcal{G}$ on $\mathbb{T}^2$, there is a positive function $\phi : \mathbb{T}^2 \to (0, \infty)$ and a flat Riemannian metric $\mathcal{G}_o$ on $\mathbb{T}^2$ such that

$$\mathcal{G} = \phi^2 \mathcal{G}_o \qquad \text{on } \mathbb{T}^2.$$

Now, the torus acts isometrically on itself as a group translations of the flat metric. To distinguish between the two, we denote the arbitrary "point on $\mathbb{T}^2$" by q and the arbitrary "translation of $\mathbb{T}^2$" by T. The area element (for the flat metric $\mathcal{G}_o$) of the points will be denoted, as usual, by $dA_{\mathcal{G}_o}(q)$, and of the translations by $d\mu(T)$.

Let $\omega : \mathbb{S}^1 \to \mathbb{T}^2$ be any loop in $\mathbb{T}^2$, T any translation. Then,

$$L_{\mathcal{G}}(\omega) = \int_{\mathbb{S}^1} |\omega'(t)|_{\mathcal{G}}\, dt = \int_{\mathbb{S}^1} (\phi \circ \omega)(t) |\omega'(t)|_{\mathcal{G}_o}\, dt,$$

which implies

$$L_{\mathcal{G}}(T \cdot \omega) = \int_{\mathbb{S}^1} (\phi \circ (T \cdot \omega))(t) |(T \cdot \omega)'(t)|_{\mathcal{G}_o}\, dt = \int_{\mathbb{S}^1} (T^*\phi)(\omega(t)) |\omega'(t)|_{\mathcal{G}_o}\, dt.$$

If we restrict ω to noncontractible loops in $\mathbb{T}^2$, then $T \cdot \omega$ is noncontractible for all translations T, which implies

$$\ell_{\mathcal{G}} \leq \int_{\mathbb{S}^1} (T^*\phi)(\omega(t))|\omega'(t)|_{\mathcal{G}_o}\, dt$$

for all translations T.

Now, average the inequality over all translations $T \in \mathbb{T}^2$. Then,

$$\begin{aligned}
\ell_{\mathcal{G}} &\leq \frac{1}{\mu(\mathbb{T}^2)} \int_{\mathbb{T}^2} d\mu(T) \int_{\mathbb{S}^1} (T^*\phi)(\omega(t))|\omega'(t)|_{\mathcal{G}_o}\, dt \\
&= \int_{\mathbb{S}^1} dt\, \frac{1}{\mu(\mathbb{T}^2)} \int_{\mathbb{T}^2} (T^*\phi)(\omega(t))|\omega'(t)|_{\mathcal{G}_o}\, d\mu(T) \\
&= \Phi \int_{\mathbb{S}^1} |\omega'(t)|_{\mathcal{G}_o}\, dt, \\
&= \Phi L_{\mathcal{G}_o}(\omega),
\end{aligned}$$

where Φ is the constant given by

$$\Phi = \frac{1}{\mu(\mathbb{T}^2)} \int_{\mathbb{T}^2} (T^*\phi)(q)\, d\mu(q) = \frac{1}{A_{\mathcal{G}_o}(\mathbb{T}^2)} \int_{\mathbb{T}^2} \phi(q)\, dA_{\mathcal{G}_o}(q);$$

that is,

$$\ell_{\mathcal{G}} \leq \Phi L_{\mathcal{G}_o}(\omega). \tag{V.1.2}$$

If we minimize (V.1.2) over all noncontractible loops ω, then we obtain $\ell_{\mathcal{G}} \leq \Phi \ell_{\mathcal{G}_o}$. The Cauchy–Schwarz inequality implies

$$\frac{{\ell_{\mathcal{G}}}^2}{{\ell_{\mathcal{G}_o}}^2} \leq \Phi^2 \leq \frac{1}{A_{\mathcal{G}_o}(\mathbb{T}^2)} \int_{\mathbb{T}^2} \phi^2(q)\, dA_{\mathcal{G}_o}(q) = \frac{A_{\mathcal{G}}}{A_{\mathcal{G}_o}};$$

therefore,

$$\frac{{\ell_{\mathcal{G}}}^2}{A_{\mathcal{G}}} \leq \frac{{\ell_{\mathcal{G}_o}}^2}{A_{\mathcal{G}_o}}. \tag{V.1.3}$$

We conclude that the systolic ratio ℓ^2/A is maximized by a flat Riemannian metric on the torus, with ℓ^2/A maximal only if it is flat. The question is: Which metric is maximal among all the flat ones?

We may always multiply the Riemannian metric by a constant to normalize the area to equal to 1. So, assume we have a flat torus of area equal to 1 and systolic length ℓ. We identify the torus with its fundamental domain – a parallelogram in $\mathbb{R}^2$, and we may assume the lattice in $\mathbb{R}^2$ is generated by

$$\mathbf{e}_1 = \ell\mathbf{i}, \quad \mathbf{e}_2 = \alpha\ell\mathbf{i} + \frac{1}{\ell}\mathbf{j}, \qquad \alpha \in (-1/2, 1/2].$$

This describes the full collection of isometrically distinct flat tori of area equal to 1 and with systolic length ℓ.

Then,

$$\ell^2 \leq \alpha^2\ell^2 + \frac{1}{\ell^2} \quad \Rightarrow \quad (1-\alpha^2)\ell^2 \leq \frac{1}{\ell^2} \quad \Rightarrow \quad \ell^4 \leq \frac{1}{1-\alpha^2} \leq \frac{4}{3},$$

that is, $\ell^2 \leq 2/\sqrt{3}$, which is the claim.

We have equality if and only if

$$\alpha = 1/2, \qquad \text{and} \qquad |\mathbf{e}_1| = |\mathbf{e}_2| = \sqrt{\frac{2}{\sqrt{3}}},$$

which implies

$$\cos\theta(\mathbf{e}_1, \mathbf{e}_2) = \frac{\ell^2}{2}\frac{\sqrt{3}}{2} = \frac{1}{2}, \quad \Rightarrow \quad \theta = \frac{\pi}{3},$$

which implies the claim. ■

Second Proof That the Optimal Metric is Flat. (M. Gromov (1996)) Again, $\mathcal{G}_o$ is a flat metric on the torus, with the lattice generated by

$$\mathbf{e}_1 = \ell\mathbf{i}, \;\; \mathbf{e}_2 = \alpha\ell\mathbf{i} + \frac{1}{\ell}\mathbf{j}, \qquad \alpha \in (-1/2, 1/2], \qquad \ell = \ell_{\mathcal{G}_o},$$

where $\mathbf{e}_1$ is the element of the lattice closest to the origin, and $\mathbf{e}_2$ is the next closest. So, we have normalized the area of the flat metric to be equal to 1.

One now considers the family of closed geodesics, determined by the projection of

$$t \mapsto \gamma_s(t) = s\mathbf{e}_2 + t\mathbf{e}_1, \qquad 0 \leq t, s \leq 1$$

to $\mathbb{T}^2_{\mathcal{G}_o}$ (in short, the horizontal segments of length ℓ starting at points on $\mathbf{e}_2$). Then,

$$dA_{\mathcal{G}_o} = |\mathbf{e}_1||\mathbf{e}_2| \sin\theta(\mathbf{e}_1, \mathbf{e}_2)\, ds\, dt = ds\, dt.$$

Assume we are given a metric $\mathcal{G}$ on $\mathbb{T}^2$ by

$$\mathcal{G} = \phi^2\mathcal{G}_o \qquad \text{on } \mathbb{T}^2.$$

Then, the Cauchy–Schwarz inequality implies

$$\begin{aligned} A_{\mathcal{G}}(\mathbb{T}^2) &= \int_0^1 ds \int_0^1 \phi^2(t,s)\, dt \geq \int_0^1 ds \left\{ \int_0^1 \phi(t,s)\, dt \right\}^2 \\ &= \int_0^1 \{L_{\mathcal{G}}{}^2(\gamma_s)/\ell^2\}\, ds \geq \ell_{\mathcal{G}}{}^2/\ell^2, \end{aligned}$$

which implies $\ell_{\mathcal{G}}{}^2/A_{\mathcal{G}}$ is maximized in its conformal class by the flat metric. ■

Before continuing with surfaces, we give an estimate for higher dimensional tori.

Theorem V.1.2. (M. Gromov (1996)) *For higher dimensional flat torus* $\mathbb{T}^n$, $n \gg 2$, *with Riemannian metric* $\mathcal{G}_o$, *we have*

$$\frac{{\ell_{\mathcal{G}_o}}^n}{V_{\mathcal{G}_o}} = o(n^{(n+1)/2}).$$

If $V_{\mathcal{G}_o} = 1$ *then* $\ell_{\mathcal{G}_o} \leq \text{const}.n^{1/2}$.

Proof. We write the torus as $\mathbb{T}^n = \mathbb{R}^n/\Gamma$, where Γ is a lattice in $\mathbb{R}^n$. Let $\ell = \ell_{\mathcal{G}_o}$, $V = V_{\mathcal{G}_o}$.

For every $R > 0$ satisfying $V(\mathbb{B}^n(R)) \geq V(\mathbb{T}^n)$, we must have $R \geq \ell/2$. Indeed, for any $\rho > 0$ satisfying $V(\mathbb{B}^n(\rho)) > V(\mathbb{T}^n)$, the covering map $p : \mathbb{R}^n \to \mathbb{T}^n$ restricted to $\mathbb{B}^n(\rho)$ cannot be a diffeomorphism. So, there exist two points $x_1, x_2 \in \mathbb{B}^n(\rho)$ so that $p(x_1) = p(x_2)$, which implies $|x_1 - x_2| \geq \ell$ (see Exercise IV.23). But $|x_1 - x_2| \leq 2\rho$, which implies the claim.

Then, for large $n \gg 1$, we have, by Stirling's formula (Olver (1974, p. 88)),

$$\frac{\ell^n}{V} \leq \frac{2^n}{\boldsymbol{\omega}_\mathbf{n}} = \frac{2^n n\Gamma(n/2)}{2\pi^{n/2}} \sim \frac{2^n n(n/2)^{n/2}(2\pi)^{1/2}}{2\pi^{n/2}e^{n/2}(n/2)^{1/2}} = \sqrt{\pi n}\left(\frac{2n}{\pi e}\right)^{n/2},$$

which implies the claim. ∎

Definition. Recall, from surface topology, that any compact oriented surface M may be realized as a 2–sphere with $g(M)$ handles attached. The number $g(M)$ is referred to as the *genus of* M.

Theorem V.1.3. (M. Gromov (1996)) *Let* M *be a compact surface of constant Gauss curvature equal to* -1, $g \gg 1$, *where* g *denotes the genus of* M. *Then,*

$$\frac{\ell^2}{A} \leq \text{const.}\left\{\frac{\ln g}{\sqrt{g}}\right\}^2.$$

Proof. We may realize M as $M = \mathbb{H}^2/\Gamma$, where $\mathbb{H}^2$ is the hyperbolic plane of constant Gauss curvature equal to -1, and Γ is a discrete subgroup of the isometries of $\mathbb{H}^2$. (See Theorem IV.2.1 and Exercise IV.23.)

By the Gauss–Bonnet theorem (V.2.15), below, we have

$$A(M) = 4\pi\{g(M) - 1\}.$$

Therefore, again, as in the previous proof, the systolic length $\ell/2 \leq R$ for every R satisfying $A(\mathbb{B}^2_{-1}(R)) \geq A(M)$, that is $\ell/2 \leq R$ for every R satisfying

$$2\pi\{\cosh R - 1\} \geq 4\pi\{g-1\}, \quad \Rightarrow \quad 2\pi\{\cosh \ell/2 - 1\} \geq 4\pi\{g-1\}.$$

For $g \gg 1$, we have

$$e^{\ell/2} \geq (1-\epsilon)4g, \quad \Rightarrow \quad \ell \geq \text{const.} \ln g = \text{const.}\frac{\ln g}{\sqrt{g}}\sqrt{g} \geq \text{const.}\frac{\ln g}{\sqrt{g}}A^{1/2},$$

which implies the theorem. ∎

Theorem V.1.4. (J. Hebda (1982)) *For all compact orientable 2–dimensional Riemannian manifolds M of genus ≥ 1, we have*

$$\ell^2(M) \leq 2\,A(M).$$

Proof. Let $\phi : \mathbb{S}^1 \to M$ denote the minimizing homotopically nontrivial geodesic in M of systolic length $\ell = \ell(M)$, $|\phi'| = \ell/2\pi$, and fix $p \in \phi(\mathbb{S}^1)$, and set $\phi(0) = p$.

Pick any $r < \ell/2$. Then, $\pi_1(\overline{B(p;r)}; p)$ is trivial. If not, there exists a minimal homotopically nontrivial geodesic loop in $\overline{B(p;r)}$ based at p. Since the two halves of the loop are geodesics emanating from p, their total length is $\leq 2r < \ell$; therefore, the shortest closed homotopically nontrivial geodesic is shorter, which is impossible. Thus, the loop must be null-homotopic, a contradiction. So $\pi_1(\overline{B(p;r)}; p)$ is trivial.

Next, $\phi(\pi)$ is the cut point of p along ϕ. Assume the opposite; then $d(p, \phi(t)) < \ell/2$ for all $t \in \mathbb{S}^1$. Now, on the one hand, for any $t_o \in (0, 2\pi)$, a minimizing geodesic joining p to $\phi(t_o)$, $\overline{p\phi(t_o)}$ cannot be homotopic to *both* $\phi|[0, t_o]$ and $\phi|[t_o, 2\pi]$ (otherwise, the full loop would be null-homtopic). But, on the other, the length of $\phi|[0, t_o] \cdot \overline{\phi(t_o)\phi(2\pi)}$ is strictly less than ℓ, which implies $\phi|[0, t_o] \cdot \overline{\phi(t_o)p}$ is null-homotopic, and the same is true for $\phi|[t_o, 2\pi] \cdot \overline{\phi(t_o)p}$, which implies a contradiction.

We claim that

$$A(M) \geq A(p; \ell/2) \geq \ell^2/2.$$

To prove the claim, consider ϕ written as $\gamma : [-\ell/2, \ell/2]$, $|\gamma'| = 1$, $\gamma(0) = p$. For any $0 < r < \ell/2$, consider $\gamma|[-r, r]$; then $\gamma(-r)$ and $\gamma(r)$ are *not* cut points of p.

Now, each component of $S(p;r)$ is homeomorphic to the image (possibly degenerate) of a circle. If $\gamma(-r)$ and $\gamma(r)$ belong to the same component of $S(p;r)$, it must be a nontrivial component. Each "half of the component" is homotopic to $\gamma|[-r, r]$, which implies its length $\geq 2r$ (if not, replace $\gamma|[-r, r]$ by that "half-component"). Therefore, each component of $S(p; r)$ has length $\geq 4r$.

Assume $\gamma(-r)$ and $\gamma(r)$ belong to different components of $S(p; r)$. If one of these components has length strictly less than ℓ, then it is null-homotopic and bounds a disk in M, which implies γ would intersect $S(p; r)$ more than twice, which is impossible. Therefore, each such component has length $\geq \ell \geq 2r$, which implies the two components of $S(p; r)$ have length $\geq 4r$.

In sum, $L(p; r) \geq 4r$ for all $r < \ell/2$, which implies

$$A(p; \ell/2) \geq \int_0^{\ell/2} 4r\, dr = \ell^2/2,$$

which is the theorem. ∎

Remark V.1.1. Gromov (1983) improved Hebda's theorem by proving a general estimate that implied

$$\frac{\ell^2(M)}{A(M)} \leq \frac{64}{4\sqrt{g(M)} + 27},$$

where $g(M)$ denotes the genus of M. In particular,

$$\sup_{M: g(M) = g} \frac{\ell^2(M)}{A(M)} \to 0, \qquad \text{as } g \to \infty.$$

More recently, Katz–Sabourau (2005) generalized Theorem V.1.3 to all compact surfaces; namely,

$$\sup_{M: g(M) = g} \frac{\ell^2(M)}{A(M)} \leq \frac{1}{\pi} \left\{ \frac{\ln g}{\sqrt{g}} \right\}^2 (1 + o(1)), \qquad \text{as } g \to \infty.$$

See Katz (2005) for a broad survey of recent systolic inequalities.

§V.2. Gauss–Bonnet Theory of Surfaces

When speaking of Euclidean space $\mathbb{R}^n$ of any dimension n, we always consider its canonical orientation $\mathfrak{e}_1 \wedge \cdots \wedge \mathfrak{e}_n$, where $\{\mathfrak{e}_1, \ldots, \mathfrak{e}_n\}$ denotes the canonical basis of $\mathbb{R}^n$. When speaking of $\mathbb{S}^n$ we always assume it is endowed with the orientation inherited from its natural imbedding in $\mathbb{R}^{n+1}$.

When given a Riemannian manifold M with orientation, then for any domain D with smooth boundary C, we orient C as in §III.7.

For the rest of this section, M will be a 2–dimensional oriented connected Riemannian manifold.

Notation. In what follows, we let ι denote the rotation of tangent spaces to M by $\pi/2$ radians.

Consider an open subset U of M with orthonormal frame field $\{e_1, e_2\}$ and associated dual coframe field of 1–forms $\{\omega^1, \omega^2\}$. Thus, $\omega^1 \wedge \omega^2$ is the area form associated to the Riemannian measure dA as in §III.7; and from §I.8, we have the connection 1–forms ${\omega_j}^k$ given by

$$ {\omega_j}^k(\xi) = \langle \nabla_\xi e_j, e_k \rangle, $$

satisfying

$$ {\omega_j}^k = -{\omega_k}^j, \qquad d\omega^j = \sum_k \omega^k \wedge {\omega_k}^j. $$

Of course, since $\dim M = 2$, we only have the one nonvanishing connection form

$$ {\omega_1}^2 = -{\omega_2}^1. $$

Also, recall that if ${\Omega_j}^k$ denotes the curvature 2–form given by

$$ {\Omega_j}^k(X, Y) = \omega^k(R(X, Y)e_j), $$

then

$$ \text{(V.2.1)} \quad d{\omega_j}^k = \sum_\ell {\omega_j}^\ell \wedge {\omega_\ell}^k - {\Omega_j}^k, \quad \text{and} \quad {\Omega_j}^k(X, Y) = \langle R(X, Y)e_j, e_k \rangle. $$

Therefore, in our 2–dimensional situation, (V.2.1) becomes

$$ \text{(V.2.2)} \qquad d{\omega_1}^2 = -\mathcal{K}\omega^1 \wedge \omega^2 = -d{\omega_2}^1, $$

where $\mathcal{K}$ denotes the Gauss curvature, by Theorem II.1.

Definition. Given an immersed C^2 path $\gamma : (\alpha, \beta) \to M$, we define the *geodesic curvature of* γ by

$$ \kappa_g = \frac{\langle \nabla_t \gamma', \iota\gamma' \rangle}{|\gamma'|^3}. $$

Thus, κ_g is the quotient of the second fundamental form of the immersion, relative to the unit normal vector $\iota\gamma'/|\gamma'|$, by the first fundamental form. (In particular, if the orientation of γ is reversed, the effect on the geodesic curvature is to multiply it by -1.)

If the image of the immersion γ is in our neighborhood U above, then

$$ \gamma' = |\gamma'|\{\xi^1 e_{1|\gamma} + \xi^2 e_{2|\gamma}\} $$

for functions $\xi^1(t), \xi^2(t)$; so, the map

$$t \mapsto \gamma'(t) \mapsto (\xi^1(t), \xi^2(t))$$

is a map from $(\alpha, \beta) \to \mathbb{S}^1$, which has a lift $\theta : (\alpha, \beta) \to \mathbb{R}$, which implies

$$\gamma' = |\gamma'|\{(\cos\theta)e_1\circ\gamma + (\sin\theta)e_2\circ\gamma\}.$$

Then,

$$\begin{aligned}\nabla_t\gamma' &= \frac{\{|\gamma'|\}'}{|\gamma'|}\gamma' + |\gamma'|\theta'\{-(\sin\theta)e_1\circ\gamma + (\cos\theta)e_2\circ\gamma\}\\ &\quad + |\gamma'|\{(\cos\theta)\nabla_t(e_1\circ\gamma) + (\sin\theta)\nabla_t(e_2\circ\gamma)\}\\ &= \frac{\{|\gamma'|\}'}{|\gamma'|}\gamma' + \theta'\iota\gamma' + |\gamma'|\{(\cos\theta){\omega_1}^2(\gamma')e_2\circ\gamma + (\sin\theta){\omega_2}^1(\gamma')e_1\circ\gamma\}\\ &= \frac{\{|\gamma'|\}'}{|\gamma'|}\gamma' + \{\theta' + {\omega_1}^2(\gamma')\}\iota\gamma',\end{aligned}$$

which implies

$$\kappa_g|\gamma'| = \theta' + {\omega_1}^2(\gamma'). \tag{V.2.3}$$

Therefore, given a domain D in U, with compact closure and smooth boundary C, we have from (V.2.2) and (V.2.3)

$$-\iint_D \mathcal{K}\,dA = \iint_D d{\omega_1}^2 = \int_C {\omega_1}^2 = \int_C \kappa_g\,ds - d\theta,$$

that is,

$$\int_C \kappa_g\,ds + \iint_D \mathcal{K}\,dA = \int_C d\theta, \tag{V.2.4}$$

where ds is the 1–dimensional Riemannian measure of C (thus, $\kappa_g\,ds$ is a differential form along the oriented C). Now, C is a compact 1–manifold; so, it consists of a finite union of imbedded circles. We, therefore, have the existence of an integer k for which

$$\int_C d\theta = 2\pi k.$$

We now comment that k is independent of the Riemannian metric on U, if U is the domain of a chart x on M. Indeed, for $p \in U$, let $G(p)$ denote the matrix of the Riemannian metric given by

$$G(p) = (g_{ij}(p)), \qquad g_{ij}(p) = \langle \partial_i|_p, \partial_j|_p\rangle, \tag{V.2.5}$$

$i, j = 1, 2$, and define the family G^ϵ of Riemannian metrics on U by

$$G^\epsilon = (1-\epsilon)G + \epsilon I,$$

where I denotes the identity matrix. Then, k depends continuously on ϵ; since k is always integral, it remains constant. Therefore, when U is diffeomorphic to a subset of the Euclidean plane, it suffices to consider the case where the plane is endowed with its flat canonical Riemannian metric.

The Umlaufsatz

Or "the theorem of turning tangents."

Theorem V.2.1. (G. N. Watson (1916)) *If C consists of precisely one imbedded circle in $\mathbb{R}^2$ then*

$$k = 1. \tag{V.2.6}$$

Proof. (H. Hopf (1935)) Assume that C is given by the orientation preserving $\Gamma : \mathbb{S}^1(L/2\pi) \to \mathbb{R}^2$, parameterized with respect to arc length, where $\mathbb{S}^1(L/2\pi)$ denotes the circle of radius $L/2\pi$ and L denotes the length of C. Set

$$\gamma(t) = \Gamma((L/2\pi)\mathcal{E}^{2\pi it/L}) \tag{V.2.7}$$

(see Corollary IV.1.1). Then, γ is L–periodic, with unit speed.

Let T be the triangle in $\mathbb{R}^2$ given by

$$T = \{(x^1, x^2) \in \mathbb{R}^2 : 0 \le x^1 \le x^2 \le L\},$$

and map $v : T \to \mathbb{S}^1$ by

$$v(t, s) = \begin{cases} \{\gamma(s) - \gamma(t)\}/|\gamma(s) - \gamma(t)| & 0 < s - t < L \\ \gamma'(t) & s = t \\ -\gamma'(0) = -\gamma'(L) & s = L, t = 0 \end{cases}.$$

Then the homotopy lifting lemma implies the existence of a continuous lift of v to $\Theta : T \to \mathbb{R}$, that is, Θ satisfies

$$v = \mathcal{E}^{i\Theta}.$$

Thus,

$$2\pi k = \Theta(L, L) - \Theta(0, 0) = \{\Theta(L, L) - \Theta(0, L)\} + \{\Theta(0, L) - \Theta(0, 0)\}.$$

To facilitate the evaluation of each of the parentheses, we may pick γ so that

$$\gamma^2(t) \ge \gamma^2(0) \tag{V.2.8}$$

for all t (see Figure V.1). Then,

$$\gamma'(0) = \mathfrak{e}_1. \tag{V.2.9}$$

One now easily checks that each of the above parentheses is equal to π. ∎

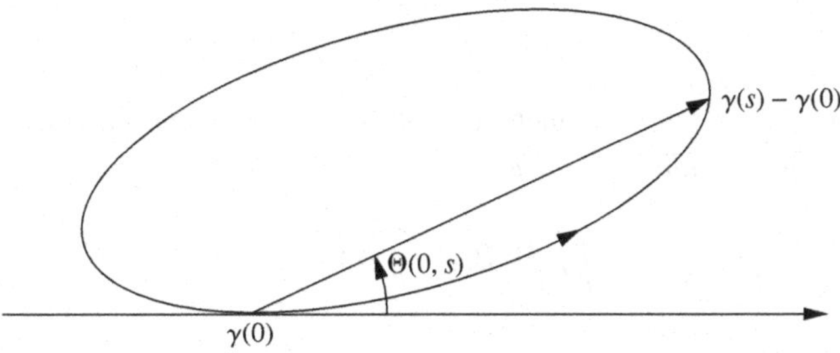

Figure V.1. For the Umlaufsatz.

Applications of the Umlaufsatz

Note that, although the closed curve C was given in the statement of the Umlaufsatz as the boundary of a domain, the argument only uses the fact that C is an imbedded circle in the plane, with a given orientation – counterclockwise. If C is simply given to be an imbedded circle Γ, then γ may be chosen to satisfy (V.2.8), and then (V.2.9) is replaced by

$$\gamma'(0) = \pm\mathfrak{e}_1.$$

One then has (V.2.6) replaced by

$$k = \pm 1,$$

with $\mathfrak{e}_1 \mapsto k = 1$ and $-\mathfrak{e}_1 \mapsto k = -1$.

Also, the argument of the Umlaufsatz only requires that $\Gamma \in C^1$.

Theorem V.2.2. *For M with closure diffeomorphic to the closed unit disk $\overline{\mathbb{B}^2}$ in $\mathbb{R}^2$, we have*

$$\int_{\partial M} \kappa_g \, ds + \iint_M \mathcal{K} \, dA = 2\pi. \tag{V.2.10}$$

For M diffeomorphic to the sphere $\mathbb{S}^2$, we have

$$\iint_M \mathcal{K} \, dA = 4\pi. \tag{V.2.11}$$

For M with closure diffeomorphic to the closed annulus in $\mathbb{R}^2$, we have

$$\int_{\partial M} \kappa_g \, ds + \iint_M \mathcal{K} \, dA = 0. \tag{V.2.12}$$

One then has:

Theorem V.2.3. (Gauss–Bonnet theorem) *For M diffeomorphic to the sphere with g attached handles, we have*

$$\iint_M \mathcal{K}\, dA = 4\pi(1-g). \tag{V.2.13}$$

Proof of Theorems V.2.2 and V.2.3. (W. Blaschke (1967, pp. 163–167)[1]) Equation (V.2.10) is a direct consequence of the Umlaufsatz applied to (V.2.4).

To obtain (V.2.11), one applies (V.2.10) to the two domains bounded by the preimage of a great circle of $\mathbb{S}^2$ under the diffeomorphism. Each domain orients this preimage opposite to the other, so the sum of the boundary integrals of the geodesic curvature vanishes, and (V.2.11) follows immediately.

For (V.2.12), we note that the Umlaufsatz applied to the total boundary of an annulus implies

$$\int d\theta = 0,$$

since the two boundary curves have opposite orientation.

We note that the same consideration implies that if D is a domain in $\mathbb{R}^2$ diffeomorphic to a disk with h holes, then

$$\int_{\partial D} d\theta = 2\pi(1-h).$$

Then, for (V.2.13), we remind the reader that for connected, oriented 2–dimensional manifolds M and $\mathcal{M}$, to say that *M is obtained by attaching a handle to $\mathcal{M}$* is to say that there exist two Jordan curves γ_1, γ_2 in M such that $M \setminus (\{\gamma_1\} \cup \{\gamma_2\})$ consists of two open connected submanifolds Ω_1, Ω_2 such that Ω_1 is diffeomorphic to $\mathcal{M}$ with two closed disks removed, and Ω_2 is diffeomorphic to a cylinder (see Figure V.2).

Thus, we may pick a point p in M with neighborhood Ω diffeomorphic to an open 2–disk, such that $M \setminus \Omega$ is diffeomorphic to a closed 2–disk with g handles attached. One now applies the above to obtain the desired result. ■

[1] It seems that not only does the proof belong to Blaschke, but also the Gauss–Bonnet theorem is in fact "Blaschke's theorem" (personal communication from R. Osserman based on his own research and conversations with S. S. Chern). See Petersen (1999, pp. 298–300) for some historical background.

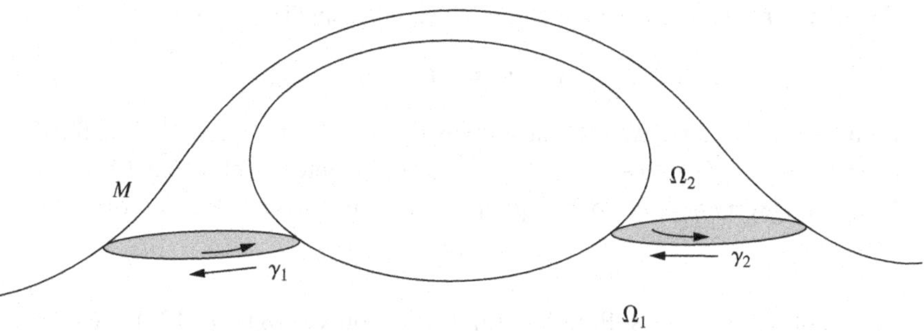

Figure V.2. Attaching a handle.

Theorem V.2.4. *The 2–dimensional torus has genus equal to* 1. *Therefore, for any Riemannian metric on a torus* M *we have*

$$\iint_M K\,dA = 0. \tag{V.2.14}$$

For any compact surface of constant curvature -1, *we have*

$$\iint_M K\,dA = 4\pi\{g(M) - 1\}. \tag{V.2.15}$$

In particular, $g(M) \geq 2$.

We now consider D with boundary C possessing corners.

Note that (V.2.6) is valid if we only assume that C is given by orientation preserving $\Gamma : \mathbb{S}^1(L/2\pi) \to M \in C^1 \cap D^2$, parameterized with respect to arc length, in which case the Umlaufsatz implies

$$\int_C d\theta = 2\pi.$$

Now consider D diffeomorphic to the 2–disk with compact closure, and with open neighborhood $U \supset \overline{D}$ diffeomorphic to a subset of $\mathbb{R}^2$; and assume the boundary C of D is given by the orientation preserving $\Gamma : \mathbb{S}^1(L/2\pi) \to M \in D^2, |\Gamma'| = 1$ at points where $\Gamma \in C^1$. Since the discontinuities of Γ' are isolated, we may define at all points $q = \gamma(t) \in \mathbb{S}^1(L/2\pi)$ the one-sided limits

$$\Gamma'_-(q) = \lim_{\tau\uparrow t} \gamma'(\tau), \qquad \Gamma'_+(q) = \lim_{\tau\downarrow t} \gamma'(\tau),$$

where $\gamma(\tau) = \Gamma((L/2\pi)\mathcal{E}^{2\pi i\tau})$, and

$$q = (L/2\pi)\mathcal{E}^{2\pi i t}.$$

To require that C has, at worst, corners (i.e., no cusps) is to require that

$$\Gamma'_{+}(q) \neq -\Gamma'_{-}(q)$$

for all $q \in \mathbb{S}^1(L/2\pi)$. We then have for every $p = \Gamma(q) \in C$, the well-defined *exterior angle* $\alpha(p) \in (-\pi, \pi)$ defined as the oriented angle from $\Gamma'_{-}(q)$ to $\Gamma'_{+}(q)$, for every p. (Of course, points of continuity of Γ' are described by $\alpha(p) = 0$.)

Theorem V.2.5. (Gauss–Bonnet formula; O. Bonnet (1848, p. 124)) *Given M, D, and C, as described above, we have*

$$\text{(V.2.16)} \qquad \int_{\partial D} \kappa_g \, ds + \iint_D \mathcal{K} \, dA = 2\pi - \sum_{j=1}^{k} \alpha(p_j),$$

where $p_1, \ldots, p_k$ are the corners of C.

Proof. Let γ be given by $\gamma(\tau) = \Gamma((L/2\pi)\mathcal{E}^{2\pi i \tau})$. By rotating $\mathbb{S}^1(L/2\pi)$ first, if necessary, we may assume that γ is C^2 at $t = 0$ and $t = L$. Set

$$q_j = (L/2\pi)\mathcal{E}^{2\pi i t_j/L}, \qquad p_j = \Gamma(q_j) = \gamma(t_j),$$

$j = 1, \ldots, k$, where $t_1 < \cdots < t_k \in (0, L)$, and $t_0 = 0$, $t_{k+1} = L$.

Then, certainly,

$$\int_C \kappa_g \, ds = \sum_{j=0}^{k} \int_{t_j}^{t_{j+1}} \kappa_g(\gamma(t)) \, dt.$$

We smooth out the corners as follows: There exist $\epsilon_0 > 0$, and Riemann normal coordinates $x_j : U_j \to \mathbb{R}^2$ centered at p_j, $j = 1, \ldots, k$, such that

$$\gamma(\mathbb{R}) \cap U_j = \gamma((-\epsilon_0 + t_j, t_j + \epsilon_0))$$

for all j. For each ϵ in $(0, \epsilon_0)$, let $\widehat{\gamma_{\epsilon,j}}$ denote the smaller circular arc in $\mathbb{R}^2$, parameterized with respect to arc length, connecting $(x_j \circ \gamma)(-\epsilon + t_j)$ to$(x_j \circ \gamma)(t_j + \epsilon)$, and tangent to $x_j \circ \gamma | (-\epsilon_0 + t_j, t_j + \epsilon_0)$ at the points $(x_j \circ \gamma)(-\epsilon + t_j)$ and $(x_j \circ \gamma)(t_j + \epsilon)$; and set

$$\gamma_{\epsilon,j} = {x_j}^{-1} \circ \widehat{\gamma_{\epsilon,j}},$$

for each $j = 1, \ldots, k$. Replace $\gamma | (-\epsilon + t_j, t_j + \epsilon)$ by $\gamma_{\epsilon,j}$, $j = 1, \ldots, k$. Call the new domain, resulting from the change of boundary, D_ϵ. Then

$$\int_{\partial D_\epsilon} \kappa_g \, ds + \iint_{D_\epsilon} \mathcal{K} \, dA = 2\pi.$$

Then, one verifies that, as $\epsilon \downarrow 0$,

$$\int_{\partial D_\epsilon} \kappa_g \, ds \to \int_{\partial D} \kappa_g \, ds + \sum_{j=1}^{k} \alpha(p_j)$$

and

$$\iint_{D_\epsilon} \mathcal{K} \, dA \to \iint_D \mathcal{K} \, dA.$$

This completes the proof. ■

Corollary V.2.1. (C. F. Gauss (1827, pp. 29ff)) *If C consists of three geodesic segments, with interior angles $\beta(p_j)$ (from $\Gamma'_+(q_j)$ to $-\Gamma'_-(q_j)$), $j = 1, 2, 3$, then*

$$\iint_D \mathcal{K} \, dA = \sum_{j=1}^{3} \beta(p_j) - \pi. \tag{V.2.17}$$

Now consider compact M with a triangulation, each of whose closed triangles is in a domain diffeomorphic to a subset of $\mathbb{R}^2$. Then,

$$\begin{aligned}\iint_M \mathcal{K} \, dA &= \sum_{\text{faces}} \{2\pi - \sum \text{exterior angles}\} \\ &= \sum_{\text{faces}} \{-\pi + \sum \text{interior angles}\} \\ &= 2\pi V - \pi F,\end{aligned}$$

where V denotes the number of vertices, and F denotes the number of faces, of the triangulation. Let E denote the number of edges of the triangulation. Then, since M is compact,

$$3F = 2E;$$

so

$$-F = 2(F - E),$$

which implies

Theorem V.2.6. (W. Blaschke (1967, pp. 163–167)) *For M compact, with triangulation as above, we have*

$$\iint_M \mathcal{K} \, dA = 2\pi\{V - E + F\} = 2\pi \chi(M), \tag{V.2.18}$$

where $\chi(M)$ denotes the Euler characteristic of M.

In particular, if M is compact orientable with no boundary, then

$$\chi(M) = 2 - 2g(M).$$

Theorem V.2.7. *More generally, for M orientable, with with compact closure and smooth boundary, we have*

$$\int_{\partial M} \kappa_g \, ds + \iint_M \mathcal{K} \, dA = 2\pi \chi(M). \tag{V.2.19}$$

If the boundary ∂M has corners with exterior angles $\{\alpha_j,\ j = 1, \ldots, k\}$, then

$$\int_{\partial M} \kappa_g \, ds + \iint_M \mathcal{K} \, dA = 2\pi \chi(M) - \sum_{j=1}^{k} \alpha_j. \tag{V.2.20}$$

§V.3. The Collar Theorem

Definition. A *Riemann surface* will be an oriented 2–dimensional Riemannian manifold of constant Gauss curvature equal to -1.

Let M be a compact Riemann surface, and let γ denote a simply closed geodesic in M of length ℓ. Then, B. Randol's (1979) collar theorem states that the distance from γ to its focal cut locus $\geq$ arcsinh csch $\ell/2$ (see §III.6 for definitions and notation). Furthermore, the area inside the focal cut locus, $A(\nu D_\gamma)$, is greater than or equal to 2ℓcsch $\ell/2$.

First, the reader should refer to Exercises IV.12–IV.14 and Exercise IV.23. Their content and arguments will be used throughout the proof.

Second, note that because the Riemann surface has negative curvature, the Gauss–Bonnet Theorem (V.2.15) implies that the genus of M is greater than or equal to 2.

Notation. For any point p in M, we let ι_p denote the rotation of $\pi/2$–radians in the tangent space M_p.

To prove Randol's collar theorem, we first require a lemma from hyperbolic trigonometry (see Ratcliffe (1994, p. 96)).

Lemma V.3.1. *Let $ABCD$ denote a geodesic quadrilateral in $\mathbb{H}^2$, the hyperbolic plane, with A, B, C all right angles, and $\measuredangle D = \phi$.*

Assume the length of AB and BC are r and t, respectively. Then,

$$\cos \phi = (\sinh r)(\sinh t).$$

Theorem V.3.1. (B. Randol (1979)) *Let M be a compact Riemann surface with simple closed geodesic γ of length $\ell > 0$, and let $\operatorname{inj}_\gamma$ denote the distance from γ to its focal cut locus. Then,*

$$\operatorname{inj}_\gamma \geq \operatorname{arcsinh} \operatorname{csch} \ell/2. \tag{V.3.21}$$

Corollary V.3.1. *Let $C_\gamma = \{q \in M : \ d(q, \gamma) < \operatorname{inj}_\gamma\}$ denote the cylinder "inside the focal cut locus." Then,*

$$A(C_\gamma) \geq 2\ell \operatorname{csch} \ell/2.$$

Proof. For the area, we have

$$A(C_\gamma) = 2\ell \int_0^{\operatorname{inj}_\gamma} \cosh t \, dt \geq 2\ell \int_0^{\operatorname{arcsinh} \operatorname{csch} \ell/2} \cosh t \, dt = 2\ell \operatorname{csch} \ell/2.$$

■

Proof of Theorem V.3.1. Let $\gamma : \mathbb{S}^1(\ell/2\pi) \to M, |\gamma'| = 1$, be a simple closed geodesic of length ℓ.

Klingenberg's lemma can be adapted here to show the existence of two distinct geodesic segments $\sigma_j : [0, \operatorname{inj}_\gamma] \to M$, $|\sigma'| = 1$, $\sigma_j(0)$ on γ, for both $j = 1, 2$, satisfying

$$\sigma_1(\operatorname{inj}_\gamma) = \sigma_2(\operatorname{inj}_\gamma), \qquad \text{and} \qquad {\sigma_1}'(\operatorname{inj}_\gamma) = -{\sigma_2}'(\operatorname{inj}_\gamma).$$

So, the union of the two geodesic segments is a smooth geodesic segment that can be written as $\sigma : [0, 2\operatorname{inj}_\gamma] \to M$, $|\sigma'| = 1$, $\sigma(0) = \sigma_1(0) := p_1$, $\sigma(2\operatorname{inj}_\gamma) = \sigma_2(0) := p_2$. We set $q = \sigma(\operatorname{inj}_\gamma)$. Note that Figure V.3 contains the two possibilities of how σ intersects γ at p_2.

Assume that the base geodesic γ is parameterized so that $\gamma(0) = p_1$ and that γ is oriented so that $\sigma'(0) = \iota_{p_1} \cdot \gamma'(0)$.

Next, consider the geodesic τ in M for which $\tau(0) = q$, and $\tau'(0) = -\iota_q \cdot \sigma'(\operatorname{inj}_\gamma)$. We claim that *if* (V.3.21) *is false, then τ is a geodesic loop that is freely homotopic to γ.*

Indeed, lift the geodesics γ, σ to geodesics $\tilde{\gamma}$, $\tilde{\sigma}$ in the universal cover of M, $\mathbb{H}^2$, starting at some lift $\tilde{p}$ of p_1. As in M, $\tilde{\sigma}'(0) = \tilde{\iota}_{\tilde{p}} \cdot \tilde{\gamma}'(0)$. Let Σ be the geodesic in M such that $\Sigma(0) = \gamma(-\ell/2) = \gamma(\ell/2)$ and $\Sigma'(0) = \iota_{\Sigma(0)} \cdot (\gamma'(\ell/2)) = \iota_{\Sigma(0)} \cdot (-\gamma'(-\ell/2))$. When γ is lifted, starting at p_1, to $\tilde{\gamma}$ by the

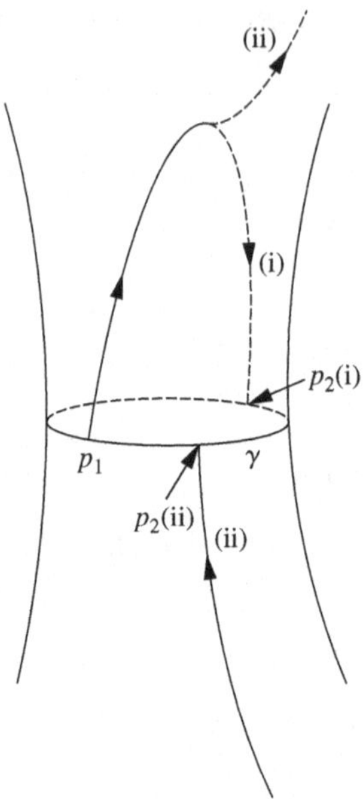

Figure V.3. The focal cut point.

covering to $\mathbb{H}^2$, the lifts of $\gamma(-\ell/2)$, $\gamma(\ell/2)$ are now distinct points, thereby determining two lifts $\tilde{\Sigma}_1$, $\tilde{\Sigma}_2$ of Σ, starting at $\tilde{\gamma}(-\ell/2)$, $\tilde{\gamma}(\ell/2)$, respectively. The lift, now, of the geodesic τ, $\tilde{\tau}$, starting at $\tilde{\tau}(0) = \tilde{\sigma}(\mathrm{inj}_\gamma)$ will be oriented so that $\tilde{\tau}'(0) = -\tilde{\iota}_{\tilde{\sigma}(\mathrm{inj}_\gamma)} \cdot \tilde{\sigma}'(\mathrm{inj}_\gamma)$. By Lemma V.3.1, if (V.3.21) is false, then $\tilde{\tau}$ will have to intersect $\tilde{\Sigma}_1$ and $\tilde{\Sigma}_2$. By symmetry, the intersection must be at the same arc length along $\tilde{\Sigma}_1$ and $\tilde{\Sigma}_2$ (see Figure V.4). The respective points of intersection $\tilde{q}_1$ and $\tilde{q}_2$ will then project, under the covering, to the same point in M. Therefore, $\tilde{\tau}$ projects to the loop τ in M, which is freely homotopic to γ. We distinguish two possibilities: Let ξ denote the continuous unit vector field

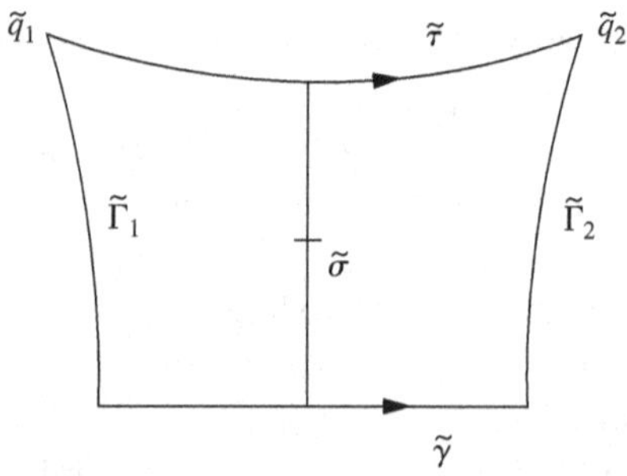

Figure V.4. The lifts to $\mathbb{H}^2$.

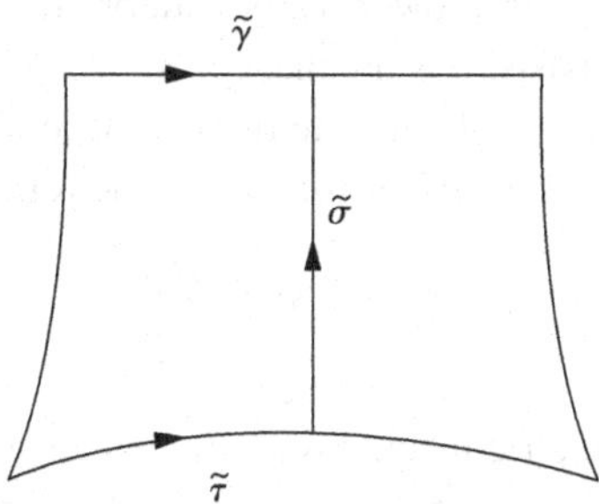

Figure V.5. The lifts to $\mathbb{H}^2$.

along γ, orthogonal to γ, for which $\sigma'(0) = \xi(p_1)$. The two possibilities are: (i) $\sigma'(2\text{inj}_\gamma) = -\xi(p_2)$ and (ii) $\sigma'(2\text{inj}_\gamma) = \xi(p_2)$.

Assume (i) $\sigma'(2\text{inj}_\gamma) = -\xi(p_2)$. Go through the same lifting construction as above, except that, now, start at p_2. Since the lift of σ starting at p_2 is oriented opposite the lift starting at p_1, the lift of τ now appears with the orientation opposite to that obtained by starting at p_1. The result is that τ is freely homotopic to γ^{-1}, which implies γ is freely homotopic to γ^{-1}, which contradicts Preissmann's theorem (Exercise IV.14).

Assume (ii) $\sigma'(2\text{inj}_\gamma) = \xi(p_2)$. This time, lifting the picture starting at p_2 will result, as in Figure V.5, by cutting M along γ in a homotopy between the new boundaries. This is impossible since the genus of M is greater than 2. ■

§V.4. The Isoperimetric Problem: Introduction

Let M be a Riemannian manifold. As usual, M always refers to the interior, independent of whether M has, or does not have, a nonempty boundary. The *isoperimetric problem* is to find, for any given v less than the volume of M, the region of volume v with minimal area of its boundary. The analytic aspect of the problem is to write this minimal area as a function of v, namely,

Definition. The *isoperimetric profile* $v \mapsto I_M(v)$ *of* M, defined for every $v \in (0, V(M))$ (the volume of M might be infinite), is given by

$$I_M(v) = \inf\{A(\partial D) : V(D) = v, \; D \subset\subset M\},$$

where D varies over relatively compact open submanifolds of M with smooth boundary in M. Such open subsets of M will simply be referred to as *smooth regions*. (In particular, they can have at most finitely many components.)

One, naturally, seeks explicit knowledge of the function $I_M(v)$. Also, one wants to know if, given any v, a region Ω exists for which $V(\Omega) = v$ and $A(\partial\Omega) = I_M(v)$; such a region is called an *isoperimetric region* or *minimizer* for short.

Furthermore, one also wants know if the boundary of the isoperimetric region is smooth. Finally, if it exists, it is unique.

A weaker version of the problem is to offer an accessible lower bound $\phi(v)$ for $I_M(v)$, that is, an *isoperimetric inequality* of the form

$$A(\partial D) \geq \phi(V(D)) \tag{V.4.1}$$

for all relatively compact smooth regions D.

The classical isoperimetric inequality in Euclidean space of all dimensions represents the best all worlds, namely, for any such region in $\mathbb{R}^n$, one has

$$\frac{A(\partial D)}{V(D)^{1-1/n}} \geq \frac{A(\mathbb{S}^{n-1})}{V(\mathbb{B}^n)^{1-1/n}} = \frac{\mathbf{c_{n-1}}}{\boldsymbol{\omega}_\mathbf{n}{}^{1-1/n}}. \tag{V.4.2}$$

One has equality in (V.4.2) if and only if D is a disk in $\mathbb{R}^n$. (See the proofs in §§VI.1 and VI.2.) Since the quotient on the right-hand side of (V.4.2) remains the same for disks of any radius r in $\mathbb{R}^n$, the inequality constitutes a complete solution of the isoperimetric problem. It says that, for any given volume v, the disks of volume v are the only minimizers of the bounding area, and the isoperimetric profile $I_M(v)$ is known explicitly, namely,

$$\mathbf{c_{n-1}} = n\boldsymbol{\omega}_\mathbf{n}, \quad \Rightarrow \quad I_{\mathbb{R}^n}(v) = n\boldsymbol{\omega}_\mathbf{n}{}^{1/n} v^{1-1/n}.$$

For the general situation, we quote a broad existence theorem.

Theorem V.4.1. (Main theorem for M without boundary) *If M^n is compact, or covers a compact, then, for any v, $0 < v < V(M)$, there exists $\Omega_v \subset\subset M$ whose boundary in M, $\Sigma_v = \partial\Omega_v$ minimizes area among smooth regions of volume v, that is, $A(\Sigma_v) = I_M(v)$. Moreover, except for a singular set of Hausdorff dimension at most $n - 8$, the boundary Σ_v of any minimizer is a smooth imbedded hypersurface with constant mean curvature.*

Of course, the theorem applies to the standard simply connected spaces of constant sectional curvature – Euclidean space, the sphere, and hyperbolic space of all dimensions.

Theorem V.4.2. (Main theorem for M with boundary) *Assume, in the above, M has boundary and $\overline{M^n}$ is compact (the boundary may be $C^{1,1}$). Then, again, the minimizer Ω_v exists, with boundary Σ_v. Σ_v might include some of the boundary ∂M, in which case $\Sigma_v \in C^{1,1}$ in a neighborhood of ∂M. The mean curvature $\mathfrak{h} = \mathfrak{h}_v$ is constant on the set of all smooth points of $\Sigma_v \cap M$. The mean curvature of $\Sigma_v \cap \partial M$ must satisfy $\mathfrak{h} \leq \mathfrak{h}_v$.*

Note that, in this case, $A(\overline{\Omega}_v \cap \partial M)$ is included in the area $A(\Sigma_v)$. Also, the mean curvature of Σ_v is always relative to the *interior* unit normal vector field along Σ_v. Finally, the mean curvature of $\Sigma_v \cap \partial M$ is defined in a weak sense. This formulation of the main theorems is from A. Ros (2001); see the references there for the proof of the theorem. The proof of $C^{1,1}$ regularity in a neighborhood of the boundary is in B. White (1991) and E. Stredulinsky–W. P. Ziemer (1997).

Let Ω be an arbitrary region with compact closure and smooth boundary in M, and assume a 1–parameter family of regions is given by perturbing the boundary $\partial\Omega$ by

$$\Phi(\epsilon; w) = \exp \epsilon\nu_{|w}, \qquad w \in \partial\Omega,$$

where ν denotes the *exterior* unit normal vector field along $\partial\Omega$. Let Ω_ϵ be the region with boundary $\partial\Omega_\epsilon = \Phi_\epsilon(\partial\Omega)$, where $\Phi_\epsilon(w) = \Phi(\epsilon; w)$. Let ν_ϵ denote the exterior unit normal vector field along $\partial\Omega_\epsilon$, for every ϵ. Set $V(\epsilon) = V(\Omega_\epsilon)$ and $A(\epsilon) = A(\partial\Omega_\epsilon)$. Then, the standard variational formulae (Exercises III.14 and III.33) read as:

$$\text{(V.4.3)} \qquad V'(\epsilon) = A(\epsilon), \qquad A'(\epsilon) = \int_{\partial\Omega_\epsilon} \mathfrak{h}_\epsilon \, dA,$$

where $\mathfrak{h}_\epsilon$ is the mean curvature relative to $-\nu_\epsilon$, and

$$\text{(V.4.4)} \qquad A''(\epsilon) = \int_{\partial\Omega_\epsilon} \{\mathfrak{h}_\epsilon{}^2 - \|\mathfrak{B}_\epsilon\|^2 - \langle \mathfrak{Ric}\,\nu_\epsilon, \nu_\epsilon\rangle\} \, dA,$$

where $\langle\ ,\ \rangle$ denotes the Riemannian inner product in M, $\|\mathfrak{B}_\epsilon\|^2$ denotes the Gram–Schmidt norm of the second fundamental form $\mathfrak{B}_\epsilon$ of $\partial\Omega_\epsilon$ relative to ν_ϵ, and $\mathfrak{Ric}$ denotes the Ricci curvature of M, .

Theorem V.4.3. (C. Bavard–P. Pansu (1986)) *Let Ω be an isoperimetric region with smooth boundary, and constant mean curvature $\mathfrak{h}$ with respect to the unit interior normal along the boundary. Then, the isoperimetric profile I_M has weak (in the sense of Calabi – see below) left- and right- first derivatives and second derivative satisfying*

$$\frac{D^+ I_M}{dv} \le \mathfrak{h} \le \frac{D^- I_M}{dv}, \qquad v = V(\Omega),$$

and

$$\frac{D^2 I_M}{dv^2} \le -\frac{1}{I_M{}^2(v)} \int_{\partial\Omega} \{\|\mathfrak{B}\|^2 + \langle \mathfrak{Ric}\,\nu, \nu\rangle\} \, dA, \qquad \text{at } v = V(\Omega).$$

If the Ricci curvature of M is bounded from below, then

$$\frac{D^2}{dv^2}\left\{ {I_M}^2(v) + \left(\inf_M \mathfrak{Ric}\right) v^2 \right\} \le 0 \qquad \text{at } v = V(\Omega).$$

In particular, the function

$$v \mapsto {I_M}^2(v) + \left(\inf_M \mathfrak{Ric}\right) v^2$$

is concave, which implies $I_M(v)$ is locally Lipschitz.

Proof. Given an arbitrary function $f(x)$, to say that it satisfies

$$\frac{D^+ f}{dx}(x_o) \le C_1 \le \frac{D^- f}{dx}(x_o) \qquad \text{and} \qquad \frac{D^2 f}{dx^2}(x_o) \le C_2$$

at x_o in the sense of Calabi is to say that there exists a smooth function $\phi(x)$ defined on some neighborhood of x_o such that

$$\begin{aligned} f(x) \le \phi(x) \ \forall\, x, \qquad & \phi(x_o) = f(x_o), \\ \phi'(x_o) = C_1, \qquad & \phi''(x_o) = C_2. \end{aligned}$$

Let $\epsilon(v)$ denote the inverse function of $V = V(\epsilon)$, then

$$\frac{dA}{dv} = \frac{A'}{A} \qquad \Rightarrow \qquad \frac{d^2 A}{dv^2} = \frac{1}{A}\left\{\frac{A''}{A} - \left(\frac{A'}{A}\right)^2\right\}.$$

Therefore, if $\partial\Omega$ has constant mean curvature, then

$$\text{(V.4.5)} \qquad \left.\frac{d^2 A}{dv^2}\right|_{\epsilon=0} = -\frac{1}{A^2}\int_{\partial\Omega} \{\|\mathfrak{B}\|^2 + \langle \mathfrak{Ric}\,\nu, \nu\rangle\}\, dA.$$

Then, for ϵ in a neighborhood of 0, we have $V(\epsilon)$ in a neighborhood of $V(\Omega)$, and

$$\begin{aligned} I_M(v) &\le A(v) \qquad \text{for } v \text{ in a neighborhood of } V(\Omega), \\ I_M(V(\Omega)) &= A(V(\Omega)). \end{aligned}$$

Then (V.4.5) implies

$$\left.\frac{d^2 A^2}{dv^2}\right|_{\epsilon=0} \le -2 \inf_M \mathfrak{Ric},$$

which implies

$$\left.\frac{d^2}{dv^2}\left\{A^2(v) + \left(\inf_M \mathfrak{Ric}\right) v^2\right\}\right|_{\epsilon=0} \le 0,$$

which implies the claim. ∎

Remark V.4.1. The theorem remains valid when M has a boundary, because we can use a variation of $\partial\Omega$ compactly supported on $\partial\Omega \cap M$ with infinitesmal variational vector field η (in the notation of Exercise III.14) pointing into Ω.

§V.5. Surfaces with Curvature Bounded from Above

Here, M is a surface. We use A and L for V and A, respectively, and write the isoperimetric function as $I(a)$.

The Wirtinger inequality (Exercise III.42) gives rise to a proof of the classical isoperimetric inequality in the plane $\mathbb{R}^2$. The argument is from A. Hurwitz (1901).

Theorem V.5.1. *Let Ω be a domain in $\mathbb{R}^2$ bounded by the C^1 Jordan curve Γ, and let the area of Ω be denoted by A, the length of Γ by L. Then,*

$$L^2 \geq 4\pi A,$$

with equality if and only if Ω is a disk.

Proof. Let $\mathbf{x}$ denote the position vector of the point $x \in \mathbb{R}^2$. One identifies the position vector $\mathbf{x}$ with the tangent vector $\Im_{\mathbf{x}}\mathbf{x} \in (\mathbb{R}^n)_x$; then, the divergence theorem (III.7.8) implies

$$2A = \int_\Gamma \langle \mathbf{x}, \nu \rangle \, ds$$

where ds denotes arc length along Γ. Cauchy's inequality, for vectors and for integrals, then, in turn, imply

$$\begin{aligned} 2A &= \int_\Gamma \langle \mathbf{x}, \nu \rangle \, ds \leq \int_\Gamma |\mathbf{x}| \, ds \leq \left\{ \int_\Gamma |\mathbf{x}|^2 \, ds \right\}^{1/2} \left\{ \int_\Gamma 1^2 \, ds \right\}^{1/2} \\ &= L^{1/2} \left\{ \int_\Gamma |\mathbf{x}|^2 \, ds \right\}^{1/2}, \end{aligned}$$

that is,

$$2A \leq L^{1/2} \left\{ \int_\Gamma |\mathbf{x}|^2 \, ds \right\}^{1/2}.$$

Next, translate the curve $\mathbf{x}$ so that

$$\int_\Gamma \mathbf{x}(s) \, ds = 0.$$

Note that $A(\Omega)$, $L(\Gamma)$, and $\mathbf{x}'(s)$ (for every s) are invariant under the translation. Therefore, we can apply Wirtinger's inequality to obtain

$$2A \leq \frac{L^{3/2}}{2\pi}\left\{\int_\Gamma |\mathbf{x}'|^2\,ds\right\}^{1/2}.$$

We are thinking of $\mathbf{x}$ as $\mathbf{x} = \mathbf{x}(s)$, where s is arc length; so $|\mathbf{x}'| = 1$ on all of Γ, which implies the theorem.

The case of equality follows easily. ■

The Theorems of Carleman and Weil

The first generalization of this inequality was given by T. Carleman (1921).

Theorem V.5.2. *Let Ω be a 2–dimensional domain in a minimal surface Σ in $\mathbb{R}^3$, bounded by the C^1 Jordan curve Γ. Then,*

$$L^2 \geq 4\pi A,$$

with equality if and only if Ω is a flat totally geodesic disk in $\mathbb{R}^3$.

Proof. The proof is the same as above, except that now uses Jellet's formula (Exercise III.30(b)) to prove

$$2A = \int_\Gamma \langle \mathbf{x}, \nu\rangle\,ds.$$

The rest of the argument is the same. ■

Theorem V.5.3. (A. Weil (1926)) *If M is a simply connected surface with complete Riemannian metric of nonpositive Gauss curvature one also has, for Ω a 2–dimensional domain in M bounded by the C^1 Jordan curve Γ, the inequality*

$$L^2 - 4\pi A \geq 0,$$

with equality if and only if Ω is a geodesic disk of Gauss curvature identically equal to 0.

Remark V.5.1. The Weil theorem is a generalization of the Carleman theorem since every minimal surface has nonpositive Gauss curvature – use (II.2.6).

Lemma V.5.2. *Let $p \in M$, $\{e_1, e_2\}$ be an orthonormal basis of M_p, and $y : M \to \mathbb{R}^2$ normal coordinates on M determined by $\{p; e_1, e_2\}$. Then, $\{p; e_1, e_2\}$*

may be chosen so that

$$\int_\Gamma y^j \, ds = 0.$$

Proof. Fix $p_o \in M, \{E_1, E_2\}$ an orthonormal frame at p_o, and parallel translate the frame $\{E_1, E_2\}$ along every geodesic emanating from p_o and thereby obtain a differentiable orthonormal frame field $\{e_1, e_2\}$ on all of M (we are using the simple connectedness and nonpositivity of the Gauss curvature). For every $p \in M$, let $y_p : M \to \mathbb{R}^2$ denote the Riemann normal coordinates on all of M determined by $\{e_1, e_2\}$ at p. Let $(y_p)^j$, $j = 1, 2$ denote the coordinate functions of y_p. Then, the vector field

$$Y(p) = \sum_{j=1}^{2} \left\{ \int_\Gamma (y_p)^j \, ds \right\} e_j(p)$$

is a continuous vector field on M. Restrict Y to a geodesic disk B in M that contains Ω. Then, the nonpositivity of the Gauss curvature implies that B is convex (see Exercise II.20); therefore, on the boundary of B, Y points into B. The Brouwer fixed-point theorem then implies that Y has a zero in B. ∎

Proof of Theorem V.5.3. Let p be given by the lemma, y given by the normal coordinates based at p, and consider the vector field on M given by

$$X = \sum_{j=1}^{2} y^j \frac{\partial}{\partial y^j}.$$

From Exercise III.8 and the relation

$$Xf = t\frac{\partial f}{\partial t}$$

for any function f, where t denotes distance from p, one deduces

$$\operatorname{div} X(\exp t\xi) = 2 + t \left\{ \frac{\partial \sqrt{\mathbf{g}}(t;\xi)}{\partial t} - \frac{1}{t} \right\},$$

where ξ denotes any unit tangent vector in M_p. Since the Gauss curvature is nonpositive, the Sturm comparison theorem (Exercise II.22) implies $\operatorname{div} X \geq 2$.

Therefore,

$$2A \leq \iint_\Omega \operatorname{div} X \, dA \leq L^{1/2} \left\{ \int_\Gamma \{|y^1|^2 + |y^2|^2\} \, ds \right\}.$$

We assume the boundary is parameterized with respect to arc length and that the point p satisfies the lemma. Then,

$$2A \leq \frac{L^{3/2}}{2\pi} \left\{ \int_{\Gamma} \{((y^1)')^2 + ((y^2)')^2\}\, ds \right\}^{1/2} = L^2/2\pi,$$

which implies the claim. The case of equality is handled easily. ■

Remark V.5.2. The inequality $L^2 - 4\pi A \geq 0$ can be extended to all compact regions in the complete simply connected M of nonpositive Gaussian curvature. Indeed, it is certainly true for every region Ω diffeomorphic to a disjoint union of disks. If the region $\mathcal{D}$ is obtained from the disjoint union of disks, Ω, by the removal of a finite number of diffeomorphic disks, then $A(\mathcal{D}) < A(\Omega)$. But $L(\partial\mathcal{D}) > L(\partial\Omega)$. Since the inequality is valid for Ω, it is automatically valid for $\mathcal{D}$. The case of equality is easily handled.

Remark V.5.3. Finally, E. F. Beckenbach–T. Rado (1933) noted that if $L^2 \geq 4\pi A$ for all simply connected domains on a surface then the Gauss curvature is nonpositive. The argument is an easy calculation. Fix any $x \in M$. Consider the geodesic disk $B(x;\epsilon)$, for $\epsilon > 0$ sufficiently small, with area $A(x;\epsilon)$ and boundary length $L(x;\epsilon)$. Then, one uses Corollary II.8.1 to show

$$L(x;\epsilon) = 2\pi\epsilon\left\{1 - \frac{\epsilon^2\mathcal{K}}{6} + O(\epsilon^3)\right\}, \quad A(x;\epsilon) = \pi\epsilon^2\left\{1 - \frac{\epsilon^2\mathcal{K}}{12} + O(\epsilon^3)\right\},$$

which implies

$$L^2(x;\epsilon)/A(x;\epsilon) = 4\pi\left\{1 - \frac{3\epsilon^2\mathcal{K}}{12} + O(\epsilon^3)\right\}.$$

Therefore, $L^2(x;\epsilon)/A(x;\epsilon) \geq 4\pi$ for all sufficiently small $\epsilon > 0$ implies that $\mathcal{K}$ is nonpositive. ■

The Bol–Fiala Inequalities

We now are given an oriented complete 2–dimensional *real analytic* Riemannian manifold M, and a relatively compact domain $\mathcal{D}$ bounded by $\mathcal{C}$, a finite number of simple closed real analytic Jordan curves, each having positive length (this excludes, for example, deleting a point from the domain). Each component curve of $\mathcal{C}$ is given by a periodic map $\omega : \mathbb{R} \to M$, parameterized with respect to arc length.

Let q denote a point in $\mathcal{C}$, as well as an arc length parameter along each component of $\mathcal{C}$. Also, let T denote the unit velocity vector field along $\mathcal{C}$.

Let $\mathtt{N}$ denote the interior unit normal vector field along $\mathcal{C}$, and ι denote the rotation of tangent vectors by $\pi/2$–radians. The orientation of $\mathcal{C}$ is chosen as usual, that is, so that $\mathtt{N} = \iota\mathtt{T}$. In particular,

$$\nabla_q \mathtt{T} = \kappa_g \mathtt{N},$$

where κ_g denotes the geodesic curvature of ω.

Consider Fermi coordinates on M, given by

$$v(t;q) = \exp t\mathtt{N}(q), \qquad t \in \mathbb{R}, \quad q \in \mathcal{C}.$$

Then, standard calculation yields (see Gauss' lemma – Theorem I.6.1)

$$|\partial_t v| = 1, \qquad \nabla_t \partial_t v = 0, \qquad \langle \partial_t v, \partial_q v\rangle = 0.$$

Also, $\partial_q v$ satisfies Jacobi's equation (see §III.6)

$$\nabla_t{}^2 \partial_q v + \mathcal{K}\partial_q v = 0,$$

where $\mathcal{K}$ denotes the Gauss curvature of M, along each geodesic $\gamma_q(t) := v(t;q)$, with initial conditions

$$\partial_q v(0;q) = \mathtt{T}(q), \quad \nabla_t \partial_q v(0;q) = \nabla_q \mathtt{N}(q) = -\kappa_g \mathtt{T}(q).$$

One can write

$$\partial_q v := \eta\,(-\iota(\partial_t v));$$

therefore,

$$ds^2 = dt^2 + \eta^2\, dq^2$$

for the Riemannian metric, and η satisfies the *scalar* Jacobi equation:

$$\partial_t{}^2\eta + \mathcal{K}\eta = 0, \quad \text{with initial conditions} \quad \eta(0;q) = 1,\ (\partial_t\eta)(0;q) = -\kappa_g(q).$$

Notation. For every $q \in \mathcal{C}$, let $P(q)$ denote the first *positive* zero of $t \mapsto \partial_q v(t;q)$; that is, *$v(P(q);q)$ is the first focal point of $\mathcal{C}$ along the geodesic* $\gamma_q(t) = v(t;q)$, $t > 0$. If no such zero exists, then we set $P(q) = +\infty$.

Since η and $\partial_t\eta$ cannot vanish simultaneously, then $P(q)$, when finite, is a real analytic function (by the implicit function theorem), with $P'(q)$ bounded on any strip $0 \le t \le r$.

Immediate results are:

Proposition V.5.1. *If, for any given $r_o > 0$, the equation $P(q) = r_o$ has infinitely many solutions in $\mathcal{C}$, then $\mathcal{D}$ is a disk.*

Similarly, if $P'(q) = 0$ has infinitely many solutions, then either (i) $\mathcal{D}$ is a disk, or (ii) for every accumulation point q_o of solutions of $P'(q) = 0$ we have $P(q_o) = +\infty$.

Let Z_o denote the set of r for which $P(q) = r$ has at least one solution q_o satisfying $P'(q_o) = 0$ at $P(q_o) = r$; then, Z_o is discrete.

Proposition V.5.2. (Hüllcurvensatz) *For each component of $\mathcal{C}$, along the path $q \mapsto v(P(q); q)$ where $P(q)$ is finite, the Riemannian metric reads as*

$$ds = |P'(q)|\, dq;$$

so when $P' \neq 0$ on an interval $[q_1, q_2]$, we have that the length of the path along the first focal locus, from q_1 to q_2, is given by $|P(q_2) - P(q_1)|$.

Definition. Given the geodesic $\gamma_q(t) = v(t; q)$, we let

$$\rho(q) = \sup \{\tau : d(\mathcal{C}, \gamma_q(t)) = t \ \forall\, t \in (0, \tau]\}.$$

So, $\rho(q) < +\infty$ is the distance from q to the focal cut point $\gamma_q(\rho(q))$ of $\mathcal{C}$ (we usually just say, *cut point of $\mathcal{C}$*) along γ_q.

Proposition V.5.3. *A cut point $p = v(\rho(q); q)$ of $\mathcal{C}$ is either the first focal point of $\mathcal{C}$ along γ_q or is the intersection of at least two geodesic arcs minimizing distance from $\mathcal{C}$ to p. Note that these two arcs need not emanate from the same component of the boundary.*

So, $\rho(q) \leq P(q)$. Furthermore, if $\rho(q_o) = P(q_o) < \infty$, then $P'(q_o) = 0$.

Proof. The proof of the first claim is standard (see §III.2 for the discussion of the cut locus of a point), so we only consider the last claim.

If $P'(q_o) \neq 0$, then there exists q close to q_o with $P(q) < P(q_o)$. One can then travel from q on $\mathcal{C}$ to $v(P(q_o); q_o)$ by going along the geodesic $\gamma_q(t)$ from q to $\gamma_q(P(q))$ and then traveling along the focal locus to $v(P(q_o); q_o)$. By Proposition V.5.2, the full path has length $P(q_o)$, which implies $\gamma_{q_o}(t)$ does not minimize distance from $\mathcal{C}$ to $\gamma_{q_o}(P(q_o))$, which is a contradiction. ∎

Definition. A cut point p of $\mathcal{C}$ along the geodesic γ_q is *normal* if it is not a focal point of $\mathcal{C}$ along γ_q (i.e., $\rho(q) < P(q)$), and it is the endpoint of precisely two distance minimizing arcs (one of them γ_q) from $\mathcal{C}$ to p.

A cut point of $\mathcal{C}$, which is not normal, will be referred to as *anormal*.

Proposition V.5.4. *Let p_o be a normal cut point of $\mathcal{C}$ along the two geodesics γ_{q_1} and γ_{q_2} at distance ρ_o. Then, the function $\rho = \rho(q)$ is real analytic on neighborhoods of q_1 and q_2.*

Also, there exists a local real analytic function $\mu(q)$, from a neighborhood of q_1 to a neighborhood of q_2, satisfying

$$\mu(q_1) = q_2, \qquad v(\rho(q); q) = v(\rho(q); \mu(q)).$$

Furthermore, the locus $v(\rho(q); q)$ near q_1 (or q_2) bisects the angle between the minimizing geodesics γ_q and $\gamma_{\mu(q)}$ meeting at $v(\rho(q); q)$.

Assume q_1 and q_2 are in the same component of $\mathcal{C}$, and $q_1 < q_2 < q_1 + \ell$, where ℓ is the length of their common component. Then, the map $q \mapsto \mu(q)$ is strictly decreasing.

Proof. Let U_1 be a neighborhood of (ρ_o, q_1) such that $v_1 := v|U_1$ is a diffeomorphism, let U_2 be a neighborhood of (ρ_o, q_2) such that $v_2 := v|U_2$ is a diffeomorphism and satisfying $v_1(U_1) = v_2(U_2)$. Set

$$(\tau, \sigma) = ({v_2}^{-1} \circ v_1)(t, s)$$

(so, we let s replace q in a neighborhood of q_1, and let σ replace q in a neighborhood of q_2). Then write for $\phi = {v_2}^{-1} \circ v_1$,

$$\tau := \phi_1(t, s), \qquad \sigma := \phi_2(t, s).$$

If $\gamma_s(t)$, $\gamma_\sigma(\tau)$ denote two distinct geodesics producing the normal cut point of $\mathcal{C}$ at $\gamma_s(\rho) = \gamma_\sigma(\rho)$, then

$$\rho = \phi_1(\rho, s), \qquad \sigma = \phi_2(\rho, s).$$

Therefore, solve $t = \phi_1(t, s)$ for t as a function of s on some neighborhood of $s = q_1$. To use the implicit function theorem, it suffices to show that $\partial\phi_1/\partial t < 1$ at $(t, s) = (\rho_o, q_1)$. But,

$$(\partial\phi_1/\partial t)(\rho_o, q_1) = \langle \partial_t v(\rho_o, q_1), \partial_\tau v(\rho_o, q_2)\rangle < 1,$$

which is the first claim. The function $\mu(q)$ is given by $\mu(q) = \phi_2(\rho(q), q)$.

Thus, the focal cut locus is given, locally by $q \mapsto v(\rho(q); q)$. Its velocity vector is given by

$$\rho'(q)\partial_t v_{|v(\rho(q);q)} + \partial_q v_{|v(\rho(q);q)} = \rho'(q){\gamma_q}'(\rho(q)) + \partial_q v_{|\gamma_q(\rho(q))};$$

so, the cosine of its angle with ${\gamma_q}'(\rho(q))$ is $\rho'(q)$. And the same is true for the geodesic $\gamma_{\mu(q)}$, since $\rho(\mu(q)) = \rho(q)$. This implies the focal cut locus is the angle bisector of the intersecting geodesics.

For the last claim, it suffices to note that the arcs $\omega|[q_1, q_2]$, $\gamma_{q_1}|[0, \rho_o]$, and $\gamma_{q_2}|[0, \rho_o]$ are the boundary of a bounded domain in $\mathcal{D}$. ■

Corollary V.5.1. *In the above, we have $\rho'(q) = 0$ if and only if $\gamma_q{}'(\rho(q)) = -\gamma_{\mu(q)}{}'(\rho(q))$, that is, the two geodesics meet smoothly at the cut point of C.*

If $\rho'(q) \equiv 0$ on one component of C, then either $\mathcal{D}$ is a geodesic disk or a geodesic annulus.

Proposition V.5.5. *The function $q \mapsto \rho(q)$ is continuous.*

If $v(\rho(q_o); q_o)$ is a normal cut point of C, then q_o has a neighborhood for which $v(\rho(q); q)$ is a normal cut point of C.

Proof. For the first claim, see the proof of Theorem III.2.1. It is easily adapted to our situation. So, we only consider the second claim.

If there is a sequence of $q_j \to q_o$ such that all $v(\rho(q_j); q_j)$ are all focal points of C, then so is $v(\rho(q_o); q_o)$.

If one has a sequence $q_j \to q_o$ so that all $v(\rho(q_j); q_j)$ are all nonfocal, anormal cut points of C, then the "normality" of $v(\rho(q_o); q_o)$ comes from the strict decrease of the number of distinct geodesics associated with each q_j, which implies that the exponential map – more precisely, v – is not locally 1–1 on the preimage of $v(\rho(q_o); q_o)$, which implies (see Exercise II.27) that $v(\rho(q_o); q_o)$ is a focal cut point along some geodesic and is therefore anormal. ■

The arguments of Propositions V.5.4 and V.5.5 also imply:

Proposition V.5.6. *Let v_o be a nonfocal, anormal cut point of C. Then there exist a finite number of q–values, $q_1, q_2, \ldots q_n$, $n \geq 3$, satisfying*

$$v_o = v(\rho(q_j); q_j), \qquad j = 1, \ldots n.$$

Furthermore, for each j, $\rho(q)$ is C^1 on a small closed interval ending at q_j and real analytic on the interior of that interval; and the set of (normal!) cut points of C in a neighborhood of v_o consist of n arcs of class C^1 ending at v_o. At each point of these arcs, the arc bisects the angle of the geodesics meeting there.

Corollary V.5.2. *The anormal cut points of C are isolated. In particular, there are at most a finite number of them.*

Definition. A number $t > 0$ is called an *exceptional value* if there exists $q \in C$ such that $v(t; q)$ is an anormal cut point of C, or is a normal cut point of C with $\rho'(q) = 0$. Otherwise, t is a *nonexceptional value.*

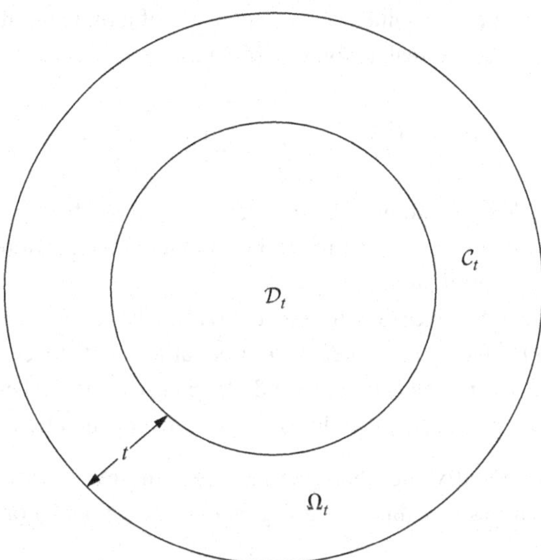

Figure V.6. Parallel-coordinate domains in $\mathcal{D}$.

Corollary V.5.3. *The set of exceptional values is finite.*

For any given r_o, if $\rho(q)$ is not identically equal to r_o on a component of $\mathcal{C}$, then the set of solutions to $\rho(q) = r_o$ is finite.

Proof. The first claim is obvious.

For the second, first recall that if $\rho(q) \equiv r_o$ on a component, then $\mathcal{D}$ must be a geodesic disk or annulus, and $\rho(q) \equiv r_o$ everywhere on $\mathcal{C}$.

So, assume $\rho(q)$ is not identically equal to r_o, but is equal to r_o at infinitely many points of $\mathcal{C}$. Then, $\rho(q) = r_o$ at infinitely many normal cut points of $\mathcal{C}$. But since $\rho(q)$ is real analytic away from nonfocal, anormal cut points of $\mathcal{C}$, this implies $\rho(q) \equiv r_o$ on a component, which implies $\rho(q) \equiv r_o$ everywhere, which we assumed is not the case. ∎

Notation. For any $t > 0$, define

$$\mathcal{C}_t = \{p \in \mathcal{D} : \ d(p, \mathcal{C}) = t\},$$
$$\Omega_t = \{p \in \mathcal{D} : \ d(p, \mathcal{C}) < t\},$$
$$\mathcal{D}_t = \{p \in \mathcal{D} : \ d(p, \mathcal{C}) > t\}$$

(see Figure V.6).

Let t be a nonexceptional value. Then, $\mathcal{C}_t$ consists of a finite number of simply closed curves, piecewise real analytic, written as

$$p_t(q) = v(t; q), \qquad q \in \bigcup_{j=1}^{m(t)} [\alpha_j(t), \beta_j(t)],$$

where the $m(t)$ denotes the number of intervals required for $\mathcal{C}_t$. The intervals $\{[\alpha_j(t), \beta_j(t)] : j = 1, \ldots, m(t)\}$ might be associated with different boundary components of the original $\mathcal{C}$.

The value of $\alpha_j(t)$ *increases* with respect to t, and the value of $\beta_j(t)$ *decreases* with respect to t. Once t is greater than or equal to the distance from $\mathcal{C}$ to its closest cut point, the points $v(t, \alpha_j(t))$ and $v(t, \beta_j(t))$ are normal cut points, and both functions $\alpha_j(t)$ and $\beta_j(t)$ are local analytic inverse functions of $t = \rho(q)$.

If t_o is an exceptional value, then one has the same result with some of the simply closed curves possibly being degenerate (i.e., a point) or two of them touching.

To consider what happens when $t \uparrow t_o$, first note that $\lim \alpha_j(t) := \alpha_o$ and $\lim \beta_j(t) := \beta_o$ exist as $t \uparrow t_o$. What might happen is that $\alpha_o = \beta_o$, in which case the "interval" disappears as t goes above t_o. Or, the interval $[\alpha_o, \beta_o]$ contains new points of the focal cut locus, in which case the number of "intervals" increases as t goes above t_o. By Corollary V.5.3, there can only be a finite number of new intervals.

Assume that at t_o the limit of the set

$$\text{“}\lim_{t \uparrow t_o}\text{''} [\alpha_j(t), \beta_j(t)] = [\alpha_o, \delta_o] \cup [\delta_o, \beta_o].$$

Even if α_o denotes an anormal cut point, we may assume that $\alpha_j(t)$ has an increasing continuous extension beyond t_o. If not, then $\alpha_j(t)$ would have a jump discontinuity at $t = t_o$, which implies there exists an interval on which $\rho(q) = t_o$, which would imply that $\mathcal{D}$ is a disk. This is the easiest case. A corresponding remark applies for β_o.

Thus, for δ_o, there exist continuous functions $\mathfrak{D}(t)$, $\mathfrak{D}^*(t)$ defined for $t > t_o$ so that

$$\lim_{t \downarrow t_o} \mathfrak{D}(t) \uparrow \delta_o \qquad \text{and} \qquad \lim_{t \downarrow t_o} \mathfrak{D}^*(t) \downarrow \delta_o,$$

where $\mathcal{C}_t \cap [\alpha_o, \beta_o] = [\alpha_j(t), \mathfrak{D}(t)] \cup [\mathfrak{D}^*(t), \beta_j(t)]$.

Notation. We let τ denote the furthest distance from $\mathcal{C}$ into $\mathcal{D}$, that is,

$$\tau = \sup \{d(q, \mathcal{C}) : q \in \mathcal{D}\}.$$

Note that τ is the *inradius of* $\mathcal{D}$. That is, there exists $q_o \in \mathcal{D}$ such that $d(q_o, \mathcal{D}) = \tau$; $B(q_o; \tau) \subseteq \mathcal{D}$; and for any $t > \tau$, $q \in \mathcal{D}$, the disk $B(q; t)$ is *not* contained in $\mathcal{D}$.

Theorem V.5.4. (G. Bol (1941) and F. Fiala (1940)) *For every value t, let $L(t)$ and $A(t)$ denote*

$$L(t) = L(\mathcal{C}_t), \qquad A(t) = A(\mathcal{D}_t).$$

Then $L(t)$ and $A(t)$ are continuous with respect to t. Moreover, at all nonexceptional values of t we have

$$A'(t) = -L(t). \tag{V.5.1}$$

If the Gauss curvature $\mathcal{K}$ satisfies $\mathcal{K} \leq \kappa$, then for all nonexceptional values of t, we have

$$L'(t) \leq -2\pi\chi(\mathcal{D}) + \kappa A(\mathcal{D}_t). \tag{V.5.2}$$

Equations (V.5.1) *and* (V.5.2) *are valid even if $\mathcal{D}$ has a finite number of components.*

Proof. We work only with $L(t)$. The argument for $A(t)$ is easier.

For any $t \in [0, \tau]$, $\mathcal{C}_t$ is a finite union of circles (some of them may be degenerate, that is, points), given by

$$\mathcal{C}_t = v(t; \mathfrak{C}_t), \qquad \mathfrak{C}_t = \bigcup_{j=1}^{m(t)} [\alpha_j(t), \beta_j(t)]$$

(notation for the intervals as described above). Then,

$$L(t) = \int_{\mathcal{C}_t} ds = \int_{\mathfrak{C}_t} \eta(t; q)\, dq,$$

which implies

$$\begin{aligned} |L(t) - L(t_o)| &= \left| \int_{\mathfrak{C}_t} \eta(t; q)\, dq - \int_{\mathfrak{C}_{t_o}} \eta(t_o; q)\, dq \right| \\ &\leq \int_{\mathfrak{C}_t \cap \mathfrak{C}_{t_o}} |\eta(t; q) - \eta(t_o; q)|\, dq + \sup_{[0,\tau]\times C} |\eta(t; q)| \int_{\mathfrak{C}_t \triangle \mathfrak{C}_{t_o}} dq, \end{aligned}$$

which implies that $L(t)$ is continuous.

As there are at most only a finite number of exceptional values of t, a similar argument shows that $L(t)$ is real analytic everywhere, except possibly at the exceptional values.

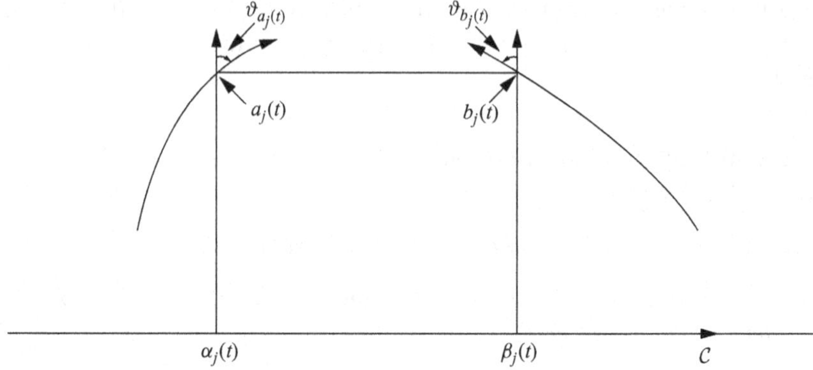

Figure V.7. The corners of C_t.

For the nonexceptional value t, the loci

$$a_j(t) := v(t; \alpha_j(t)), \qquad b_j(t) := v(t; \beta_j(t)), \qquad j = 1, \ldots m(t),$$

consist of normal cut points along C_t, and are locally real analytic paths with respect to t. Let

$$\vartheta_{a_j(t)} = \measuredangle[\partial_t v \to a_j{}'(t)], \qquad \vartheta_{b_j(t)} = \measuredangle[\partial_t v \to b_j{}'(t)]$$

(see Figure V.7). Then

$$\begin{aligned} a_j{}' &= \partial_t v + \alpha_j{}'\partial_q v = \partial_t v + (-\alpha_j{}'\eta)\iota\partial_t v, \\ b_j{}' &= \partial_t v + \beta_j{}'\partial_q v = \partial_t v + (-\beta_j{}'\eta)\iota\partial_t v, \end{aligned}$$

which implies

$$\alpha_j{}'(t)\eta(t; \alpha_j(t)) = -\tan\,\vartheta_{a_j(t)}, \qquad \beta_j{}'(t)\eta(t; \beta_j(t)) = -\tan\,\vartheta_{b_j(t)}.$$

Therefore,

$$\begin{aligned} L(t) &= \sum_{j=1}^{m(t)} \int_{\alpha_j(t)}^{\beta_j(t)} \eta(t; q)\, dq, \\ L'(t) &= \sum_{j=1}^{m(t)} \left\{ \beta_j{}'(t)\eta(t; \beta_j(t)) - \alpha_j{}'(t)\eta(t; \alpha_j(t)) + \int_{\alpha_j(t)}^{\beta_j(t)} (\partial_t \eta)(t; q)\, dq, \right\} \\ &= \sum_{j=1}^{m(t)} \left\{ -\tan\,\vartheta_{b_j(t)} + \tan\,\vartheta_{a_j(t)} + \int_{\alpha_j(t)}^{\beta_j(t)} -\kappa_g(t; q)\eta(t; q)\, dq, \right\}, \end{aligned}$$

where $\kappa_g(t; q)\eta(t; q)$ is the geodesic curvature of C_t at $v(t; q)$. We now apply the Gauss–Bonnet formula (V.2.20), and note that the exterior angle at the j–th corner is given by $\vartheta_{b_j(t)} - \vartheta_{a_j(t)}$, that $\vartheta_{b_j(t)}$ is positive and $\vartheta_{a_j(t)}$ is negative, and that $\chi(\Omega_t)$ is nonpositive. We then have, by (V.2.19) and (V.2.20),

$$\begin{aligned}
L'(t) &= \sum_{j=1}^{m(t)} \left\{\vartheta_{b_j(t)} - \tan\vartheta_{b_j(t)} - (\vartheta_{a_j(t)} - \tan\vartheta_{a_j(t)})\right\} + 2\pi\chi(\Omega_t) \\
&\qquad - \int_{\partial\mathcal{D}} \kappa_g\, ds - \iint_{\Omega_t} \mathcal{K}\, dA \\
&\le - \int_{\partial\mathcal{D}} \kappa_g\, ds - \iint_{\Omega_t} \mathcal{K}\, dA \\
&= -2\pi\chi(\mathcal{D}) + \iint_{\mathcal{D}_t} \mathcal{K}\, dA,
\end{aligned}$$

that is,

$$L'(t) \le -2\pi\chi(\mathcal{D}) + \iint_{\mathcal{D}_t} \mathcal{K}\, dA. \qquad \blacksquare$$

Theorem V.5.5. *If $\mathcal{D}$ is a relatively compact region in the complete real analytic surface M, with Gauss curvature $\mathcal{K}$ less or equal to the constant κ, then*

$$L^2(\partial\mathcal{D}) \ge 4\pi\chi(\mathcal{D})A(\mathcal{D}) - \kappa A^2(\mathcal{D}). \tag{V.5.3}$$

If $\mathcal{D}$ has n components, all of which are, topologically, disks, then

$$L^2(\partial\mathcal{D}) \ge 4\pi n A(\mathcal{D}) - \kappa A^2(\mathcal{D}). \tag{V.5.4}$$

Proof. Set $A(t) = A(\mathcal{D}_t)$. Then,

$$L(t)L'(t) \le -2\pi\chi(\mathcal{D})L(t) + \kappa L(t)A(t).$$

Integrate from 0 to τ; then,

$$\frac{1}{2}\{L^2(\tau) - L^2(0)\} \le 2\pi\chi(\mathcal{D})\{A(\tau) - A(0)\} - \frac{\kappa}{2}\{A^2(\tau) - A^2(0)\},$$

that is,

$$-\frac{1}{2}L^2(\partial\mathcal{D}) \le -2\pi\chi(\mathcal{D})A(\mathcal{D}) + \frac{\kappa}{2}A^2(\mathcal{D}),$$

which implies (V.5.3). $\blacksquare$

Remark V.5.4. If M is a Cartan–Hadamard surface, that is, it is complete, simply connected, with $\mathcal{K} \le \kappa \le 0$ – and is therefore diffeomorphic to the plane – then, one has

$$L^2(\partial\mathcal{D}) \geq 4\pi A(\mathcal{D}) - \kappa A^2(\mathcal{D})$$

for all regions in M. Indeed, it is certainly true for every region Ω diffeomorphic to a disjoint union of disks. If the region $\mathcal{D}$ is obtained from Ω by the removal of a finite number of diffeomorphic disks, then $A(\mathcal{D}) < A(\Omega)$, which implies (since $\kappa \leq 0$)

$$4\pi A(\mathcal{D}) - \kappa A^2(\mathcal{D}) < 4\pi A(\Omega) - \kappa A^2(\Omega).$$

But, $L(\partial\mathcal{D}) > L(\partial\Omega)$. Since the inequality is valid for Ω, it is automatically valid for $\mathcal{D}$. The case of equality is easily handled.

Proposition V.5.7. (Y. D. Burago (1978)) *Assume M is complete, and homeomorphic to the plane, with $\mathcal{K} \leq \kappa$. Then, for $p_o \in \mathcal{D}$ satisfying $d(p_o, \partial\mathcal{D}) = \tau$, τ the inradius of $\mathcal{D}$, we have*

$$\operatorname{inj} p_o \geq \pi/\sqrt{\kappa}.$$

(If $\kappa \leq 0$, then $\pi/\sqrt{\kappa}$ denotes $+\infty$.)

Proof. (C. Bavard (1984)) If $\kappa \leq 0$, then $\operatorname{inj} M = +\infty$ by the Cartan–Hadamard theorem (Theorem IV.1.3). So, assume $\kappa > 0$. First, note that, by the Morse–Schönberg Theorem (Theorem II.6.3), the inequality $\mathcal{K} \leq \kappa$ on all of M implies that the first conjugate point of the initial point of a geodesic cannot occur prior to distance $\pi/\sqrt{\kappa}$ along the geodesic.

Let q_o denote the closest cut point of p_o in M and assume q_o is not conjugate to p_o. Then, by Klingenberg's lemma (Theorem III.2.4), one has a geodesic loop starting and ending at p_o of length $2(\operatorname{inj} p_o)$, and q_o bisects the curve. Let $\mathfrak{D}$ denote the disk bounded by the loop. Then, one can give an argument based on those of Propositions V.5.4 and V.5.5 that there is a geodesic emanating from q_o into $\mathfrak{D}$ with a conjugate point q in $\mathfrak{D}$. In particular, $d(q, q_o) \geq \pi/\sqrt{\kappa}$.

For every $w \in \partial\mathcal{D}$, consider a minimizing geodesic connecting w to q and intersecting the geodesic loop at x. Then,

$$d(w, q) = d(w, x) + d(x, q), \qquad d(p_o, q_o) = d(p_o, x) + d(x, q_o),$$

which implies

$$d(w, q) + d(p_o, q_o) \geq d(w, p_o) + d(q, q_o),$$

which implies

$$d(w, q) + \operatorname{inj} p_o \geq \tau + \pi/\sqrt{\kappa}.$$

Therefore, by minimizing $d(w, q)$ over $\partial\mathcal{D}$, we obtain

$$\tau + \operatorname{inj} p_o \geq d(\partial\mathcal{D}, q) + \operatorname{inj} p_o \geq \tau + \pi/\sqrt{\kappa},$$

which implies the claim. ∎

Notation. For any $q \in M$ and $r > 0$, we write $A(q;r) = A(B(q;r))$. In the complete simply connected 2–dimensional space form of constant curvature κ, we let $\mathbb{B}_\kappa(r)$ denote a metric disk of radius r, $A_\kappa(r)$ its area, and $L_\kappa(r)$ the length of its boundary circle.

Corollary V.5.4. *By the Günther–Bishop Theorem* (Theorem III.4.1), *we also have* $A(\mathcal{D}) \geq A_\kappa(\tau)$.

Lemma V.5.2. (R. Osserman (1979)) *Consider the real numbers* A, a, L, ℓ, *and* κ, *satisfying* $A, a > 0$, $L, \ell \geq 0$, *and*

$$0 = \ell^2 - 4\pi a + \kappa a^2, \tag{V.5.5}$$

$$\ell L \geq 2\pi(a + A) - \kappa a A, \tag{V.5.6}$$

If $\kappa A < 4\pi$, *then* $\ell > 0$, *and*

$$L^2 - 4\pi A + \kappa A^2 \geq \left\{\frac{2\pi}{\ell}(A - a)\right\}^2. \tag{V.5.7}$$

Proof. From (V.5.5), we have $4\pi - \kappa a \geq 0$. Therefore, $4\pi - \kappa A > 0$ implies

$$\frac{1}{a} + \frac{1}{A} > \frac{\kappa}{2\pi},$$

which implies from (V.5.6) that $\ell > 0$. Now, square both sides of (V.5.6) and use (V.5.5). ∎

Theorem V.5.6. (R. Osserman (1979)) *Assume* $\mathcal{D}$ *is, topologically a disk, with boundary* $\mathcal{C}$. *Then*

$$L^2 - 4\pi A + \kappa A^2 \geq \left\{\frac{2\pi}{L_k(\tau)}(A - A_\kappa(\tau))\right\}^2 \geq 0. \tag{V.5.8}$$

We have $L^2 - 4\pi A + \kappa A^2 = 0$ *if and only if* $\mathcal{D} = \mathbb{B}_\kappa(\tau)$.

Proof. Set

$$\mathcal{A}(t) = A(\Omega_t) = A - A(\mathcal{D}_t).$$

Then, (V.5.2) reads as:

$$\mathcal{A}'' + \kappa\mathcal{A} \leq -2\pi + \kappa A, \qquad \mathcal{A}(0) = 0, \quad \mathcal{A}'(0) = L.$$

We want to compare $\mathcal{A}(t)$ to the solution $\mathcal{B}(t)$ of:

$$\mathcal{B}'' + \kappa\mathcal{B} = -2\pi + \kappa A, \qquad \mathcal{B}(0) = 0, \quad \mathcal{B}'(0) = L.$$

To this end, consider b_ϵ given by:

$$b_\epsilon{}'' + \kappa b_\epsilon = -2\pi + \kappa A, \qquad b_\epsilon(0) = 0, \quad b_\epsilon{}'(0) = L + \epsilon,$$

and $a_\epsilon := b_\epsilon - \mathcal{A}$; thus,

$$a_\epsilon{}'' + \kappa a_\epsilon \geq 0, \qquad a_\epsilon(0) = 0, \quad a_\epsilon{}'(0) = \epsilon.$$

A standard Sturmian argument (Exercise II.22) implies $a_\epsilon(t) \geq \epsilon \mathbf{S}_\kappa(t)$, which implies

$$b_\epsilon(t) \geq \mathcal{A}(t) + \epsilon \mathbf{S}_\kappa(t) \qquad \forall \epsilon > 0,$$

which implies $\mathcal{B}(t) \geq \mathcal{A}(t)$ $\forall t \in [0, \pi/\sqrt{\kappa}]$. One checks explicitly that

$$\mathcal{B}(t) = \frac{(-2\pi + \kappa A)}{2\pi} A_\kappa(t) + \frac{L}{2\pi} L_\kappa(t);$$

therefore,

$$LL_\kappa(t) \geq 2\pi(\mathcal{A}(t) + A_\kappa(t)) - \kappa A A_\kappa(t).$$

Let $\tau_\kappa = \min\{\tau, \pi/\sqrt{\kappa}\}$. Then, Burago's result (Proposition V.5.7) implies

$$A = \mathcal{A}(\tau) \geq A(p_o; \tau) \geq A_\kappa(\tau_\kappa)$$

(where $p_o \in \mathcal{C}_\tau$), with equality in the first inequality if and only if $\mathcal{D} = B(p_o; \tau)$, and in the second inequality if and only if $\tau \leq \pi/\sqrt{\kappa}$ and $\mathcal{K}|B(p_o; \tau) = \kappa$.

On the other hand, our interest in $\mathcal{D}$ is only for $A < 4\pi/\kappa$ – otherwise, the inequality yields no information. For this case, we have, by Corollary V.5.4, $\tau \leq \pi/\sqrt{\kappa}$, which implies

$$LL_\kappa(\tau) \geq 2\pi(A + A_\kappa(\tau)) - \kappa A A_\kappa(\tau),$$

which implies (V.5.8).

The case of equality follows easily. ∎

Remark V.5.5. The benefit of the Bonnesen type estimate is that it preserves the characterization of the case of equality even when approximating the differentiable case by the analytic one.

§V.6. The Isoperimetric Problem on the Paraboloid of Revolution

The Curvature Flow: Background

The beginning of our discussion seems modest enough. M. Gage (1983), in the spirit of ideas and methods of R. Hamilton (1982) for deforming Riemannian metrics along their (Ricci) curvature, proved the following:

Given $\mathbf{x}(t)$ a parameterization of a C^2 Jordan curve in $\mathbb{R}^2$, with *inward* pointing unit normal vector field $\mathbf{n}(t)$, and curvature $\kappa(t)$ with respect to $\mathbf{n}(t)$. Deform $\mathbf{x}(t)$ by the *heat equation*:

$$\frac{\partial \mathbf{x}_\epsilon(t)}{\partial \epsilon} = \kappa_\epsilon(t)\mathbf{n}_\epsilon(t), \qquad \mathbf{x}_0(t) = \mathbf{x}(t),$$

where κ_ϵ denotes the curvature of the deformed curve $\mathbf{x}_\epsilon$, and $\mathbf{n}_\epsilon(t)$ denotes the unit normal vector field along $\mathbf{x}_\epsilon$. If s denotes arc length along each curve $t \mapsto \mathbf{x}_\epsilon(t)$, then the differential equation reads as the usual heat equation:

$$\frac{\partial \mathbf{x}}{\partial \epsilon} = \frac{\partial^2 \mathbf{x}}{\partial s^2}.$$

Let $\mathcal{D}_\epsilon$ the domain bounded by $\mathbf{x}_\epsilon$, and $A(\epsilon) = A(\mathcal{D}_\epsilon)$, $L(\epsilon) = L(\mathbf{x}_\epsilon)$. Then,

$$\frac{\partial}{\partial \epsilon}\left(\frac{L^2}{A}\right) = -\frac{2L}{A}\left\{\int_{\mathbf{x}_\epsilon} {\kappa_\epsilon}^2\, ds - \frac{\pi L}{A}\right\}.$$

If $\mathbf{x}$ is convex then

$$\int_{\mathbf{x}} \kappa^2\, ds \geq \frac{\pi L}{A}.$$

Also, if $\mathbf{x}(t)$ is convex, then $\mathbf{x}_\epsilon$ is convex for all ϵ, which implies the quotient L^2/A is decreasing with respect to ϵ. Note that, by the Umlaufsatz (Theorem V.2.1), $A'(\epsilon) = -2\pi$.

Gage (1984) then proved that a family of C^2 closed convex curves $\mathbf{x}_\epsilon$, which satisfies the heat equation for $0 < \epsilon < \epsilon_o$ and for which $\lim_{\epsilon \to \epsilon_o} A(\epsilon) = 0$, also satisfies $L^2(\epsilon)/A(\epsilon) \to 4\pi$ as $\epsilon \to \epsilon_o$ – the result *uses* the isoperimetric inequality. Also, the normalized curves $\mathbf{y}_\epsilon = \sqrt{\pi/A(\epsilon)}\,\mathbf{x}_\epsilon$ converge in the Hausdorff metric (see Chavel (2001, p. 53ff), Burago–Burago–Ivanov (2001, p. 252ff)) to the unit circle.

This was followed by the proof of M. Gage and R. Hamilton (1986) that if $\mathbf{x}$ is a convex curve in $\mathbb{R}^2$, then the heat equation shrinks $\mathbf{x}$ to a point in finite time. The curve remains convex and becomes circular in the sense that (a) the ratio of the inscribed to circumscribed radius $\to 1$; (b) $\max \kappa/\min \kappa \to 1$; (c) the higher order derivatives of κ go to 0 uniformly; (d) if the initial curve is convex, but contains straight line segments, then the straight line segments disappear immediately as the curve evolves, and the succeeding curves are all strictly convex.

Finally, for the Euclidean plane, M. A. Grayson (1987) extended the Gage–Hamilton result to nonconvex curves of bounded curvature. He showed that the curve becomes convex in finite time, without developing any singularities.

Before giving the statement of the theorem for the curvature flow on surfaces, we first recall, from Exercise II.20:

Definition. Let M be a complete Riemannian manifold. A set A in M is *convex* if, for any $p, q \in A$, there exists a geodesic $\gamma_{pq} \subseteq A$ such that γ_{pq} is the unique minimizer in M connecting p to q.

Given $A \subset M$, define the *convex hull of* A, conv A, to be the smallest convex set containing A.

Theorem V.6.1. (The curvature flow theorem for surfaces) (M. A. Grayson (1989b)) *Let M be a 2–dimensional Riemannian manifold that is convex at ∞, that is, the convex hull of every compact set is compact. Then there exists $\epsilon_\infty \in (0, \infty]$ for which a solution of the heat equation exists for $\epsilon \in (0, \epsilon_\infty)$.*

If ϵ_∞ is finite, then the deformation converges to a point. If $\epsilon_\infty = \infty$, the geodesic curvature $\to 0$ in the sense of (b) *and* (c) *above (in additon to $\kappa \to 0$).*

Grayson notes that, for $\epsilon_\infty = \infty$, one does not rule out the possibility that the evolution is accumulating to an infinite set of closed geodesics. One can say: (i) they all have the same length; (ii) any two intersect at least once; and (iii) the number of intersections between any two is independent of the geodesics.

The Isoperimetric Inequality

Going in the direction opposite to that of Gage, Topping (1998) indicated how the curvature flow might be used to *prove* isoperimetric inequalities:

Assume, in Grayson's theorem, that

$$\iint_{B(p;R)} \mathcal{K}^+ \, dA < 2\pi \quad \forall\, p \in M,\ R > 0, \qquad \mathcal{K} \leq \kappa,$$

where $\mathcal{K}^+ = \max\{\mathcal{K}, 0\}$. Let C be a simple closed curve, bounding a domain $\mathcal{D}$. For the argument that follows, assume that M is diffeomorphic to a plane, and C converges to a point under the curvature flow. Then,

$$L^2 \geq 4\pi A - \kappa A^2.$$

Proof. The formula for the first variation of the area of the deformed $\mathcal{D}_\epsilon$ and the length of the deformed $\mathcal{C}_\epsilon$ is given by (see Exercise III.14)

$$A'(\epsilon) = -\int_{\mathcal{C}_\epsilon} \kappa_{g,\epsilon}\, ds = -2\pi + \iint_{\mathcal{D}_\epsilon} \mathcal{K}\, dA < 0, \tag{V.6.1}$$

$$L'(\epsilon) = -\int_{\mathcal{C}_\epsilon} {\kappa_{g,\epsilon}}^2\, ds \le -\frac{1}{L(\epsilon)}\left\{\int_{\mathcal{C}_\epsilon} \kappa_{g,\epsilon}\, ds\right\}^2, \tag{V.6.2}$$

where $\kappa_{g,\epsilon}$ denotes the geodesic curvature of $\mathcal{C}_\epsilon$ (the last inequality in (V.6.2) is a consequence of the Cauchy–Schwarz inequality). The function $A(\epsilon)$ is strictly decreasing; so, we may set $\epsilon = \epsilon(a)$ to be the inverse function of $A(\epsilon)$.

$$\frac{1}{2}\frac{dL^2}{da} \ge \int_{\mathcal{C}_\epsilon} \kappa_{g,\epsilon}\, ds = 2\pi - \iint_{\mathcal{D}_\epsilon} \mathcal{K}\, dA \ge 2\pi - \kappa a,$$

which implies the inequality, by integrating from $a = 0$ to $a = A(\mathcal{D})$. ■

The Paraboloid of Revolution

The second development of using the curvature flow to *prove* isoperimetric inequalities was the solution by I. Benjamini–J. Cao (1996) of the isoperimetric problem for the 2–dimensional paraboloid of revolution. We discuss it in more detail.

Theorem V.6.2. *Assume M is diffeomorphic to the plane, with Riemannian metric a complete surface of revolution about the vertex $o \in M$. Let B_R denote the geodesic disk of radius R centered at o. Assume*

$$\iint_{B_R} \mathcal{K}^+\, dA < 2\pi, \qquad \forall\, R > 0, \tag{V.6.3}$$

and assume that the Gauss curvature of M, $\mathcal{K} = \mathcal{K}(R)$ is a decreasing function of R. Given Ω a relatively compact region in M, then for $A(B_R) = A(\Omega)$, we have $L(\partial B_R) \le L(\partial\Omega)$ with equality if and only if Ω is isometric to B_R. The result applies to the paraboloid of revolution.

Remark V.6.1. The condition (V.6.3) implies that, for any disk $\Omega \subset M$, we have $\int_{\partial\Omega} \kappa_g\, ds > 0$. Indeed, the Gauss–Bonnet formula implies

$$\int_{\partial\Omega} \kappa_g\, ds = 2\pi - \iint_M \mathcal{K}\, dA \ge 2\pi - \iint_M \mathcal{K}^+\, dA > 0. \tag{V.6.4}$$

One might say that the condition (V.6.3) implies that every curve that bounds a disk is, *on average*, convex. Once $\int \mathcal{K}^+\, dA > 2\pi$, the convexity breaks down. Just consider a metric disk of radius greater than $\pi/2$ on the unit sphere.

Proof. We break up the proof of Theorem V.6.2 into a succession of steps.

Step 1. *Start with an arbitrary M, a 2–dimensional Riemannian manifold.* Let Ω_ϵ be a smooth 1–parameter family of domains obtained by the curvature flow applied to the domain $\partial\Omega$, for all ϵ. Then, for the general Ω, the inequality (V.6.2) implies

$$\frac{d\,L^2}{da} \geq 2\left\{2\pi\,\chi(\Omega_\epsilon) - \iint_{\partial\Omega_\epsilon} \mathcal{K}\,dA\right\}. \tag{V.6.5}$$

Assume, in addition, that M is complete, simply connected, and satisfies (V.6.3). Then, M must be the plane or the sphere. But Gauss–Bonnet implies that M must be diffeomorphic to the plane. So, *henceforth*, M is diffeomorphic to the plane.

Now *assume, also, that M is a surface of revolution about a point o.* Note that the geodesic curvature is constant on ∂B_R for all R. So (V.6.4) implies that each ∂B_R has positive geodesic curvature.

Step 2. We claim that B_R is convex for every $R > 0$. If not, there exist points p and q in ∂B_R such that the interior of a minimizing geodesic pq connecting them is contained in $M \setminus \overline{B_R}$. Then, an open topological disk Δ in $M \setminus \overline{B_R}$ is bounded by pq and an arc σ of ∂B_R. For every point z in the interior of pq, consider the geodesic γ_z emanating from it orthogonally into Δ, and the first distance d_z along it at which γ_z hits σ. Then, d_z assumes a maximum value, at which point γ_z is perpendicular to σ. This implies that the geodesic curvature of ∂B_R relative to the interior of Δ is nonnegative, which contradicts the fact that relative to the interior of B_R it is strictly positive.

In particular, M is convex at infinity.

By Grayson's theorem (Theorem V.6.1), the curvature flow deforms a Jordan curve to a point or a closed geodesic. If it flows to a geodesic; but the geodesic will not have positive geodesic curvature, which is impossible. Therefore, the curvature flow deforms a Jordan curve to a point.

Step 3. If $\Omega \subset\subset M$ is a domain (i.e., Ω is connected) with C^2 boundary and $\chi(\Omega) \leq 0$, then one can find a domain D diffeomorphic to the disk so that

$$A(D) = A(\Omega), \qquad L(\partial D) \leq L(\partial\Omega).$$

The proof goes as follows: First, there is a simply connected domain Ω^* from which Ω is obtained by deleting a finite number of disks with smooth boundary. Replace Ω by Ω^*. Then,

$$L(\partial\Omega^*) \leq L(\partial\Omega) \qquad \text{and} \qquad A(\Omega^*) \geq A(\Omega).$$

Now apply the curvature flow to $\partial\Omega^*$. Because Ω^* is a disk, (V.6.1) implies, with Grayson's theorem, that $A(\epsilon) \downarrow 0$ and $L(\epsilon)$ is decreasing. Then stop ϵ at ϵ_o satisfying $A(\epsilon_o) = A(\Omega)$.

Step 4. *The final added assumption*: *the Gauss curvature is a decreasing function of* R. Start with the case where Ω is a domain. By Step 3, we may assume Ω is bounded by a single Jordan curve.

Set $a_o = A(\Omega) = A(B_R)$. One can use the curvature flow to obtain a family of domains

$$\{\Omega_a : \Omega = \Omega_{a_o} \text{ and } A(\Omega_a) = a, \ \forall\, 0 < a \leq a_o\}.$$

Proceed as follows: Apply the curvature flow to Ω to obtain a family of domains Σ_ϵ converging to a point as $\epsilon \to$ some $\epsilon_o < +\infty$. Let $A(\epsilon) = A(\Sigma_\epsilon)$, and $\epsilon(a)$ its inverse function (by (V.6.1) the function $A(\epsilon)$ is strictly decreasing). Then, set $\Omega_a = \Sigma_{\epsilon(a)}$.

Also consider the function $\rho = \rho(a)$ satisfying $A(B_{\rho(a)}) = a$, that is, $\rho(a)$ is the radius of geodesic disk about o having area equal to a. Then, $A(\Omega_a) = A(B_{\rho(a)}) = a$, and (V.6.5) implies

$$\begin{aligned}\frac{d\,L^2(\partial\Omega_a)}{d\,a} &\geq 2\left\{2\pi - \iint_{\Omega_a} \mathcal{K}\,dA\right\}\\ &\geq 2\left\{2\pi - \iint_{B_{\rho(a)}} \mathcal{K}\,dA\right\} = \frac{d\,L^2(\partial B_{\rho(a)})}{d\,a},\end{aligned}$$

the second inequality uses $\mathcal{K}(R)\downarrow$, and the last equality uses the fact that κ_r, the geodesic curvature along ∂B_r, is constant for each r. Since

$$\liminf_{a\downarrow 0} L(\partial\Omega_a) \geq 0 = \lim_{a\downarrow 0} L(\partial B_{\rho(a)}),$$

we integrate the inequality

$$\frac{d\,L^2(\partial\Omega_a)}{d\,a} \geq \frac{d\,L^2(\partial B_{\rho(a)})}{d\,a}$$

from 0 to a_o to obtain

$$L^2(\partial\Omega) = L^2(\partial\Omega_{a_o}) \geq L^2(\partial B_{\rho(a_o)}) = L^2(\partial B_R),$$

which is the claim.

If $L(\partial\Omega) = L(\partial B_R)$, then

$$\frac{d\,L^2(\partial\Omega_a)}{d\,a} = \frac{d\,L^2(\partial B_{\rho(a)})}{d\,a}$$

for every $a \in [0, a_o]$, which implies

$$\iint_{\Omega_a} \mathcal{K}\,dA = \iint_{B_{\rho(a)}} \mathcal{K}\,dA$$

for every $a \in [0, a_o]$. Since $\mathcal{K}(R)$ is nonincreasing, this implies that if B_ρ is the smallest disk, centered at o and containing Ω, then the Gauss curvature $\mathcal{K}$ must be constant on B_ρ. This would then easily imply that Ω is a disk isometric to B_R.

Remark V.6.2. In the case of equality above, if the Gauss curvature is *strictly* decreasing, for example, then Ω *is* B_R.

Continuation of the Proof of Theorem V.6.2. Assume Ω has more than one component – we may assume each of which is diffeomorphic to a disk – and then subject Ω to the curvature flow. If the components never meet, then the argument is similar to the above. Should components touch, for some ϵ, then one can replace the "offending" domains with fewer domains having the same area and lower boundary, as follows:

Step 5. Consider the case where $\Omega = \Omega_1 \cup \Omega_2$, where Ω_1 and Ω_2 are domains, with disjoint closures, each diffeomorphic to a disk. Subject $\partial\Omega_1$ and $\partial\Omega_2$ simultaneously to the curvature flow, to obtain the families of domains $(\Omega_1)_\epsilon$ and $(\Omega_2)_\epsilon$. Let ϵ_1 be the time for Ω_1 to deform to a point, and the same for ϵ_2 and Ω_2.

First *assume* that $0 < \epsilon_2 \le \epsilon_1$ and that $\overline{(\Omega_1)_\epsilon} \cap \overline{(\Omega_2)_\epsilon} = \varnothing$ for all $\epsilon \in [0, \epsilon_2)$. Then, set

$$\Omega_\epsilon = \begin{cases} (\Omega_1)_\epsilon \cup (\Omega_2)_\epsilon & 0 \le \epsilon \le \epsilon_2 \\ (\Omega_1)_\epsilon & \epsilon_2 \le \epsilon \le \epsilon_1 \end{cases},$$

and $A(\epsilon) = A(\Omega_\epsilon)$. Let $\epsilon(a)$ denote the inverse function of $A(\epsilon)$, and set

$$D_a = \Omega_{\epsilon(a)}.$$

Then, $A(D_a) = a$ for all $a \in [0, A(\Omega)]$, and

$$\begin{aligned} \frac{d\,L^2(\partial D_a)}{d\,a} &\ge 2\int_{\partial D_a} \kappa_g\,ds \\ &= \begin{cases} 2\left(2\pi - \iint_{D_a} \mathcal{K}\,dA\right) & 0 \le a \le A(\epsilon_2) \\ 2\left(4\pi - \iint_{D_a} \mathcal{K}\,dA\right) & A(\epsilon_2) \le a \le A(\Omega_1) + A(\Omega_2) \end{cases} \\ &> 2\left(2\pi - \iint_{B_{\rho(a))}} \mathcal{K}\,dA\right) \\ &= \frac{d\,L^2(\partial B_{\rho(a)})}{d\,a}. \end{aligned}$$

If we set $a_1 = A(\Omega_1)$ and $a_2 = A(\Omega_2)$, then we have

$$L(\partial\Omega_1) + L(\partial\Omega_1) \ge L(B_{\rho(a_1+a_2)}),$$

which implies the inequality in this case. The case of equality is now easy.

Step 6. As in Step 5, consider the case where $\Omega = \Omega_1 \cup \Omega_2$, where Ω_1 and Ω_2 are domains, with disjoint closures, each diffeomorphic to a disk; and subject $\partial\Omega_1$ and $\partial\Omega_2$ simultaneously to the curvature flow, to obtain the families of domains $(\Omega_1)_\epsilon$ and $(\Omega_2)_\epsilon$. Except that, now, *assume* one has ϵ_o such that ϵ_o is the lowest value of ϵ for which $\overline{(\Omega_1)_\epsilon} \cap \overline{(\Omega_2)_\epsilon} \neq \emptyset$.

One first backs off a bit to $\epsilon^* < \epsilon_o$, and constructs a *connected* Ω^* satisfying

$$a^* := A(\Omega^*) > A((\Omega_1)_{\epsilon^*}) + A((\Omega_2)_{\epsilon^*}) := a_{\epsilon^*}, \tag{V.6.6}$$

$$L(\partial\Omega^*) < L(\partial(\Omega_1)_{\epsilon^*}) + L(\partial(\Omega_2)_{\epsilon^*}). \tag{V.6.7}$$

We describe how to do this below in Step 7. If Ω^* is not diffeomorphic to a disk, replace by a domain that is, as in Step 3, and call the new domain Ω^*.

Now, deform Ω^*, using the curvature flow, to obtain the family of domains $(\Omega^*)_a$, where $A((\Omega^*)_a) = a$, for all $0 \le a \le a^*$.

Set $a_1 = A(\Omega_1)$, $a_2 = A(\Omega_2)$ as above, and define

$$D_a = \begin{cases} (\Omega^*)_a & 0 \le a \le a_{\epsilon^*} \\ (\Omega_1)_{\epsilon(a)} \cup (\Omega_2)_{\epsilon(a)} & a_{\epsilon^*} \le a \le a_1 + a_2 \end{cases},$$

where $\epsilon(a)$ is the inverse function of $A(\epsilon) = A((\Omega_1)_\epsilon) + A((\Omega_2)_\epsilon), 0 \le \epsilon \le \epsilon^*$.

Then, $L(0) = 0$,

$$\frac{d\,L^2(\partial D_a)}{d\,a} \ge \frac{d\,L^2(\partial B_{\rho(a)})}{d\,a},$$

as in Step 4. By integrating the inequality, we obtain

$$\lim_{a \uparrow a_{\epsilon^*}} L^2(\partial D_a) \ge L^2(\partial B_{\rho(a_{\epsilon^*})}).$$

Since $a_{\epsilon^*} < a^*$ and since $L(\partial(\Omega^*)_a)$ is increasing on $[0, a^*]$, (V.6.6) and (V.6.7) imply

$$\lim_{a \downarrow a_{\epsilon^*}} L^2(\partial D_a) > \lim_{a \uparrow a_{\epsilon^*}} L^2(\partial D_a) \ge L^2(\partial B_{\rho(a_{\epsilon^*})}).$$

For $a \in (a_{\epsilon^*}, a_1 + a_2)$, we have

$$\begin{aligned} \frac{d\,L^2(\partial D_a)}{d\,a} &\ge 2\int_{\partial D_a} \kappa_g\, ds \\ &= 2\left(4\pi - \iint_{D_a} \mathcal{K}\, dA\right) \\ &> 2\left(2\pi - \iint_{B_{\rho(a)}} \mathcal{K}\, dA\right) \\ &= \frac{d\,L^2(\partial B_{\rho(a)})}{d\,a}, \end{aligned}$$

as above. If we integrate the inequality from a_{ϵ^*} to $a_1 + a_2$, we obtain

$$L(\partial\Omega_1) + L(\partial\Omega_1) \geq L(B_{\rho(a_1+a_2)}),$$

which is the claim.

Step 7. We now show how to replace $(\Omega_1)_{\epsilon^*} \cup (\Omega_2)_{\epsilon^*}$ with Ω^*, satisfying (V.6.6) and (V.6.7).

Let p_o be any point where $(\Omega_1)_{\epsilon_o}$ touches $(\Omega_2)_{\epsilon_o}$. Let $\partial((\Omega_1)_\epsilon)$ and $\partial((\Omega_2)_\epsilon)$ be parameterized by closed paths $(\omega_2)_\epsilon$ and $(\omega_2)_\epsilon$, respectively, and assume that both $(\omega_1)_{\epsilon_o}$ and $(\omega_2)_{\epsilon_o}$ are parameterized with respect to arc length, such that $(\omega_1)_\epsilon(0) = (\omega_2)_{\epsilon_o}(0) = p_o$. For small $\tau > 0$, consider the arcs

$$(\omega_1)_{\epsilon_o - \tau^2}|[-\tau, \tau] \qquad \text{and} \qquad (\omega_2)_{\epsilon_o - \tau^2}|[-\tau, \tau],$$

and replace them by the two minimizing geodesics connecting

$$(\omega_1)_{\epsilon_o-\tau^2}(-\tau) \text{ to } (\omega_2)_{\epsilon_o-\tau^2}(-\tau) \qquad \text{and} \qquad (\omega_1)_{\epsilon_o-\tau^2}(\tau) \text{ to } (\omega_2)_{\epsilon_o-\tau^2}(\tau).$$

For sufficiently small $\tau > 0$, the triangle inequality and Taylor's formula imply that the resulting Ω^* satisfies (V.6.6) and (V.6.7).

Step 8. If one starts with more than two components of Ω, then the argument is a generalization of Steps 5–7 (see the original paper Benjamimi-Cao (1996)). ■

§V.7. Notes and Exercises

Systolic Inequalities

Note V.1. The study of the shortest homotopically nontrivial closed geodesic was initiated by C. Loewner (unpublished) for the 2–torus, and continued in Pu (1962) for the projective plane and the Möbius strip. For Riemann surfaces (of higher genus), the notion of extremal length was introduced for homology in Ahlfors–Beurling (1950) and worked out in Accola (1960), Blatter (1961). The generalizations to higher dimensions were formulated in Berger (1972a, 1972b); and completely new insights were embarked on in Gromov (1983). Also see Katz (1983, 2005). Finally, also see Gromov's notes (1996), and the English edition of his book (1999).

The Gauss–Bonnet Theorem

Note V.2. We refer the reader to Milnor (1965) and Berger–Gostiaux (1988, pp. 253ff) for discussions of degree and index of singularities of vector fields.

Note V.3. The Umlaufsatz actually states that if $\Gamma : \mathbb{S}^1 \to \mathbb{R}^2$ is an imbedding of $\mathbb{S}^1$ in the plane, $T : \mathbb{S}^1 \to \mathbb{S}^1$ given by $T = \Gamma'/|\Gamma'|$, then the degree of T is equal to ± 1, depending on the orientation of Γ. As is well known, the degree of a smooth mapping from a compact manifold M to a compact manifold N of the same dimension depends only on the smooth homotopy class of the mapping; so, given a 1–parameter family Γ_ϵ of imbeddings of $\mathbb{S}^1$ in the plane, the degree of the associated T_ϵ remains constant. Furthermore, a famous theorem of H. Hopf (1926a) states that when $N = \mathbb{S}^n$, $n = \dim M$, then any two smooth maps $M \to N$ of the same degree are smoothly homotopic. Thus, given Γ_0, Γ_1 two imbeddings of the circle in the plane with associated T_0, T_1 possessing the same degree, there is a smooth homotopy T_ϵ, $\epsilon \in [0, 1]$ taking T_0 to T_1.

Exercise V.1. Prove the following result of H. Whitney (1937): Given Γ_0, Γ_1 two imbeddings of the circle in the plane with associated T_0, T_1 possessing the same degree, there is a smooth homotopy Γ_ϵ, $\epsilon \in [0, 1]$ of imbeddings of $\mathbb{S}^1$ into the plane taking Γ_0 to Γ_1.

Exercise V.2. Consider an oriented Riemannian 2–manifold M with a local positively oriented frame field $\{e_1, e_2\}$ on a neighborhood U diffeomorphic to a subset of $\mathbb{R}^2$. Let X denote a vector field on M with isolated zero at $p \in U$, and let $\mathfrak{x}_1 = X/|X|$ be the associated unit vector field on $U \setminus \{p\}$. Let $\mathfrak{x}_2$ denote the vector field on $U \setminus \{p\}$ obtained by rotating $\mathfrak{x}_1$ by $\pi/2$ radians. For any circle C (in local or polar coordinates) about p, one can write

$$\mathfrak{x}_1 = (\cos\theta)e_1 + (\sin\theta)e_2, \qquad \mathfrak{x}_2 = -(\sin\theta)e_1 + (\cos\theta)e_2.$$

Show that if $\omega_j{}^k$ denotes the connection 1–form of the frame field $\{e_1, e_2\}$ on U, and $\tau_j{}^k$ denotes the connection 1–form of the frame field $\{\mathfrak{x}_1, \mathfrak{x}_2\}$ on $U \setminus \{p\}$, then

$$\tau_1{}^2 = d\theta + \omega_1{}^2.$$

Note that

$$\int_C d\theta = 2\pi(\text{index } X \text{ at } p).$$

Exercise V.3. (H. Hopf (1956, pp. 107–118))

(a) Given an orientable compact Riemannian 2–manifold M with smooth vector field X whose set of zeros is the subset $\{p_1, \ldots, p_\ell\}$ of M. Show that

$$\int_M \mathcal{K}\, dA = 2\pi \sum_{j=1}^{\ell} (\text{index } X \text{ at } p_j).$$

This recaptures the Poincaré–Hopf theorem (Hopf (1926b)) that the sum of the indices of the singularities of a vector field on a compact differentiable manifold is equal to its Euler characteristic.

(b) Also prove that if M is imbedded in $\mathbb{R}^3$ with Gauss map $\mathbf{n} : M \to \mathbb{S}^2$ (as described in Exercise III.19), show that the degree of $\mathbf{n}$ satisfies

$$\deg \mathbf{n} = \chi(M)/2.$$

Note V.4. Higher dimensional versions of the Gauss–Bonnet theorem were given, for submanifolds of higher dimensional Euclidean space, by C. B. Allendoerfer (1940) and W. Fenchel (1940). Allendoerfer–Weil (1943) considered the case of Riemannian polyhedra, but, again used imbedding methods. The first intrinsic proof of the higher dimensional Gauss–Bonnet theorem was given by S.S. Chern (1944).

Exercise V.4. Prove Theorem V.2.7, the Gauss–Bonnet formula for manifolds with boundary. That is, show that if M is a 2–dimensional orientable Riemannian manifold with compact closure and differentiable boundary, then

$$\int_{\partial M} \kappa_g \, ds + \iint_M \mathcal{K} \, dA = 2\pi \chi(M).$$

On Randol's Collar Theorem

Note V.5. I. Chavel and E. A. Feldman (1978a) considered the result for negative variable Gauss curvature on an oriented compact surface. They assumed that the Gauss curvature $\mathcal{K}$ satisfied

$$-1 \leq \mathcal{K} \leq -\delta^2 < 0$$

and γ a simple closed geodesic of length ℓ. Then, the estimate for the distance to the cut locus of γ is as above,

$$\operatorname{inj}_\gamma \geq \operatorname{arcsinh} \operatorname{csch} \ell/2,$$

and the area estimate is given by

$$A(C_\gamma) \geq (2\ell/\delta) \sinh(\delta \operatorname{arcsinh} \operatorname{csch} \ell/2).$$

When $\delta = -1$, their estimates coincide with those of Randol.

The estimate for $\operatorname{inj}_\gamma$ only uses $-1 \leq \mathcal{K} \leq 0$, and the fact that the genus of M is ≥ 2. If the genus of M were equal to 1, then possibility (ii) in the proof of Randol's theorem could occur – for the torus $\mathbb{T}^2$.

Buser (1978) also has a differential geometric generalization of Randol's estimate. For early results, see Keen (1974) and Halpern (unpublished). More developments can be found in Basmajian (1972).

The Theorems of Carleman and Weil

Note V.6. Carleman's argument was based on complex analysis. The first to free the original proofs from such considerations was W. T. Reid (1959), who based his argument on Jellet's formula.

Definitive results for minimal varieties of arbitrary co-dimension in Euclidean spaces were obtained, via geometric measure theory, by F. Almgren (1986).

One advantage of the above argument is that there are no restrictions on the topology of Ω, only that it be bounded by one curve. See Chavel (1978) for the details, and for the literature between Carleman (1921) and Chavel (1978).

The proof of Lemma V.5.2 is adapted from Weinberger (1956).

I. Chavel (1978) combined Jellet's formula with A. Hurwitz (1901), and Lemma V.5.2, to give a unified generalization of the Carleman and Weil result to higher dimensions ($n > 2$). The result is a weak isoperimetric inequality, since the analogue of Wirtinger's inequality is Rayleigh's principle; and the lowest eigenvalue of the boundary depends on more than just its $(n-1)$–area.

The approach here to Carleman's theorem was extended further in P. Li–R. Schoen–S. T. Yau (1984) and J. Choe (1990).

Note V.7. P. Hartman (1964) extended the Bol–Fiala inequalities to the case of minimal differentiability assumptions by direct analysis of the continuity and differentiability of the function $L(t)$. Namely, except for a set of measure 0 (in t), the curves Γ_t are piecewise smooth, and the function $L(t)$ is well-defined and C^1, with $L(t)$ integrable with respect to t. Furthermore, Fiala's differential inequalities for $L'(t)$ remain valid off this singular set of measure 0. A more recent extended discussion of these results can be found in Shiohama–Shioya–Tanaka (2003, Chapter 4)

I. Chavel and E. A. Feldman (1980) extended the inequalities to the differentiable case by approximating the differentiable case by the real analytic one.

It seems that the first to consider the isoperimetric inequality on manifolds besides Euclidean space was F. Bernstein (1905), wherein he considered the isoperimetric problem for domains in the standard 2–sphere bounded by convex curves. The Bol–Fiala result is generalization of his result. Subsequent generalizations of the Bol–Fiala inequalities were given by A. D. Alexandrov (1948) and A. Huber (1954), and summarized in a unified presentation in Y. D. Burago–V. A. Zalgaller (1988). Further generalization was given in P. Topping (1998, 1999).

Note V.8. Higher dimensions. C. Croke (1984) proved the following: Assume M is a 4–dimensional Riemannian manifold with smooth boundary, so that $\overline{M}$ is compact. Assume that M has nonpositive sectional curvature, and every geodesic in M minimizes distance from its initial point until it hits the boundary. Then,

$$\frac{A(\partial M)}{V(M)^{3/4}} \geq \frac{A(\mathbb{S}^3)}{V(\mathbb{B}^4)^{3/4}},$$

with equality if and only if M is isometric to a disk in $\mathbb{R}^n$.

B. Kleiner (1992) then considered the 3–dimensional case: Let M be a 3–dimensional complete simply connected Riemannian manifold, with sectional curvature less than or equal to a constant $\kappa \leq 0$, and let $\Omega \subset\subset M$ with smooth boundary, and $\mathbb{B}_\kappa(\Omega)$ the disk in the 3–dimensional model space of constant curvature κ, having the same volume as Ω. Then,

$$A(\partial\Omega) \geq A(\partial(\mathbb{B}_\kappa(\Omega))), \tag{V.7.1}$$

with equality if and only if Ω is isometric to $\mathbb{B}_\kappa(\Omega)$.

The corresponding inequalities in higher dimensions, known as the *Aubin conjecture* (1976), that simply connected domains D on manifolds of all dimensions ≥ 2 with nonpositive Riemannian sectional curvature satisfy (V.4.2), namely

$$\frac{A(\partial D)}{V(D)^{1-1/n}} \geq \frac{A(\mathbb{S}^{n-1})}{V(\mathbb{B}^n)^{1-1/n}} = n\omega_{\mathbf{n}}{}^{1/n},$$

remains open.

Curvature Flow in Higher Dimensions

Note V.9. For higher dimensions, G. Huisken (1984) proved the following: Let $\mathbf{x} : M^{n-1} \to \mathbb{R}^n$ be an $(n-1)$–dimensional strictly convex hypersurface in $\mathbb{R}^n$. Then, the heat equation

$$\frac{\partial \mathbf{x}_\epsilon}{\partial \epsilon} = \mathfrak{h}_\epsilon \mathbf{n}_\epsilon, \qquad \mathbf{x}_0 = \mathbf{x}$$

has a smooth solution on a finite time interval $0 < \epsilon < \epsilon_o$, and $\mathbf{x}_\epsilon$ converges to a point as P_o as $\epsilon \to \epsilon_o$. Also, with a specific change of parameter, $\epsilon \mapsto \tau$, $\mathbf{x}_\epsilon$ has a homothetic expansion $\mathbf{y}_\tau$ about P_o , of area $A(M)$, so that $\mathbf{y}_\tau \to$ sphere of area $A(M)$ as $\tau \to \infty$ in the C^∞ convegence. M. A. Grayson (1989a) and S. B. Angenent (1992) showed that the theorem is false if M is not convex.

Further Isoperimetric Inequalities on Surfaces

Note V.10. An elementary approach to the isoperimetric problem for compact surfaces is presented in Hass–Morgan (1996), followed by application of its methods to the isoperimetric problem for surfaces of revolution of decreasing Gaussian curvature, beyond the paraboloid of revolution, in Morgan–Hutchings–Howards (2000).

In H. Howards–M. Hutchings–F. Morgan (1999), one can find results on circular cylinders, flat tori and Klein bottles, and circular cones; in C. Adams–F. Morgan (1999), hyperbolic surfaces; and in M. Ritoré (2001), general surfaces of nonnegative curvature. In A. Ros (2005), one can find a number of 3–dimensional examples, especially $\mathbb{P}^3$—3–dimensional real projective space.

VI

Isoperimetric Inequalities (Constant Curvature)

This chapter continues the discussion initiated in §V.4, the isoperimetric problem, with the emphasis on results valid for all dimensions $n \geq 1$ (the case $n = 1$ is easy to work out and is included to cover all formulae.) As mentioned in (V.4.2), the main result, to be proved below, is

$$\text{(VI.0.1)} \qquad \frac{A(\partial\Omega)}{V(\Omega)^{1-1/n}} \geq \frac{A(\mathbb{S}^{n-1})}{V(\mathbb{B}^n)^{1-1/n}} = \frac{\mathbf{c_{n-1}}}{\boldsymbol{\omega}_{\mathbf{n}}{}^{1-1/n}} = n\boldsymbol{\omega}_{\mathbf{n}}{}^{1/n},$$

where Ω is any domain in $\mathbb{R}^n$, A denotes $(n-1)$–dimensional measure, and V denotes n–dimensional measure. Equality is achieved if and only if Ω is an n–disk. Furthermore, the inequality is invariant under similarities of $\mathbb{R}^n$.

As soon as one expands the problem to the model spaces of constant sectional curvature, that is, to spheres and hyperbolic spaces, one has no self-similarities of the Riemannian spaces in question. And, if the disks on the right-hand side of (VI.0.1) are to have radius r, then the right-hand side of the inequality in (VI.0.1) is no longer independent of the value of r. So, the analytic formulation becomes more involved. For $n = 2$, the Bol–Fiala inequality (see §V.5) reads as follows: If $M = \mathbb{M}^2_\kappa$, then the isoperimetric inequality becomes

$$\text{(VI.0.2)} \qquad L^2 \geq 4\pi A - \kappa A^2.$$

By Remark V.5.4, the result is valid for all *regions* with compact closure and smooth boundary, when $\kappa \leq 0$, with equality if and only if the region is a geodesic disk. When $\kappa > 0$, the result is only proved when the region is a pairwise disjoint union of topological disks (again, with equality if and only if the region is a geodesic disk).

In this chapter, we investigate the standard space forms of constant sectional curvature of all dimensions, the Euclidean spaces, spheres, and hyperbolic spaces. Instead of giving a definitive all embracing method to definitively prove the theorem, we prefer to give a variety of arguments illustrating the wealth of techniques associated with the inequality.

A variety of techniques were presented for Euclidean space in Chavel (2001), with a full proof using Steiner symmetrization for sets with finite perimeter. Here, we do not aim for such generality; although the proofs below (excluding the solution of the Neumann problem) of the isoperimetric inequality are valid for compacta with quite general boundaries, the characterization of equality is limited to relatively compacta with C^2 boundary, in order not to take the discussion to far afield.

The proofs of the isoperimetric inequality in simply connected space of constant sectional curvature are consequences of the Brunn–Minkowski inequality (see §VI.1 for Euclidean space and §VI.3 for hyperbolic spaces and spheres). The two arguments given are classical, whereas for Euclidean space we included (§VI.2) a recent striking proof of X. Cabré (2000, 2003) which presupposes the solution of a Neumann problem in Euclidean space. Only then do we consider characterization of the case of equality.

The characterization of equality for the isoperimetric inequality on spheres (§VI.6) is achieved by a far-reaching method of M. Gromov (1986), which was first employed for compact manifolds with strictly positive Ricci curvature (including the spheres, of course). The argument then appeared in M. Berger (2003, pp. 63–66) applied to Euclidean space, and is valid for hyperbolic spaces as well. These latter two cases are presented separately (§VI.5), as a "warm-up," as they do not use the full power of the method. In this manner, the student might come to a better appreciation of the different components of Gromov's argument.

A limited bibliography for background, in general, and symmetrization arguments, various differential geometric arguments, and Gromov's argument in Euclidean space using Stokes' theorem, in particular, is given in Notes VI.1–VI.3 at the end of the chapter.

§VI.1. The Brunn–Minkowski Theorem

For $n \geq 1$, we present here the "integrated version" of the isoperimetric inequality, more popularly known as the Brunn–Minkowski inequality. More specifically, let $M = \mathbb{M}_\kappa$ be the simply connected n–dimensional space of constant sectional curvature, $n \geq 1$. To each compact subset X of $\mathbb{M}_\kappa$, associate the closed metric disk D in $\mathbb{M}_\kappa$ with the same volume as that of X. For every $\epsilon > 0$, let $[X]_\epsilon$ denote the closed set of points in $\mathbb{M}_\kappa$ whose distance from X is less than or equal to ϵ, that is,

$$[K]_h = \{x \in \mathbb{M}_\kappa : d(x, K) \leq h\}.$$

Note that X is a metric disk if and only if $[X]_\epsilon$ is a metric disk for all $\epsilon > 0$.

Definition. Let K be a compact subset of $\mathbb{M}_\kappa$. Define its *Minkowski area*, Mink (K) by

$$\text{Mink}\,(K) = \liminf_{h \downarrow 0} \frac{\mathbf{v}_n([K]_h) - \mathbf{v}_n(K)}{h}.$$

Thus, $K \mapsto \text{Mink}\,(K)$ is a functional defined on collection compacts sets K – not on the boundary of K. Furthermore, the Minkowski area is generalization of Riemannian $(n-1)$–dimensional area in that:

Proposition VI.1.1. *When $K = \overline{\Omega}$ in $\mathbb{M}_\kappa$ with $\partial\Omega \in C^1$, we have*

$$\text{Mink}\,(K) = A(\partial\Omega).$$

Remark VI.1.1. However, note that when $n = 1$, then Mink $\{x_o\} = 2$ at the same time that $\mathbf{v}_1(\{x_o\}) = 0$.

Theorem VI.1.1. *Let X and Y be bounded measurable subsets of $\mathbb{R}^n$. Then,*

$$V(X+Y)^{1/n} \geq V(X)^{1/n} + V(Y)^{1/n}. \tag{VI.1.1}$$

Proof. Assume X and Y consist of one "box" each, with no intersections (if they have an intersection then we may translate one of them – this will have no effect on the volume). Then,

$$V(X) = \prod_{j=1}^{n} \alpha_j, \qquad V(Y) = \prod_{j=1}^{n} \beta_j,$$

and $X + Y$ is then a box with

$$V(X+Y) = \prod_{j=1}^{n} (\alpha_j + \beta_j).$$

The arithmetic–geometric mean in inequality then implies

$$\left\{\frac{\prod \alpha_j}{\prod(\alpha_j + \beta_j)}\right\}^{1/n} + \left\{\frac{\prod \beta_j}{\prod(\alpha_j + \beta_j)}\right\}^{1/n}$$
$$\leq \frac{1}{n}\sum_{j=1}^{n} \frac{\alpha_j}{\alpha_j + \beta_j} + \frac{1}{n}\sum_{j=1}^{n} \frac{\beta_j}{\alpha_j + \beta_j} = 1,$$

which implies the inequality (VI.1.1).

Now assume that we have the inequality when X and Y are each the union of non-overlapping boxes (except maybe at the boundary) such that the total number of boxes is $\leq k - 1, k > 2$. So, consider the case when the union of X

and Y consists of k boxes. Then, either X or Y has at least 2 boxes. Assume it is X.

Then there exists a hyperplane in $\mathbb{R}^n$ that divides $\mathbb{R}^n$ into two closed half-spaces H^+ and H^-, such that any box in X is either in H^+ or H^-, and that each half-space has at least one box in X. Then, one can translate the boxes of Y so that P divides Y into Y^+ and Y^- satisfying

$$\frac{V(X^+)}{V(X)} = \frac{V(Y^+)}{V(Y)}, \qquad \frac{V(X^-)}{V(X)} = \frac{V(Y^-)}{V(Y)}.$$

If X consists of ℓ boxes, then Y consists of $k - \ell$ boxes, and both Y^+ and Y^- have at most $k - \ell$ boxes. But each of X^+ and X^- have strictly less than ℓ boxes. So, both $X^+ \cup Y^+$ and $X^- \cup Y^-$ have strictly less than k boxes, each. By hypothesis, the theorem is valid for $X^+ \cup Y^+$ and $X^- \cup Y^-$. Therefore,

$$\begin{aligned}
V(X+Y) &\geq V(X^+ + Y^+) + V(X^- + Y^-)\\
&\geq \left\{V(X^+)^{1/n} + V(Y^+)^{1/n}\right\}^n + \left\{V(X^-)^{1/n} + V(Y^-)^{1/n}\right\}^n\\
&= V(X^+)\left\{1 + \left(\frac{V(Y^+)}{V(X^+)}\right)^{1/n}\right\}^n + V(X^-)\left\{1 + \left(\frac{V(Y^-)}{V(X^-)}\right)^{1/n}\right\}^n\\
&= V(X^+)\left\{1 + \left(\frac{V(Y)}{V(X)}\right)^{1/n}\right\}^n + V(X^-)\left\{1 + \left(\frac{V(Y)}{V(X)}\right)^{1/n}\right\}^n\\
&= \frac{V(X^+)}{V(X)}\left\{V(X)^{1/n} + V(Y)^{1/n}\right\}^n + \frac{V(X^-)}{V(X)}\left\{V(X)^{1/n} + V(Y)^{1/n}\right\}^n\\
&= \left\{V(X)^{1/n} + V(Y)^{1/n}\right\}^n.
\end{aligned}$$

which is (III.2.2), when X and Y consist of a finite number of boxes.

For arbitrary bounded measurable sets X and Y, one approximates them from within by finite unions of nonoverlapping boxes. ■

Corollary VI.1.1. (The isoperimetric inequality for Minkowski area) *If* $Y = \mathbb{B}^n$*, the unit n–disk in* $\mathbb{R}^n$*, then* $X + \epsilon\mathbb{B}^n = [X]_\epsilon$*, which implies*

(VI.1.2) $$V([X]_\epsilon)^{1/n} \geq V(X)^{1/n} + \epsilon{\boldsymbol{\omega}_\mathbf{n}}^{1/n},$$

which implies

(VI.1.3) $$\text{Mink}\,(X) \geq n{\boldsymbol{\omega}_\mathbf{n}}^{1/n}V(X)^{1-1/n} = \frac{A(\mathbb{S}^{n-1})}{V(\mathbb{B}^n)^{1-1/n}}V(X)^{1-1/n}.$$

If Ω *is a relatively compact region with* C^1 *boundary, then*

(VI.1.4) $$A(\partial\Omega)) \geq n{\boldsymbol{\omega}_\mathbf{n}}^{1/n}V(\Omega)^{1-1/n}.$$

Remark VI.1.2. Note that even if one has an easy characterization of equality in (VI.1.2), it cannot guarantee that the argument itself will pass to the limiting result (VI.1.3). However, D. Ohman (1955) proved that if A is nonconvex, then there exists a constant c so that

$$V([X]_\epsilon) \geq \left\{V(X)^{1/n} + \epsilon {\boldsymbol{\omega}_\mathbf{n}}^{1/n}\right\}^n + c\epsilon.$$

See Burago–Zalgaller (1988, pp. 71–74). In particular, if A is nonconvex, one cannot have equality in the classical isoperimetric inequality. In the case where A is convex, one can give the classical proof (based on calculus) to characterize equality in the isoperimetric inequality.

§VI.2. Solvability of a Neumann Problem in $\mathbb{R}^n$

First note that, for Euclidean space, one may reduce the collection of competing regions to domains (that is, connected regions). Indeed, if $\Omega = \Omega_1 \cup \cdots \cup \Omega_k$ is a decomposition of Ω into domains then Minkowski's inequality (applied to the characteristic functions of the components of Ω) implies (see Lemma VIII.3.1)

$$\sum_{j=1}^{k} \{V(\Omega_j)\}^{(n-1)/n} \leq \left\{\sum_{j=1}^{k} V(\Omega_j)\right\}^{(n-1)/n},$$

with equality if and only if Ω is connected. Therefore, knowledge of the isoperimetric inequality for domains implies its validity for relatively compact regions, with the same characterization of the case of equality.

Therefore, consider a domain $D \subset\subset \mathbb{R}^n$, with C^2 boundary. The argument of X. Cabré (2000, 2003) relies on the solvability of the Neumann problem

$$\Delta u = 1 \quad \text{on } D \qquad\qquad \frac{\partial u}{\partial \nu} = c \quad \text{on } \partial D.$$

A necessary condition is:

$$V(D) = \iint_D \Delta u \, dV = \int_{\partial D} \partial u/\partial \nu \, dA = cA(\partial D).$$

So, c must given by $c = V/A$. Standard arguments from Fredholm and regularity theory (see Gilbarg–Trudinger (1977)) imply that there is indeed a solution for $c = V/A$.

Now define

$$\Gamma^+ := \{y \in D : \; u(x) \geq u(y) + (\nabla u)(y) \cdot (x - y) \, \forall \, x \in D\}.$$

That is, the tangent hyperplane to the graph of u, at the point $(y, u(y))$, supports the graph in the upper half-space defined by the tangent hyperplane. At every point of Γ^+, the Hessian of u is positive semidefinite, so all of its eigenvalues are nonnegative.

We normally think of ∇u as a vector field on D, but we may also think of it as a mapping $\nabla u : D \to \mathbb{R}^n$. Then, the Hessian of u is the Jacobian matrix of the mapping ∇u, which implies

$$\frac{V(D)}{n^n} \geq \frac{V(\Gamma^+)}{n^n} = \iint_{\Gamma^+} \left[\frac{\Delta u}{n}\right]^n dV = \iint_{\Gamma^+} \left[\frac{\text{tr Hess}\, u}{n}\right]^n dV = \cdots$$

$$\cdots \geq \iint_{\Gamma^+} \det \text{Hess}\, u \, dV \geq V\left[(\nabla u)(\Gamma^+)\right]$$

(the last inequality might come from many to one points of the mapping). It remains to show

$$(\nabla u)(\Gamma^+) \supseteq \mathbb{B}^n(c). \tag{VI.2.1}$$

If we do so, then we would have $V/n^n \geq \omega_{\mathbf{n}} c^n = \omega_{\mathbf{n}} V^n/A^n$, which is $A \geq n\omega_{\mathbf{n}}{}^{1/n} V^{1-1/n}$, the isoperimetric inequality (VI.0.1).

To prove (VI.2.1), we argue as follows: For any $\xi \in \mathbb{R}^n$, consider a hyperplane in $\mathbb{R}^{n+1}$ moving up the x^{n+1}–axis from $-\infty$, with normal vector $\xi - \mathbf{e}_{n+1}$. Then, there is a first point at which the hyperplane intersects – and therefore touches – the graph of $u : \overline{D} \to \mathbb{R}^n$. If that first point of contact is over ∂D, then the slope of the hyperplane is $\geq c$, which implies the slope of the normal $\leq 1/c$, which implies $|\xi| \geq c$.

So, for any ξ, satisfying $|\xi| < c$, the hyperplane must hit the graph at some $(y, u(y))$, $y \in D$, which implies $\xi = (\nabla u)(y)$; so, $y \in \Gamma^+$, and this implies the (VI.2.1).

If we have equality in the isoperimetric inequality, then: (1) $(\nabla u)(\Gamma^+) = \mathbb{B}^n(c)$, the n–disk of radius c; (2) ∇u is a diffeomorphism; (3) $\Gamma^+ = D$, which implies u is convex; and (4) we have equality in the arithmetic–geometric mean inequality.

Therefore

$$\frac{\partial^2 u}{\partial x^{j2}} = \frac{1}{n}, \qquad \frac{\partial^2 u}{\partial x^j \partial x^k} = 0 \ \forall j \neq k,$$

which implies

$$\frac{\partial u}{\partial x^j} = \frac{x^j}{n} + \alpha_j,$$

that is,

$$(\nabla u)(x) = \frac{x}{n} + \alpha.$$

Therefore, for all $w \in \partial D$, we have

$$|w + n\alpha| = nc,$$

so ∂D is a sphere.

§VI.3. Fermi Coordinates in Constant Sectional Curvature Spaces

We review here some basic information from §III.6 to be used in the sections that follow.

Let M be an n–dimensional simply connected space of constant sectional curvature κ, $\mathfrak{M}$ an $(n-1)$–dimensional submanifold of M, $q \in \mathfrak{M}$, ξ a unit vector at q orthogonal to $\mathfrak{M}$, $\gamma_\xi(t)$ the geodesic with initial velocity vector ξ.

Let η a principal direction of the second fundamental form of $\mathfrak{M}$ relative to ξ, that is, an eigenvector of $\mathfrak{A}^\xi : \mathfrak{M}_q \to \mathfrak{M}_q$ with eigenvalue λ. Then, the transverse vector field $Y(t)$ along the geodesic $\gamma_\xi(t)$ satisfying

$$Y(0) = \eta, \qquad \nabla_t Y(0) = -\lambda\eta,$$

is given by

$$Y(t) = \{\mathbf{C}_\kappa - \lambda \mathbf{S}_\kappa(t)\}\boldsymbol{\eta}(t), \tag{VI.3.1}$$

where $\boldsymbol{\eta}(t)$ is the parallel vector field along $\gamma_\xi(t)$ satisfying $\boldsymbol{\eta}(0) = \eta$. Let $\beta_{\kappa,\lambda}$ denote the first positive zero of $Y(t)$, should it exist, that is, the first positive solution to

$$\frac{\mathbf{C}_\kappa(\beta_{\kappa,\lambda})}{\mathbf{S}_\kappa(\beta_{\kappa,\lambda})} = \lambda. \tag{VI.3.2}$$

If we let $\mathcal{A}(t;\xi)$ denote the matrix solution to Jacobi's equation along γ, pulled back to $\xi^\perp$, as in §III.1:

$$\mathcal{A}'' + \mathcal{R}\mathcal{A} = 0,$$

subject to the initial conditions

$$\mathcal{A}(0;\xi)|\mathfrak{M}_q = I, \qquad \mathcal{A}'(0;\xi) = -\mathfrak{A}^\xi,$$

then for an orthonormal basis $e_1, \ldots, e_{n-1}$ of $\mathfrak{M}_q$ of principal directions of $\mathfrak{A}^\xi$, we have

$$\mathcal{A}(t;\xi) = \{\mathbf{C}_\kappa - \lambda_j \mathbf{S}_\kappa(t)\}e_j, \qquad j = 1, \ldots, n-1,$$

and

$$\det \mathcal{A}(t;\xi) = \prod_{j=1}^{n-1} \{\mathbf{C}_\kappa(t) - \lambda_j \mathbf{S}_\kappa(t)\}.$$

Also, recall that if $c_\nu(\xi)$ denotes the distance to the focal cut point of $\mathfrak{M}$ along γ_ξ, that is,

$$c_\nu(\xi) := \sup\{t > 0 : d(\mathfrak{M}, \gamma_\xi(t)) = t\},$$

then

(VI.3.3) $$c_\nu(\xi) \leq \min_{j=1,\dots,n-1} \beta_{\kappa,\lambda_j}.$$

Example. Let $\mathfrak{M} = \partial\Omega$, where Ω is a domain in M with compact closure and smooth boundary. Construct Fermi coordinates in Ω based on $\partial\Omega$. So, for every $q \in \partial\Omega$ let ξ_q denote the *interior* unit vector at q, orthogonal to $\partial\Omega$, and

$$E(t;q) = \exp_q t\xi_q.$$

If $\lambda_1(q), \dots, \lambda_{n-1}(q)$ denote the principal curvatures of the second fundamental form of $\partial\Omega$ at q, with respect to ξ_q, then, by (III.6.3), the volume of Ω is given by

(VI.3.4) $$V(\Omega) = \int_{\partial\Omega} dA_q \int_0^{c_\nu(q)} \prod_{j=1}^{n-1} \{\mathbf{C}_\kappa(t) - \lambda_j(q)\mathbf{S}_\kappa(t)\}\, dt.$$

Note that when Ω is an n–disk of radius r, $\mathbb{B}(r)$, the formula (VI.3.4) reduces to

(VI.3.5) $$V_\kappa(r) = A_\kappa(r) \int_0^r \{\mathbf{C}_\kappa(t) - (\mathbf{C}_\kappa(r)/\mathbf{S}_\kappa(r))\mathbf{S}_\kappa(t)\}^{n-1}\, dt.$$

Example. Let $M = \mathbb{M}^n_\kappa$, $\mathfrak{M} = \mathbb{M}^{n-1}_\kappa$ be a totally geodesic hypersurface of constant sectional curvature κ. Then,

$$\mathcal{A}(t;\xi) = \mathbf{C}_\kappa(t) I, \qquad \det \mathcal{A}(t;\xi) = {\mathbf{C}_\kappa}^{n-1}(t).$$

Let $\mathfrak{M}^t$ denote the *t–parallel hypersurface in M*, that is,

$$\mathfrak{M}^t = E(t;\mathfrak{M}); \quad \Rightarrow \quad d(x,\mathfrak{M}) = |t| \qquad \forall\ x \in \mathfrak{M}^t.$$

Then, $\mathfrak{M}^t$ is totally umbilic in M with principal curvature (relative to $\gamma_\xi{}'(t)$) equal to $\lambda = \kappa\mathbf{S}_\kappa(t)/\mathbf{C}_\kappa(t)$ (compare Exercises II.19 and III.25), and is therefore an $(n-1)$–dimensional Riemannian manifold with intrinsic constant sectional curvature $K = \kappa + \kappa^2\mathbf{S}_\kappa^2(t)/\mathbf{C}_\kappa^2(t)$ (by (II.2.6)). Thus,

$$\mathfrak{M}^t = \mathbb{M}^{n-1}_{\kappa+\kappa^2\mathbf{S}_\kappa^2(t)/\mathbf{C}_\kappa^2(t)}.$$

By (III.6.3), the volume element on M is given by

(VI.3.6) $$dV(t;q) = \mathbf{C}_\kappa{}^{n-1}(t)\,dt\,dA(q), \qquad t \in \mathbb{R}\ q \in \mathfrak{M}.$$

§VI.4. Spherical Symmetrization and Isoperimetric Inequalities

Symmetrization arguments are the earliest of those initiating the modern study of isoperimetric inequalities, starting with Steiner (1838), and continuing in the work of Schwarz (1884) and Caratheodory–Study (1909). As the title suggests, it relies heavily on the symmetries of the ambient Riemannian space, and is therefore most useful in the classical space forms of constant sectional curvature $\mathbb{M}_\kappa$.

Definition. Let M denote any metric space. The metric on M induces the Hausdorff metric δ on the space $\mathfrak{X}$ of nonempty compact subsets of M, given by

$$\delta(X,Y) = \min\{\rho :\ X \subseteq [Y]_\rho, Y \subseteq [X]_\rho\}.$$

The Blaschke selection theorem (see Chavel (2001, pp. 55ff)) states that if M is a space in which closed and bounded sets are compact, then $\mathfrak{X}$ is complete. Also, if M is compact, then $\mathfrak{X}$ is compact.

Definition. For any bounded subset Y of M, we let $r(Y)$ denote the *circumradius of* Y, that is,

$$r(Y) = \min\{\rho :\ Y \subseteq \overline{B(x;\rho)} \text{ for some } x \in M\}.$$

A *circumdisk* of Y is a disk of radius $r(Y)$ containing Y.

The set function $X \mapsto r(X)$ is continuous on $\mathfrak{X}$. But, one only has the set function $X \mapsto V(X)$ is upper semicontinuous on $\mathfrak{X}$, that is,

$$\limsup_{k\to\infty} V(X^k) \le V(X),$$

when X^k is a sequence in $\mathfrak{X}$ for which $\delta(X^k, X) \to 0$ as $k \to \infty$.

Definition. Given any compact X in M, set

$$\mathfrak{U}(X) = \{Y \in \mathfrak{X} :\ V(Y) = V(X),\ V([Y]_\epsilon) \le V([X]_\epsilon)\ \forall\, \epsilon > 0\}.$$

Lemma VI.4.1. *Assume M is one of the simply connected constant curvature spaces. Given $X \in \mathfrak{X}$, there exists an element $Y \in \mathfrak{U}(X)$ with minimal circumradius.*

Proof. If X itself realizes the minimum circumradius, then we are done. If not, then we need only consider elements of $\mathfrak{U}(X)$ with circumradius less than or equal to $r(X)$. Using the isometry group of M, we need only consider those $Y \in \mathfrak{U}(X)$, with circumdisk centered at the center o of the circumdisk of X, and satisfying $r(Y) \leq r(X)$. Then, this collection $\mathfrak{V}_o(X)$ of compact subsets of $\overline{B(o; r(X))}$ is compact in $\mathfrak{X}$. Given a sequence Y^k in $\mathfrak{V}_o(X) \cap \mathfrak{U}(X)$, such that $r(Y^k) \to \min_{\mathfrak{V}_o(X)} r$, there exists $Y \in \mathfrak{V}_o(X)$ such that $\delta(Y^k, Y) \to 0$ as $k \to \infty$. This implies $r(Y) = \min r$. So, we want to verify that $Y \in \mathfrak{U}(X)$.

Suppose we are given $\epsilon > 0$. For every $\eta > 0$, there exists $k_0 > 0$ such that $Y \subseteq [Y^k]_\eta$ for all $k \geq k_0$. Then,

$$[Y]_\epsilon \subseteq [Y^k]_{\eta+\epsilon},$$

which implies

$$V([Y]_\epsilon) \leq V([Y^k]_{\eta+\epsilon}) \leq V([X]_{\eta+\epsilon})$$

for all $\eta, \epsilon > 0$, since $Y^k \in \mathfrak{U}(X)$ for all k. This implies for all $\epsilon > 0$

$$V([Y]_\epsilon) \leq \inf_\eta V([X]_{\eta+\epsilon}) = V\left(\bigcap_{\eta>0} [X]_{\eta+\epsilon}\right) = V([X]_\epsilon),$$

the last equality since X is compact. So, to show $Y \in \mathfrak{U}(X)$, it remains to show that $V(Y) = V(X)$. From the above, we certainly have $V(Y) \leq V(X)$. But the upper semicontinuity of the Riemannian measure on $\mathfrak{U}$ implies

$$V(Y) \geq \limsup_{k\to\infty} V(Y^k) = V(X),$$

which is the claim. ∎

The Argument in Hyperbolic Space

Theorem VI.4.1. *Let M be hyperbolic space with constant curvature κ. Given $X \in \mathfrak{X} = \mathfrak{X}(M)$, let D denote the geodesic disk with volume equal to that X. Then,*

$$V([X]_\epsilon) \geq V([D]_\epsilon)$$

for all $\epsilon > 0$.

Corollary VI.4.1. *If Ω is a relatively compact region with C^1 boundary, then*

$$A(\partial\Omega) \geq A(\partial D).$$

Proof of Theorem VI.4.1. Before proceeding, we note that, for $n = 0$, the space M can be considered to be the metric space consisting of one point, the space endowed with counting measure. The theorem is then valid in this case.

The method of proof of Theorem VI.4.1 is by induction on $n = \dim M$, starting with the truth of the theorem for $n = 0$. When wishing to prove the theorem for the dimension n, one assumes that one already has the truth of the theorem in dimension $n - 1$. (We only use the induction in Lemma VI.4.2 below.)

We prove the theorem for $\kappa = -1$ (of course, it is then valid for any κ). Henceforth we consider $n \geq 1$.

First, consider a fixed $(n-1)$–dimensional hyperbolic space $\mathbb{H}^{n-1}$ totally geodesic in $\mathbb{H}^n$, and introduce Fermi coordinates based on $\mathbb{H}^{n-1}$, relative to a (parallel) normal unit tangent vector field $\boldsymbol{\xi}$ along $\mathbb{H}^{n-1}$, specifically

$$E(t;q) = \operatorname{Exp} t\boldsymbol{\xi}_q, \qquad t \in \mathbb{R}, \;\; q \in \mathbb{H}^{n-1}.$$

Then, the metric on $\mathbb{H}^n$ reads as

$$|dx|^2 = (dt)^2 + \cosh^2 t\,|dq|^2,$$

where $|dx|^2$ denotes the Riemannian metric on $\mathbb{H}^n$, and $|dq|^2$ on $\mathbb{H}^{n-1}$. Also, by (VI.3.6),

$$dV(t;q) = \cosh^{n-1} t\,dt\,dA(q), \tag{VI.4.1}$$

Let

$$\mathfrak{H}^t = E(t;\mathbb{H}^{n-1})$$

denote the t–parallel hypersurface in $\mathbb{H}^n$. Then, $\mathfrak{H}^t$ is totally umbilic in $\mathbb{H}^n$ with principal curvature $\kappa = \tanh t$ (compare Exercise III.25), and is therefore an $(n-1)$–dimensional Riemannian manifold with intrinsic constant sectional curvature $K = -1 + \tanh^2 t$ (by (II.2.6)). Thus,

$$\mathfrak{H}^t = \mathbb{H}^{n-1}_{-1+\tanh^2 t}.$$

Second, for every unit tangent vector ξ, one considers the geodesic $\gamma_\xi(t)$ (determined by the initial velocity vector ξ) and the hypersurface $\mathfrak{H}_\xi$ determined by ξ, namely, the totally geodesic $\mathbb{H}^{n-1}$ in $\mathbb{H}^n$ orthogonal to ξ. Let $\mathfrak{H}^t_\xi$ denote the hypersurface t–parallel to $\mathfrak{H}_\xi$.

As above, introduce the Fermi coordinates

$$E_\xi(t,q) = \operatorname{Exp} t\boldsymbol{\xi}_q := \lambda_t(q), \qquad t \in \mathbb{R},\ \ q \in \mathbb{H}^{n-1},$$

where $\boldsymbol{\xi}$ is a continuous extension of ξ to a unit normal vector field on $\mathfrak{H}_\xi$.

Fix $\epsilon > 0$. For any σ, there exists a real-valued function $\eta_\sigma(s)$ supported on $\{s : |s-\sigma| \le \epsilon\}$, such that, for any $x = E_\xi(\sigma, q)$, $q \in \mathfrak{H}_\xi$, the closed n–disk about x of radius ϵ is given by

$$[\{x\}]_\epsilon = \overline{B(x;\epsilon)} = \bigcup_{\{s:\,|s-\sigma|\le\epsilon\}} \lambda_s(B_{\mathfrak{H}_\xi}(q;\eta_\sigma(s))).$$

This implies that, for

$$Z \subseteq \mathfrak{H}_\xi^\sigma,$$

we have

$$[Z]_\epsilon \cap \mathfrak{H}_\xi^s = \lambda_s([\lambda_\sigma{}^{-1}(Z)]_{\eta_\sigma(s)} \cap \mathfrak{H}_\xi) := \lambda_s([\lambda_\sigma{}^{-1}(Z)]_{\eta_\sigma(s),\xi}),$$

where $[Y]_{\epsilon,\xi}$ denotes ϵ–thickening of Y in $\mathfrak{H}_\xi$. That is, we start with an $(n-1)$–dimensional subset Z of $\mathfrak{H}_\xi^\sigma$, thicken it *in* M by ϵ, and then slice it by $\mathfrak{H}_\xi^s$. The result is the same as exponentiating to $\mathfrak{H}_\xi^s$ the thickening of Z in $\mathfrak{H}_\xi$ by $\eta_\sigma(s)$. Note that $|s-\sigma| > \epsilon$ implies $[Z]_\epsilon \cap \mathfrak{H}_\xi^s = \emptyset$.

For a compact subset X in M, and $t \in \mathbb{R}$, let $X_\xi^t = X \cap \mathfrak{H}_\xi^t$; and $D_\xi^t = D_\xi^t(X)$ denote closed (intrinsic) $(n-1)$–dimensional disk in $\mathfrak{H}_\xi^t$, centered at $\gamma_\xi(t)$, with $(n-1)$–dimensional area $A(D_\xi^t) = A(X_\xi^t)$; and we set

$$\mathcal{S}_\xi(X) = \bigcup_t D_\xi^t.$$

One verifies directly that $\mathcal{S}_\xi(X)$ is compact and, by (VI.4.1), $V(\mathcal{S}_\xi(X)) = V(X)$.

Lemma VI.4.2. *Given any compact X in M, $\xi \in \mathsf{S}M$, we have $\mathcal{S}_\xi(X) \in \mathfrak{U}(X)$.*

Proof. Given X and ξ (for convenience, we drop the subscript ξ in the rest of this lemma). Fix $\epsilon > 0$. Set

$$[X]_\epsilon^t = [X]_\epsilon \cap \mathfrak{H}^t.$$

For any $t \in \mathbb{R}$, the intersection of $[X]_\epsilon$ with $\mathfrak{H}^t$, $[X]_\epsilon^t$, is composed of contributions from X^s, for $|s-t| \le \epsilon$. Therefore,

$$[X]_\epsilon^t = \bigcup_{\{s:\,|s-t|\le\epsilon\}} \lambda_t([\lambda_s{}^{-1}X^s]_{\eta_s(t)}).$$

Also, for $W = \mathcal{S}_\xi(X)$,

$$[W]^t_\epsilon = \bigcup_{\{s:\,|s-t|\le\epsilon\}} \lambda_t([\lambda_s{}^{-1}W^s]_{\eta_s(t)}).$$

Of course, $A(W^t) = A(X^t)$ by the very construction of W.

Of course, we have

$$[X]^t_\epsilon = \bigcup_{\{s:\,|s-t|\le\epsilon\}} \lambda_t([\lambda_s{}^{-1}X^s]_{\eta_s(t)}) \supseteq \lambda_t([\lambda_s{}^{-1}X^s]_{\eta_s(t)}) \quad \text{for all } \{s : |s-t| \le \epsilon\}$$

which implies

$$A([X]^t_\epsilon) = A\left(\bigcup_{\{s:\,|s-t|\le\epsilon\}} \lambda_t([\lambda_s{}^{-1}X^s]_{\eta_s(t)})\right) \ge \sup_{\{s:\,|s-t|\le\epsilon\}} A(\lambda_t([\lambda_s{}^{-1}X^s]_{\eta_s(t)})),$$

that is,

$$\text{(VI.4.2)} \qquad A([X]^t_\epsilon) \ge \sup_{\{s:\,|s-t|\le\epsilon\}} A(\lambda_t([\lambda_s{}^{-1}X^s]_{\eta_s(t)})).$$

For $W = \Sigma(X)$, we have equality in (VI.4.2), since each $\lambda_t([\lambda_s{}^{-1}W^s]_{\eta_s(t)})$ is an $(n-1)$–disk in $\mathfrak{H}^t$ centered at $\gamma(t)$.

Now assume the Theorem VI.4.1 is true for dimension $n-1$. Then, it is true for every hyperbolic space $\mathfrak{H}^t$, which implies

$$A(\lambda_t([\lambda_s{}^{-1}X^s]_{\eta_s(t)})) \ge A(\lambda_t([\lambda_s{}^{-1}W^s]_{\eta_s(t)}))$$

for all $|t-s| \le \epsilon$, which implies

$$\begin{aligned} A([X]^t_\epsilon) &\ge \sup_{\{s:\,|s-t|\le\epsilon\}} A(\lambda_t([\lambda_s{}^{-1}X^s]_{\eta_s(t)})) \\ &\ge \sup_{\{s:\,|s-t|\le\epsilon\}} A(\lambda_t([\lambda_s{}^{-1}W^s]_{\eta_s(t)})) \\ &= A([W]^t_\epsilon). \end{aligned}$$

We therefore have $A([X]^t_\epsilon) \ge A([W]^t_\epsilon)$ for all $t \in \mathbb{R}$. By (VI.4.1), we have

$$V([X]_\epsilon) \ge V([W]_\epsilon).$$

■

Lemma VI.4.3. *Consider $X \in \mathfrak{X}$, which is not a disk. Let $r = r(X)$, and $\overline{B(o;r)}$ the circumdisk of X. Then there exists a finite number of unit tangent vectors $\xi_1, \ldots, \xi_k$ at o such that*

$$r(\mathcal{S}_{\xi_k}(\mathcal{S}_{\xi_{k-1}}(\ldots(X)\ldots))) < r(X).$$

Proof. Given X not a disk, any symmetrization determined by a unit tangent vector at o leaves $\overline{B(o;r)}$ invariant. Since X is not all of $\overline{B(o;r)}$, then $\mathcal{S}_\xi(X)$

is not all of $\overline{B(o;r)}$ for any ξ and does not contain the complete boundary $S(o;r) = \partial\overline{B(o;r)}$. The idea of what follows is to decrease the intersection of the image of successive symmetrizations of X with $S(o;r)$, until it becomes empty. But then the circumradius will have become strictly less than r.

To carry out the argument, one simply notes:

(i) If Y is a closed subset of $\overline{B(o;r)}$, and $x \in \{S(o;r) \setminus Y\}$, then $x \notin \mathcal{S}_\xi(Y)$ for all ξ.

(ii) Let Y be a closed subset of $\overline{B(o;r)}$, and G a relatively open subset of $S(o;r)$, which does not intersect Y. Since G is open, it contains a spherical cap subtending an angle $\alpha > 0$ in $(\mathbb{H}^n)_o$, the tangent space to $\mathbb{H}^n$ at o. If $\alpha > \pi$, then G contains antipodal points, and one symmetrizes relative to the geodesic they determine to finish the job. On the other hand, if $0 < \alpha < \pi$, then one can find a ξ for which there exists a relatively open subset G_ξ of $S(o;r)$, disjoint from $\mathcal{S}_\xi(Y)$, which contains a spherical cap subtending an angle $5\alpha/4$ in $(\mathbb{H}^n)_o$.

This suffices to prove the lemma, and with it, Theorem VI.4.1. ■

The Argument in the Sphere

To carry out the argument on the sphere, one notes that, for every unit tangent vector ξ, one considers the geodesic segment $\gamma_\xi(t)$ for $t \in [-\pi/2, \pi/2]$, and the "equator" $\mathfrak{S}_\xi$ determined by ξ, namely, the totally geodesic $\mathbb{S}^{n-1}$ in $\mathbb{S}^n$ orthogonal to ξ.

Then, one considers Fermi coordinates based on $\mathfrak{S}_\xi$,

$$E_\xi(t;q) = \operatorname{Exp} t\boldsymbol{\xi}_q, \qquad t \in \mathbb{R},\ q \in \mathfrak{S}_\xi,$$

where $\boldsymbol{\xi}$ is the continuous extension of ξ to a normal unit field along $\mathfrak{S}_\xi$. Then, the metric on $\mathbb{S}^n$ reads as

$$|dx|^2 = (dt)^2 + \cos^2 t\,|dq|^2,$$

where $|dx|$ denotes the Riemannian metric on $\mathbb{S}^n$ and $|dq|^2$ on $\mathfrak{S}_\xi$. The volume element, in these coordinates, is given by (see (VI.3.6))

$$dV(t;q) = \cos^{n-1} t\,dt\,dA(q). \tag{VI.4.3}$$

Let

$$\mathfrak{S}^t = E_\xi(t;\mathbb{S}^{n-1})$$

denote the *t–parallel hypersurface in* $\mathbb{S}^n$, that is

$$d(x, \mathfrak{S}_\xi) = |t| \qquad \forall\ x \in \mathfrak{S}^t.$$

Then, $\mathfrak{S}^t$ is totally umbilic in $\mathbb{S}^n$ with principal curvature $\kappa = \tan t$, and is therefore an $(n-1)$–dimensional Riemannian manifold with intrinsic constant sectional curvature $K = 1 + \tan^2 t$. Thus,

$$\mathfrak{S}^t = \mathbb{S}^{n-1}_{1+\tan^2 t}.$$

Now one can work through the argument as above.

§VI.5. M. Gromov's Uniqueness Proof – Euclidean and Hyperbolic Space

Let M denote n–dimensional Euclidean space or hyperbolic space of constant curvature -1, $\mathbb{B}(r)$ the disk of radius r, with boundary $\mathbb{S}(r)$. We know that $\mathbb{B}(r)$ is *a* solution to the isoperimetric problem for domains Ω satisfying $V(\Omega) = V(\mathbb{B}(r))$, that is, $A(\partial\Omega) \geq A(\mathbb{S}(r))$ for all such Ω. Here, we consider the uniqueness question in the category of domains with C^2 boundary. We show that if $V(\Omega) = V(\mathbb{B}(r))$, $A(\partial\Omega) = A(\mathbb{S}(r))$, $\partial\Omega \in C^2$, then Ω is isometric to $\mathbb{B}(r)$.

Of course, one can invoke the Theorem V.4.1 to claim that $\partial\Omega$ is regular except for, at worst, a singular set of codimension ≥ 7, in which case the argument presented below is valid as well. But, for purposes of the exposition, we find it simpler to stay in the more restricted category of C^2.

First, let $\mathsf{V}(r)$ denote the volume of $\mathbb{B}(r)$, $\mathsf{A}(r)$ the area of $\mathbb{S}(r)$, and $\mathsf{R}(v)$ the inverse function of $\mathsf{V}(r)$. Then,

$$\frac{\mathsf{A}'(r)}{\mathsf{V}'(r)} = \frac{\mathsf{A}'(r)}{\mathsf{A}(r)} = \mathfrak{h}_r,$$

where $\mathfrak{h}_r$ denotes the mean curvature of $\mathbb{S}(r)$ relative to the *inward* pointing normal of $\mathbb{S}(r)$ into $\mathbb{B}(r)$. So,

$$\mathsf{V}(r) = \omega_{\mathbf{n}} r^n = \mathbf{c}_{\mathbf{n-1}} r^n/n, \qquad \mathsf{A}(r) = \mathbf{c}_{\mathbf{n-1}} r^{n-1}, \qquad \mathfrak{h}_r = (n-1)/r$$

for Euclidean space, and

$$\mathsf{V}(r) = \mathbf{c}_{\mathbf{n-1}} \int_0^r \sinh^{n-1} t\, dt, \quad \mathsf{A}(r) = \mathbf{c}_{\mathbf{n-1}} \sinh^{n-1} r, \quad \mathfrak{h}_r = (n-1)\coth r$$

for hyperbolic space.

In both cases, we let $\mathcal{I}(v)$ denote the isoperimetric profile of M; therefore,

$$\mathcal{I}(v) = \mathsf{A}(\mathsf{R}(v)).$$

For any isoperimetric region Ω with C^2 boundary, the mean curvature of $\partial\Omega$ (relative to the interior of Ω) is constant, $\mathfrak{h}_\Omega$, and for any C^2 variation

$\Omega_t, t \in (-t_o, t_o)$, of Ω, we have

$$\frac{A(\partial(\Omega_t))}{\mathcal{I}(V(\Omega_t))} \geq 1, \qquad \frac{A(\partial(\Omega))}{\mathcal{I}(V(\Omega))} = 1,$$

which implies

$$0 = \frac{d}{dt}\frac{A(\partial(\Omega_t))}{\mathcal{I}(V(\Omega_t))}\bigg|_{t=0} = \mathfrak{h}_\Omega - \frac{\mathsf{A}'(r)}{\mathsf{V}'(r)} = \mathfrak{h}_\Omega - \mathfrak{h}_{\mathsf{R}(V(\Omega))},$$

by Exercises III.16 and III.14 (the first variation of volume and area), and the chain rule. That is,

$$\mathfrak{h}_\Omega = \mathfrak{h}_{\mathsf{R}(V(\Omega))} \tag{VI.5.1}$$

for any isoperimetric region Ω in M.

Next, consider $M = \mathbb{R}^n$. Let Ω be an isoperimetric region in $\mathbb{R}^n$, with $V(\Omega) = \mathsf{V}(\mathsf{r})$. Construct Fermi coordinates in Ω based on $\partial\Omega$. For every $q \in \partial\Omega$, let ξ_q denote the *interior* unit vector at q, orthogonal to $\partial\Omega$, and

$$E(t; q) = \exp_q t\xi_q.$$

If $\lambda_1(q), \ldots, \lambda_{n-1}(q)$ denote the principal curvatures of the second fundamental form of $\partial\Omega$ at q, with respect to ξ_q (recall that the mean curvature is the sum of the principal curvatures), then, by (VI.3.3),

$$c_\nu(q) \leq 1/\max_{j=1,\ldots,n-1} \lambda_j(q).$$

Also, the volume of Ω is given by (see (VI.3.4))

$$V(\Omega) = \int_{\partial\Omega} dA_q \int_0^{c_\nu(q)} \prod_{j=1}^{n-1}\{1 - \lambda_j(q)t\}\,dt.$$

The arithmetic–geometric mean inequality implies

$$\begin{aligned}
V(\Omega) &\leq \int_{\partial\Omega} dA_q \int_0^{1/\max\lambda_j(q)} \prod_{j=1}^{n-1}\{1 - \lambda_j(q)t\}\,dt \\
&\leq \int_{\partial\Omega} dA_q \int_0^{1/\max\lambda_j(q)} \left[1 - \frac{\mathfrak{h}_\Omega t}{n-1}\right]^{n-1} dt \\
&\leq \int_{\partial\Omega} dA_q \int_0^{(n-1)/\mathfrak{h}_\Omega} \left[1 - \frac{\mathfrak{h}_\Omega t}{n-1}\right]^{n-1} dt \\
&= A(\partial\Omega) \int_0^{(n-1)/\mathfrak{h}_{\mathsf{r}}} \left[1 - \frac{\mathfrak{h}_{\mathsf{r}} t}{n-1}\right]^{n-1} dt \\
&= \mathsf{A}(\mathsf{r}) \int_0^{\mathsf{r}} \left[1 - \frac{t}{\mathsf{r}}\right]^{n-1} dt \\
&= \mathsf{V}(\mathsf{r})
\end{aligned}$$

the last equality follows from (VI.3.5). Therefore, we have equality in the arithmetic–geometric mean inequality, which implies every point of $\partial\Omega$ is umbilic, so $\partial\Omega$ is a sphere and Ω a disk. ∎

In hyperbolic space, the argument is the same, only the formulae are different. Let Ω be an isoperimetric region in $\mathbb{H}^n$, with $V(\Omega) = \mathsf{V}(\mathsf{r})$. Construct Fermi coordinates in Ω based on $\partial\Omega$. For every $q \in \partial\Omega$, let ξ_q denote the *interior* unit vector field at q, orthogonal to $\partial\Omega$, and

$$E(t; q) = \exp_q t\xi_q.$$

If $\lambda_1(g), \ldots, \lambda_{n-1}(q)$ denote the principal curvatures of the second fundamental form of $\partial\Omega$ at q, with respect to ξ_q, then

$$c_\nu(q) \le \operatorname{arctanh}(1/\max_{j=1,\ldots,n-1} \lambda_j(q)).$$

Also, the volume of Ω is given by

$$V(\Omega) = \int_{\partial\Omega} dA_q \int_0^{c_\nu(q)} \prod_{j=1}^{n-1} \{\cosh t - \lambda_j(q) \sinh t\}\, dt,$$

and

$$\begin{aligned}
V(\Omega) &\le \int_{\partial\Omega} dA_q \int_0^{\operatorname{arctanh}(1/\max \lambda_j(q))} \prod_{j=1}^{n-1} \{\cosh t - \lambda_j(q) \sinh t\}\, dt \\
&\le \int_{\partial\Omega} dA_q \int_0^{\operatorname{arctanh}(1/\max \lambda_j(q))} \left[\cosh t - \frac{\mathfrak{h}_\Omega}{n-1} \sinh t\right]^{n-1} dt \\
&\le \int_{\partial\Omega} dA_q \int_0^{\operatorname{arctanh}((n-1)/\mathfrak{h}_\Omega)} \left[\cosh t - \frac{\mathfrak{h}_\Omega}{n-1} \sinh t\right]^{n-1} dt \\
&= \int_{\partial\Omega} dA_q \int_0^{\operatorname{arctanh}((n-1)/\mathfrak{h}_\mathsf{r})} \left[\cosh t - \frac{\mathfrak{h}_\mathsf{r}}{n-1} \sinh t\right]^{n-1} dt \\
&= \int_{\partial\Omega} dA_q \int_0^{\mathsf{r}} \left[\cosh t - \frac{\mathfrak{h}_\mathsf{r}}{n-1} \sinh t\right]^{n-1} dt \\
&= A(\Omega) \int_0^{\mathsf{r}} \left[\cosh t - \frac{\mathfrak{h}_\mathsf{r}}{n-1} \sinh t\right]^{n-1} dt \\
&= A(\mathsf{r}) \int_0^{\mathsf{r}} [\cosh t - \coth \mathsf{r} \sinh t]^{n-1}\, dt \\
&= V(\mathsf{r}).
\end{aligned}$$

Therefore, we have equality in the arithmetic–geometric mean inequality, which implies every point of $\partial\Omega$ is umbilic; so, $\partial\Omega$ is a sphere and Ω a disk. ∎

§VI.6. The Isoperimetric Inequality on Spheres

In this section, we prove a theorem of M. Gromov, a special case of which is the solution of the isoperimetric problem on spheres.

Theorem VI.6.1. (M. Gromov (1986)) *Let M be an n–dimensional compact Riemannian manifold whose Ricci curvature satisfies*

$$\mathrm{Ric} \geq (n-1)\kappa > 0,$$

and $\mathbb{M}_\kappa$ the n–sphere of constant sectional curvature equal to $\kappa > 0$. Then, the Bishop theorem (III.4.4) *states that*

$$\beta := \frac{V(M)}{V(\mathbb{M}_\kappa)} \leq 1.$$

To any given $\Omega \subseteq M$, we associate the geodesic disk D in $\mathbb{M}_\kappa$ satisfying

$$V(\Omega) = \beta V(D).$$

Then,

$$A(\partial\Omega) \geq \beta A(\partial D), \tag{VI.6.1}$$

with equality in (VI.6.1) *if and only if M is isometric to the sphere $\mathbb{M}_\kappa$, and Ω is isometric to the geodesic disk D.*

Proof.

Step 1. We let Γ vary over all compact $(n-1)$–submanifolds of M that divide M into two domains M_1, M_2 for which

$$V(M_1) = V(\Omega);$$

and we consider the variational problem of minimizing the area of Γ, $A(\Gamma)$, over this collection of Γ. Let A_0 be the minimum value and assume that we have a differentiable Γ_0 for which $A_0 = A(\Gamma_0)$ (see Step 5). Then (see Exercise (III.18)), Γ_0 has constant mean curvature λ, with the unit normal vector field ξ along Γ_0 pointing into the domain M_1^0 with volume equal to that of Ω.

Step 2. As above, introduce the Fermi coordinates as $E : (-\infty, +\infty) \times \Gamma_0 \to M$ given by

$$E(t; q) = \exp t\xi_{|q}.$$

When discussing the focal cut locus, it is best to write

$$c(q) = c_\nu(\xi_{|q}), \qquad c(-q) = c_\nu(-\xi_{|q}),$$

where c_ν denotes the distance to the cut point of Γ_o.

Step 3. For $f \in L^1(M)$, one has in Fermi coordinates based on the submanifold Γ_0,

$$\text{(VI.6.2)} \qquad \int_M f\,dV = \int_{\Gamma_0} dA(q) \int_{-c(-q)}^{c(q)} f(E(t;q)) \det \mathcal{A}(t;q)\,dt,$$

by (III.6.3).

Now the Bishop comparison theorem (Theorem III.6.1) in this context reads as follows: Let $\beta^+_{\kappa,\lambda}$ denote the first positive zero of $(\mathbf{C}_\kappa - \lambda\mathbf{S}_\kappa)(t)$ (should it exist – otherwise, we set $\beta^+_{\kappa,\lambda} = +\infty$), and $-\beta^-_{\kappa,\lambda}$ the first negative zero of $(\mathbf{C}_\kappa - \lambda\mathbf{S}_\kappa)(t)$ (should it exist – otherwise we set $\beta^-_{\kappa,\lambda} = \infty$). Then,

$$\text{(VI.6.3)} \qquad c(-q) \le \beta^-_{\kappa,\lambda}, \qquad c(q) \le \beta^+_{\kappa,\lambda},$$

and

$$\text{(VI.6.4)} \qquad \det \mathcal{A}(t;q) \le (\mathbf{C}_\kappa - \lambda\mathbf{S}_\kappa)^{n-1}(t)$$

(by the arithmetic–geometric mean inequality) for all $t \in [-c(-q), c(q)]$. We have equality in (VI.6.4) at a given $\tau \in [-c(-q), c(q)] \setminus \{0\}$ if and only if

$$\text{(VI.6.5)} \qquad \mathcal{A}(t;q) = (\mathbf{C}_\kappa - \lambda\mathbf{S}_\kappa)(t)I$$

for all t in the interval connecting τ to 0, in which case we have

$$A^{\xi|_q} = \lambda I, \qquad \mathcal{R}(t) = \kappa I$$

on the interval connecting τ to 0.

Step 4. We are now ready to finish the proof of the theorem. First, let r_D denote the radius of D.

Let $f = I_{M_1^0}$. Then,

$$\begin{aligned} V(\Omega) = V(M_1^0) &= \int_{\Gamma_0} dA(q) \int_0^{c(q)} \det \mathcal{A}(t;q)\,dt \\ &\le A(\Gamma_0) \int_0^{\beta^+_{\kappa,\lambda}} (\mathbf{C}_\kappa - \lambda\mathbf{S}_\kappa)^{n-1}(t)\,dt. \end{aligned}$$

If Ω is a disk of radius r in $\mathbb{M}_\kappa$, then we have equality above, which yields

$$V_\kappa(r) = A_\kappa(r) \int_0^r (\mathbf{C}_\kappa - \lambda\mathbf{S}_\kappa)^{n-1}(t)\,dt.$$

We conclude for general Ω,

$$V(\Omega) \le A(\Gamma_0) \frac{V_\kappa(\beta^+_{\kappa,\lambda})}{A_\kappa(\beta^+_{\kappa,\lambda})}.$$

Let r_D denote the radius of D. Note that the function $r \mapsto A_\kappa(r)/V_\kappa(r)$ is strictly decreasing. Therefore, if $\mathrm{r}_D \geq \beta^+_{\kappa,\lambda}$ then

$$A(\Gamma_0) \geq \frac{V(\Omega)}{V_\kappa(\beta^+_{\kappa,\lambda})} A_\kappa(\beta^+_{\kappa,\lambda}) \geq \frac{V(\Omega)}{V_\kappa(\mathrm{r}_D)} A_\kappa(\mathrm{r}_D) = \beta A_\kappa(\mathrm{r}_D).$$

On the other hand, if $\mathrm{r}_D < \beta^+_{\kappa,\lambda}$, then $\mathrm{r}_D > \beta^-_{\kappa,\lambda}$. Then, one has

$$A(\Gamma_0) \geq \frac{V(M \setminus \overline{\Omega})}{V_\kappa(\beta^-_{\kappa,\lambda})} A_\kappa(\beta^-_{\kappa,\lambda}) \geq \frac{V(\Omega)}{V_\kappa(\mathrm{r}_D)} A_\kappa(\mathrm{r}_D) = \beta A_\kappa(\mathrm{r}_D).$$

This implies (VI.6.1).

Step 5. Our proof assumed that Γ_0 has no singularities. However, the main existence–regularity theorem for the isoperimetric problem (see Theorem V.4.1) implies the validity of (VI.6.2), even when there are singularities.

First, the singularities of Γ are restricted to subsets of submanifolds of codimension greater than or equal to 7, and therefore the collection of singularities has $(n-1)$–measure equal to 0. So, if Γ_0^* denotes the regular points of Γ_0, then

$$\int_{\Gamma_0} dA(q) \cdots = \int_{\Gamma_0^*} dA(q) \cdots .$$

Moreover (see Gromov (1980)), for any point x in $M \setminus \Gamma_0$, there exists a point $q_x \in \Gamma_0^*$ for which

$$x = E(d(x, q_x), q_x).$$

That is, any point in M not in Γ is in the image of the exponential map of the normal bundle over points of Γ_0^*. Indeed, such a q_x that minimizes distance from x to Γ_0 certainly exists in Γ_0. Let y be the midpoint of a minimizing geodesic segment connecting x to q_x, and consider the closed metric disk $\overline{B(y; d(x, q_x)/2)}$. Then,

$$\Gamma_0 \cap \overline{B(y; d(x; q_x)/2)} = \{q_x\}$$

(since, otherwise, one would have a broken minimizing geodesic from x to Γ_0). Furthermore, the bounding metric sphere $S(y; d(x, q_x)/2)$ is smooth at q_x. We conclude that the tangent cone of Γ_0 at x is contained in a half-space at x, which implies (Almgren (1976)) Γ_0 is regular at x.

Step 6. One easily checks that equality implies $\beta = 1$, which implies M is isometric to $\mathbb{M}_\kappa$ by the Toponogov–Cheng theorem (Theorem III.4.6). ■

§VI.7. Notes and Exercises

Note VI.1. Bibliographic sampler on isoperimetric inequalities. The literature is quite large, and a good start is in the following: Burago–Zalgaller (1988), Gromov (1986, 1981, Chapter 6) Osserman (1978, 1979), and A. Ros (2005).

Symmetrization

Note VI.2. The modern study of isoperimetric inequalities, inaugurated in the papers of J. Steiner (1838), H. A. Schwarz (1884), and C. Caratheodory–E. Study (1909), was originally restricted to 2– and 3–dimensional Euclidean space. The isoperimetric problem for manifolds beyond the Euclidean plane and space was first solved for domains in the 2–sphere bounded by convex curves, by F. Bernstein (1905); and a unified proof for the simply connected spaces of constant sectional curvature of all dimensions (which might be viewed as a summary statement of all the previous work) using spherical symmetrization was given in E. Schmidt (1948). See also the extensive treatment in Y. D. Burago–V. A. Zalgaller (1988).

Yet another symmetrization argument of J. Steiner (1842) was, more recently, revived by W. Y. Hsiang (1991), H. Howard–M. Hutchings–F. M. Morgan (1999), and A. Ros (2001). See our comments in Chavel (1984, p. 11ff).

Still another approach to symmetrization, two-point symmetrization, was initiated, it seems, by L. V. Ahlfors (1973), and used on the n–sphere by A. Baernstein–B. A. Taylor (1976). Also, see the notes of Y. Benjamini (1983).

For current presentations of symmetrization methods, see E. H. Lieb–M. Loss (1996) and A. Baernstein (2004). For the Brunn–Minkowski inequalities see the recent R. J. Gardner (2002). For a complete proof of the isoperimetric inequality, using Steiner symmetrization, valid for compacta and domains with finite perimeter, and with characterization of the case of equality, see Chavel (2001).

Other Methods

Note VI.3. The classical isoperimetric inequality in $\mathbb{R}^n$ also has the advantage of possessing a veritable wealth of techniques (especially in the plane), each of which "represents" some subfield of analysis and/or geometry. Many of the 2–dimensional arguments are presented in Burago–Zalgaller (1988, Chapter 1). We only sampled two of higher dimensional arguments in our presentation, here. In Chavel (2001, Chapter II), we present a sampling of other arguments associated with the isoperimetric inequality, most of which are related to uniqueness.

(i) We note there (see also Exercise III.18) that the classical Euler–Lagrange equation for an extremal area subject to constant volume constraint is that the mean curvature of the boundary of the extremal domain have constant mean curvature.

(ii) We give F. Almgren's (1976) proof that any solution to the isoperimetric problem in $\mathbb{R}^n$, with C^2 boundary must be an n–disk.

(iii) We then give A. D. Alexandrov's (1962) proof that *any* domain in $\mathbb{R}^n$ with C^2 boundary of constant mean curvature must be isometric to a disk. These arguments complement the argument of Gromov presented here.

(iv) Finally, we give, there, M. Gromov's (1986) argument proving the isoperimetric inequality for domains with C^1 boundary, *as a consequence of Stokes' theorem*!

The Elementary Version of Steiner Symmetrization

We describe the basic idea of Steiner symmetrization (Steiner (1838)). In general, symmetrization arguments are always at the heart of isoperimetric inequalities in space forms of constant sectional curvature, since they give immediate expression to the intuition that the more symmetric a set, the "closer" the set is to the solution of the isoperimetric problem.

The basic idea is as follows: We work in $\mathbb{R}^2$. Given two differentiable functions $f_j(x)$, $j = 1, 2$ defined on the interval $[a, b]$ in $\mathbb{R}$, with $f_1(x) \leq f_2(x)$ for all $x \in [a, b]$, consider the domain

$$\Omega =: \{(x, y) : \ f_1(x) \leq y \leq f_2(x), \qquad a \leq x \leq b\}.$$

To the domain Ω, we associate the symmetrized domain Ω^* obtained by considering the function

$$f^*(x) = \frac{f_2(x) - f_1(x)}{2}$$

and setting

$$\Omega^* =: \{(x, y) : \ -f^*(x) \leq y \leq f^*(x), \ a \leq x \leq b\}.$$

Exercise VI.1. Prove

$$A(\Omega) = A(\Omega^*), \qquad L(\partial\Omega) \geq L(\partial\Omega^*).$$

Also show that, if Ω is given by

$$\Omega =: \{(x, y) : \ f_{2j-1}(x) \leq y \leq f_{2j}(x), \ j = 1, \ldots, k \ \ a \leq x \leq b\},$$

where $f_1(x) \le f_2(x) \le \ldots \le f_{2k-1}(x) \le f_{2k}(x)$ for all x, with $f_{2j} \not\equiv f_{2j+1}$ for at least one j, and the symmetrization Ω^* of Ω is defined by setting

$$f^*(x) = \frac{1}{2} \sum_{j=1}^{k} \{f_{2j}(x) - f_{2j-1}(x)\},$$

and

$$\Omega^* =: \{(x, y) : -f^*(x) \le y \le f^*(x),\ a \le x \le b\},$$

then show that

$$A(\Omega) = A(\Omega^*), \qquad L(\partial\Omega) > L(\partial\Omega^*).$$

So, the symmetrization preserves the area and decreases the length.

Note VI.4. See Pólya–Szegö (1951) and Kawohl (1985) for extended discussions emphasizing the connection of symmetrization methods to analysis.

The Faber–Krahn Inequality

We give below an application of symmetrization to analysis. It involves the application of "geometric" isoperimetric inequalities to isoperimetric inequalities for eigenvalues. The inequality is as follows: Let $M = \mathbb{M}_\kappa$ be the simply connected space form of constant sectional curvature κ. To each open set Ω, consisting of a finite union of relatively compact domains with smooth boundary, we associate the disk D in M with $V(\Omega) = V(D)$. Then, the isoperimetric inequality implies $A(\partial\Omega) \ge A(\partial D)$, with equality if and only if Ω is congruent to D.

But first we have to quote the coarea formula (Theorem VIII.3.3 below; also Theorem III.5.2 and Exercise III.12).

Coarea Formula. *Let Ω be a domain in M with compact closure and $f : \overline{\Omega} \to \mathbb{R}$ a function in $C^0(\overline{\Omega}) \cap C^\infty(\Omega)$, with $f \mid \partial\Omega = 0$. For any regular value t of $|f|$, we let*

$$\Gamma(t) = |f|^{-1}[t], \qquad A(t) = A(\Gamma(t)),$$

and dA_t denote the $(n-1)$–dimensional Riemannian measure on $\Gamma(t)$. Then,

$$dV_{|\Gamma(t)} = \frac{dA_t\, dt}{|\text{grad}\, f|},$$

and for any function $\phi \in L^1(\Omega)$, we have

$$\int_\Omega \phi |\mathrm{grad}\, f|\, dV = \int_0^\infty dt \int_{\Gamma(t)} \phi\, dA_t.$$

Recall that the critical values, in $\mathbb{R}$, of f have Lebesgue measure equal to 0, by Sard's theorem (Narasimhan (1968, p. 19ff)). The regular values of f, R_f, are open in $\mathbb{R}$, and for $t \in R_f$ the preimage $f^{-1}[t] \cap \Omega$ is an $(n-1)$–submanifold in M with $f^{-1}[t] \cap \Omega$ compact. For any $t \in R_f$, we write

$$\Omega(t) = \{x : |f|(x) > t\}, \qquad V(t) = V(\Omega(t)).$$

Theorem. (C. Faber (1923) and E. Krahn (1925)) *Let Ω be a bounded domain in M, and let D be the disk in M satisfying $V(\Omega) = V(D)$. Then,*

$$\lambda^*(\Omega) \geq \lambda_\kappa(D),$$

where λ^* denotes the fundamental tone, and λ_κ the lowest eigenvalue, relative to the Dirichlet eigenvalue problem, *of the domain indicated. If Ω also has smooth boundary, then one has equality if and only if Ω is isometric to D.*

The theorem answers in the affirmative a conjecture of Lord Rayleigh (1877, §210). We sketch the proof, with some details relegated to exercises. (The reader should check §III.7 for background.)

Exercise VI.2. Consider the collection of functions

$$H^{\mathrm{n}\partial} := \{\phi \in C_c(\Omega) : \partial(\mathrm{supp}\,\phi) \in C^\infty,\ (\phi|\mathrm{int\,supp}\,\phi) \in C^\infty,$$
$$\phi|\ \mathrm{int\,supp}\,\phi \text{ only has nondegenerate critical points}\}.$$

Then, $H^{\mathrm{n}\partial}$ is dense in $\mathfrak{H}(\Omega)$ (Aubin (1982, p. 40)). Show that to prove the first claim of the theorem, it suffices to show that for any $\phi \in H^{\mathrm{n}\partial}$, $\Omega^\phi := \mathrm{int\,supp}\,\phi$, for the geodesic disk D^ϕ in M satisfying $V(\Omega^\phi) = V(D^\phi)$, we have

$$\mathcal{D}[\phi, \phi]/\|\phi\|^2 \geq \lambda_\kappa(D^\phi).$$

Exercise VI.3. Let $f = |\phi|$, $\phi \in H^{\mathrm{n}\partial}$. By Lemma VIII.3.2, the Rayleigh quotients of ϕ and $|\phi|$ are the same. Prove that, for f, the volume function $V(t)$ is continuous.

Now the idea of the proof is to associate to f a function $F : \overline{D} \to \mathbb{R}$ for which $F|\partial D = 0$, and for which

$$\iint_\Omega |\text{grad}\, f|^2\, dV \geq \iint_D |\text{grad}\, F|^2\, dV, \qquad \iint_\Omega f^2\, dV = \iint_D F^2\, dV.$$

This would then prove the claim.

To this end, let $T = \max\ f|\Omega$, and for $t \in [0, T]$ let $D(t)$ be the geodesic disk (fix the center o once and for all) in M with $V(t) = V(D(t))$. So, if $\mathfrak{r}(t)$ denotes the radius of $D(t)$, then

$$V(t) = V_\kappa(\mathfrak{r}(t)).$$

Also, set $r_0 = \mathfrak{r}(0)$; in particular, $D = \mathbb{B}_\kappa(o; r_0)$. Now, the function $\mathfrak{r} : [0, T] \to [0, r_0]$ is in $C^0([0, T]) \cap C^\infty(R_f \cap (0, T))$ and is strictly decreasing. We let $\psi : [0, r_0] \to [0, T]$ be the inverse function of $\mathfrak{r}$, and define $F : \overline{D} \to \mathbb{R}$ by

$$F = \psi \circ r, \qquad r(x) = d(o, x).$$

Exercise VI.4.

(a) Verify

$$V'(t) = V_\kappa{}'(\mathfrak{r}(t))\mathfrak{r}'(t) = A_\kappa(\mathfrak{r}(t))\mathfrak{r}'(t), \qquad 1 = \psi'(\mathfrak{r}(t))\mathfrak{r}'(t),$$

and

$$\iint_\Omega f^2\, dV = \iint_D F^2\, dV.$$

(b) Prove

$$\iint_\Omega |\text{grad}\, f|^2\, dV \geq \int_0^T A^2(t) \left\{ \int_{\Gamma(t)} |\text{grad}\, f|^{-1}\, dA_t \right\}^{-1} dt.$$

(c) Prove

$$\int_0^T A^2(t) \left\{ \int_{\Gamma(t)} |\text{grad}\, f|^{-1}\, dA_t \right\}^{-1} dt \geq -\int_0^T A_\kappa(\mathfrak{r}(t))(\mathfrak{r}'(t))^{-1}\, dt.$$

(d) Prove

$$\iint_D |\text{grad}\, F|^2\, dV = -\int_0^T A_\kappa(\mathfrak{r}(t))(\mathfrak{r}'(t))^{-1}\, dt.$$

This, then, implies the first claim of the theorem. To consider the case of equality in the theorem, one could not obtain (easily, if at all) a characterization using a collection of functions dense in $\mathfrak{H}(\Omega)$. (For that, one would need a Bonnesen inequality – see Note 6.) So, we assume Ω has smooth boundary, in which case $\lambda^*(\Omega) = \lambda_1(\Omega)$, and we may work directly with a positive eigenfunction f of $\lambda_1(\Omega)$. Of course, $f|\partial\Omega = 0$. Similarly, Ω might be a nodal domain of an eigenfunction ϕ (i.e., Ω is a connected component of $\{\phi \neq 0\}$) of a larger domain (e.g., a compact manifold). Then, as we showed in §III.9, $f = |\phi|\Omega|$ realizes $\lambda^*(\Omega)$. So, again, we work with the specific f.

Exercise VI.5. Let f be the eigenfunction under consideration. If p is a point in Ω such that grad f vanishes at p, show that for Riemann normal coordinates $x : U(p) \to \mathbb{R}^n$ about p, we have

$$\frac{\partial^2 f}{\partial x^{j2}} < 0$$

for at least one of the $j \in \{1, \ldots, n\}$. Thus, the set of points for which grad $f = 0$ is contained in an $(n-1)$–manifold. In particular, for every t, $V(\Gamma(t)) = 0$, which implies $V = V(t)$ is continuous with respect to t.

Now one can argue as above. If $\lambda_1(\Omega) = \lambda_\kappa(D)$, then the argument shows that $A(t) = A_\kappa(\mathfrak{r}(t))$ for all regular values of t. But, then, for every such t, one has $\Omega(t)$ isometric to $D(t)$. Sard's theorem implies that $\Omega = \Omega(0)$ is isometric to $D(0) = D$. ∎

Note VI.5. The Faber–Krahn inequality was generalized, in Chavel–Feldman (1980), to simply connected domains on surfaces with Gauss curvature bounded from above.

Note VI.6. A different generalization and application of the Faber–Krahn inequality was given in Bérard–Meyer (1982). Let M be compact with Ricci curvature bounded from below by $(n-1)\kappa$, with $\kappa > 0$. Set $\beta = V(M)/V(\mathbb{M}_\kappa)$, and given any domain Ω in M, let D be the metric disk in $\mathbb{M}_\kappa$ with $V(\Omega) = \beta V(D)$.

Exercise VI.6.

(a) Prove $\lambda^*(\Omega) \geq \lambda_\kappa(D)$. If Ω has smooth boundary, characterize equality.

(b) Prove the Obata theorem (Obata (1962)). Namely, we know that $\lambda_2(M) \geq n\kappa$ (see Exercise III.45). Show that $\lambda_2 = n\kappa$ if and only if M is isometric to $\mathbb{M}_\kappa$.

Other Developments

Note VI.7. Start, for convenience, in the Euclidean plane $\mathbb{R}^2$. A *Bonnesen inequality* is a sharpened form of the isoperimetric inequality. Namely, given the domain Ω, one looks for a nonnegative invariant $B(\Omega) \geq 0$ of the domain Ω (i) that vanishes if and only if Ω is a disk, and (ii) that satisfies

$$L^2 - 4\pi A \geq B.$$

This not only contains an immediate proof of the isoperimetric inequality, but it also contains an immediate characterization of equality.

Of course, one can easily formulate such questions for surfaces with curvature bounded from above and for higher dimensional Euclidean space. The first such result was F. Bernstein's proof (1905) of the isoperimetric inequality on the 2–sphere. Osserman's (1979) sharpened form of the Bol–Fiala inequality (Theorem V.5.6) was of Bonnesen type.

Note VI.8. A different approach to the isoperimetric inequality in Euclidean space, using geometric measure theory, is given in Almgren (1986). It is built on the following:

Theorem. *Let M be a k–dimensional submanifold of $\mathbb{R}^n$, $k < n$, without boundary, and assume that the length of the mean curvature vector on all of M is less than or equal to that of $\mathbb{S}^k$. Then, the $V_k(M) \geq \mathbf{c_k}$ (where V_k denotes k–dimensional volume), with equality if and only if M is congruent to $\mathbb{S}^k$.*

Note VI.9. For fuller discussion of analytic isoperimetric inequalities related to geometric considerations, see Chavel (1984, Chapter IV; 2001, Chapters VI–VIII).

VII

The Kinematic Density

In this chapter, we discuss integration over the unit tangent bundle of a given Riemannian manifold. The geodesic flow of the Riemannian metric acts on the unit tangent bundle, and one of its salient features is the existence of a natural measure on the unit tangent bundle, called *the kinematic density* or *the Liouville measure*, which is invariant under the action of the geodesic flow. Furthermore, the integral of a function on the unit tangent bundle can be calculated by first integrating the function relative to the kinematic density over each of the fibers ($(n-1)$–spheres) in the unit tangent bundle and then integrating the resulting function on the base manifold relative to its Riemannian measure. The measure on the fibers is the natural measure on spheres induced by Lebesgue measure on the tangent spaces.

We could present the kinematic density by simply writing it as the local product measure of the natural measure on tangent spheres and the Riemannian measure on the base manifold, and then verifying that it is invariant relative to the geodesic flow. However, we prefer a different route, one that detours through the formalism of classical analytical mechanics. This affords an opportunity to connect the discussion to an extremely important collection of ideas, important historically and in current research. We do not pursue this connection here to any extent, rather, we concentrate on Riemannian results that emerged from these notions.

Our presentation of the formalism of analytic mechanics is different from that usually presented in classical mechanics (Goldstein (1950) and Arnold (1980)). By this, I mean that in classical mechanics, (i) one starts with a Lagrangian functional associated with some dynamical system in $\mathbb{R}^3$; (ii) in generalized coordinates in configuration space (i.e., the view of the variables of the system as a differentiable manifold), the Lagrangian becomes a function defined on the state space (the tangent bundle); (iii) under the canonical transformations of the data to generalized momenta in phase space (the cotangent bundle), the energy function associated with the Lagrangian becomes a Hamiltonian, relative to which the dynamical system satisfies the associated Hamilton equations. But, we shall proceed in the opposite direction, namely, (a) we start with the cotangent bundle and note the existence of a canonical 1–form on the cotangent bundle whose exterior derivative, the canonical 2–form, is nondegenerate. This determines, after any choice of (Hamiltonian) function on the cotangent bundle, the Hamiltonian system of differential equations of the induced flow. The important point is that the canonical forms are determined by the differentiable structure of the original base manifold – nothing else. (b) Given any (Lagrangian) function on the tangent bundle, one then constructs an associated bundle map of the tangent bundle to the cotangent bundle to pull the canonical forms from the cotangent bundle back to the tangent bundle. (c) Once these forms exist on the tangent bundle, they and the energy function associated with the Lagrangian determine the flow in the tangent bundle whose integral curves project to the solutions of the associated classical Euler–Lagrange equations.

From there, we restrict to the Riemannian case and interpret the above in this situation. As mentioned previously, our emphasis will be on Riemannian results. We present the analytic aspects of the solution to the Blaschke conjecture, and the first steps in the dynamical theory of manifolds with no conjugate points. In the last section, we discuss Santalo's formula, of interest in its own right, in its application to the Blaschke conjecture (see, e.g., Exercise VII.13(c)) and to isoperimetric inequalities on Riemann manifolds – to be discussed in the following chapter.

A note on our presentation of the classical mechanics. We have done so by calculating in local coordinates. Such treatments are out of fashion these days – there is an almost ideological prejudice against such treatments. But then that very fact, that local coordinates are no longer "in," may constitute a recommendation for a second look for at least one topic.

§VII.1. The Differential Geometry of Analytical Dynamics

For any vector space V with bilinear form B, one has the natural homomorphism (as in the case of an inner product)

$$\theta_B : V \to V^*$$

given by

$$(\theta_B(\xi))(\eta) = B(\xi, \eta).$$

If E is finite dimensional, then θ_B is an isomorphism if and only if B is nondegenerate (i.e., $B(\xi, \eta) = 0$ for all $\eta \in E$ implies $\xi = 0$).

Definition. A *symplectic* 2–form Ω on a differentiable manifold M is a closed (i.e., $d\Omega = 0$) nondegenerate differentiable differential 2–form on M.

Let M be any manifold endowed with the symplectic 2–form Ω. Then, with any function f on M, we associate the vector field (analogous to the gradient) X_f on M given by

$$X_f = -{\theta_\Omega}^{-1}\, df$$

that is,

$$\text{(VII.1.1)} \qquad -\,df = \theta_\Omega(X_f) = \mathrm{i}(X_f)\Omega.$$

Proposition VII.1.1. (a) *The function f is constant along the integral curves of X_f, and* (b) *the symplectic 2–form Ω is invariant relative to the flow of X_f.*

Proof. For (a), we simply have

$$X_f f = df(X_f) = -\Omega(X_f, X_f) = 0,$$

which implies the claim; and for (b), we have (see §III.7), by (III.7.1),

$$\mathcal{L}_{X_f}\Omega = \{d \circ \mathrm{i}(X_f) + \mathrm{i}(X_f) \circ d\}\Omega = d \circ \mathrm{i}(X_f)\Omega = -d^2 f = 0. \qquad ■$$

The Canonical Symplectic Form on Cotangent Bundles

Our standard example is given by considering our perennial n–dimensional manifold M, with respective tangent and cotangent bundles TM and TM^*, and respective projection maps

$$\pi_1 : TM \to M, \qquad \pi_2 : TM^* \to M.$$

Definition. We define the *canonical* 1*–form on* TM^*, ω, by

$$\omega_{|\tau} = {\pi_2}^*\tau,$$

where $\omega_{|\tau}$ denotes the 1–form ω evaluated at $\tau \in TM^*$. We define the *canonical* 2*–form on* M, Ω, by

$$\Omega = d\omega.$$

Proposition VII.1.2. *The* 2*–form Ω on TM^* is symplectic.*

Proof. Clearly, Ω is closed. To show that Ω is nondegenerate, we consider a local chart $x : U \to \mathbb{R}^n$, with associated charts

$$(q^1, \ldots, q^n; \dot{q}^1, \ldots, \dot{q}^n), \qquad (q^1, \ldots, q^n; p_1, \ldots, p_n)$$

on ${\pi_1}^{-1}[U]$ in TM, and on ${\pi_2}^{-1}[U]$ in TM^*, respectively, defined as follows: For $\xi \in {\pi_1}^{-1}[U]$ we have

$$\xi = \sum_{j=1}^{n} \xi^j \frac{\partial}{\partial x^j} = \sum_{j=1}^{n} dx^j(\xi)\frac{\partial}{\partial x^j};$$

so we set

$$q^j = x^j \circ \pi_1, \qquad \dot{q}^j = dx^j,$$

where dx^j is viewed here as a *function* on ${\pi_1}^{-1}[U]$. For $\tau \in {\pi_2}^{-1}[U]$, we have

$$\tau = \sum_{j=1}^{n} \tau_j \, dx^j = \sum_{j=1}^{n} \left(\frac{\partial}{\partial x^j}\right)^{**} (\tau)\, dx^j;$$

so we set

$$q^j = x^j \circ \pi_2, \qquad p_j = \left(\frac{\partial}{\partial x^j}\right)^{**}.$$

Then,

$$\tau = \sum_j p_j(\tau)\, dx^j$$

implies

$$\omega_\tau = {\pi_2}^* \tau = \sum_j \{p_j \, dq^j\}_{|\tau},$$

which implies

$$\Omega = d\omega = \sum_{j=1}^{n} dp_j \wedge dq^j,$$

which is nondegenerate. ∎

Given any function, a *Hamiltonian* $H : TM^* \to \mathbb{R}$, then to calculate X_H we set

$$X_H = \sum_j \alpha^j \frac{\partial}{\partial q^j} + \beta^j \frac{\partial}{\partial p_j}.$$

Then,

$$dH = \sum_j \frac{\partial H}{\partial q^j} dq^j + \frac{\partial H}{\partial p_j} dp_j = -\mathrm{i}(X_H)\Omega = \sum_j \alpha^j dp_j - \beta^j dq^j.$$

So,

$$X_H = \sum_j \frac{\partial H}{\partial p_j}\frac{\partial}{\partial q^j} - \frac{\partial H}{\partial q^j}\frac{\partial}{\partial p_j},$$

and the differential equation of the flow of X_H on TM^* is given by

$$\frac{dq^j}{dt} = \frac{\partial H}{\partial p_j}, \qquad \frac{dp_j}{dt} = -\frac{\partial H}{\partial q^j}.$$

Lagrangian-Induced Symplectic Forms on Tangent Bundles

Now, the symplectic form on TM^* is determined by the differentiable structure on M. What about symplectic forms on TM?

We proceed as follows: To any function, a *Lagrangian* $L : TM \to \mathbb{R}$, one associates a bundle map $FL : TM \to TM^*$ defined by

$$\{FL(\xi)\}(\eta) = \frac{d}{dt}(L(\xi + t\eta))\bigg|_{t=0}$$

for all $\eta \in M_p$, where $p \in M$, $\xi \in M_p$. Then, one sets

$$\Theta = FL^*\Omega.$$

Certainly, Θ is closed. If FL has maximal rank, then Θ will also be nondegenerate and, therefore, symplectic. In what follows, we will explicitly calculate Θ for a natural class of examples.

One defines the *action* $A : TM \to \mathbb{R}$ and *energy* $E : TM \to \mathbb{R}$ functions on TM by

$$A(\xi) = \{FL(\xi)\}(\xi), \qquad E = A - L.$$

When FL has maximal rank, then Θ is nondegenerate; then, one can study the flow of the vector field X on TM determined by Θ and E.

We calculate the above quantities in local coordinates. Let $x : U \to \mathbb{R}^n$ be a chart on M, and $(q; \dot{q}) : {\pi_1}^{-1}[U] \to \mathbb{R}^{2n}$, $(q; p) : {\pi_2}^{-1}[U] \to \mathbb{R}^{2n}$ the associated charts on TM, TM^*, respectively. Then,

$$L(\xi + t\eta) = L(q(\xi + t\eta); \dot{q}(\xi + t\eta)) = L(q(\xi); \dot{q}(\xi + t\eta)),$$

which implies

$$\frac{d}{dt}L(\xi + t\eta)\bigg|_{t=0} = \sum_j \frac{\partial L}{\partial \dot{q}^j}(\xi)\dot{q}^j(\eta) = \left\{\sum_j \frac{\partial L}{\partial \dot{q}^j}(\xi)\,dx^j\right\}(\eta);$$

so $FL(\xi)$ is the 1–form at $\pi_1(\xi)$ given by

$$FL(\xi) = \sum_j \frac{\partial L}{\partial \dot{q}^j}(\xi)\, dx^j,$$

which implies

$$A(\xi) = \sum_j \frac{\partial L}{\partial \dot{q}^j}\dot{q}^j, \qquad E = \sum_j \frac{\partial L}{\partial \dot{q}^j}\dot{q}^j - L. \tag{VII.1.2}$$

To calculate Θ, we have the map FL given in the local coordinates by

$$q^j \mapsto q^j, \qquad \dot{q}^j \mapsto p_j(\xi) = \frac{\partial L}{\partial \dot{q}^j}(\xi).$$

We conclude

$$\Theta = FL^*\Omega = FL^* \sum_j dp_j \wedge dq^j = \sum_j d\left(\frac{\partial L}{\partial \dot{q}^j}\right) \wedge dq^j. \tag{VII.1.3}$$

So, Θ is the symplectic form of TM induced by the Lagrangian L. Given the energy function E, derived from L above, we calculate the vector field X on TM determined by Θ and E, as in equation (VII.1.1), namely,

$$-\,dE = \mathrm{i}(X)\Theta. \tag{VII.1.4}$$

First set

$$X = \sum_j \gamma^j \frac{\partial}{\partial q^j} + \delta^j \frac{\partial}{\partial \dot{q}^j}. \tag{VII.1.5}$$

Then, from (VII.1.2), we have

$$-\,dE = \sum_j \frac{\partial L}{\partial q^j}\, dq^j - \dot{q}^j\, d\left(\frac{\partial L}{\partial \dot{q}^j}\right); \tag{VII.1.6}$$

on the other hand, from (VII.1.3) and (VII.1.5), one easily has

$$\mathrm{i}(X)\Theta = \sum_j \left\{X\left(\frac{\partial L}{\partial \dot{q}^j}\right)\right\} dq^j - \gamma^j d\left(\frac{\partial L}{\partial \dot{q}^j}\right). \tag{VII.1.7}$$

We claim that (VII.1.6) and (VII.1.7) imply

$$\gamma^j = \dot{q}^j \tag{VII.1.8}$$

for all j, that is, the vector field X actually represents a second-order differential equation on M (see §I.3), and

$$X\left(\frac{\partial L}{\partial \dot{q}^j}\right) = \frac{\partial L}{\partial q^j} \tag{VII.1.9}$$

for all j. Well,

$$d\left(\frac{\partial L}{\partial \dot{q}^j}\right) = \sum_k \left\{\frac{\partial^2 L}{\partial q^k \partial \dot{q}^j}\, dq^k + \frac{\partial^2 L}{\partial \dot{q}^k \partial \dot{q}^j}\, d\dot{q}^k\right\}.$$

But, to say that Θ is nondegenerate is to say that the matrix $\mathfrak{L}$, given by

$$\mathfrak{L}_{kj} = \frac{\partial^2 L}{\partial \dot{q}^k \partial \dot{q}^j}$$

is nonsingular. Then, this implies the claim. Equation (VII.1.9) is referred to as the *Euler–Lagrange equation.*

To determine δ^j in (VII.1.5), one uses (VII.1.8) and (VII.1.9):

$$\frac{\partial L}{\partial q^j} = X\left(\frac{\partial L}{\partial \dot{q}^j}\right) = \sum_k \dot{q}^k \frac{\partial^2 L}{\partial q^k \partial \dot{q}^j} + \delta^k \frac{\partial^2 L}{\partial \dot{q}^k \partial \dot{q}^j},$$

which implies

$$\text{(VII.1.10)} \qquad \sum_k \delta^k \frac{\partial^2 L}{\partial \dot{q}^k \partial \dot{q}^j} = \frac{\partial L}{\partial q^j} - \sum_k \dot{q}^k \frac{\partial^2 L}{\partial q^k \partial \dot{q}^j}.$$

One can now use the nonsingularity of $\mathfrak{L}$ to determine δ^j.

Newton's Equations in Riemannian Geometry

The classical example from Newtonian mechanics goes as follows: Let M be a Riemannian manifold, and $V : M \to \mathbb{R}$ a function on M. For the Lagrangian $L : TM \to \mathbb{R}$, we pick

$$\text{(VII.1.11)} \qquad L(\xi) = \frac{|\xi|^2}{2} - (V \circ \pi_1)(\xi),$$

so, in the local coordinates on TM, we have

$$L(\xi) = \frac{1}{2} \sum_{j,k} g_{jk}(q)\dot{q}^j \dot{q}^k - V(q).$$

The function V is referred to as a *potential function of the force field* $-\operatorname{grad} V$ *on* M.

Theorem VII.1.1. *The vector field X represents "Newton's equation of motion" on M for the force field $-\operatorname{grad} V$. That is, if Ψ_t denotes the flow of X on TM, and for any $\xi \in TM$, we set*

$$\gamma_\xi(t) = \pi_1 \circ \Psi_t(\xi),$$

then

$$\text{(VII.1.12)}\qquad \gamma_\xi{}'(t) = \Psi_t(\xi), \qquad \nabla_t \gamma_\xi{}' = -(\operatorname{grad} V)\circ\gamma_\xi .$$

Thus, when V is constant on M, X is the vector field of the geodesic flow. In other words, the projection of the trajectories of X to M yield the paths given by the "law of inertia" – geodesics.

Proof. The first equality in (VII.1.12) is (VII.1.8); so, we concentrate on proving the second.

We give the local calculation. First,

$$\frac{\partial L}{\partial q^j} = \frac{1}{2}\sum_{r,s} \frac{\partial g_{rs}}{\partial q^j}\dot q^r \dot q^s - \frac{\partial V}{\partial q^j}, \qquad \frac{\partial L}{\partial \dot q^k} = \sum_s g_{ks}\dot q^s,$$

which implies

$$\frac{\partial^2 L}{\partial q^j \partial \dot q^k} = \sum_s \frac{\partial g_{ks}}{\partial q^j}\dot q^s, \qquad \frac{\partial^2 L}{\partial \dot q^j \partial \dot q^k} = g_{jk}.$$

If we substitute into (VII.1.10), we obtain

$$\begin{aligned}\sum_s \delta^s g_{sj} &= \frac{1}{2}\sum_{r,s} \frac{\partial g_{rs}}{\partial q^j}\dot q^r \dot q^s - \frac{\partial V}{\partial q^j} - \sum_{r,s}\frac{\partial g_{js}}{\partial q^r}\dot q^r \dot q^s \\ &= \frac{1}{2}\sum_{r,s}\left\{\frac{\partial g_{rs}}{\partial q^j} - \frac{\partial g_{js}}{\partial q^r} - \frac{\partial g_{rj}}{\partial q^s}\right\}\dot q^r \dot q^s - \frac{\partial V}{\partial q^j};\end{aligned}$$

so,

$$\delta^j = -\sum_{r,s} \Gamma_{rs}{}^j \dot q^r \dot q^s - \sum_s g^{js}\frac{\partial V}{\partial q^s},$$

which implies the theorem. ■

The Liouville Theorems

Our goal now is to calculate the forms

$$\Theta^n, \qquad \vartheta \wedge \Theta^{n-1},$$

where

$$\vartheta = FL^*\omega, \qquad \Theta = FL^*\Omega$$

(ω is the canonical 1–form on TM^*, and the exponent of the forms indicates the number of times the form is wedged with itself). We first note

Lemma VII.1.1. *Given a vector space V and*

$$\{\sigma^1, \ldots, \sigma^{2n}\} \subseteq V^*,$$

then there exists a (not necessarily positive) constant $c(n, k)$, so that

$$\left(\sum_{j=1}^{n} \sigma^j \wedge \sigma^{j+n}\right)^k = c(n,k) \sum_{j_1<\ldots<j_k} \sigma^{j_1} \wedge \cdots \wedge \sigma^{j_k} \wedge \sigma^{j_1+n} \wedge \cdots \wedge \sigma^{j_k+n}.$$

Therefore, for the canonical 1–form and 2–form on TM^*,

$$\omega = \sum_j p_j \, dq^j, \qquad \Omega = \sum_j dp_j \wedge dq^j,$$

we have

$$\Omega^n = c(n)\, dp_1 \wedge \cdots \wedge dp_n \wedge dq^1 \wedge \cdots \wedge dq^n \tag{VII.1.13}$$

and

$$\omega \wedge \Omega^{n-1} = (-1)^n c(n, n-1) \left\{ \sum_j (-1)^j p_j \, dp_1 \wedge \cdots \wedge \widehat{dp_j} \wedge \cdots \wedge dp_n \right\} \wedge$$

$$\wedge dq^1 \wedge \cdots \wedge dq^n. \tag{VII.1.14}$$

Given L as in (VII.1.11), then

$$\{FL(\xi)\}(\eta) = \langle \xi, \eta \rangle,$$

which implies that the coordinate p_j at the point $FL(\xi)$ in TM^* is given by

$$p_j(FL(\xi)) = \frac{\partial L}{\partial \dot{q}^j}(\xi) = \sum_r g_{jr}(q) \dot{q}^r. \tag{VII.1.15}$$

Therefore, if we let $\mathfrak{m}(\dot{q}; q)$ denote the volume form of $M_{\pi_1(q)}$ at $\dot{q}$, associated with Lebesgue measure on $M_{\pi_1(q)}$, and σ the *local* volume form of the Riemannian measure on M, then (VII.1.13) implies

$$\begin{aligned} \Theta^n &= c(n) g \, d\dot{q}^1 \wedge \cdots \wedge d\dot{q}^n \wedge dq^1 \wedge \cdots \wedge dq^n \\ &= c(n)\, \mathfrak{m}(\dot{q}; q) \wedge \pi_1{}^* \sigma(q), \end{aligned}$$

where g, as usual, denotes $\det(g_{ij})$. We conclude:

Theorem VII.1.2. *The form Θ^n is the local product, up to constant multiple, of the Riemannian volume form on M with the Lebesgue volume form on* *the tangent space to M. In particular, TM is orientable, independent of the orientability of M.*

Thus, for any Θ^n–integrable function F on TM, we have "integration over the fibers"

$$\int_{TM} F\,\Theta^n = c(n)\int_M dV(p)\int_{M_p}\{F|M_p\}\,d\mathfrak{m}_p,$$

where $d\mathfrak{m}_p$ denotes the canonical Lebesgue measure on M_p, for any $p \in M$.

Also, since Ω is invariant with respect to the flow of X_H for any Hamiltonian H on TM^*, we also have Θ, and therefore Θ^n, invariant with respect to the flow Ψ_t of X on TM.

The same type of theorem is true of $\vartheta \wedge \Theta^{n-1}$, although the details are messier. They go as follows:

First, we have from (VII.1.14) and (VII.1.15)

$$\begin{aligned}
&\vartheta \wedge \Theta^{n-1}\\
&\quad = \mathfrak{c}(n)\sum_{j,r}(-1)^j g_{jr}\dot{q}^r \cdot\\
&\qquad \cdot \sum_{r_1,\ldots,\widehat{r_j},\ldots,r_n} (g_{1r_1}\,d\dot{q}^{r_1})\wedge\cdots\wedge\widehat{(g_{jr_j}\,d\dot{q}^{r_j})}\wedge\cdots\wedge(g_{nr_n}\,d\dot{q}^{r_n})\wedge\\
&\qquad \wedge dq^1\wedge\cdots\wedge dq^n\\
&\quad = \mathfrak{c}(n)\sum_j(-1)^j\sum_{r_1,\ldots,r_n} g_{1r_1}\cdots g_{nr_n}\dot{q}^{r_j}\,d\dot{q}^{r_1}\wedge\cdots\wedge\widehat{d\dot{q}^{r_j}}\wedge\cdots\wedge d\dot{q}^{r_n}\wedge\\
&\qquad \wedge dq^1\wedge\cdots\wedge dq^n.
\end{aligned}$$

So, we study

$$d\dot{q}^{r_1}\wedge\cdots\wedge\widehat{d\dot{q}^{r_j}}\wedge\cdots\wedge d\dot{q}^{r_n}$$

where r_j has the value ℓ. If $\ell < j$, then

$$\begin{aligned}
&d\dot{q}^{r_1}\wedge\cdots\wedge\widehat{d\dot{q}^{r_j}}\wedge\cdots\wedge d\dot{q}^{r_n}\\
&\quad = d\dot{q}^{r_1}\wedge\cdots\wedge d\dot{q}^{r_{\ell-1}}\wedge\widehat{d\dot{q}^{\ell}}\wedge d\dot{q}^{r_\ell}\wedge\cdots\wedge d\dot{q}^{r_{j-1}}\wedge d\dot{q}^{r_{j+1}}\wedge\cdots\wedge d\dot{q}^{r_n}\\
&\quad = \varepsilon^{1\cdots(\ell-1)\ell(\ell+1)\cdots j(j+1)\cdots n}_{r_1\cdots r_{l-1}\ell r_l\cdots r_{j-1}r_{j+1}\cdots r_n}\,d\dot{q}^1\wedge\cdots\wedge\widehat{d\dot{q}^\ell}\wedge\cdots\wedge d\dot{q}^n\\
&\quad = (-1)^{\ell-j}\varepsilon^{1\cdots n}_{r_1\cdots r_n}\,d\dot{q}^1\wedge\cdots\wedge\widehat{d\dot{q}^\ell}\wedge\cdots\wedge d\dot{q}^n,
\end{aligned}$$

where ε denotes the parity of the indicated permutation; and a corresponding argument holds for $\ell > j$. Therefore,

$$\begin{aligned}\vartheta \wedge \Theta^{n-1} &= \mathfrak{c}(n) \sum_j (-1)^j \sum_{r_1,\dots,r_n} g_{1r_1} \cdots g_{nr_n} \dot{q}^{r_j}\, d\dot{q}^{r_1} \wedge \cdots \wedge \widehat{d\dot{q}^{r_j}} \\ &\quad \wedge \cdots \wedge d\dot{q}^{r_n} \wedge dq^1 \wedge \cdots \wedge dq^n \\ &= \mathfrak{c}(n) \sum_{r_1,\dots,r_n} \varepsilon^{1\cdots n}_{r_1\cdots r_n} g_{1r_1} \cdots g_{nr_n} \cdot \\ &\quad \cdot \sum_\ell (-1)^\ell \dot{q}^\ell\, d\dot{q}^1 \wedge \cdots \wedge \widehat{d\dot{q}^\ell} \wedge \cdots \wedge d\dot{q}^n \wedge dq^1 \wedge \cdots \wedge dq^n \\ &= \mathfrak{c}(n) g \sum_\ell (-1)^\ell \dot{q}^\ell\, d\dot{q}^1 \wedge \cdots \wedge \widehat{d\dot{q}^\ell} \wedge \cdots \wedge d\dot{q}^n \wedge dq^1 \wedge \cdots \wedge dq^n \\ &= \mathfrak{c}(n) \{\sqrt{g} \sum_\ell (-1)^\ell \dot{q}^\ell\, d\dot{q}^1 \wedge \cdots \wedge \widehat{d\dot{q}^\ell} \wedge \cdots \wedge d\dot{q}^n\} \wedge \pi_1^* \sigma.\end{aligned}$$

Lemma VII.1.2. *For any p in the domain U of our chart, the form*

$$\left\{ \sqrt{g} \sum_j (-1)^j \dot{q}^j\, d\dot{q}^1 \wedge \cdots \wedge \widehat{d\dot{q}^j} \wedge \cdots \wedge d\dot{q}^n \right\} \Bigg| \mathsf{S}_p$$

is, up to sign, the $(n-1)$–volume form on S_p associated with μ_p the Euclidean $(n-1)$–measure on the tangent sphere S_p.

Proof. The simplest way to derive the result is to assume that the coordinates q^j, $j = 1, \dots, n$ are orthonormal at p (nothing is lost by such an assumption). Then, the problem is to show that the $(n-1)$–form

$$\tau = \sum_j (-1)^j x^j\, dx^1 \wedge \cdots \wedge \widehat{dx^j} \wedge \cdots \wedge dx^n$$

on $\mathbb{R}^n$ is, when restricted to the unit sphere $\mathbb{S}^{n-1}$, the volume form of $\mathbb{S}^{n-1}$.

But this may be easily handled by noting that τ is invariant with respect to the orthogonal group acting on $\mathbb{R}^n$. Since the orthogonal group acts on the unit sphere by isometries, then both the volume form on $\mathbb{S}^{n-1}$ and $\tau \mid \mathbb{S}^{n-1}$ are invariant $(n-1)$–forms on $\mathbb{S}^{n-1}$. But, at $(1, 0, \dots, 0)$, we have

$$\tau|\mathbb{S}^{n-1} = -dx^2 \wedge \cdots \wedge dx^n,$$

which is minus the volume form of $\mathbb{S}^{n-1}$ at $(1, 0, \dots, 0)$, since $\{dx^2, \dots, dx^n\}$ is the coframe dual to a positively oriented orthonormal frame at $(1, 0, \dots, 0)$. The invariance of both forms under the action of the orthogonal group on $\mathbb{S}^{n-1}$ implies that $\tau|\mathbb{S}^{n-1}$ is minus the volume form on all of $\mathbb{S}^{n-1}$. ■

Theorem VII.1.3. *The form* $(\vartheta \wedge \Theta^{n-1})_{|\mathsf{S}M}$ *is the local product, up to constant multiple, of the Riemannian volume form on* M *with the Lebesgue* $(n-1)$*–volume form on the unit tangent spheres of* M*. In particular,* $\mathsf{S}M$ *is orientable, independent of the orientability of* M*.*

Thus, for any $((\vartheta \wedge \Theta^{n-1})_{|\mathsf{S}M})$*–integrable function* f *on* $\mathsf{S}M$*, we have "integration over the fibers"*

$$\int_{\mathsf{S}M} f\,\vartheta \wedge \Theta^{n-1} = \mathfrak{c}(n) \int_M dV(p) \int_{\mathsf{S}_p} \{f|M_p\}\,d\mu_p.$$

Finally, if we let μ *denote the measure on* $\mathsf{S}M$*, associated with* $(n-1)$*–form* $\mathfrak{c}(n)^{-1}\vartheta \wedge \Theta^{n-1}$*, that is,* μ *is the local product of the canonical measure on unit tangent spheres with the Riemannian measure on the base manifold, then* $V = 0$ *on all of* M *implies that* μ *invariant with respect to the geodesic flow* Ψ_t *of* X *on* TM*.*

Proof. We only need consider the invariance of μ with respect to the geodesic flow. Well, the first thing one must note is that if $V = 0$, then X is always tangent to $\mathsf{S}M \subseteq TM$, since the geodesic flow maps $\mathsf{S}M$ to $\mathsf{S}M$.

Then,

$$\mathcal{L}_X \left((\vartheta \wedge \Theta^{n-1})_{|\mathsf{S}M}\right) = \left(\mathcal{L}_X(\vartheta \wedge \Theta^{n-1})\right)_{|\mathsf{S}M} = (dL \wedge \Theta^{n-1})_{|\mathsf{S}M} = 0,$$

which implies the claim. ■

We summarize:

Definition. Let M be an n–dimensional Riemannian manifold, and $\mathsf{S}M$ the unit tangent bundle of M with natural projection $\pi : \mathsf{S}M \to M$.

The geodesic flow on $\mathsf{S}M$ is denoted by Φ_t and is given by

$$\Phi_t(\xi) = \gamma_\xi{}'(t),$$

where γ_ξ denotes the geodesic with initial point $\pi(\xi)$ and initial velocity vector ξ. The *kinematic density* or *Liouville measure* $d\mu$ on $\mathsf{S}M$ is given by

$$\int_{\mathsf{S}M} F(\xi)\,d\mu(\xi) = \int_M dV(x) \int_{\mathsf{S}_x} F(\xi)\,d\mu_x(\xi);$$

Theorem VII.1.4. (Liouville's theorem) *The measure* $d\mu$ *on* $\mathsf{S}M$ *is invariant with respect to the geodesic flow.*

§VII.2. The Berger–Kazdan Inequalities

The first inequality is strictly analytic, so we temporarily suspend any reference to geometric data. Our discussion here closely follows our presentation in §V.1 and §V.2 of Chavel (1984).

The setting will be an N–dimensional inner product space V over $\mathbb{R}$, with given family of self-adjoint linear transformations $\mathcal{R}(t) : V \to V$, where t ranges over $\mathbb{R}$. One associates with $\mathcal{R}(t)$ the matrix Jacobi equation

(VII.2.1) $$\mathsf{A}'' + \mathcal{R}(t)\mathsf{A} = 0,$$

where each $\mathsf{A}(t)$ is a linear transformation of V. Of course, for each fixed $\xi \in V$, the vector function

$$\eta(t) = \mathsf{A}(t)\xi$$

is a solution of the vector Jacobi equation

$$\eta'' + \mathcal{R}(t)\eta = 0.$$

Notation. In what follows, we let $\mathcal{A}(t)$ denote the solution of (VII.2.1) determined by the initial conditions

$$\mathcal{A}(0) = 0, \qquad \mathcal{A}'(0) = I,$$

where I denotes the identity transformation of V. For every $s \in \mathbb{R}$, we let $\mathcal{C}_s(t)$ denote the solution of (VII.2.1) determined by the initial conditions

$$\mathcal{C}_s(s) = 0, \qquad \mathcal{C}_s{}'(s) = I.$$

Proposition VII.2.1. *For every t, we have*

(VII.2.2) $$\mathcal{A}^*\mathcal{A}' = \mathcal{A}'^*\mathcal{A}.$$

Assume that $\mathcal{I}$ is an interval in $\mathbb{R}$ such that $\mathcal{A}(t)$ is invertible for all $t \in \mathcal{I}$, $t \neq 0$. Then, for any $s \in \mathcal{I}$, we have the representation formula for all $t \in \mathcal{I}$,

(VII.2.3) $$\mathcal{C}_s(t) = \mathcal{A}(t)\left\{\int_s^t (\mathcal{A}^*\mathcal{A})^{-1}(\tau)\,d\tau\right\}\mathcal{A}^*(s).$$

Proof. Recall from §III.4 that, for any linear transformations $A(t), B(t) : V \to V$, depending differentiably on t, their associated Wronskian $\mathsf{W}(t)$ is defined by

$$\mathsf{W}(A, B) := A'^*B - A^*B',$$

and for solutions A, B of (VII.2.1), we have $\mathsf{W}(A, B)$ is constant. One then easily has, for our $\mathcal{A}(t)$,

$$\mathsf{W}(\mathcal{A}, \mathcal{A}) = 0,$$

which implies (VII.2.2).

Furthermore, if we set

$$\mathcal{U} := \mathcal{A}'\mathcal{A}^{-1},$$

then

$$\mathcal{U}^* - \mathcal{U} = (\mathcal{A}^{-1})^* \mathsf{W}(\mathcal{A}, \mathcal{A})\mathcal{A}^{-1} = 0;$$

so, $\mathcal{U}$ is self-adjoint.

Denote the right-hand side of (VII.2.3) by $\mathcal{B}_s(t)$. Then, the self-adjointness of $\mathcal{A}'\mathcal{A}^{-1}$ implies that $\mathcal{B}_s$ is a solution of Jacobi's equation. Certainly, $\mathcal{B}_s(s) = 0$. Furthermore,

$$\mathcal{B}_s{}' = \left\{\mathcal{A}' \int_s (\mathcal{A}^*\mathcal{A})^{-1}(\tau)\, d\tau + (\mathcal{A}^{-1})^*\right\} \mathcal{A}^*(s),$$

which implies $\mathcal{B}_s{}'(s) = I$. ■

Theorem VII.2.1. (J. L. Kazdan (1978)) *If $\mathcal{A}(t)$ is invertible for all $t \in (0, \pi)$, then for any C^2 positive function $m(t)$ on $(0, \pi)$ satisfying*

(VII.2.4) $$m(\pi - t) = m(t),$$

on $(0, \pi)$, we have

(VII.2.5)
$$\int_0^\pi ds \int_s^\pi m(t-s) \det \mathcal{C}_s(t)\, dt \geq \int_0^\pi ds \int_s^\pi m(t-s) \sin^N(t-s)\, dt.$$

Equality in (VII.2.5) *is achieved if and only if*

$$\mathcal{R}(t) = I$$

on all of $[0, \pi]$, *that is, if and only if*

$$\mathcal{C}_s(t) = \sin(t-s) I$$

on all of $[0, \pi]$.

Proof. We proceed in a series of steps.

Step 1. We use the representation formula (VII.2.3), to which we apply the following version of *Jensen's inequality*:

If $F = F(B)$ is a strictly convex function defined on $\mathfrak{C}$, the convex set of positive definite self-adjoint linear transformations of V, and ν is any positive measure on $\mathbb{R}$, then

$$\text{(VII.2.6)} \quad F\left\{\frac{1}{\nu((\alpha,\beta))}\int_\alpha^\beta B(\tau)\,d\nu(\tau)\right\} \le \frac{1}{\nu((\alpha,\beta))}\int_\alpha^\beta F(B(\tau))\,d\nu(\tau)$$

for any $B : [\alpha, \beta] \to \mathfrak{C}$, with equality in (VII.2.6) if and only if $B(\tau)$ is a constant function on $[\alpha, \beta]$.

To apply the inequality, one sets

$$F(B) = (\det B)^{-1}, \qquad \phi(t) = \{\det \mathcal{A}(t)\}^{1/N},$$

and

$$B(\tau) = \phi^2(\tau)(\mathcal{A}^*\mathcal{A})^{-1}(\tau), \qquad d\nu(\tau) = \phi^{-2}(\tau)\,d\tau.$$

To check that the function $B \mapsto F(B)$ is strictly convex, restrict F to some line $B(t)$ in $\mathfrak{C}$, that is, $B''(t) = 0$. Use the formula for differentiating determinants (Proposition II.8.2) to imply

$$\frac{d^2}{dt^2}(\det B)^{-1} = (\det B)^{-1}\{(\operatorname{tr} B^{-1}B')^2 + \operatorname{tr}(B^{-1}B')^2\}.$$

Now, $B^{-1}B'$ is self-adjoint with respect to the quadratic form $\mathcal{B}(x, x) = Bx \cdot x$ (where $\cdot$ denotes the inner product in V), from which one concludes that $(d^2/dt^2)(\det B)^{-1}$ is nonnegative and is equal to 0 if and only if $B' = 0$.

Step 2. So, we may now apply the Jensen inequality. Then, we obtain

$$\det \int_s^t (A^*A)^{-1}(\tau)\,d\tau \ge \left\{\int_s^t \phi^{-2}(\tau)\,d\tau\right\}^N,$$

which, in turn, implies from (VII.2.3)

$$\text{(VII.2.7)} \qquad \det C_s(t) \ge \left\{\phi(t)\phi(s)\int_s^t \phi^{-2}(\tau)\,d\tau\right\}^N.$$

We consider the case of equality in (VII.2.7). This is characterized by

$$\text{(VII.2.8)} \qquad \mathcal{A}^*\mathcal{A}(\tau) = \phi^2(\tau)I$$

for all τ. Then, (VII.2.2) and (VII.2.8) imply

$$\mathcal{A}^*\mathcal{A}' = \phi'\phi I,$$

which implies

$$\mathcal{A}' = \phi'\phi(\mathcal{A}^*)^{-1} = \frac{\phi'}{\phi}\mathcal{A},$$

which implies

$$\mathcal{A} = \phi I$$

on all of $(0, \pi)$. Certainly, if $\mathcal{A}$ is a scalar multiple of I at each t, then one has equality in (VII.2.7).

Step 3. The next step is to apply Hölder's inequality to the functions

$$f = \{\det \mathcal{C}_s(t)\}^{1/N}, \qquad h = \sin^{N-1}(t-s),$$

with respective conjugate exponents

$$p = N, \qquad q = N/(N-1),$$

and measure ϵ on

$$\Omega_2 := \{(t, s) \in \mathbb{R}^2 : s \le t \le \pi,\ 0 \le s \le \pi\}$$

given by

$$d\epsilon = m(t-s)\,dt\,ds.$$

Then Hölder's inequality and (VII.2.7) combine to imply

$$\text{(VII.2.9)} \qquad \int_0^\pi ds \int_s^\pi m(t-s) \det \mathcal{C}_s(t)\,dt$$
$$\ge \{G(\phi)\}^N \left\{\int_0^\pi ds \int_s^\pi m(t-s) \sin^N(t-s)\,dt\right\}^{1-N},$$

where

$$G(\phi) = \int_0^\pi \phi(s)\,ds \int_s^\pi m(t-s) \sin^{N-1}(t-s)\phi(t)\,dt \int_s^t \phi^{-2}(\tau)\,d\tau.$$

Equality is achieved in (VII.2.9) if and only if

$$\sin^N(t-s) = \det \mathcal{C}_s(t), \qquad \mathcal{A}(t) = \phi(t)I,$$

that is, if and only if

$$\text{(VII.2.10)} \qquad \mathcal{A}(t) = \sin t\ I$$

on $[0, \pi]$.

Step 4. Thus, we are led to the study of $G(\phi)$. Note that if $\phi(t) = \sin t$, then

$$\int_s^t \phi^{-2}(\tau)\,d\tau = \int_s^t \frac{d\tau}{\sin^2 \tau} = \frac{\sin(t-s)}{(\sin t)(\sin s)},$$

which implies

$$G(\sin) = \int_0^\pi ds \int_s^\pi m(t-s) \sin^N(t-s)\,dt.$$

Therefore, the inequality (VII.2.5) is a consequence of the inequality

$$\text{(VII.2.11)} \qquad G(\phi) \ge G(\sin).$$

We shall prove (VII.2.11) under the hypothesis that $\phi(t)$ has the form $\phi(t) = t^\alpha(\pi - t)^\beta h(t)$, where $0 \le \alpha, \beta \le 1$, and $h(t)$ is positive and continuous on all of $[0, \pi]$.

Let

$$\Omega_3 := \{(\tau, t, s) \in \mathbb{R}^3 : s \le \tau \le t,\ s \le t \le \pi,\ 0 \le s \le \pi\}$$

and consider the measure σ on Ω_3 given by

$$d\sigma = \frac{(\sin s)(\sin t)}{\sin^2 \tau} m(t-s)\sin^{N-1}(t-s)\, d\tau dt ds.$$

Note that $G(\sin) = \sigma(\Omega_3)$.

Since $\phi(t)$ is of the form described above, we may write $\phi(t)$ as

$$\phi(t) = e^{u(t)} \sin t,$$

where $u(t)$ is continuous on $(0, \pi)$. Then, the usual form of Jensen's inequality implies

$$\text{(VII.2.12)} \quad G(\phi) \ge G(\sin)\exp\left\{\frac{1}{G(\sin)}\int_{\Omega_3} \{u(s) + u(t) - 2u(\tau)\}\, d\sigma\right\}.$$

So, (VII.2.11) will be a consequence of

$$\text{(VII.2.13)} \qquad \int_{\Omega_3} \{u(s) + u(t) - 2u(\tau)\}\, d\sigma = 0$$

for all u. Note that we have yet to use the symmetry hypothesis (VII.2.4) for $m(t)$. We show that (VII.2.13) is valid for all u under consideration if and only if $m(t) = m(\pi - t)$ for all $t \in [0, \pi]$.

Step 5. First one employs some manipulation to rewrite (VII.2.13) as

$$\text{(VII.2.14)} \qquad \int_0^\pi u(t) f(t) (\sin t)^{-2}\, dt = 0$$

for all u, where $f(t)$ is a C^2 function on $[0, \pi]$, satisfying $f(0) = 0$, and

$$\{(\sin t)^{-2} f'(t)\}' = (\sin t)^{-2}\{(\sin^3 t)[\rho(t) - \rho(\pi - t)]\}',$$

$$\{(\sin t)^{-2} f'(t)\}\,|_{t=0} = \int_0^\pi (\cos s)\{\rho(s) - \rho(\pi - s)\}\, ds,$$

where

$$\rho(t) := m(t)\sin^{N-1} t, \qquad \rho(t) - \rho(\pi - t) = \{m(t) - m(\pi - t)\}\sin^{N-1} t.$$

If $m(t) = m(\pi - t)$ for all $t \in [0, \pi]$, then $f = 0$, and (VII.2.14) is valid for all u. Conversely, if (VII.2.14) is valid for all u, then $f = 0$ and $\rho(t) = \text{const.}(\sin t)^{-3}$, which implies the constant is 0.

This concludes the proof of the inequality (VII.2.5). One can now easily deal with the case of equality. ∎

Before proceeding to the geometric applications of Kazdan's inequality, we first prove:

Proposition VII.2.2. *For all $n \geq 1$, we have*

$$\mathbf{c_n} = \mathbf{c_{n-1}} \int_0^\pi \sin^{n-1} t\, dt, \tag{VII.2.15}$$

and

$$\pi \mathbf{c_n}/2\mathbf{c_{n-1}} = \int_0^\pi ds \int_0^{\pi-s} \sin^{n-1} t\, dt. \tag{VII.2.16}$$

Proof. Equation (VII.2.15) is just (III.3.12). See the discussion in §III.3.

To prove (VII.2.16), we have

$$\begin{aligned}\pi \mathbf{c_n}/\mathbf{c_{n-1}} &= \int_0^\pi ds \int_0^\pi \sin^{n-1} t\, dt \\ &= \int_0^\pi ds \int_0^{\pi-s} \sin^{n-1} t\, dt + \int_0^\pi ds \int_{\pi-s}^\pi \sin^{n-1} t\, dt,\end{aligned}$$

and

$$\begin{aligned}\int_0^\pi ds \int_{\pi-s}^\pi \sin^{n-1} t\, dt &= \int_0^\pi ds \int_{\pi-s}^\pi \sin^{n-1}(\pi - t)\, dt \\ &= \int_0^\pi ds \int_0^s \sin^{n-1} t\, dt \\ &= \int_0^\pi ds \int_0^{\pi-s} \sin^{n-1} t\, dt\end{aligned}$$

(by the change of variable $s \mapsto \pi - s$), which implies the claim. ∎

Theorem VII.2.2. (M. Berger (1980)) *If M is a compact Riemannian manifold of dimension $n \geq 1$, then*

$$V(M) \geq \mathbf{c_n}\{\text{inj}\, M/\pi\}^n, \tag{VII.2.17}$$

with equality in (VII.2.17) *if and only if M is isometric to the standard n–sphere with radius equal to* inj M.

Proof. We first refer the reader to §§III.1–3 for background and notation.

For convenience, normalize the Riemannian metric on M, so that inj $M = \pi$ (namely, change the Riemannian metric on M by multiplying the length of every element of TM by $\pi/\text{inj}\, M$).

Step 1. First, note that for all $r \in [0, \pi/2]$, and all $\xi \in \mathsf{S}M$, we have empty intersection of the geodesic disks $B(\gamma_\xi(0); r)$ and $B(\gamma_\xi(\pi); \pi - r)$. Therefore,

$$V(M) \geq V(\gamma_\xi(0); r) + V(\gamma_\xi(\pi); \pi - r). \tag{VII.2.18}$$

Next note that

$$\begin{aligned}\int_{\mathsf{S}M} V(\pi(\xi); r)\, d\mu(\xi) &= \int_M dV(p) \int_{\mathsf{S}_p} V(\pi(\xi); r)\, d\mu_p(\xi)\\ &= \mathbf{c_{n-1}} \int_M V(p; r)\, dV(p),\end{aligned}$$

and Liouville's theorem implies

$$\begin{aligned}\int_{\mathsf{S}M} V(\gamma_\xi(\pi); \pi - r)\, d\mu(\xi) &= \int_{\mathsf{S}M} V(\pi \circ \Phi_\pi(\xi); \pi - r)\, d\mu(\xi)\\ &= \int_{\mathsf{S}M} V(\pi(\xi); \pi - r)\, d\mu(\xi)\\ &= \mathbf{c_{n-1}} \int_M V(p; \pi - r)\, dV(p).\end{aligned}$$

If we integrate (VII.2.18) over all of $\mathsf{S}M$, we obtain

$$\begin{aligned}\mathbf{c_{n-1}} V(M)^2 &= \mu(\mathsf{S}M) V(M)\\ &\geq \int_{\mathsf{S}M} \{V(\gamma_\xi(0); r) + V(\gamma_\xi(\pi); \pi - r)\}\, d\mu(\xi)\\ &= \mathbf{c_{n-1}} \int_M \{V(p; r) + V(p; \pi - r)\}\, dV(p),\end{aligned}$$

that is,

$$V(M)^2 \geq \int_M \{V(p; r) + V(p; \pi - r)\}\, dV(p). \tag{VII.2.19}$$

Step 2. One easily verifies

$$\int_M V(p; r)\, dV(p) = \int_0^r dt \int_{\mathsf{S}M} \sqrt{\mathbf{g}}(t; \xi)\, d\mu(\xi), \tag{VII.2.20}$$

for all $r \in [0, \pi]$, and, by interchanging the order of integration, we obtain

$$\begin{aligned}&\int_0^{\pi/2} dr \int_0^r \sqrt{\mathbf{g}}(t; \xi)\, dt + \int_0^{\pi/2} dr \int_0^{\pi - r} \sqrt{\mathbf{g}}(t; \xi)\, dt\\ &\quad = \int_0^{\pi/2} \{(\pi - t)\sqrt{\mathbf{g}}(t; \xi) + t\sqrt{\mathbf{g}}(\pi - t; \xi)\}\, dt\end{aligned} \tag{VII.2.21}$$

for all $\xi \in \mathsf{S}M$.

Now integrate (VII.2.19) over $r \in [0, \pi/2]$. Then, (VII.2.20) and (VII.2.21) imply

$$
\begin{aligned}
(\pi/2)V(M)^2 &\geq \int_0^{\pi/2} dr \left\{ \int_0^r dt \int_{SM} \sqrt{\mathbf{g}}(t;\xi)\, d\mu(\xi) \right. \\
&\quad \left. + \int_0^{\pi - r} dt \int_{SM} \sqrt{\mathbf{g}}(t;\xi)\, d\mu(\xi) \right\} \\
&= \int_0^{\pi/2} dt \int_{SM} \{(\pi - t)\sqrt{\mathbf{g}}(t;\xi) + t\sqrt{\mathbf{g}}(\pi - t;\xi)\}\, d\mu(\xi).
\end{aligned}
$$

But,

$$
\begin{aligned}
(\pi - t) \int_{SM} \sqrt{\mathbf{g}}(t;\xi)\, d\mu(\xi) &= \int_0^{\pi - t} dr \int_{SM} \sqrt{\mathbf{g}}(t;\xi)\, d\mu(\xi) \\
&= \int_0^{\pi - t} dr \int_{SM} \sqrt{\mathbf{g}}(t;\Phi_r\xi)\, d\mu(\xi)
\end{aligned}
$$

by Liouville's theorem, and, similarly,

$$
t \int_{SM} \sqrt{\mathbf{g}}(\pi - t;\xi)\, d\mu(\xi) = \int_0^t dr \int_{SM} \sqrt{\mathbf{g}}(\pi - t;\Phi_r\xi)\, d\mu(\xi).
$$

We, therefore, have

$$
\begin{aligned}
(\pi/2)V(M)^2 \geq \int_0^{\pi/2} dt \int_{SM} d\mu(\xi) &\left\{ \int_0^{\pi - t} \sqrt{\mathbf{g}}(t;\Phi_r\xi)\, dr \right. \\
&\left. + \int_0^t \sqrt{\mathbf{g}}(\pi - t;\Phi_r\xi)\, dr \right\}.
\end{aligned}
\tag{VII.2.22}
$$

Step 3. We now apply Kazdan's inequality (VII.2.5) for $N = n - 1$, and

$$
\mathcal{C}_s(t) = \mathcal{A}(t - s;\Phi_s\xi), \qquad \det \mathcal{C}_s(t) = \sqrt{\mathbf{g}}(t - s;\Phi_s\xi).
$$

Inequality (VII.2.5) can then be written as

$$
\int_0^{\pi} ds \int_0^{\pi - s} m(r)\sqrt{\mathbf{g}}(r;\Phi_s\xi)dr \geq \int_0^{\pi} ds \int_0^{\pi - s} m(r) \sin^{n-1} r dr.
\tag{VII.2.23}
$$

For the function $m(t)$, we pick

$$
m = \delta_t + \delta_{\pi - t},
\tag{VII.2.24}
$$

the sum of the Dirac distribution at t and the Dirac distribution at $\pi - t$. Inequality (VII.2.5) is valid for this choice of m, since (VII.2.5) is valid for all positive

C^2 functions $m = m(t)$ on $[0, \pi]$ satisfying (VII.2.4). One easily obtains

$$\int_0^{\pi-t} \sqrt{\mathbf{g}}(t; \Phi_r\xi)\, dr = \int_0^{\pi} dr \int_0^{\pi-r} \delta_t(s)\sqrt{\mathbf{g}}(s; \Phi_r\xi)\, ds,$$

and

$$\int_0^{t} \sqrt{\mathbf{g}}(\pi - t; \Phi_r\xi)\, dr = \int_0^{\pi} dr \int_0^{\pi-r} \delta_{\pi-t}(s)\sqrt{\mathbf{g}}(s; \Phi_r\xi)\, ds.$$

Therefore, (VII.2.22) and (VII.2.23) imply

$$\begin{aligned}(\pi/2)V(M)^2 &\geq \int_0^{\pi/2} dt \int_{\mathsf{S}M} d\mu(\xi) \int_0^{\pi} ds \int_0^{\pi-s} \{\delta_t + \delta_{\pi-t}\}(r)\sin^{n-1} r\, dr \\ &= \int_{\mathsf{S}M} d\mu(\xi) \int_0^{\pi/2} dt \int_0^{\pi} ds \int_0^{\pi-s} \{\delta_t + \delta_{\pi-t}\}(r)\sin^{n-1} r\, dr \\ &= \mathbf{c_{n-1}} V(M) \int_0^{\pi/2} dt \int_0^{\pi} ds \int_0^{\pi-s} \{\delta_t + \delta_{\pi-t}\}(r)\sin^{n-1} r\, dr.\end{aligned}$$

that is,

$$(\pi/2)V(M) \geq \mathbf{c_{n-1}} \int_0^{\pi/2} dt \int_0^{\pi} ds \int_0^{\pi-s} \{\delta_t + \delta_{\pi-t}\}(r)\sin^{n-1} r\, dr.$$

(VII.2.25)

Step 4. We now finish the proof of the theorem. First, one checks that if M is isometric to $\mathbb{S}^n$, then *all* the above inequalities are equalities. This implies that the right-hand side of (VII.2.25) is equal to $(\pi/2)\mathbf{c_n}$; so, (VII.2.25) is to be rewritten as

$$(\pi/2)V(M) \geq (\pi/2)\mathbf{c_n},$$

and (VII.2.17) follows.

If we have equality in (VII.2.17), then we have equality in (VII.2.23), with m as given in (VII.2.24). We conclude, from the case of equality in Kazdan's inequality, that

$$\mathcal{A}(t; \xi) = \{\sqrt{\mathbf{g}}(t; \xi)\}^{1/(n-1)} I$$

for all $(t; \xi) \in [0, \pi] \times \mathsf{S}M$, and

$$\sqrt{\mathbf{g}}(r - s; \xi) = \sin^{n-1}(r - s)$$

almost everywhere on $\{(r, s) : 0 \leq s \leq r, 0 \leq r \leq \pi\}$, with respect to the measure

$$d\epsilon = \{\delta_t + \delta_{\pi-t}\}(r - s)\, dr\, ds,$$

for all $\xi \in \mathsf{S}M$. Thus,

$$\sqrt{\mathrm{g}}(t;\xi) = \sin^{n-1} t, \qquad \sqrt{\mathrm{g}}(\pi - t;\xi) = \sin^{n-1}(\pi - t)$$

for all $t \in [0, \pi/2]$. Thus, we have

$$\mathcal{A}(t;\xi) = \sin t\, I$$

on all of $[0, \pi] \times \mathsf{S}M$, which implies M has constant sectional curvature equal to 1. The universal covering of M is (by Theorem IV.1.4) $\mathbb{S}^n$. But since $V(M) = V(\mathbb{S}^n)$, we must have M isometric to $\mathbb{S}^n$. ∎

The Blaschke Conjecture

Here is the basic definition.

Definition. A complete Riemannian manifold M is referred to as a *wiedersehnsraum* if there exists a constant $\kappa > 0$ such that for every $p \in M$, $\exp|\mathsf{B}(p;\pi/\sqrt{\kappa})$ has maximal rank, and $\exp_* T(\mathsf{S}(p;\pi/\sqrt{\kappa})) = 0$.

The reader will recall (for sure) that wiedersehnsräume were the topic of Lemma IV.1.2, in which it was proved that,
(i) for every $p \in M$, the image $\exp(\mathsf{S}(p;\pi/\sqrt{\kappa}))$ in M consists of precisely one point. Thus, a map $Q : M \to M$ is well defined by

$$Q(p) = \exp(\mathsf{S}(p;\pi/\sqrt{\kappa})). \tag{VII.2.26}$$

(ii) One has

$$Q^2 = \mathrm{id}_M,$$

(iii) Q is an isometry of M.
(iv) every unit speed geodesic γ on M is periodic, with period equal to $2\pi/\sqrt{\kappa}$.
(v) M is diffeomorphically covered by the sphere.

W. Blaschke (1967) had conjectured that any simply connected wiedersehnsraum is, in fact, *isometric* to a sphere and had derived the above properties (i)–(v). L. W. Green (1963) verified the conjecture in the 2–dimensional case (see our discussion in Exercise VII.13). Later, A. Weinstein (1974) showed that, for M, a simply connected even dimensional wiedersehnsraum one has its volume equal to the volume of the standard sphere whose geodesics have the same length as the common length of those in M. With the Berger–Kazdan inequalities, the Blaschke conjecture was thereby settled for even dimensions. Later,

C. T. Yang (1980) settled the Blaschke conjecture by explicitly calculating the volume of wiedersehnsräume in odd dimensions.

For additional discussion, see Berger–Kazdan (1980), Kazdan (1982), and Yang (1990). For more general discussions of Riemannian manifolds, all of whose geodesics are closed, see Besse (1978).

§VII.3. On Manifolds with No Conjugate Points

We start with a partial converse to the Morse–Schönberg theorem (Theorem II.14).

Theorem VII.3.1. (L. W. Green–M. Berger (Green (1963))) *Let M be a compact Riemannian manifold, $\kappa > 0$, and assume that, for any unit tangent vector $\xi \in \mathsf{S}M$, the point $\gamma_\xi(t) = \exp t\xi$ is not conjugate to $\gamma_\xi(0)$ along γ_ξ for all $t < \pi/\sqrt{\kappa}$ (i.e., the distance between conjugate points, when they exist, along any geodesic is $\geq \pi/\sqrt{\kappa}$). Then, the integral of the scalar curvature S over M* (see §II.1) *satisfies*

$$\text{(VII.3.1)} \qquad \frac{1}{V(M)}\int_M S\,dV \leq n(n-1)\kappa,$$

with equality if and only if M has constant sectional curvature κ.

Proof. To any $\xi \in \mathsf{S}M$, we associate its geodesic

$$\gamma_\xi(t) = \exp t\xi = (\pi \circ \Phi_t)(\xi), \qquad \gamma_\xi{}'(t) = \Phi_t(\xi),$$

where π denotes the natural projection of $\mathsf{S}M$ to M (we hope there will be no confusion with the various uses of π) and Φ_t the geodesic flow. Then, given any vector field $X \in C^\infty$ along $\gamma_\xi|[-\pi/2\sqrt{\kappa}, \pi/2\sqrt{\kappa}]$, pointwise orthogonal to γ_ξ, and vanishing at $\gamma_\xi(-\pi/2\sqrt{\kappa})$ and $\gamma_\xi(\pi/2\sqrt{\kappa})$, we have by Jacobi's criterion (Theorem II.10)

$$\text{(VII.3.2)} \qquad 0 \leq \int_{-\pi/2\sqrt{\kappa}}^{\pi/2\sqrt{\kappa}} \{|\nabla_t X|^2 - \langle R(\gamma_\xi{}', X)\gamma_\xi{}', X\rangle\}\,dt,$$

with equality if and only if X is a Jacobi field on γ_ξ.

Given any $\xi \in \mathsf{S}M$, $p = \pi(\xi)$. Complete ξ to an orthonormal basis $\{e_1, \ldots, e_{n-1}, \xi\}$ of M_p, and let $\{E_1(t), \ldots, E_{n-1}(t), \Phi_t\xi\}$ be the parallel orthonormal frame field along γ_ξ determined by the initial data $\{e_1, \ldots, e_{n-1}, \xi\}$. Set

$$X_j(t) = (\cos\sqrt{\kappa}t)E_j(t);$$

then

$$\begin{aligned}0 &\le \sum_{j=1}^{n-1} \int_{-\pi/2\sqrt{\kappa}}^{\pi/2\sqrt{\kappa}} \{|\nabla_t X_j|^2 - \langle R(\gamma_\xi{}', X_j)\gamma_\xi{}', X_j\rangle\}\, dt \\ &= \sum_{j=1}^{n-1} \int_{-\pi/2\sqrt{\kappa}}^{\pi/2\sqrt{\kappa}} \{\kappa \sin^2 \sqrt{\kappa}t - (\cos^2 \sqrt{\kappa}t)\mathcal{K}(E_j(t), \Phi_t\xi)\}\, dt \\ &= \int_{-\pi/2\sqrt{\kappa}}^{\pi/2\sqrt{\kappa}} \{(n-1)\kappa \sin^2 \sqrt{\kappa}t - (\cos^2 \sqrt{\kappa}t)\mathrm{Ric}\,(\Phi_t\xi, \Phi_t\xi)\}\, dt,\end{aligned}$$

with equality if and only if each X_j, $j = 1, \ldots, n-1$, is a Jacobi field. In such a case, one concludes that $\mathcal{K}(E_j(t), \Phi_t\xi) = \kappa$ for all $t \in [-\pi/2\sqrt{\kappa}, \pi/2\sqrt{\kappa}]$ for all $j = 1, \ldots, n-1$, that is, all 2–planes containing $\Phi_t\xi$, $t \in [-\pi/2\sqrt{\kappa}, \pi/2\sqrt{\kappa}]$, have sectional curvature equal to κ.

Lemma VII.3.1. *Let V be a real n–dimensional inner product space, $\beta : V \times V \to \mathbb{R}$ a bilinear form, S the unit sphere in V with canonical $(n-1)$–measure $d\epsilon$. Then,*

$$\int_{\mathsf{S}} \beta(\xi, \xi)\, d\epsilon(\xi) = \mathbf{c_{n-1}} \frac{\mathrm{tr}\,\beta}{n}. \tag{VII.3.3}$$

Proof. First diagonalize β with respect to the inner product, that is, pick an orthonormal basis $\{e_1, \ldots, e_n\}$ of V for which $\beta(\xi, \xi) = \sum_{j=1}^n \lambda_j (\xi^j)^2$ whenever $\xi = \sum_{j=1}^n \xi^j e_j$. Then, $\mathrm{tr}\,\beta = \lambda_1 + \cdots + \lambda_n$, and

$$\int_{\mathsf{S}} \beta(\xi, \xi)\, d\epsilon(\xi) = \sum_{j=1}^n \lambda_j \int_{\mathsf{S}} (\xi^j)^2\, d\epsilon(\xi).$$

From the symmetry of S, we have

$$\int_{\mathsf{S}} (\xi^j)^2\, d\epsilon(\xi)$$

independent of $j = 1, \ldots, n$. One concludes

$$\int_{\mathsf{S}} (\xi^j)^2\, d\epsilon(\xi) = \frac{\mathbf{c_{n-1}}}{n},$$

from which one obtains (VII.3.3). ■

Conclusion of the Proof of Theorem VII.3.1. Set

$$f_\kappa(t, \xi) = (n-1)\kappa \sin^2 \sqrt{\kappa}t - (\cos^2 \sqrt{\kappa}t)\mathrm{Ric}\,(\Phi_t\xi, \Phi_t\xi),$$

and define the function $\mathcal{F}_\kappa$ on $\mathsf{S}M$ by

$$\mathcal{F}_\kappa(\xi) = \int_{-\pi/2\sqrt{\kappa}}^{\pi/2\sqrt{\kappa}} f_\kappa(t, \xi)\, dt.$$

Then, $\mathcal{F}_\kappa \geq 0$ on all of $\mathsf{S}M$, which implies

$$\begin{aligned} 0 &\leq \int_{\mathsf{S}M} \mathcal{F}_\kappa(\xi)\, d\mu(\xi) \\ &= \int_{\mathsf{S}M} d\mu(\xi) \int_{-\pi/2\sqrt{\kappa}}^{\pi/2\sqrt{\kappa}} f_\kappa(t, \xi)\, dt \\ &= \int_{-\pi/2\sqrt{\kappa}}^{\pi/2\sqrt{\kappa}} dt \int_{\mathsf{S}M} f_\kappa(t, \xi)\, d\mu(\xi) \\ &:= J_1 - J_2, \end{aligned}$$

where

$$J_1 = \int_{-\pi/2\sqrt{\kappa}}^{\pi/2\sqrt{\kappa}} dt \int_{\mathsf{S}M} (n-1)\kappa \sin^2 \sqrt{\kappa} t\, d\mu(\xi) = (n-1)\kappa \mathbf{c_{n-1}} V(M)\pi/2\sqrt{\kappa},$$

and

$$\begin{aligned} J_2 &= \int_{-\pi/2\sqrt{\kappa}}^{\pi/2\sqrt{\kappa}} dt \int_{\mathsf{S}M} \cos^2 \sqrt{\kappa} t\, \mathrm{Ric}\,(\Phi_t\xi, \Phi_t\xi)\, d\mu(\xi) \\ &= \int_{-\pi/2\sqrt{\kappa}}^{\pi/2\sqrt{\kappa}} \cos^2 \sqrt{\kappa} t\, dt \int_{\mathsf{S}M} \mathrm{Ric}\,(\Phi_t\xi, \Phi_t\xi)\, d\mu(\xi) \\ &= \int_{-\pi/2\sqrt{\kappa}}^{\pi/2\sqrt{\kappa}} \cos^2 \sqrt{\kappa} t\, dt \int_{\mathsf{S}M} \mathrm{Ric}\,(\xi, \xi)\, d\mu(\xi) \\ &= \frac{\pi}{2\sqrt{\kappa}} \int_{\mathsf{S}M} \mathrm{Ric}\,(\xi, \xi)\, d\mu(\xi) \\ &= \frac{\pi}{2\sqrt{\kappa}} \int_M dV(x) \int_{\mathsf{S}_x} (\mathrm{Ric}\,(\xi, \xi)|\mathsf{S}_x)\, d\mu_x(\xi) \\ &= \frac{\pi}{2\sqrt{\kappa}} \int_M \frac{\mathbf{c_{n-1}} S}{n}\, dV(x), \end{aligned}$$

to go from the second to the third lines, one uses the invariance of the $d\mu$ relative to the geodesic flow. So,

$$0 \leq J_1 - J_2 = \frac{\pi \mathbf{c_{n-1}}}{2\sqrt{\kappa}} \left\{ (n-1)\kappa V(M) - \frac{1}{n} \int_M S\, dV \right\},$$

which is the inequality (VII.3.1).

If we have equality in (VII.3.1), then $\mathcal{F}_\kappa = 0$ on all of $\mathsf{S}M$. But this will then imply $\mathcal{K} = \kappa$ on all of M. ∎

Theorem VII.3.2. (E. Hopf (1948), L. W. Green (1958)) *If M above has no conjugate points, then*

$$\int_M S\,dV \le 0. \tag{VII.3.4}$$

Furthermore, one has equality in (VII.3.4) *if and only if M has sectional curvature identically equal to* 0.

If M is a compact Riemannian 2–dimensional manifold, diffeomorphic to a torus, and possessing no conjugate points, then the Gauss curvature vanishes identically on M.

Proof. (Following Berger (1965, pp. 273–276)) The final claim is a direct consequence of the Gauss–Bonnet Theorem (Theorems V.2.3, V.2.6), since it implies that one has equality in (VII.3.4). So, the real issue is inequality (VII.3.4) and the characterization of equality in (VII.3.4).

The inequality (VII.3.4) follows from that fact that if M has no conjugate points, then (VII.3.1) is valid for all positive κ. Simply let $\kappa \downarrow 0$.

We now consider the case of equality in (VII.3.4), but first, we require some preliminaries.

Let γ be a geodesic in a Riemannian manifold, and $\mathcal{I}$ an interval containing $t = 0$, such that any solution to Jacobi's equation along γ vanishes at most once on $\mathcal{I}$ (said casually, the interval $\mathcal{I}$, or the geodesic segment $\gamma\,|\mathcal{I}$, has no conjugate points). In what follows, we view the Jacobi equation as a matrix equation in a single real vector space, most conveniently, in $V = M_{\gamma(0)}$ as described toward the end of §III.1.

Lemma VII.3.2. *We let $\mathcal{A}(t)$ denote the matrix solution to Jacobi's equation determined by the initial data*

$$\mathcal{A}(0) = 0, \qquad \mathcal{A}'(0) = I;$$

and for any $T \in \mathcal{I} \setminus \{0\}$, we let $\mathcal{D}_T$ denote the matrix solution of Jacobi's equation determined by the boundary data

$$\mathcal{D}_T(0) = I, \qquad \mathcal{D}_T(T) = 0.$$

Then, for $t, T \in \mathcal{I} \setminus \{0\}$, $tT > 0$, we have

$$\mathcal{D}_T(t) = \mathcal{A}(t) \int_t^T (\mathcal{A}^*\mathcal{A})^{-1}(s)\,ds. \tag{VII.3.5}$$

Proof. The argument is the same as that of Proposition VII.2.1. ■

Furthermore, from the above argument we have

$$\text{(VII.3.6)} \qquad {\mathcal{D}_T}'(t) = \mathcal{A}'(t) \int_t^T (\mathcal{A}^*\mathcal{A})^{-1}(s)\, ds - (\mathcal{A}^{-1})^*(t),$$

and

$$\mathsf{W}(\mathcal{A}, \mathcal{D}_T) = I.$$

Lemma VII.3.3. (L.W. Green (1958)) *Assume M is a complete Riemannian manifold with a geodesic γ, with no two points of γ conjugate to each other along γ. Then,*

$$\mathcal{D} := \lim_{T\uparrow+\infty} \mathcal{D}_T$$

exists, and

$$\mathcal{D}' = \lim_{T\uparrow+\infty} {\mathcal{D}_T}'.$$

The convergence of $\mathcal{D}_T \to \mathcal{D}$ and ${\mathcal{D}_T}' \to \mathcal{D}'$ is uniform on compact subsets of $\mathbb{R}$, and $\mathcal{D}$ is nonsingular on all of $\mathbb{R}$.

In particular, for any $e \in \gamma'(0)^\perp$ and $T > 0$, consider the solution Y_T to Jacobi's equation along γ satisfying

$$Y_T(0) = e, \qquad Y_T(T) = 0.$$

Then, Y_T converges as $T \uparrow +\infty$ uniformly on compact subsets of $\mathbb{R}$ to a nowhere vanishing solution Y_e of Jacobi's equation satisfying

$$Y_e(0) = e.$$

Proof. First, we note that by evaluating $\mathsf{W}(\mathcal{D}_T, \mathcal{D}_T)(t)$ at $t = T$, one has

$$\mathsf{W}(\mathcal{D}_T, \mathcal{D}_T) = 0,$$

and by then evaluating $\mathsf{W}(\mathcal{D}_T, \mathcal{D}_T)(t)$ at $t = 0$, one has the self-adjointness of $\mathcal{D}_T$.

Next, for $T > \sigma$, we have

$$\mathcal{D}_T - \mathcal{D}_\sigma = \mathcal{A} \int_\sigma^T (\mathcal{A}^*\mathcal{A})^{-1}(s)\, ds,$$

$${\mathcal{D}_T}' - {\mathcal{D}_\sigma}' = \mathcal{A}' \int_\sigma^T (\mathcal{A}^*\mathcal{A})^{-1}(s)\, ds,$$

$${\mathcal{D}_T}'(0) - {\mathcal{D}_\sigma}'(0) = \int_\sigma^T (\mathcal{A}^*\mathcal{A})^{-1}(s)\, ds.$$

Thus, $\mathcal{D}_T{}'(0)$, viewed as a self-adjoint linear transformation depending on T, is strictly increasing with respect to T.

To show that $\mathcal{D}_T{}'(0)$ is bounded from above, fix $\alpha < 0$; and for any $v \in V$, let

$$\eta_{_T}(t) = \begin{cases} \mathcal{D}_\alpha(t)v & \alpha \le t \le 0 \\ \mathcal{D}_T(t)v & 0 \le t \le T \end{cases}.$$

Then,

$$\begin{aligned} 0 &< I(\eta_{_T}, \eta_{_T}) \\ &= \int_\alpha^T \{|\eta_{_T}{}'|^2 - \langle R(\gamma_\xi{}', \eta_{_T})\gamma_\xi{}', \eta_{_T}\rangle\}(s)\,ds \\ &= \langle \eta_{_T}, \eta_{_T}{}'\rangle(0-) - \langle \eta_{_T}, \eta_{_T}{}'\rangle(0+) - \int_\alpha^T \langle \eta_{_T}, \eta_{_T}{}'' + R(\gamma_\xi{}', \eta_{_T})\gamma_\xi{}'(s)\rangle\,ds \\ &= \langle \mathcal{D}_\alpha{}'(0)v, v\rangle - \langle \mathcal{D}_T{}'(0)v, v\rangle. \end{aligned}$$

One now has easily the claims of the lemma. ∎

Remark VII.3.1. One also has from the proof of the lemma

$$(\mathcal{D} - \mathcal{D}_T)(t) = A(t)\int_T^\infty (\mathcal{A}^*\mathcal{A})^{-1}(s)\,ds, \tag{VII.3.7}$$

and

$$\mathcal{D}(t) = A(t)\int_t^\infty (\mathcal{A}^*\mathcal{A})^{-1}(s)\,ds. \tag{VII.3.8}$$

Conclusion of the Proof of Theorem VII.3.2. (Berger (1965)) Given equality in (VII.3.4), we have by Fatou's lemma

$$\liminf_{\kappa\downarrow 0} \mathcal{F}_\kappa(\xi) = 0$$

for almost all $\xi \in \mathsf{S}M$.

Assume we are considering such a ξ. Then, for any parallel vector field $E(t)$ along γ_ξ, pointwise orthogonal to γ_ξ, we have for

$$X_\kappa(t) = \cos\sqrt{\kappa}t\,E(t)$$

the limit

$$0 = \liminf_{\kappa\downarrow 0} \int_{-\pi/2\sqrt{\kappa}}^{\pi/2\sqrt{\kappa}} \{|\nabla_t X_\kappa|^2 - \langle R(\gamma_\xi{}', X_\kappa)\gamma_\xi{}', X_\kappa\rangle\}\,dt.$$

Let $Y_{j,\kappa}$, $j = 1, \ldots, n-1$, denote the Jacobi field along γ_ξ such that

$$Y_{j,\kappa}(0) = E_j(0), \qquad Y_{j,\kappa}(\pi/2\sqrt{\kappa}) = 0;$$

then, by Lemma VII.3.3, we have

$$Y_j := \lim_{\kappa\downarrow 0} Y_{j,\kappa}, \qquad \nabla_t Y_j = \lim_{\kappa\downarrow 0} \nabla_t Y_{j,\kappa}$$

uniformly on compact subsets of $\mathbb{R}$, with $\{Y_1, \ldots, Y_{n-1}\}$ pointwise linearly independent on all of $\mathbb{R}$ and with

$$Y_j(0) = E_j(0).$$

Determine functions $\alpha_{j,\kappa}$ for which X_κ above is given by

$$X_\kappa = \sum_{j=1}^{n-1} \alpha_{j,\kappa} Y_{j,\kappa}.$$

Then, Theorem II.5.4 implies

$$0 = \liminf_{\kappa\downarrow 0} \int_{-\pi/2\sqrt{\kappa}}^{\pi/2\sqrt{\kappa}} |\sum_j \alpha_{j,\kappa}{}' Y_{j,\kappa}|^2 \geq \int_{\mathbb{R}} \liminf_{\kappa\downarrow 0} |\sum_j \alpha_{j,\kappa}{}' Y_{j,\kappa}|^2;$$

so

$$0 = \liminf_{\kappa\downarrow 0} |\sum_j \alpha_{j,\kappa}{}' Y_{j,\kappa}|^2. \tag{VII.3.9}$$

On the other hand, since $X_\kappa \to E$ and $Y_{j,\kappa} \to Y_j$ as $\kappa \downarrow 0$ uniformly on compact subsets of $\mathbb{R}$, there exist functions α_j, $j = 1, \ldots, n-1$, such that $\alpha_{j,\kappa} \to \alpha_j$ uniformly on compact subsets of $\mathbb{R}$; and since $\nabla_t X_\kappa \to 0$ and $\nabla_t Y_{j,\kappa} \to \nabla_t Y_j$ as $\kappa \downarrow 0$, we also have $\alpha_{j,\kappa}{}' \to \alpha_j{}'$ uniformly on compact subsets of $\mathbb{R}$. Thus, (VII.3.9) implies $\alpha_j{}' = 0$ on all of $\mathbb{R}$. In particular,

$$E = \sum_j \alpha_j Y_j,$$

where α_j, $j = 1, \ldots, n-1$, are constants. We conclude that all parallel vector fields along γ_ξ are Jacobi fields. This implies, directly from Jacobi's equation, that all sectional curvatures of 2–planes containing velocity vectors of γ_ξ all vanish. This is true for almost all $\xi \in \mathsf{S}M$. Continuity of the sectional curvature implies that it vanishes identically on all of M. ■

Second Proof of Theorem VII.3.2. (L. W. Green (1958)) Again, we use Lemma VII.3.3, except that now we vary the initial point of the geodesic. So, $\mathcal{D}_T$ and $\mathcal{D}$ are now replaced by matrix solutions $\mathcal{D}_{T;\xi}$ and $\mathcal{D}_\xi$, respectively, of the Jacobi equation

$$\text{(VII.3.10)} \qquad \nabla_t{}^2 A + \mathsf{R}_\xi(t) A = 0,$$

where $\mathsf{R}_\xi(t)$ is the self-adjoint map of $\gamma_\xi{}'(t)^\perp = (\Phi_t \xi)^\perp$ given by

$$\mathsf{R}_\xi(t)\eta = R(\gamma_\xi{}'(t), \eta)\gamma_\xi{}'(t).$$

Thus, $\mathcal{D}_{T;\xi}(t)$ and $\mathcal{D}_\xi(t)$ are linear transformations of $(\Phi_t \xi)^\perp$. The initial data for $\mathcal{D}_{T;\xi}$ are given by

$$\mathcal{D}_{T;\xi}(0) = I, \qquad \mathcal{D}_{T;\xi}(T) = 0;$$

and, of course,

$$\mathcal{D}_\xi = \lim_{T \uparrow +\infty} \mathcal{D}_{T;\xi}.$$

We now comment on ∇_t in (VII.3.10). Given *any* 1–parameter family of linear transformations

$$\mathfrak{a}_\xi(t) : (\Phi_t \xi)^\perp \to (\Phi_t \xi)^\perp$$

we define its covariant derivative

$$\nabla_t \mathfrak{a}_\xi(t) : (\Phi_t \xi)^\perp \to (\Phi_t \xi)^\perp$$

in the obvious way, namely, given $\eta \in (\Phi_t \xi)^\perp$ let X be a parallel vector field along γ_ξ for which $X(t) = \eta$. Then, $(\nabla_t \mathfrak{a}_\xi)(\eta)$ is defined by

$$(\nabla_t \mathfrak{a}_\xi)(\eta) = \nabla_t(\mathfrak{a}_\xi X).$$

One then has a natural notion of parallel translation of $\mathfrak{a}_\xi$ along γ_ξ.

We denote all parallel translation along γ_ξ, from $\gamma_\xi(t)$ to $\gamma_\xi(s)$, by $\tau_{t,s;\xi}$; for convenience, $\tau_{s;\xi} = \tau_{0,s;\xi}$.

Note that one easily verifies

$$\mathcal{D}_{T-s;\Phi_s\xi}(t) = \mathcal{D}_{T;\xi}(s+t) \circ \tau_{s,t+s;\xi} \circ \mathcal{D}_{T;\xi}(s)^{-1},$$

which implies

$$\mathcal{D}_{\Phi_s\xi}(t) = \mathcal{D}_\xi(s+t) \circ \tau_{s,t+s;\xi} \circ \mathcal{D}_\xi(s)^{-1}.$$

Therefore, if we set

$$U_\xi(t) = (\nabla_t \mathcal{D}_\xi)(t) \circ \mathcal{D}_\xi(t)^{-1},$$

then we have

$$U_{\Phi_s\xi}(t) = U_\xi(t+s).$$

Moreover, $U_\xi(t)$ is a self-adjoint solution of the matrix Riccati equation

$$\nabla_t U + U^2 + \mathsf{R}_\xi(t) = 0$$

along γ_ξ (see the proof of Theorem III.4.3). As in the proof of Bishop's theorem (Theorem III.4.3), we have

$$(\mathrm{tr}\, U_\xi)' + \frac{(\mathrm{tr}\, U_\xi)^2}{n-1} + \mathrm{tr}\, \mathsf{R}_\xi \le 0. \tag{VII.3.11}$$

We consider $U = U_\xi(t)$ as a function on $\mathbb{R} \times \mathsf{S}M$. Fix $s > 0$ and integrate (VII.3.11) over $[0, s] \times \mathsf{S}M$. Of course, one must first verify the measurability of U – an easy matter.[1] Indeed, for any $T > 0$,

$$U_{T;\xi} = (\nabla_t \mathcal{D}_{T;\xi}) \circ \mathcal{D}_{T;\xi}{}^{-1}$$

varies continuously with respect to $\xi \in \mathsf{S}M$. Now, let $T \uparrow +\infty$.

So, we consider the integration of (VII.3.11) over $[0, s] \times \mathsf{S}M$. First,

$$\begin{aligned}\int_0^s dt \int_{\mathsf{S}M} (\mathrm{tr}\, U_\xi)'(t)\, d\mu(\xi) &= \int_{\mathsf{S}M} \{\mathrm{tr}\, U_\xi(s) - \mathrm{tr}\, U_\xi(0)\}\, d\mu(\xi) \\ &= \int_{\mathsf{S}M} \mathrm{tr}\, U_{\Phi_s\xi}(0)\, d\mu(\xi) - \int_{\mathsf{S}M} \mathrm{tr}\, U_\xi(0)\, d\mu(\xi) \\ &= 0\end{aligned}$$

by Liouville's theorem. Furthermore, as in the proof of Theorem VII.3.1,

$$\begin{aligned}\int_{\mathsf{S}M} \mathrm{tr}\, \mathsf{R}_\xi(t)\, d\mu(\xi) &= \int_{\mathsf{S}M} \mathrm{Ric}\,(\Phi_t\xi, \Phi_t\xi)\, d\mu(\xi) \\ &= \int_{\mathsf{S}M} \mathrm{Ric}\,(\xi, \xi)\, d\mu(\xi) \\ &= \frac{\mathbf{c}_{n-1}}{n} \int_M S\, dV,\end{aligned}$$

again, we have used the Liouville theorem. Thus, we obtain

$$\frac{\mathbf{c}_{n-1}}{n} \int_M S\, dV \le -\frac{1}{s} \int_0^s dt \int_{\mathsf{S}M} \frac{(\mathrm{tr}\, U_\xi)^2}{n-1}(t)\, d\mu(\xi) \le 0. \tag{VII.3.12}$$

This implies, again, (VII.3.4).

If we have equality in (VII.3.4), then we have equality in (VII.3.11) on all of $(0, +\infty)$. Since $U_\xi(t)$ is self-adjoint, this implies that $U_\xi(t)$ is a scalar multiple of the identity transformation of $(\Phi_t\xi)^\perp$. Then, equality in (VII.3.12) implies $U_\xi(t) = 0$ for all $t > 0$ for almost all $\xi \in \mathsf{S}M$. In particular, $\mathsf{R}_\xi(t) = 0$ for all $t > 0$ for almost all $\xi \in \mathsf{S}M$. But R_ξ is continuous with respect to ξ. Thus R_ξ vanishes identically on $\mathsf{S}M$, and M is flat. ■

[1] But, see Remark VII.3.3.

Remark VII.3.2. The *E. Hopf conjecture* states that any Riemannian metric on an n–dimensional torus, $n \geq 2$, that has no conjugate points must be flat. The conjecture was proved in Burago–Ivanov (1994). Regrettably, any detailed discussion here would take us too far afield.

Remark VII.3.3. In E. Hopf's (1948), he claimed that he had a proof that U_ξ was continuous with respect to ξ (although he did not publish it). However, a counterexample is given in Ballman–Brin–Burns (1987).

Remark VII.3.4. One thinks of $\mathcal{D}_\xi$ as a "contracting," or *stable,* nonsingular solution of the matrix Jacobi equation on all of γ_ξ – "contracting" in the direction of ξ. One might denote $\mathcal{D}_\xi$, more precisely, as $\mathcal{D}_\xi^-$. Of course, one can construct a corresponding "expanding", or *unstable,* solution $\mathcal{D}_\xi^+$ to the matrix Jacobi equation along γ_ξ, namely, $\mathcal{D}_\xi^+ = \mathcal{D}_{-\xi}$. The corresponding logarithmic derivative solutions of the matrix Riccati equation are then denoted by U_ξ^- and U_ξ^+, respectively.

§VII.4. Santalo's Formula

We are given a Riemannian manifold M and a relatively compact domain Ω in M with smooth boundary. For any $\xi \in \mathsf{S}\Omega$, we set

$$\tau(\xi) = \sup\{\tau > 0 : \gamma_\xi(t) \in \Omega \ \forall\, t \in (0, \tau)\},$$

that is, when $\tau(\xi)$ is finite, then $\gamma_\xi(\tau(\xi))$ will be the first point on the geodesic to hit the boundary of Ω. We also set

$$\ell(\xi) = \inf\{c(\xi), \tau(\xi)\},$$

where $c(\xi)$ denotes the distance from $\pi(\xi)$ to its cut point along γ_ξ, and

$$\mathsf{U}\Omega = \{\xi \in T\Omega : c(\xi) \geq \tau(\xi)\}.$$

We now consider the boundary $\partial\Omega$ of Ω. Let $\overline{\nu}$ denote the *inward* unit normal vector field along $\partial\Omega$, and let $\mathsf{S}^+\partial\Omega$ denote the collection of inward pointing unit vectors along $\partial\Omega$, that is,

$$\mathsf{S}^+\partial\Omega = \{\xi \in \mathsf{S}\overline{\Omega}|\partial\Omega : \langle \xi, \overline{\nu}_{\pi(\xi)} \rangle > 0\},$$

with measure

$$d\sigma(\xi) = d\mu_{\pi(\xi)}(\xi) dA(\pi(\xi)),$$

where dA denotes the $(n-1)$–measure on $\partial\Omega$.

One checks that $\tau = \tau(\xi)$ is lower semicontinuous on $\mathsf{S}\Omega \cup \mathsf{S}^+\partial\Omega$.

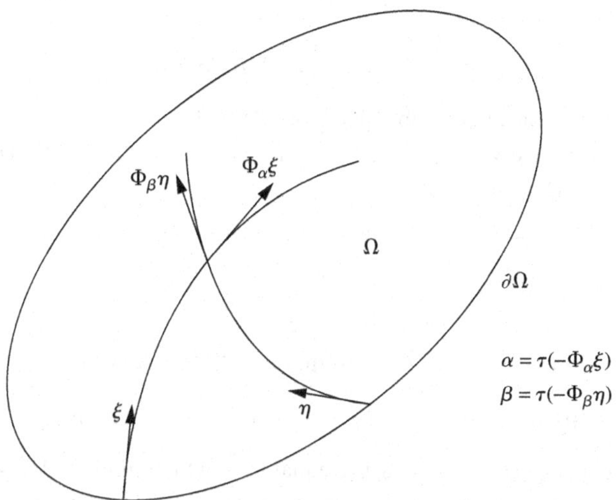

Figure VII.1. For Santalo's formula.

Theorem VII.4.1. (Santalo's formula (1976, pp. 336ff)) *For all integrable F on $\mathsf{S}\Omega$, we have*

$$\text{(VII.4.1)} \qquad \int_{-\mathsf{U}\Omega} F\,d\mu = \int_{\mathsf{S}^+\partial\Omega} \langle \xi, \overline{\nu}_{\pi(\xi)}\rangle\, d\sigma(\xi) \int_0^{\ell(\xi)} F(\Phi_t\xi)\,dt.$$

Furthermore, if $\tau < +\infty$ on all of $\mathsf{S}\Omega$, then we also have

$$\text{(VII.4.2)} \qquad \int_{\mathsf{S}\Omega} F\,d\mu = \int_{\mathsf{S}^+\partial\Omega} \langle \xi, \overline{\nu}_{\pi(\xi)}\rangle\, d\sigma(\xi) \int_0^{\tau(\xi)} F(\Phi_t\xi)\,dt.$$

Proof. Map

$$(t,\xi) \mapsto \Phi_t\xi, \qquad (t,\xi) \in (0,\infty) \times \mathsf{S}^+\partial\Omega$$

(see Figure VII.1). Then (in what follows, we let s denote distance from $\partial\Omega$),

$$\begin{aligned}
d\mu(\Phi_t\xi) &= (\Phi_t)_* d\mu(\xi) \\
&= (\Phi_t)_* d\mu_{\pi(\xi)}\, dV(\pi(\xi)) \\
&= (\Phi_t)_* d\mu_{\pi(\xi)}\, ds(\pi(\xi))\, dA(\pi(\xi)) \\
&= (\Phi_t)_* \frac{ds}{dt}(\xi)\, dt\, d\sigma(\xi) \\
&= (\Phi_t)_* \langle \xi, \overline{\nu}_{\pi(\xi)}\rangle\, dt\, d\sigma(\xi) \\
&= \langle \xi, \overline{\nu}_{\pi(\xi)}\rangle\, (\Phi_t)_* dt\, d\sigma(\xi) \\
&= \langle \xi, \overline{\nu}_{\pi(\xi)}\rangle\, dt\, d\sigma(\xi),
\end{aligned}$$

that is,

$$d\mu(\Phi_t\xi) = \langle \xi, \overline{\nu}_{\pi(\xi)}\rangle \, dt \, d\sigma(\xi). \tag{VII.4.3}$$

We note that the third line is obtained from (III.6.3).

Since Ω is relatively compact in M, we have $\ell < \infty$ on all of $\mathsf{S}\Omega$. Therefore, with any $\eta \in \mathsf{S}\Omega$ we associate

$$t = \ell(-\eta) \qquad \text{and} \qquad \xi = -\gamma_{-\eta}{}'(\ell(-\eta));$$

of course,

$$t < \ell(\xi).$$

Therefore, the map that takes (t, ξ) to $\Phi_t\xi$ is a diffeomorphism

$$\{(t, \xi) : 0 < t < \ell(\xi), \ \xi \in \mathsf{S}^+\partial\Omega\} \to -\mathsf{U}\Omega \setminus N,$$

where N denotes a set of μ–measure equal to 0, which implies (VII.4.1). (We have to leave out a set N, since $-\mathsf{U}\Omega$ contains $N := \{\Phi_{\ell(\xi)}\xi : \xi \in \mathsf{S}^+\partial\Omega\}$.) This implies (VII.4.1).

To obtain (VII.4.2), we note that since $\tau < \infty$ on all of $\mathsf{S}\Omega$, the map which takes (t, ξ) to $\Phi_t\xi$ is a diffeomorphism

$$\{(t, \xi) : 0 < t < \tau(\xi), \ \xi \in \mathsf{S}^+\partial\Omega\} \to \mathsf{S}\Omega. \qquad \blacksquare$$

Proposition VII.4.1. *For any $e \in \mathbb{S}^{n-1}$, we have*

$$\int_{H_e} \langle \xi, e\rangle \, dV(\xi) = \frac{\mathbf{c_{n-2}}}{n-1}, \tag{VII.4.4}$$

where H_e denotes the hemisphere of $\mathbb{S}^{n-1}$ centered at e.

Proof. By direct calculation: Write $\xi = (\cos\theta)e + (\sin\theta)\eta$, where η varies over the equator of e (that is, $\theta = \pi/2$). Then,

$$\begin{aligned}\int_{H_e} \langle \xi, e\rangle \, dV(\xi) &= \int_{\mathbb{S}^{n-2}} d\mu_{n-2}(\eta) \int_0^{\pi/2} \cos\theta \sin^{n-2}\theta \, d\theta \\ &= \mathbf{c_{n-2}} \int_0^{\pi/2} \cos\theta \sin^{n-2}\theta \, d\theta \\ &= \frac{\mathbf{c_{n-2}}}{n-1}.\end{aligned} \qquad \blacksquare$$

Definition. For every $x \in M$, we let U_x denote the subset of S_x given by

$$\mathsf{U}_x = (\pi|\mathsf{U}\Omega)^{-1}[x], \qquad \omega_x := \frac{\mu_x(\mathsf{U}_x)}{\mathbf{c_{n-1}}}, \qquad \omega := \inf_{x\in\Omega} \omega_x.$$

It is common to refer to ω_x as *the visibility angle at* x.

In particular,

$$\mu(-\mathsf{U}\Omega) = \mu(\mathsf{U}\Omega)) = \mathbf{c_{n-1}} \int_\Omega \omega_x \, dV(x) \geq \mathbf{c_{n-1}}\omega V(\Omega).$$

For any $w \in \partial\Omega$, set

$$\mathsf{S}^+_w = (\pi|\mathsf{S}^+\partial\Omega)^{-1}[w].$$

Theorem VII.4.2. (C. Croke (1980)) *Let $d(\Omega)$ denote the diameter of Ω. Then,*

$$\frac{A(\partial\Omega)}{V(\Omega)} \geq \frac{(n-1)\mathbf{c_{n-1}}\omega}{\mathbf{c_{n-2}}d(\Omega)}, \tag{VII.4.5}$$

with equality in (VII.4.5) *if Ω is a hemisphere of constant positive curvature, in which case $\omega = 1$ and $d(\Omega)$ is the diameter of the hemisphere.*

Proof. We have

$$\begin{aligned} \omega\mathbf{c_{n-1}}V(\Omega) &\leq \mu(\mathsf{U}\Omega) \\ &= \int_{\mathsf{S}^+\partial\Omega} \ell(\xi)\langle\xi, \overline{\nu}_{\pi(\xi)}\rangle \, d\sigma(\xi) \\ &= \int_{\partial\Omega} dA(w) \int_{\mathsf{S}^+_w} \ell(\xi)\langle\xi, \overline{\nu}_w\rangle \, d\mu_w(\xi) \\ &\leq \frac{d(\Omega)\mathbf{c_{n-2}}A(\partial\Omega)}{(n-1)}, \end{aligned}$$

by (VII.4.4), that is,

$$\omega\mathbf{c_{n-1}}V(\Omega) \leq \frac{d(\Omega)\mathbf{c_{n-2}}A(\partial\Omega)}{(n-1)},$$

which implies (VII.4.5).

One easily verifies the equality when Ω is a hemisphere of constant positive curvature. ∎

Theorem VII.4.3. (C. Croke (1980)) *Let M be a compact n–dimensional Riemannian manifold, all of whose Ricci curvatures are bounded below by $(n-1)\kappa$. Let $d(M)$ denote the diameter of M. If Γ is any $(n-1)$–dimensional compact submanifold of M dividing M into open submanifolds M_1, M_2, satisfying $\partial M_1 = \partial M_2 = \Gamma$, then, setting $\Omega = M_1$, we have*

$$\omega_x \geq V(M_2) \Big/ \mathbf{c_{n-1}} \int_0^{d(M)} \mathbf{S}_\kappa{}^{n-1} \tag{VII.4.6}$$

for all $x \in M_1$.

Proof. Santalo's formula (VII.4.1) is valid even when Ω is a finite disjoint union of regular domains.

To prove (VII.4.6) for $x \in M_1$, let

$$\Theta_x = \{q \in M : q = \gamma_\xi(t),\ t \in (0, c(\xi)],\ \xi \in \mathsf{U}_x\}.$$

Then, any point $q \in M_2$ has a unit speed minimizing geodesic connecting x to q. This geodesic must hit the boundary of M_1 (to reach a point in M_2); therefore, its initial velocity vector is in U_x. We conclude $M_2 \subseteq \Theta_x$, which implies

$$\begin{aligned} V(M_2) &\le \int_{\mathsf{U}_x} d\mu_x(\xi) \int_0^{c(\xi)} \sqrt{\mathbf{g}}(t;\xi)\,dt \\ &\le \int_{\mathsf{U}_x} d\mu_x(\xi) \int_0^{c(\xi)} \mathbf{S}_\kappa{}^{n-1}(t)\,dt \\ &\le \omega_x \mathbf{c_{n-1}} \int_0^{d(M)} \mathbf{S}_\kappa{}^{n-1}, \end{aligned}$$

the right-hand side of the first line is precisely equal to $V(\Theta_x)$; and the second line is a consequence of the Bishop comparison theorem (Theorem III.4.3). ■

§VII.5. Notes and Exercises

The Kinematic Density Via Moving Frames

Our treatment of the Liouville measure was presented from the perspective of Hamiltonian mechanics, namely, given any manifold, the differentiable structure alone determined a canonical 1–form on the cotangent bundle, with its associated symplectic 2–form and Hamiltonian differential equation. Then, given *any* Lagrangian on the tangent bundle, one has a natural (depending on the Lagrangian) method of bringing the data from the cotangent bundle to the tangent bundle. Given a Riemannian metric, one picks the Lagrangian to be half the norm squared of vectors in the tangent bundle, and lo and behold, one obtains the geodesic flow on the tangent bundle.

In what follows, we give an explicit Riemannian approach, with the calculations using the method of moving frames. To fix the data, let M be an n–dimensional Riemannian manifold, with projection maps

$$\pi_1 : TM \to M, \qquad \pi_2 : TM^* \to M$$

of the tangent and cotangent bundles, respectively. Recall the natural isomorphism $\theta : TM \to TM^*$ given by

$$(\theta(\xi))(\eta) = \langle \xi, \eta \rangle,$$

and the canonical 1–form ω on TM^* given by

$$\omega_{|\tau} = \pi_2{}^*\tau,$$

which determines the 1–form ϑ on TM (the same as in §VII.1) given by

$$\vartheta = \theta^*\omega.$$

Then, ϑ is the fundamental object in what follows.

Now, simply refer to π_1 as π. Given a chart $x : U \to \mathbb{R}^n$ on an open set U in M, fix a frame field $\{e_1, \ldots, e_n\}$ on U, with coframe field $\{\omega^1, \ldots, \omega^n\}$ and matrix of connection 1–forms $({\omega_j}^k)$.

Use the chart x and the frame field $\{e_1, \ldots, e_n\}$ to determine a chart $y : \pi^{-1}[U] \to \mathbb{R}^{2n}$ given by

$$y^j(\xi) = x^j \circ \pi(\xi), \qquad y^{j+n}(\xi) = \langle \xi, e_j \rangle.$$

Then, of course,

$$\pi_*\partial^y_j = \partial^x_j, \qquad \pi_*\partial^y_{j+n} = 0.$$

For

$$e_j = \sum_k {A_j}^k \partial^x_k,$$

set

$$E_j = \sum_k ({A_j}^k \circ \pi)\partial^y_k,$$

and consider the local forms on $\pi^{-1}[U]$ given by

$$\tau^j = \pi^*\omega^j, \qquad {\tau_j}^k = \pi^*{\omega_j}^k.$$

Exercise VII.1.

(a) Show that

$$\vartheta_{|\pi^{-1}[U]} = \sum_j y^{j+n}\tau^j,$$

and

$$\Theta = d\vartheta = \sum_j \left\{ dy^{j+n} + \sum_k y^{k+n}{\tau_k}^j \right\} \wedge \tau^j.$$

(b) For the vector field $\mathcal{G}$ of the geodesic flow on $\pi^{-1}[U]$ show that

$$\mathcal{G} = \sum_j y^{j+n}E_j - \sum_k y^{k+n}{\omega_k}^j \partial^y_{j+n},$$

where ${\omega_k}^j$ is viewed here, and, in what follows, as a real-valued function on TM.

(c) Show that

$$\tau^j(\mathcal{G}) = y^{j+n}, \qquad {\tau_k}^j(\mathcal{G}) = {\omega_k}^j, \qquad dy^{j+n}(\mathcal{G}) = -\sum_k y^{k+n}{\omega_k}^j.$$

Exercise VII.2.

(a) Show that for the function L on TM, given by

$$L(\xi) = |\xi|^2 = \sum_j (y^{j+n})^2,$$

we have

$$\mathcal{G}L = 0.$$

So, the geodesic flow takes the unit bundle $\mathsf{S}M$ to itself.

(b) Show that

$$\mathrm{i}(\mathcal{G})\vartheta = L, \qquad \mathrm{i}(\mathcal{G})\Theta = -dL, \qquad \mathcal{L}_{\mathcal{G}}\vartheta = \frac{1}{2}dL, \qquad \mathcal{L}_{\mathcal{G}}\Theta = 0.$$

(c) Show that

$$(\vartheta \wedge \Theta^{n-1})_{|\mathsf{S}M}$$

is, up to a constant depending only on n, the Liouville measure on $\mathsf{S}M$, and is invariant with respect to the geodesic flow on $\mathsf{S}M$.

(d) Show that Θ, restricted to any complementary subspace of $\mathcal{G}$ in $T\mathsf{S}M$, is nondegenerate.

(e) Show there is no closed codimension 1 submanifold of $T\mathsf{S}M$ that is transverse to $\mathcal{G}$.

The Differential of the Geodesic Flow

Continue the previous discussion. Let Φ_t denote the geodesic flow, with vector field $\mathcal{G}$. Let $\pi : TM \to M$ denote the standard projection of TM to M, and $\overline{\pi} : TTM \to TM$ the standard projection to TM from its tangent bundle. Assume M is complete.

Consider the map $\mathcal{T} : TTM \to TM$ given as follows: For any $p \in M$, $\xi \in M_p$, and $\mathfrak{x} \in (TM)_\xi$, let $Z(\epsilon)$ be a path in TM satisfying

$$Z(0) = \xi, \qquad Z'(0) = \mathfrak{x},$$

so Z' is the velocity vector of a path in TM. Define

$$\mathcal{T}\mathfrak{x} = (\nabla_\epsilon Z)(0),$$

where the covariant differentiation is along the path $\pi \circ Z$.

Exercise VII.3.

(a) Show that $\mathcal{T}$ is well defined, that is, show that it only depends on $\mathfrak{x}$; more particularly, calculate, using moving frames with the above notation,

$$\mathcal{T}\mathfrak{x} = \sum_j \left\{ dy^{j+n}(\mathfrak{x}) + \sum_k y^{k+n}(\overline{\pi}\circ\mathfrak{x})\omega_k{}^j(\xi) \right\} e_j,$$

where $y^{k+n}(\overline{\pi} \circ \mathfrak{x})$ denotes the k–th coordinate of $\pi \circ \mathfrak{x}$, and $\omega_k{}^j(\xi)$ denotes the action of the 1–form $\omega_k{}^j$ on ξ.

(b) Show that the kernel of $\mathcal{T}$ is transverse to the kernel of π_*.

(c) Define the *Sasaki metric* (Sasaki (1958, 1962)) on TM by

$$|\mathfrak{x}|_{\mathsf{S}}{}^2 = |\pi_*\mathfrak{x}|^2 + |\mathcal{T}\mathfrak{x}|^2.$$

Since the two kernels are transverse, the metric provides an orthogonal decomposition of every tangent space of TM.

Exercise VII.4. Show that $\mathcal{T}\mathcal{G} = 0$.

Exercise VII.5. Given the path $Z(\epsilon)$ above, consider the geodesic variation

$$\Omega(t, \epsilon) = \exp tZ(\epsilon),$$

with associated Jacobi field $Y_\mathfrak{x}$ along γ_ξ given by: $Y_\mathfrak{x} = \partial_\epsilon \Omega_{|\epsilon=0}$. Show that

$$Y_\mathfrak{x}(t) = \pi_* \circ (\Phi_t)_*\mathfrak{x}, \qquad (\nabla_t Y_\mathfrak{x})(t) = \mathcal{T} \circ (\Phi_t)_*\mathfrak{x}.$$

which implies

$$|(\Phi_t)_*\mathfrak{x}|_{\mathsf{S}}{}^2 = |Y_\mathfrak{x}(t)|^2 + |\nabla_t Y_\mathfrak{x}(t)|^2.$$

We know that there exist constants a and b so that $\langle Y_\mathfrak{x}, \gamma_\xi{}' \rangle = at + b$. Show that

$$a = \frac{1}{2}(|Z|^2)'(0), \qquad b = \langle (\Phi_t)_*\mathfrak{x}, \mathcal{G}(\Phi_t\xi) \rangle_{\mathsf{S}}.$$

If we restrict ξ to $\mathsf{S}M$, and $\mathfrak{x}$ to $T\mathsf{S}M$, then $a = 0$. Henceforth, restrict $\mathfrak{x}$ to $\mathcal{G}(\xi)^\perp$. Then, $b = 0$.

Exercise VII.6. Show that the geodesic flow Φ_t is an isometry for all $t \in \mathbb{R}$ if and only if M has constant sectional curvature equal to 1.

Manifolds Without Conjugate Points

Assume M has no conjugate points, and to each $\xi \in \mathsf{S}M$, $T > 0$, consider the matrix solutions $\mathcal{D}_{\xi;T}, \mathcal{D}_{\xi;-T}$ of Jacobi's equation on γ_ξ, as described in the proof of E. Hopf's theorem. Then, the linear span of Jacobi vector fields

$$\mathcal{J}_{\xi;T} := \{\mathcal{D}_{\xi;T}\,\eta, \mathcal{D}_{\xi;-T}\,\eta : \eta \in \xi^\perp\}$$

is actually equal to $\mathcal{J}_\xi{}^\perp$, the collection of all Jacobi fields along γ_ξ pointwise orthogonal to γ_ξ. Furthermore, the subspaces

$$\mathcal{J}^-_{\xi;T} := \{\mathcal{D}_{\xi;T}\eta : \eta \in \xi^\perp\}, \qquad \mathcal{J}^+_{\xi;T} := \{\mathcal{D}_{\xi;-T}\eta : \eta \in \xi^\perp\}$$

determine respective $(n-1)$–dimensional subspaces[2] $X_{s;T}(\xi)$, $X_{u;T}(\xi)$ of $(\mathsf{S}M)_\xi$, whose direct sum is all of $\mathcal{G}(\xi)^\perp$.

The question is: what happens when $T \uparrow \infty$? Said differently, are the $(n-1)$–dimensional subspaces

$$\mathcal{J}^-_\xi := \{\mathcal{D}^-_\xi\eta : \eta \in \xi^\perp\}, \qquad \mathcal{J}^+_\xi := \{\mathcal{D}^+_\xi\eta : \eta \in \xi^\perp\}$$

of Jacobi fields in $\mathcal{J}^\perp_\xi$ transverse one to the other? The two extremes are illustrated by: $M = \mathbb{R}^n$, where the two spaces $\mathcal{J}^-_\xi$, $\mathcal{J}^+_\xi$ coincide; and by $M = \mathbb{H}^n$, where the two spaces are transverse, and hence the subspaces $X_s(\xi)$, $X_u(\xi)$ are transverse (exercise for the reader).

Exercise VII.7. Assume M has negative sectional curvature uniformly bounded away from 0. Show that $X_s(\xi)$, $X_u(\xi)$ are transverse for all $\xi \in \mathsf{S}M$.

Note VII.1. A theorem of P. Eberlein (1973) states that, when $X_s(\xi)$, $X_u(\xi)$ are transverse for all $\xi \in \mathsf{S}M$, M compactly homogeneous (e.g., M is the cover of a compact manifold), then the geodesic flow on $\mathsf{S}M$ is Anosov. (See his paper for definitions and proofs.) This then implies the earlier theorem of D. V. Anosov (1967) that a Riemannian manifold with sectional curvature bounded above and below by two negative constants is Anosov.

A Variant of E. Hopf's Theorem

The following theorem is proved in Green–Gulliver (1985).

Theorem. *Let g be a smooth Riemannian metric on $\mathbb{R}^2$ that differs from the canonical flat metric g_0 on at most a compact set. If $(\mathbb{R}^2, g)$ has no conjugate points, then it is isometric to $(\mathbb{R}^2, g_0)$.*

The Osserman–Sarnak Inequality

We sketch here the results of Osserman–Sarnak (1984) in the narrow sense, that is, only in those aspects pertinent to the Riemannian situation. See their discussion of how the inequalities that follow relate to the various entropies of the geodesic flow.

[2] "s" is for *stable* and "u" is for *unstable*.

We are given the compact n–dimensional Riemannian manifold M without conjugate points. Then, to every $\xi \in \mathsf{S}M$, we have the nonsingular stable matrix solution U_ξ of the Riccati equation defined on all of $\mathbb{R}$, as described in the second proof of Hopf's theorem. Recall that

$$U_{\Phi_s\xi}(t) = U_\xi(t+s).$$

So, we may think of $U(\xi) = U_\xi(0)$ as a matrix function on $\mathsf{S}M$. We let $U'(\xi)$ denote $(\nabla_t U_\xi)(0)$, where the covariant differentiation is along the geodesic $\gamma_\xi(t)$.

Exercise VII.8. Show that

$$\int_{\mathsf{S}M} \operatorname{tr} U'U^{-1}\, d\mu = 0.$$

Now assume that M has strictly negative curvature. With every $\xi \in \mathsf{S}M$, associate the quadratic form Q_ξ on $M_{\pi(\xi)}$ given by

$$Q_\xi(w) = \langle R(\xi, w)\xi, w\rangle,$$

with self-adjoint linear transformation K_ξ associated with $-Q_\xi$ and given by

$$K_\xi(w) = -R(\xi, w)\xi = -(\mathsf{R}_\xi(0))(w)$$

We denote the commmon spectra spec K_ξ = spec $- Q_\xi$ by

$$\{0 = \lambda_0(\xi) < \lambda_1(\xi) \leq \ldots \leq \lambda_{n-1}(\xi)\},$$

and consider

$$-\sigma(\xi) = \operatorname{tr} {K_\xi}^{1/2} = \sum_{j=1}^{n-1} \lambda_j(\xi)^{1/2}, \qquad \alpha(M) = \int_{\mathsf{S}M} \sigma(\xi)\, d\mu(\xi).$$

(We are loyal to the notations of Osserman–Sarnak (1984).)

Exercise VII.9. Prove

$$-\sigma(\xi) = \min \sum_{j=1}^{n-1} \langle K_\xi e_j, e_j\rangle^{1/2},$$

where the minimum is taken with respect to all orthonormal bases $\{e_1, \ldots, e_{n-1}\}$ of the orthogonal complement $\xi^\perp$ of ξ in $M_{\pi(\xi)}$. Show that one has equality for a choice of a particular orthonormal basis if and only if that orthonormal basis consists of eigenvectors of K_ξ.

Exercise VII.10. Show that

$$\int_{\mathsf{S}M} \operatorname{tr} U \, d\mu = \int_{\mathsf{S}M} \operatorname{tr} KU^{-1} \, d\mu.$$

Exercise VII.11.

(a) Use the Cauchy–Schwarz inequality on

$$\operatorname{tr} K^{1/2} = \operatorname{tr} (U^{-1/2}K^{1/2})U^{1/2}$$

to show

$$\operatorname{tr} K^{1/2} \leq \{\operatorname{tr} KU^{-1}\}^{1/2}\{\operatorname{tr} U\}^{1/2},$$

with equality if and only if there exists a positive function $k(\xi)$ on $\mathsf{S}M$ such that

$$K = k^2(\xi)U^2.$$

(b) Use the integral Cauchy–Schwarz inequality to show that

$$\int_{\mathsf{S}M} \operatorname{tr} K^{1/2} \, d\mu \leq \int_{\mathsf{S}M} \operatorname{tr} U \, d\mu,$$

with equality if and only if there exists a constant β such that

$$\operatorname{tr} KU^{-1} = \beta \operatorname{tr} U.$$

Exercise VII.12. (Osserman–Sarnak (1984)) Show that

$$-\alpha(M) \leq \int_{\mathsf{S}M} \operatorname{tr} U \, d\mu,$$

with equality if and only if K_ξ is parallel along the geodesic γ_ξ, for all ξ, that is, if and only if M is locally symmetric.

Aufwiedersehnsfläche

Exercise VII.13. (On the title, see Green (1963).) Give the following elementary proof (namely, L. W. Green's (1963)) of Blaschke's 2–dimensional conjecture.

(a) Consider properties (i)–(v) of wiedersehnsfläche as given. Now prove that, if M is simply connected, then the Gauss–Bonnet theorem (Theorem IV.V.2.3) (for the sphere) and Theorem VII.3.1 above imply that the area of M, $A(M)$, satisfies

$$A(M) \geq 4\pi,$$

with equality if and only if M has constant sectional curvature (in which case it is isometric to the standard sphere).

(b) Show that $c(\xi) = \operatorname{conj}(\xi)$ for all $\xi \in SM$, where $\operatorname{conj}(\xi)$ denotes distance to first conjugate point along γ_ξ.

(c) Use Santalo's formula (VII.4.1) to explicitly calculate the area of M.

An Eigenvalue Inequality

Exercise VII.14. We are given a relatively compact domain Ω with smooth boundary in an n–dimensional Riemannian manifold M, with $\tau(\xi) < +\infty$ for all $\xi \in S\Omega$. Let $\lambda = \lambda(\Omega)$ denote the lowest eigenvalue of the Dirichlet eigenvalue problem of Ω. For every $\xi \in S\Omega$ set

$$\delta(\xi) = \tau(\xi) + \tau(-\xi),$$

that is, $\delta(\xi)$ is the full length of the geodesic segment in Ω determined by ξ. Prove

$$\lambda \geq \pi^2 n \inf_{x\in\Omega} \frac{1}{\mathbf{c_{n-1}}} \int_{S_x} \frac{d\mu_x(\xi)}{\delta^2(\xi)}.$$

Note VII.2. The result of the above exercise is from Croke–Derdziński (1987), in which they also give a characterization of the case of equality (a nontrivial matter). Another estimate, for domains in $\mathbb{R}^n$, similar to the Croke–Derdziński result, is presented in Davies (1984), with an inequality of Hardy in place of the fixed-endpoint Wirtinger inequality. See his discussion (with some added detail) in Davies (1989, pp. 25–33, 56–57).

An interesting feature of both the Croke–Derdziński and Davies results is that if the domain Ω is perturbed by deleting a small disk then the lower bound is not grossly disturbed. That the Dirichlet eigenvalues are not grossly disturbed by "small" perturbations of the domain is a highly developed subject. See Chavel–Feldman (1978b, 1988) For further references to the earlier literature see Chavel–Feldman (1988). We also refer the reader to Courtois (1987, 1995).

VIII
Isoperimetric Inequalities (Variable Curvature)

We now return to isoperimetric inequalities in general complete Riemannian manifolds. Here, even if the manifold satisfies the conditions of Theorem V.4.1, to guarantee the existence of an isoperimetric region, one would have no idea how to identify it and how to decide whether it is unique. So, the focus shifts elsewhere to describe the isoperimetric profile, $\mathcal{I}(v)$, usually by providing lower bounds such as $\mathcal{I}(v) \geq \text{const.}v^{1-1/n}$ – in qualitative imitation of $\mathbb{R}^n$ (when the manifold has dimension n).

A better sense of the possibilities is given the isoperimetric inequality for a surface of constant curvature κ, namely,

$$L^2 \geq 4\pi A - \kappa A^2. \tag{VIII.0.1}$$

Note that when $\kappa < 0$, and the domain is a geodesic disk of radius r for large r, then the dominant term on the right-hand side of (VIII.0.1) is $-\kappa A^2$. More precisely, the inequality (VIII.0.1) implies *both* the inequalities

$$L/A^{1/2} \geq \sqrt{4\pi}, \qquad L/A \geq \sqrt{-\kappa},$$

and both are sharp. The first is sharp for geodesic disks of radius r as $r \downarrow 0$, and the second is sharp for geodesic disks of radius r as $r \uparrow +\infty$. This suggests that, for a more profound understanding of the relation of areas to volume in general Riemannian manifolds, one cannot remain limited to minimizing the quotient $L/A^{1/2}$ or, more generally, the quotient $A/V^{1-1/n}$ over the domains in question. Rather, it is important to consider a fuller apparatus of isoperimetric constants.

We will proceed as follows: First we give C. Croke's (1980) lower bound (Theorem VIII.1.1) for $A(\partial\Omega)/V(\Omega)^{1-1/n}$ valid in any n–dimensional Riemannian manifold, using Santalo's formula from the previous chapter. Then, we prove P. Buser's (1982) lower bound (Theorem VIII.2.1) for

$$A(\Gamma)/\min\{V(D_1), V(D_2)\}$$

complete n–dimensional Riemannian manifold, with

$$\mathrm{Ric} \geq (n-1)\kappa, \qquad \kappa \leq 0$$

on all of M, where Γ is a smooth hypersurface in $B(x;r)$ ($x \in M$, $r > 0$) dividing $B(x;r)$ into domains D_1 and D_2. Then, in §VIII.3, we introduce the full apparatus of isoperimetric constants, and study, in the following sections, the relation of isoperimetric constants of the manifold with those of any discretization of the manifold.

Before proceeding, a comment on notation. In what follows, for functions f and vector fields X on a Riemannian manifold, we denote their respective L^p–norms by

$$\|f\|_p = \left\{\int |f|^p \, dV\right\}^{1/p}, \qquad \|X\|_p = \left\{\int |X|^p \, dV\right\}^{1/p},$$

respectively.

§VIII.1. Croke's Isoperimetric Inequality

In what follows, we require the following version of the Berger–Kazdan inequality (see §VII.2): Given $\xi \in \mathsf{S}M$, then for all $\ell \in (0, c(\xi)]$, we have

$$\text{(VIII.1.1)} \qquad \int_0^\ell ds \int_0^{\ell - s} \sqrt{\mathbf{g}}(r; \Phi_s\xi)\, dr \geq \frac{\pi \mathbf{c_n}}{2\mathbf{c_{n-1}}}(\ell/\pi)^{n+1},$$

with equality if and only if

$$\mathcal{A}(t;\xi) = (\sin t\pi/\ell)I$$

for all $t \in [0, \ell]$.

This follows from (VII.2.5) with $m = 1$ in (VII.2.4), and (VII.2.15).

Theorem VIII.1.1. (C. Croke (1980)) *Let Ω be a relatively compact domain in the n–dimensional Riemannian manifold M, with $\partial\Omega \in C^\infty$. Let ω be the minimum visibility angle as defined in* §VII.4. *Then,*

$$\text{(VIII.1.2)} \qquad \frac{A(\partial\Omega)}{V(\Omega)^{1-1/n}} \geq \frac{\mathbf{c_{n-1}}}{\{\mathbf{c_n}/2\}^{1-1/n}}\,\omega^{1+1/n},$$

with equality if and only if Ω is a hemisphere of a constant sectional curvature sphere.

Proof. For every $x \in \Omega$, we have

$$V(\Omega) \geq \int_{\mathsf{S}_x} d\mu_x(\xi) \int_0^{\ell(\xi)} \sqrt{\mathbf{g}}(t;\xi)\,dt$$

(where $\ell(\xi)$ is given in §VII.4), which implies

$$\begin{aligned}
V(\Omega)^2 &\geq \int_\Omega dV(x) \int_{\mathsf{S}_x} d\mu_x(\xi) \int_0^{\ell(\xi)} \sqrt{\mathbf{g}}(t;\xi)\,dt \\
&= \int_{\mathsf{S}\Omega} d\mu(\xi) \int_0^{\ell(\xi)} \sqrt{\mathbf{g}}(t;\xi)\,dt \\
&\geq \int_{-\mathsf{U}\Omega} d\mu(\xi) \int_0^{\ell(\xi)} \sqrt{\mathbf{g}}(t;\xi)\,dt \\
&= \int_{\mathsf{S}^+\partial\Omega} \langle \xi, \overline{\nu}_{\pi(\xi)} \rangle \, d\sigma(\xi) \int_0^{\ell(\xi)} ds \int_0^{\ell(\Phi_s \xi)} \sqrt{\mathbf{g}}(t;\Phi_s\xi)\,dt,
\end{aligned}$$

by Santalo's formula (VII.4.1) applied to the function

$$F(\xi) = \int_0^{\ell(\xi)} \sqrt{\mathbf{g}}(t;\xi)\,dt.$$

Since

$$\ell(\Phi_s\xi) \geq \ell(\xi) - s,$$

we have

$$\begin{aligned}
V(\Omega)^2 &\geq \int_{\mathsf{S}^+\partial\Omega} \langle \xi, \overline{\nu}_{\pi(\xi)} \rangle \, d\sigma(\xi) \int_0^{\ell(\xi)} ds \int_0^{\ell(\xi)-s} \sqrt{\mathbf{g}}(t;\Phi_s\xi)\,dt \\
\text{(VIII.1.3)} \qquad &\geq \frac{\mathbf{c_n}}{2\pi^n \mathbf{c_{n-1}}} \int_{\mathsf{S}^+\partial\Omega} \ell(\xi)^{n+1} \langle \xi, \overline{\nu}_{\pi(\xi)} \rangle \, d\sigma(\xi) \\
&\geq \frac{\mathbf{c_n}}{2\pi^n \mathbf{c_{n-1}}} \left\{ \int_{\mathsf{S}^+\partial\Omega} \ell(\xi) \langle \xi, \overline{\nu}_{\pi(\xi)} \rangle \, d\sigma(\xi) \right\}^{n+1} \\
\text{(VIII.1.4)} \qquad &\quad \times \left\{ \int_{\mathsf{S}^+\partial\Omega} \langle \xi, \overline{\nu}_{\pi(\xi)} \rangle \, d\sigma(\xi) \right\}^{-n} \\
\text{(VIII.1.5)} \qquad &= \frac{\mathbf{c_n}}{2\pi^n \mathbf{c_{n-1}}} \left\{ \frac{n-1}{\mathbf{c_{n-2}}} \right\}^n \frac{\mu(\mathsf{U}\Omega)^{n+1}}{A(\partial\Omega)^n} \\
&\geq \frac{\omega^{n+1}\mathbf{c_n}}{2} \left\{ \frac{(n-1)\mathbf{c_{n-1}}}{\pi \mathbf{c_{n-2}}} \right\}^n \frac{V(\Omega)^{n+1}}{A(\partial\Omega)^n},
\end{aligned}$$

the inequality (VIII.1.3) is the Berger–Kazdan inequality; the inequality (VIII.1.4) is Hölder's inequality, and the equality (VIII.1.5) is (VII.4.4).

Note that if Ω is a hemisphere in a constant sectional curvature sphere, then $\omega = 1$ and we have equality in every step of the argument. In particular,

$$\frac{\mathbf{c_n}}{2}\left\{\frac{(n-1)\mathbf{c_{n-1}}}{\pi\,\mathbf{c_{n-2}}}\right\}^n = \frac{\mathbf{c_{n-1}}^n}{\{\mathbf{c_n}/2\}^{n-1}}.$$

Thus, the inequality (VIII.1.2) is valid, with equality if Ω is a hemisphere in a constant sectional curvature sphere.

It remains to assume equality in (VIII.1.2) and show that Ω is a hemisphere in a constant sectional curvature sphere.

From the continuity of $c(\xi)$ and lower semicontinuity of $\tau(\xi)$, we have given $\xi_0 \in \mathsf{S}\Omega \cup \mathsf{S}^+\partial\Omega$ for which $c(\xi_0) < \tau(\xi_0)$, the existence of a neighborhood G of ξ_0 in $\mathsf{S}\Omega \cup \mathsf{S}^+\partial\Omega$ on which $c < \tau$.

Now, assume equality in (VIII.1.2). Then, $\mathsf{U}\Omega = \mathsf{S}\Omega$, that is $c = \tau$ on all of $\mathsf{S}\Omega \cup \mathsf{S}^+\partial\Omega$. Equality in (VIII.1.4), the Hölder inequality, implies $\ell(\xi)$ is constant, say, equal to ℓ, on all of $\mathsf{S}^+\partial\Omega$. Thus,

$$\{\gamma_\xi(t) : \xi \in \mathsf{S}^+\partial\Omega,\ t \in [0, \ell]\}$$

covers all of Ω; and equality in (VIII.1.3), the Berger–Kazdan inequality, implies that Ω has constant sectional curvature equal to $(\pi/\ell)^2$. Then,

$$c(\xi) = \tau(\xi) = \ell$$

for all $\xi \in \mathsf{S}^+\partial\Omega$ implies that Ω is a hemisphere. ∎

Corollary VIII.1.1. *Given any $o \in M$, $\rho > 0$, such that* $\exp_o$ *is defined on* $\overline{B(o;\rho)}$, *then for*

$$r < \frac{1}{2}\min\left\{\inf_{x\in B(o;\rho)} \operatorname{inj} x,\ \rho\right\}$$

we have

$$\omega(o;r) := \omega(B(o;r)) = 1,$$

which implies

$$V(o;r) \geq \text{const.}_n r^n.$$

§VIII.2. Buser's Isoperimetric Inequality

Theorem VIII.2.1. (P. Buser (1982)) *Let M be a complete n–dimensional Riemannian manifold, with*

(VIII.2.1) $$\operatorname{Ric} \geq (n-1)\kappa, \qquad \kappa \leq 0$$

on all of M. Then there exists a positive constant depending on n, κ, and r, $c(n,\kappa,r)$, such that for any given $x \in M$, $r > 0$, a dividing smooth hypersurface Γ in $B(x;r)$ with $\overline{\Gamma}$ imbedded in $\overline{B(x;r)}$, and

$$B(x;r) \setminus \Gamma = D_1 \cup D_2,$$

where D_1, D_2 are open in $B(x;r)$, we have

$$\min\{V(D_1), V(D_2)\} \leq c(n,\kappa,r)A(\Gamma). \tag{VIII.2.2}$$

Proof. We shall actually prove a more general result, namely, we shall consider a domain D, in a Riemannian manifold M satisfying (VIII.2.1), for which there exist $0 < r \leq R$ and $o \in M$ such that

(a) $\exp_o$ is defined on $\overline{\mathsf{B}(o;R)}$,

(b) $B(o;r) \subseteq D \subseteq B(o;R)$,

(c) D is starlike with respect to o, that is, for any $q \in D$, any minimizing geodesic joining o to q is contained in D.

Under these conditions, we shall prove that for any $t \in (0, r/2)$, we have

$$\frac{A(\Gamma)}{\min\{V(D_1), V(D_2)\}} \geq c(n,\kappa,t,r,R), \tag{VIII.2.3}$$

where Γ ranges over smooth hypersurfaces in D for which $\overline{\Gamma}$ is imbedded in $\overline{D}$, and

$$D \setminus \Gamma = D_1 \cup D_2,$$

where D_1, D_2 are open in D. So, one obtains the best lower bound by maximizing the right-hand side of (VIII.2.3) with respect to $t \in (0, r/2)$.

Before starting the proof, we recall some notation and some comparison estimates (from Chapter III): For o in M, S_o denotes the unit tangent sphere at o, with $(n-1)$–measure $d\mu_o$. For $\xi \in \mathsf{S}_o$, $c(\xi)$ denotes the distance along the geodesic γ_ξ from o to its cut point along γ_ξ. Also recall that $C(o)$ denotes the cut locus of o.

For $\xi \in \mathsf{S}_o$, $s \in (0, c(\xi))$, we have for the n–measure on M

$$dV(\exp s\xi) = \sqrt{\mathbf{g}}(s;\xi)\, ds\, d\mu_o(\xi),$$

and for $(n-1)$–measure on the smooth points of the metric sphere $S(o;s)$

$$dA_{S(o;s)}(\exp s\xi) = \sqrt{\mathbf{g}}(s;\xi)\, d\mu_o(\xi).$$

When $M = \mathbb{M}_\kappa$, the simply connected space form of constant curvature κ, the area of the sphere $S(o;s)$, and the volume of the disk $B(o;s)$ are given respectively by

$$A_\kappa(s) = \mathbf{c_{n-1}} \mathbf{S}_\kappa{}^{n-1}(s), \qquad V_\kappa(s) = \mathbf{c_{n-1}} \int_0^s \mathbf{S}_\kappa{}^{n-1}(t)\,dt.$$

Then, Theorem III.4.3 implies

$$\text{(VIII.2.4)} \qquad \frac{d}{ds}\left\{\frac{\sqrt{\mathbf{g}}(s;\xi)}{A_\kappa(s)}\right\} \le 0,$$

from which one easily derives

$$\text{(VIII.2.5)} \qquad \frac{\sqrt{\mathbf{g}}(s_1;\xi)}{A_\kappa(s_1)} \ge \frac{\sqrt{\mathbf{g}}(s_2;\xi)}{A_\kappa(s_2)}$$

for $0 < s_1 \le s_2$,

$$\text{(VIII.2.6)} \qquad \frac{\sqrt{\mathbf{g}}(r;\xi)}{A_\kappa(r)} \ge \frac{1}{V_\kappa(R) - V_\kappa(r)} \int_r^R \sqrt{\mathbf{g}}(s;\xi)\,ds$$

for $0 \le r < R$,

$$\frac{1}{V_\kappa(r_1) - V_\kappa(r_0)} \int_{r_0}^{r_1} \sqrt{\mathbf{g}}(s;\xi)\,ds \ge \frac{1}{V_\kappa(r_2) - V_\kappa(r_1)} \int_{r_1}^{r_2} \sqrt{\mathbf{g}}(s;\xi)\,ds$$

(VIII.2.7)

for $0 \le r_0 < r_1 \le r_2$, and

$$\text{(VIII.2.8)} \qquad \frac{1}{V_\kappa(r)} \int_0^r \sqrt{\mathbf{g}}(s;\xi)\,ds \ge \frac{1}{V_\kappa(R)} \int_0^R \sqrt{\mathbf{g}}(s;\xi)\,ds$$

for $0 < r < R$.

Basic idea of the Proof.[1] First note that we may substitute

$$\text{(VIII.2.9)} \qquad A(\Gamma) \ge c(n,\kappa,t,r,R)V(D_1)$$

for (VIII.2.3), independent of whether D_1 has the minimum volume of $V(D_1)$, $V(D_2)$. Next, note that if $\Gamma = S(o;r/2)$, then pick $D_1 = D \setminus B(o;r/2)$. Then, the claim of the theorem easily follows from (VIII.2.6). Then, one adjusts the above argument to the general situation for Γ, where $D \setminus B(o;r/2)$ contains a "significant" portion of D_1 (CASE 1 in Steps 2–3). When $D \setminus B(o;r/2)$ does not contain a "significant" portion, then one gives the argument relative to a new "origin" (replacing o) in D (CASE 2 in Steps 4–6).

Step 1 of the Proof. Fix $t \in (0, r/2)$. We set, for any $s > 0$,

$$B_s = B(o;s), \qquad V_s = V(o;s).$$

[1] My thanks to P. Buser for helpful discussions of his theorem.

Given Γ, determine the two disjoint nonempty open subsets D_1, D_2 whose union is $D \setminus \Gamma$, and assume that D_1 satisfies

$$V(D_1 \cap B_{r/2}) \le \frac{1}{2} V_{r/2}. \tag{VIII.2.10}$$

(Certainly, either D_1 or D_2 must satisfy the inequality.) Then, it suffices to verify (VIII.2.9) for this particular choice of D_1.

Now, fix any $\alpha \in (0, 1)$ and consider two CASES: the FIRST:

$$V(D_1 \cap B_{r/2}) \le \alpha V(D_1),$$

and the SECOND:

$$V(D_1 \cap B_{r/2}) \ge \alpha V(D_1).$$

Step 2. CASE 1. To each $p \in D_1 \setminus C(o)$, we determine p^* to be the first point on the directed geodesic segment po, from p "back" to o, where this ray intersects Γ. If the geodesic segment po from p to o is completely contained in D_1, then we set $p^* = o$.

To each $p \in D_1 \setminus \{C(o) \cup B_{r/2}\}$ – that is,

$$p = \exp_o s\xi, \qquad r/2 \le s < c(\xi), \quad |\xi| = 1,$$

determine the geodesic segment

$$\text{rod}\, p := \{\exp_o \tau\xi : t \le \tau \le s\},$$

and the subsets $\mathcal{A}_1$, $\mathcal{A}_2$, $\mathcal{A}_3$ of D_1 by:

$$\begin{aligned}
\mathcal{A}_1 &:= \{p \in D_1 \setminus \{C(o) \cup \overline{B_{r/2}}\} : \; p^* \notin \overline{B_t}\},\\
\mathcal{A}_2 &:= \{p \in D_1 \setminus \{C(o) \cup \overline{B_{r/2}}\} : \; p^* \in \overline{B_t}\},\\
\mathcal{A}_3 &:= \{B_{r/2} \setminus B_t\} \cap \bigcup_{p \in \mathcal{A}_2} \text{rod}\, p.
\end{aligned}$$

(See Figure VIII.1.[2]) Then, (VIII.2.7) implies

$$\frac{V(\mathcal{A}_2)}{V(\mathcal{A}_3)} \le \frac{V_\kappa(R) - V_\kappa(r/2)}{V_\kappa(r/2) - V_\kappa(t)} := \gamma. \tag{VIII.2.11}$$

Also, since we are in CASE 1, we have $(1-\alpha)V(D_1) \le V(D_1 \setminus B_{r/2}) = V(\mathcal{A}_1) + V(\mathcal{A}_2)$. But, $V(\mathcal{A}_3) \le V(D_1 \cap B_{r/2}) \le \alpha V(D_1)$, which implies by (VIII.2.11)

$$(1-\alpha)V(D_1) \le V(\mathcal{A}_1) + \gamma\alpha V(D_1),$$

[2] My thanks to P. Buser and Gauthier–Villars for their permission to reprint Figure VIII.1 from Buser (1982, p. 224).

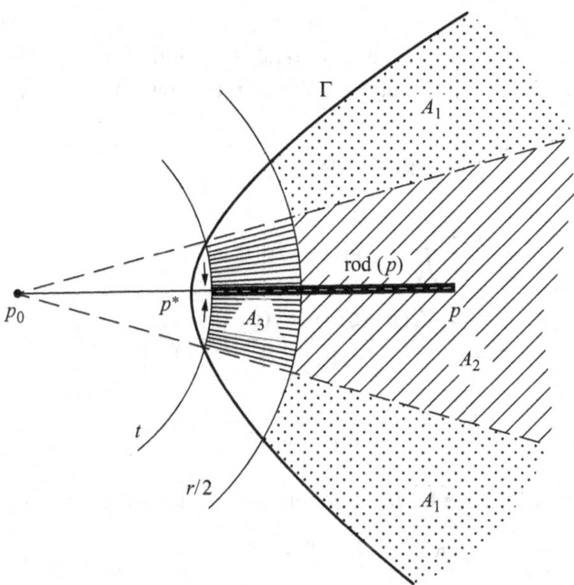

Figure VIII.1. For Buser's inequality.

that is,

$$\{1 - \alpha(1+\gamma)\}V(D_1) \leq V(\mathcal{A}_1) \tag{VIII.2.12}$$

which implies

$$\frac{A(\Gamma)}{V(D_1)} = \frac{A(\Gamma)}{V(\mathcal{A}_1)}\frac{V(\mathcal{A}_1)}{V(D_1)} \geq \frac{A(\Gamma)}{V(\mathcal{A}_1)}\{1 - \alpha(1+\gamma)\}. \tag{VIII.2.13}$$

Step 3. Now project $\nu : B_R \setminus \{C(o) \cup \{o\}\} \to \mathsf{S}_o$ by

$$\nu(\exp s\xi) = \xi.$$

Then,

$$\begin{aligned} V(\mathcal{A}_1) &= \int_{\mathsf{S}_o} d\mu_o(\xi) \int_0^{\min\{R,c(\xi)\}} I_{\mathcal{A}_1}(\exp s\xi)\sqrt{\mathbf{g}}(s;\xi)\,ds \\ &= \int_{\nu(\mathcal{A}_1)} d\mu_o(\xi) \int_{r/2}^{\min\{R,c(\xi)\}} I_{\mathcal{A}_1}(\exp s\xi)\sqrt{\mathbf{g}}(s;\xi)\,ds. \end{aligned}$$

Now, for $\xi \in \nu(\mathcal{A}_1)$, we have

$$\int_{r/2}^{\min\{R,c(\xi)\}} I_{\mathcal{A}_1}(\exp s\xi)\sqrt{\mathbf{g}}(s;\xi)\,ds = \sum_{j_\xi} \int_{\alpha'_{j_\xi}}^{\beta_{j_\xi}} \sqrt{\mathbf{g}}(s;\xi)\,ds.$$

If $\alpha'_{j_\xi} > r/2$, then set $\alpha_{j_\xi} = \alpha'_{j_\xi}$; if $\alpha'_{j_\xi} = r/2$, then set $\alpha_{j_\xi} = |\exp_o{}^{-1}(\exp_o (r/2)\xi)^*|$. Then, for all j_ξ, we have $\exp \alpha_{j_\xi}\xi \in \Gamma$, and $\nu(\mathcal{A}_1) = \nu(\Gamma_0)$ for $\Gamma_0 = \{\exp \alpha_{j_\xi}\xi : j_\xi = 1, \ldots; \xi \in \nu(\mathcal{A}_1)\}$. We therefore have for $\xi \in \nu(\mathcal{A}_1)$

$$\begin{aligned}
\sum_{j_\xi} \int_{\alpha'_{j_\xi}}^{\beta_{j_\xi}} \sqrt{\mathbf{g}}(s;\xi)\,ds
&\leq \sum_{j_\xi} \int_{\alpha_{j_\xi}}^{\beta_{j_\xi}} \sqrt{\mathbf{g}}(s;\xi)\,ds \\
&\leq \sum_{j_\xi} \frac{V_\kappa(\beta_{j_\xi}(\xi)) - V_\kappa(\alpha_{j_\xi}(\xi))}{A_\kappa(\alpha_{j_\xi}(\xi))} \sqrt{\mathbf{g}}(\alpha_{j_\xi}(\xi);\xi) \\
&\leq \sum_{j_\xi} \frac{V_\kappa(R) - V_\kappa(t)}{A_\kappa(t)} \sqrt{\mathbf{g}}(\alpha_{j_\xi}(\xi);\xi),
\end{aligned}$$

which implies, where we let $R_{\nu|\Gamma_0}$ denote the regular values of $\nu|\Gamma_0$,

$$\begin{aligned}
V(\mathcal{A}_1) &\leq \frac{V_\kappa(R) - V_\kappa(t)}{A_\kappa(t)} \int_{\nu(\Gamma_0)} \sum_{j_\xi} \sqrt{\mathbf{g}}(\alpha_{j_\xi}(\xi);\xi)\,d\mu_o(\xi) \\
&= \frac{V_\kappa(R) - V_\kappa(t)}{A_\kappa(t)} \int_{R_{\nu|\Gamma_0}} \sum_{j_\xi} \sqrt{\mathbf{g}}(\alpha_{j_\xi}(\xi);\xi)\,d\mu_o(\xi) \\
&\leq \frac{V_\kappa(R) - V_\kappa(t)}{A_\kappa(t)} A(\nu_{|\Gamma}{}^{-1}[R_{\nu|\Gamma_0}]) \\
&\leq \frac{V_\kappa(R) - V_\kappa(t)}{A_\kappa(t)} A(\Gamma),
\end{aligned}$$

which implies by (VIII.2.13)

$$\frac{A(\Gamma)}{V(D_1)} \geq \frac{A_\kappa(t)}{V_\kappa(R) - V_\kappa(t)}\{1 - \alpha(1+\gamma)\}, \tag{VIII.2.14}$$

where

$$\gamma = \frac{V_\kappa(R) - V_\kappa(r/2)}{V_\kappa(r/2) - V_\kappa(t)}.$$

Step 4. Before considering CASE 2, we first require the following:

Lemma VIII.2.1. *Set either*

$$W_0 = D_1 \cap B_{r/2}, \qquad W_1 = D_2 \cap B_{r/2},$$

or

$$W_0 = D_2 \cap B_{r/2}, \qquad W_1 = D_1 \cap B_{r/2}.$$

Then for at least one of two choices of the pair $\{W_0, W_1\}$, we have the existence of a point $w_0 \in W_0$, and a measurable subset $\mathcal{W}_1 \subseteq W_1$, such that

(i) $V(\mathcal{W}_1) \geq V(W_1)/2$,

(ii) *for each $q \in \mathcal{W}_1$ every minimizing directed geodesic segment $\overline{qw}_0$ from q to w_0 intersects Γ in a first point q^* such that*

$$d(q, q^*) \leq d(q^*, w_0).$$

Proof. Consider $W_1 \times W_0$ with the product measure. Since the cut locus of any point has n–measure equal to 0, we have, except for a possible nullset $N \subseteq W_1 \times W_0$, for each $(q, w) \in \{(W_1 \times W_0) \setminus N\}$ a *unique* minimizing geodesic $\overline{qw}$ from q to w.

Now $\overline{qw}$ is not necessarily contained in $B_{r/2}$, but since it is minimizing and has length less than r, it must be contained in B_r, that is,

$$\overline{qw} \subseteq B_r \subseteq D,$$

which implies $\overline{qw}$ must intersect Γ.

Let $V_0, V_1 \subseteq (W_1 \times W_0) \setminus N$ be given by

$$V_0 = \{(q, w) : d(q, q^\dagger) \leq d(q^\dagger, w)\},$$

and

$$V_1 = \{(q, w) : d(q, w^\dagger) \geq d(w^\dagger, w)\},$$

where $q^\dagger$ (respectively, $w^\dagger$) is the first intersection of $\overline{qw}$ (respectively, $\overline{wq}$) with Γ. Since $V_0 \cup V_1 = (W_1 \times W_0) \setminus N$, we either have

$$\mathrm{vol}_{2n}(V_0) \geq \frac{1}{2}\mathrm{vol}_{2n}(W_1 \times W_0), \qquad \text{or} \qquad \mathrm{vol}_{2n}(V_1) \geq \frac{1}{2}\mathrm{vol}_{2n}(W_1 \times W_0),$$

where vol_{2n} denotes $2n$–dimensional Riemannian measure on $W_1 \times W_0$. One now uses Fubini's theorem to obtain the claim. ■

Step 5. CASE 2. Here, we have $\alpha V(D_1) \leq V(D_1 \cap B_{r/2}) \leq (1/2)V(B_{r/2})$, which implies $\alpha V(D_1) \leq V(D_2 \cap B_{r/2})$. Therefore, however $\mathcal{W}_1$ is picked according to the conclusion of the lemma, we have

$$\alpha V(D_1) \leq 2V(\mathcal{W}_1),$$

so we bound $A(\Gamma)/V(\mathcal{W}_1)$ from below.

Center geodesic spherical coordinates at w_0, and project

$$\nu : B(o; r/2) \setminus C(w_0) \to \mathsf{S}_{w_0}$$

as described before for o. (We are keeping the same notation ν, for simplicity, even though the actual projection here is different.)

To each $\xi \in \nu(\mathcal{W}_1 \setminus C(w_0))$, we determine a collection of disjoint intervals $\{(\alpha_{\xi,j}, \beta_{\xi,j}) : j \in \mathcal{I}_\xi\}$, where $\mathcal{I}_\xi$ is a finite or countably infinite index set, as follows: For

$$q = \exp t_0\xi \in \mathcal{W}_1, \quad \text{pick} \quad \alpha_q = d(w_0, q^*),$$

where q^* is as described in the lemma, and

$$\beta_q = \sup\{t > \alpha_q : \gamma_\xi(t) \in \mathcal{W}_1 \text{ and } \gamma_\xi((\alpha_q, t)) \subseteq D \setminus \Gamma\}.$$

The collection of open intervals thus obtained is disjoint and, at most, countably infinite. Also,

$$\mathcal{W}_1 \cap \gamma_\xi((0,\infty)) \subseteq \bigcup_{j\in\mathcal{I}_\xi} \gamma_\xi((\alpha_{\xi,j}, \beta_{\xi,j})).$$

By the definition of q^*, we have

$$\beta_{\xi,j} \leq 2\alpha_{\xi,j} < r$$

for all $j \in \mathcal{I}_\xi$, $\xi \in \nu(\mathcal{W}_1 \setminus C(w_0))$.

Therefore,

$$\begin{aligned} V(\mathcal{W}_1) &\leq \int_{\nu(\mathcal{W}_1\setminus C(w_0))} d\mu_{w_0}(\xi) \sum_{j\in\mathcal{I}_\xi} \int_{\alpha_{\xi,j}}^{\beta_{\xi,j}} \sqrt{\mathbf{g}}(s;\xi)\, ds \\ &\leq \int_{\nu(\mathcal{W}_1\setminus C(w_0))} \sum_{j\in\mathcal{I}_\xi} \frac{V_\kappa(\beta_{\xi,j}) - V_\kappa(\alpha_{\xi,j})}{A_\kappa(\alpha_{\xi,j})} \sqrt{\mathbf{g}}(\alpha_{\xi,j};\xi)\, d\mu_{w_0}(\xi). \end{aligned}$$

Now,

$$\frac{V_\kappa(\beta_{\xi,j}) - V_\kappa(\alpha_{\xi,j})}{A_\kappa(\alpha_{\xi,j})} \leq \frac{V_\kappa(\beta_{\xi,j}) - V_\kappa(\beta_{\xi,j}/2)}{A_\kappa(\beta_{\xi,j}/2)} \leq \frac{V_\kappa(r) - V_\kappa(r/2)}{A_\kappa(r/2)},$$

which implies

$$V(\mathcal{W}_1) \leq \frac{V_\kappa(r) - V_\kappa(r/2)}{A_\kappa(r/2)} A(\Gamma).$$

Therefore,

$$\frac{A(\Gamma)}{V(D_1)} \geq \frac{\alpha}{2} \frac{A_\kappa(r/2)}{V_\kappa(r) - V_\kappa(r/2)}.$$

Step 6. In summary, we have

$$\frac{A(\Gamma)}{V(D_1)}$$

$$\geq \max_{t\in(0,r/2)} \max_{\alpha\in(0,1)} \left\{ \min\left[\frac{\alpha}{2} \frac{A_\kappa(r/2)}{V_\kappa(r) - V_\kappa(r/2)}, \frac{A_\kappa(t)}{V_\kappa(R) - V_\kappa(t)} \{1 - \alpha(1+\gamma)\}, \right] \right\},$$

where

$$\gamma = \frac{V_\kappa(R) - V_\kappa(r/2)}{V_\kappa(r/2) - V_\kappa(t)}.$$

This concludes the proof of (VIII.2.3). ∎

Remark VIII.2.1. Note that the argument of CASE 1 does not require the assumption (VIII.2.10).

§VIII.3. Isoperimetric Constants

M is our given n–dimensional Riemannian manifold. We let V denote n–dimensional Riemannian measure, and A denote $(n-1)$–dimensional Riemannian measure.

Definition. To each $\nu > 1$ and open submanifold Ω of M, with compact closure and smooth boundary, associate the *ν–isoperimetric quotient of* Ω, $\mathfrak{I}_\nu(\Omega)$, defined by

$$\mathfrak{I}_\nu(\Omega) = \frac{A(\partial\Omega)}{V(\Omega)^{1-1/\nu}}.$$

The *ν–isoperimetric constant of* M, $I_\nu(M)$, is defined as the infimum of $\mathfrak{I}_\nu(\Omega)$ over all Ω described above. For $\nu = \infty$, define *Cheeger's constant* $I_\infty(M)$ by

$$I_\infty(M) = \inf_\Omega \frac{A(\partial\Omega)}{V(\Omega)},$$

where Ω ranges over open submanifolds of M described above.

Remark VIII.3.1. If M is compact, then by considering $M \setminus B(x;\epsilon)$, for small $\epsilon > 0$, one can easily show that $I_\nu(M) = 0$ for all ν. A similar remark holds for the Sobolev constants considered below. We indicate the appropriate modification of the definitions in §VIII.5.

Lemma VIII.3.1. *For domains Ω_j, $j = 1, \ldots, N$, in M, and any $k \geq 1$, we have*

$$\sum_j V(\Omega_j)^{1/k} \geq \left\{\sum_j V(\Omega_j)\right\}^{1/k}.$$

Proof. The inequality is a simple application of Minkowski's inequality. Indeed, for I_j the indicator function of Ω_j, we have

$$\left\{\sum_j V(\Omega_j)^{1/k}\right\}^k = \left\{\sum_j \|I_j\|_k\right\}^k \geq \|\sum_j I_j\|_k{}^k = \cdots$$

$$\cdots = \int \left\{\sum_j I_j\right\}^k \geq \int \sum_j I_j{}^k = \sum_j \int I_j = \sum_j V(\Omega_j). \quad \blacksquare$$

Theorem VIII.3.1. (S. T. Yau (1975)) *In the definition of $I_\nu(M)$, $\nu \in (1, \infty]$, it suffices to let Ω range over open submanifolds of M which are connected.*

Proof. We only consider the case where ν is finite. The proof for $\nu = \infty$ is similar.

Let $\mathcal{I}_\nu(M)$ denote the infimum of $\Im_\nu(\Omega)$, where Ω ranges over domains (i.e., *connected* open submanifolds) in M with compact closure and smooth boundary. Then, obviously, $I_\nu(M) \leq \mathcal{I}_\nu(M)$. So, we wish to show the opposite inequality.

Let Ω be open with compact closure and smooth boundary. Then, we must show

$$A(\Omega) \geq \mathcal{I}_\nu(M) V(\Omega)^{1-1/\nu}. \tag{VIII.3.1}$$

To this end, we write

$$\partial\Omega = \bigcup_{j=1}^{k} \Gamma_j, \tag{VIII.3.2}$$

where $\Gamma_1, \ldots, \Gamma_k$ are compact connected $(n-1)$–submanifolds in M and verify (VIII.3.1) via induction on k.

If $k = 1$, then Ω is connected, and (VIII.3.1) is valid.

Assume (VIII.3.1) is valid for all open submanifolds with k boundary components, for all $k \leq k_0$, and suppose $\partial\Omega$ is given by (VIII.3.2) with $k = k_0 + 1$. If Ω is connected, then we are done. If not, we may assume that Ω may be written as the disjoint union of open sets Ω_1, Ω_2. We number the components of $\partial\Omega$ so that

$$\partial\Omega_1 = \Gamma_1 \cup \cdots \cup \Gamma_\ell, \qquad \partial\Omega_2 = \Gamma_{\ell+1} \cup \cdots \cup \Gamma_{k_0+1}.$$

Then,

$$\begin{aligned} A(\partial\Omega) &= A(\partial\Omega_1) + A(\partial\Omega_2) \\ &\geq \mathcal{I}_\nu(M)V(\Omega_1)^{1-1/\nu} + \mathcal{I}_\nu(M)V(\Omega_2)^{1-1/\nu} \\ &\geq \mathcal{I}_\nu(M)\{V(\Omega_1) + V(\Omega_2)\}^{1-1/\nu} \\ &= \mathcal{I}_\nu(M)V(\Omega)^{1-1/\nu} \end{aligned}$$

(the third line follows from Lemma VIII.3.1), which is the claim (VIII.3.1). ■

Definition. For each $\nu > 1$, define the *Sobolev constant of M*, $\mathcal{S}_\nu(M)$, by

$$\mathcal{S}_\nu(M) = \inf_f \frac{\|\text{grad}\, f\|_1}{\|f\|_{\nu/(\nu-1)}},$$

where f ranges over $C_c^\infty(M)$. Similarly, define $\mathcal{S}_\infty(M)$ by

$$\mathcal{S}_\infty(M) = \inf_f \frac{\|\text{grad}\, f\|_1}{\|f\|_1},$$

where f ranges over $C_c^\infty(M)$.

Theorem VIII.3.2. (H. Federer–W. H. Fleming (1960), Federer (1969)) *We have for all $\nu \in (1, \infty]$*

(VIII.3.3) $$I_\nu(M) = \mathcal{S}_\nu(M).$$

Proof. Again, we only consider the case of finite ν. The case of $\nu = \infty$ is referred to as *Cheeger's theorem* (Cheeger (1970b)).

Let Ω be any open submanifold of M with compact closure and smooth boundary. For sufficiently small $\epsilon > 0$, consider the function

$$f_\epsilon(x) = \begin{cases} 1 & x \in \Omega \\ (1/\epsilon)d(x, \partial\Omega) & x \in M \setminus \Omega,\ d(x, \partial\Omega) < \epsilon \\ 0 & x \in M \setminus \Omega,\ d(x, \partial\Omega) \geq \epsilon \end{cases}.$$

Then, f_ϵ is Lipschitz for every ϵ, and we may approximate (using regularization arguments – see Adams (1975, p. 29); also, see Chavel (2001, pp. 19–23)) f_ϵ by functions $\phi_{\epsilon,j} \in C_c^\infty(M)$ for which

$$\|\phi_{\epsilon,j} - f_\epsilon\|_{\nu/(\nu-1)} \to 0, \qquad \|\text{grad}\, \phi_{\epsilon,j} - \text{grad}\, f_\epsilon\|_1 \to 0,$$

as $j \to \infty$. One concludes

$$S_\nu(M) \le \frac{\|\operatorname{grad} f_\epsilon\|_1}{\|f_\epsilon\|_{\nu/(\nu-1)}}.$$

One easily sees that

$$\lim_{\epsilon \downarrow 0} \int_M |f_\epsilon|^{\nu/(\nu-1)}\, dV = V(\Omega).$$

Furthermore,

$$|\operatorname{grad} f_\epsilon| = \begin{cases} 1/\epsilon & x \in M \setminus \overline{\Omega},\ d(x, \partial\Omega) < \epsilon \\ 0 & \text{otherwise} \end{cases},$$

which implies

$$\lim_{\epsilon \downarrow 0} \int_M |\operatorname{grad} f_\epsilon|\, dV = \lim_{\epsilon \downarrow 0} \frac{V(\{x \notin \Omega : d(x, \partial\Omega) < \epsilon\})}{\epsilon} = A(\partial\Omega).$$

Thus,

$$S_\nu(M) \le \frac{A(\partial\Omega)}{V(\Omega)^{1-1/\nu}}$$

for all such Ω, from which we conclude

$$S_\nu(M) \le I_\nu(M).$$

It remains to prove the opposite inequality, that is,

$$\text{(VIII.3.4)} \qquad \int_M |\operatorname{grad} f|\, dV \ge I_\nu(M) \left\{ \int_M |f|^{\nu/(\nu-1)}\, dV \right\}^{1-1/\nu}$$

for all $f \in C_c^\infty(M)$.

To prove (VIII.3.4), we require the

Theorem VIII.3.3. (Coarea formula)[3] *Let Ω be a domain in M with compact closure and $f : \overline{\Omega} \to \mathbb{R}$ a function in $C^0(\overline{\Omega}) \cap C^\infty(\Omega)$, with $f \mid \partial\Omega = 0$. For any regular value t of $|f|$, we let*

$$\Gamma(t) = |f|^{-1}[t], \qquad A(t) = A(\Gamma(t)),$$

and dA_t denote the $(n-1)$–dimensional Riemannian measure on $\Gamma(t)$. Then,

$$\text{(VIII.3.5)} \qquad dV_{|\Gamma(t)} = \frac{dA_t\, dt}{|\operatorname{grad} f|},$$

[3] See Theorem III.5.2 and Exercise III.12.

and for any function $\phi \in L^1(\Omega)$, *we have*

$$\text{(VIII.3.6)} \qquad \int_\Omega \phi |\text{grad}\, f|\, dV = \int_0^\infty dt \int_{\Gamma(t)} \phi\, dA_t.$$

Proof. The critical values, in $\mathbb{R}$, of f have Lebesgue measure equal to 0, by Sard's theorem (Narasimhan (1968, p. 19ff)). The regular values of f, R_f, are open in $\mathbb{R}$, and for $t \in R_f$ the preimage $f^{-1}[t] \cap \Omega$ is an $(n-1)$–submanifold in M with $f^{-1}[t] \cap \Omega$ compact.

Let $(\alpha, \beta) \subseteq R_f$ and $\mu \in (\alpha, \beta)$. Then, one can construct a diffeomorphism

$$\Psi : f^{-1}[\mu] \times (\alpha, \beta) \to f^{-1}[(\alpha, \beta)]$$

for which

$$f(\Psi(q, t)) = t$$

for all $(q, t) \in f^{-1}[\mu] \times (\alpha, \beta)$. One does this as follows: On $f^{-1}[(\alpha, \beta)]$, define the vector field

$$X := \frac{\text{grad}\, f}{|\text{grad}\, f|^2},$$

ψ_t the flow determined by X; and

$$\Psi(q, t) = \psi_{t-\mu}(q).$$

Then, Ψ is the desired map. Moreover,

$$|d\Psi/dt| = |\text{grad}\, f|^{-1},$$

and $d\Psi/dt$ is always orthogonal to the level surface $f^{-1}[t]$. One immediately has (VIII.3.5) and (VIII.3.6). ∎

Lemma VIII.3.2. *Given any* $\phi \in C_c^\infty(M)$, *then*

$$\text{(VIII.3.7)} \qquad |\text{grad}\, |\phi|\,| = |\text{grad}\, \phi|$$

almost everywhere on M.

Proof. On the open set $\{\phi > 0\}$ we have $|\phi| = \phi$, and on the open set $\{\phi < 0\}$ we have $|\phi| = -\phi$, and (VIII.3.7) is certainly valid. So, we only have to consider the validity of (VIII.3.7) on $\{\phi = 0\}$ – even here, we only must consider what happens on

$$N_\phi := \{\phi = 0\} \cap \{\text{grad}\, \phi \neq 0\}.$$

But, in this case, N_ϕ is an $(n-1)$–submanifold of M, and therefore has n–measure equal to 0. ∎

Proof of (VIII.3.4). Given $f \in C_c^\infty(M)$, let

$$\Omega(t) = \{x : |f|(x) > t\}, \qquad V(t) = V(\Omega(t))$$

for $t \in R_f$. Then, the coarea formula implies

$$\int_M |\mathrm{grad}\, f|\, dV = \int_0^\infty A(t)\, dt \geq I_\nu(M) \int_0^\infty V(t)^{1-1/\nu}\, dt,$$

and

$$\begin{aligned} \int_M |f|^{\nu/(\nu-1)}\, dV &= \int_M dV \int_0^{|f|} \frac{\nu}{\nu-1} t^{1/(\nu-1)}\, dt \\ &= \frac{\nu}{\nu-1} \int_0^\infty t^{1/(\nu-1)}\, dt \int_{\Omega(t)} dV \\ &= \frac{\nu}{\nu-1} \int_0^\infty t^{1/(\nu-1)} V(t)\, dt \end{aligned}$$

the second equality is the "layer cake representation" (Lieb–Loss (1996, p. 26). So, to prove (VIII.3.4), it suffices to show

(VIII.3.8) $$\int_0^\infty V(t)^{1-1/\nu}\, dt \geq \left\{ \frac{\nu}{\nu-1} \int_0^\infty t^{1/(\nu-1)} V(t)\, dt \right\}^{1-1/\nu}.$$

To establish (VIII.3.8), set

$$F(s) = \int_0^s V(t)^{1-1/\nu}\, dt, \qquad G(s) = \left\{ \frac{\nu}{\nu-1} \int_0^s t^{1/(\nu-1)} V(t)\, dt \right\}^{1-1/\nu}.$$

Note that

$$F(0) = G(0);$$

also, since $V(s)$ is a decreasing function of s, we have

$$\begin{aligned} G'(s) &= \frac{\nu-1}{\nu} \left[\frac{\nu}{\nu-1} \right]^{1-1/\nu} \left\{ \int_0^s t^{1/(\nu-1)} V(t)\, dt \right\}^{-1/\nu} s^{1/(\nu-1)} V(s) \\ &\leq \left[\frac{\nu}{\nu-1} \right]^{-1/\nu} \left\{ \int_0^s t^{1/(\nu-1)}\, dt \right\}^{-1/\nu} s^{1/(\nu-1)} V(s)^{1-1/\nu} \\ &= V(s)^{1-1/\nu} \\ &= F'(s). \end{aligned}$$

Then, (VIII.3.8) follows immediately. ∎

Proposition VIII.3.1. *Suppose, for a given $\nu \in (1, \infty)$, we have*

$$I_\nu(M) > 0.$$

Then, for the area and volume of metric spheres and disks, we have

$$A(x;r) \geq I_\nu(M)V(x;r)^{1-1/\nu} \tag{VIII.3.9}$$

for all $x \in M$ *and* $r > 0$.

Proof. The point, of course, is even when $r > \operatorname{inj} x$. The argument is similar to the argument in the first half of the Federer–Fleming theorem (Theorem VIII.3.2).

For $\epsilon > 0$, define the function $\rho_\epsilon : [0, \infty) \to [0, 1]$ by (i) $\rho_\epsilon(s) = 1$ when $s \in [0, r]$, (ii) $\rho_\epsilon(s) = (r + \epsilon - s)/\epsilon$ when $s \in [r, r + \epsilon]$, and (iii) $\rho_\epsilon(s) = 0$ when $s > r + \epsilon$. Also, define the function $f_\epsilon : M \to \mathbb{R}$ by

$$f_\epsilon(y) = \rho_\epsilon(d(x, y)).$$

Then,

$$\begin{aligned} V(x;r)^{1-1/\nu} &\leq \|f_\epsilon\|_{\nu/(\nu-1)} \leq I_\nu(M)^{-1}\|\operatorname{grad} f_\epsilon\|_1 \\ &\leq I_\nu(M)^{-1}\frac{V(x;r+\epsilon) - V(x;r)}{\epsilon}. \end{aligned}$$

Now, let $\epsilon \downarrow 0$. Then, we conclude, from Proposition III.3.2,

$$V(x;r)^{1-1/\nu} \leq I_\nu(M)^{-1}\mathfrak{A}(x;r) \leq I_\nu(M)^{-1}A(x;r), \tag{VIII.3.10}$$

which implies the claim. ∎

Proposition VIII.3.2. *Suppose, for a given $\nu \in (1, \infty)$, we have $I_\nu(M) > 0$. Then, for the volume of metric disks, we have for all $x \in M$*

$$V'(x;r) \geq I_\nu(M)V(x;r)^{1-1/\nu} \tag{VIII.3.11}$$

(where the prime denotes differentiation with respect to r) for almost all $r > 0$.

In particular,

$$\liminf_{r\uparrow\infty} V(x;r)r^{-\nu} > 0. \tag{VIII.3.12}$$

Proof. One simply uses (VIII.3.10) in conjunction with the fact that

$$A(x;r) = V'(x;r)$$

for almost all r (Propositions III.3.2 and III.5.1) to obtain (VIII.3.11). Now integrate (VIII.3.11) to obtain (VIII.3.12). ∎

A similar argument shows:

Proposition VIII.3.3. *Suppose $I_\infty(M) > 0$. Then, for the volume of metric disks, we have for all $x \in M$*

$$V'(x;r) \geq I_\infty(M)V(x;r) \tag{VIII.3.13}$$

for almost all $r > 0$.

In particular,

$$\liminf_{r\uparrow\infty} V(x;r)e^{-I_\infty(M)r} > 0. \tag{VIII.3.14}$$

Theorem VIII.3.4. (S. T. Yau (1975)) *If M is connected, complete, simply connected, with all sectional curvatures less than or equal to $\kappa < 0$, then*

$$I_\infty(M) \geq (n-1)\sqrt{-\kappa}. \tag{VIII.3.15}$$

The inequality (VIII.3.15) *is sharp in the sense that we have equality for M the n–dimensional hyperbolic space of constant sectional curvature κ.*

Proof. The argument is a simple application of the divergence theorem, as follows:

Consider a fixed point $o \in M$, and the distance function $r : M \to (0,\infty)$ given by

$$r(x) = d(o,x).$$

Then, $r \in C^\infty$ on $M \setminus \{o\}$, and

$$\Delta r = \frac{\partial_r \sqrt{\mathbf{g}}(r;\xi)}{\sqrt{\mathbf{g}}(r;\xi)} \geq (n-1)\frac{\mathbf{C}_\kappa}{\mathbf{S}_\kappa} \circ r,$$

(by the Bishop comparison theorem (Theorem III.4.1)) which implies, for any domain Ω in M with compact closure and smooth boundary,

$$\begin{aligned} A(\partial\Omega) &\geq \int_{\partial\Omega} \langle \operatorname{grad} r, \nu\rangle \, dA \\ &= \iint_\Omega \Delta r \, dV \\ &\geq (n-1)\iint_\Omega \frac{\mathbf{C}_\kappa}{\mathbf{S}_\kappa} \circ r \, dV \\ &\geq (n-1)\sqrt{-\kappa}\, V(\Omega), \end{aligned}$$

which implies the first claim.

The last claim simply follows from considering geodesic disks and their bounding spheres, namely, one easily verifies $A_\kappa(r)/V_\kappa(r) \to (n-1)\sqrt{-\kappa}$ as $r \uparrow \infty$. ■

Theorem VIII.3.5. *Let M be an open submanifold of a Riemannian manifold N, M possessing compact closure and smooth boundary. Then,*

$$I_\infty(M) > 0.$$

The result is an easy consequence of Santalo's formula – see Theorems VII.4.2 and VII.4.3.

Remark VIII.3.2. We note that the inequality $I_\nu(M) > 0$ is only possible for $n \le \nu \le \infty$. Indeed, let $\nu < n$, and consider small metric disks $B(x;\epsilon)$, with center $x \in M$ and radius $\epsilon > 0$. Then, for the isoperimetric quotient of $B(x;\epsilon)$, we have

$$\mathfrak{I}_\nu(B(x;\epsilon)) \sim \text{const.}\epsilon^{n-1-n(1-1/\nu)} = \text{const.}\epsilon^{n/\nu-1}$$

as $\epsilon \downarrow 0$; so $I_\nu(M) = 0$ whenever $\nu < n$. So, it seems at first glance that one only has a discussion of isoperimetric constants for $\nu \ge n = \dim M$. However, we deal with modified isoperimetric constants, for $\nu < n$.

Modified Isoperimetric Constants

We now deal with the fact that the inequality $I_\nu(M) > 0$ is only possible for $\nu \ge n$. As we described, it is a strictly local phenomenon. Nevertheless, one has a simple example to illustrate the necessity of a corresponding notion for $\nu \in [1, n)$. Consider the Riemannian product $M = M_0 \times \mathbb{R}^k$, where M_0 is an $(n-k)$–dimensional *compact* Riemannian manifold. Then $I_k(M) = 0$. Yet, for extremely large domains – for example, geodesic disks of large radius – one expects the volume of these domains and the area of their boundaries to reflect k–dimensional space.

On the other hand, in discretizations of M, as described in §IV.4, the local phenomena disappear and no local difficulties occur. In our example, pick one point $x_0 \in M_0$ and consider the integer lattice $\mathbb{Z}^k$ in $\mathbb{R}^k$. Then,

$$\mathcal{G} := \{x_0\} \times \mathbb{Z}^k$$

is a discretization of $M = M_0 \times \mathbb{R}^k$, and we expect, since $\mathbb{Z}^k$ is a discretization of $\mathbb{R}^k$, that isoperimetric data of M should somehow be similar, or "equivalent," to isoperimetric data on $\mathbb{R}^k$. That is, for general M, the isoperimetric data should depend only on the coarse macroscopic geometry of M — once we assume some version of local uniformity of M.

Definition. In the variational problems for modified isoperimetric constants with parameter ρ, to be defined below, we let Ω vary over open submanifolds of M with compact closure and smooth boundary, and with *inradius greater than* ρ, that is, Ω which contain a closed metric disk of radius ρ.

For $\nu = 1$, $\rho > 0$, the *modified 1–isoperimetric constant of* M, $I_{1,\rho}(M)$, is defined as the infimum of

$$\Im_1(\Omega) = A(\partial\Omega).$$

For $\nu > 1$, $\rho > 0$, the *modified ν–isoperimetric constant of* M, $I_{\nu,\rho}(M)$, is defined as the infimum of $\Im_\nu(\Omega) = A(\partial\Omega)/V(\Omega)^{1-1/\nu}$. For $\nu = \infty$, $\rho > 0$, we define the *modified Cheeger constant* $I_{\infty,\rho}(M)$ as the infimum of $\Im_\infty(\Omega) = A(\partial\Omega)/V(\Omega)$.

Recall from Chapter IV that the Riemannian manifold M has *bounded geometry* if the Ricci curvature of M is bounded uniformly from below, and if the injectivity radius of M is bounded uniformly away from 0 on all of M.

Theorem VIII.3.6. (Chavel–Feldman (1991)) *If M is Riemannian complete with bounded geometry, then $I_{1,\rho}(M) > 0$ for every $\rho > 0$.*

Proof. Set

$$\delta = \min\{\rho, \operatorname{inj} M\}.$$

Assume we are given Ω as above, containing $B(x;\rho)$ for some $x \in M$. One easily has the existence of $z \in M$ for which $\partial\Omega \cap B(z;\delta/2)$ divides $B(z;\delta/2)$ into two open subsets for which the smaller volume is greater than or equal to $V(z;\delta/2)/3$. But, then, Buser's isoperimetric inequality states that

$$A(\partial\Omega \cap B(z;\delta/2)) \geq \text{const.}\frac{V(z;\delta/2)}{3}.$$

The corollary above implies

$$V(z;\delta/2) \geq c_n\delta^n,$$

which bounds $A(\partial\Omega)$ away from 0. ■

For the modified isoperimetric constants, the Federer–Fleming theorem goes as follows: For each $\rho > 0$, let $C^\infty_{c,\rho}(M)$ consist of those compactly supported Lipschitz functions ϕ on M, for which (i) there exists an $x \in M$ such that the preimage of max $|\phi|$, Ω_ϕ, satisfies $\Omega_\phi \supseteq \overline{B(x;\rho)}$, and (ii) $\phi \mid M\setminus\Omega_\phi \in C^\infty$. For each $\nu > 1$ and $\rho > 0$, define the *modified Sobolev constant of* M, $S_{\nu,\rho}(M)$, by minimizing the functional

(VIII.3.16)
$$f \mapsto \frac{\|\text{grad } f\|_1}{\|f\|_{\nu/(\nu-1)}},$$

where f ranges over $C^\infty_{c,\rho}(M)$. Then, one also has

(VIII.3.17)
$$I_{\nu,\rho}(M) = S_{\nu,\rho}(M)$$

by the argument of the Federer–Fleming theorem.

Similarly, one has:

Proposition VIII.3.4. *If M is Riemannian complete, and $I_{\nu,\rho}(M) > 0$, for some $\nu \in [1, \infty)$, $\rho > 0$, then*

(VIII.3.18)
$$\liminf_{r\to+\infty} V(x;r)r^{-\nu} > 0$$

for all $x \in M$. If M is Riemannian complete, and $I_{\infty,\rho}(M) > 0$, for some $\rho > 0$, then

(VIII.3.19)
$$\liminf_{r\to+\infty} V(x;r)e^{-I_{\infty,\rho}(M)r} > 0$$

for all $x \in M$.

To restate the matter, if ν_0 is the supremum of ν for which (VIII.3.18) is valid, then $I_{\nu,\rho}(M) = 0$ for all $\nu > \nu_0$, $\rho > 0$. Also, one easily verifies:

Proposition VIII.3.5. *If $I_{\nu,\rho}(M) > 0$ for some $\nu \in [1, \infty)$, $\rho > 0$ then, for any $\epsilon > 0$, $V(x;\rho+\epsilon)$ is uniformly bounded from below for all x with $d(x, \partial M) > \rho + \epsilon$. Therefore, $I_{\mu,\rho+\epsilon}(M) > 0$ for all μ in $[1, \nu)$.*

Remark VIII.3.3. Note that the proof of the proposition breaks down when $\nu = \infty$ — unless we *postulate* the existence of $\rho_0 > 0$ for which

$$V(x;\rho_0) \geq \text{const.} > 0$$

for all x (the constant independent of x). Moreover, one has a counterexample.

Consider the Riemannian metric on $M = \mathbb{R} \times \mathbb{S}$ given by

$$ds^2 = dr^2 + e^{2r}\, d\theta^2.$$

Then

$$dV = e^r\, dr\, d\theta.$$

To calculate $I_\infty(M) = S_\infty(M)$, we consider $f : M \to (0, +\infty)$, and for each $\theta \in \mathbb{S}$, we set

$$F_\theta := \{r : f(r,\theta) \neq 0\} = \bigcup_j (\alpha_j(\theta), \beta_j(\theta)).$$

Then,

$$\|\text{grad } f\|_1 \geq \int_{\mathbb{S}} d\theta \sum_j \left| \int_{\alpha_j(\theta)}^{\beta_j(\theta)} (\partial_r f) e^r \, dr \right| = \|f\|_1,$$

the last equality obtained by integration by parts; so $I_{\infty,r_0}(M) \geq I_\infty(M) = S_\infty(M) \geq 1$. On the other hand, for any $r_0 > 0$ and $\alpha \in \mathbb{R}$ one has for, $\Omega = (\alpha, \alpha + r_0) \times \mathbb{S}$,

$$A(\partial\Omega) = 2\pi e^\alpha (e^{r_0} + 1), \qquad V(\Omega) = 2\pi e^\alpha (e^{r_0} - 1).$$

One easily shows

$$I_{\infty,\rho}(M) = I_\infty(M) = 1, \qquad I_{\nu,\rho}(M) = 0$$

for all $\nu \in [1, \infty)$, $\rho > 0$. By the way, the Gauss curvature of M is identically equal to -1.

Proposition VIII.3.6. *Let D be a relatively compact domain in the Riemannian complete M with smooth boundary Γ, D' an n–dimensional Riemannian manifold with compact closure and smooth boundary Γ such that M', given by*

$$M' = \{M \backslash D\} \cup D',$$

is smooth Riemannian. If $I_{\nu,\rho}(M) > 0$ for given $\nu \in [1, \infty)$ and $\rho > 0$, then there exist $\rho' > 0$ such that $I_{\nu,\rho'}(M') > 0$.

Proof. Suppose we are given any $\alpha \in (0, \infty)$. The value of α will be fixed throughout the argument, although its precise value will be determined as we go along.

Since $I_{\nu,\rho}(M) > 0$, there exists $R > \rho$ such that

$$V(x; R)^{1-1/\nu} \geq \alpha A(\Gamma)$$

for all $x \in M$. Let δ' denote the diameter of D'. We pick

$$\rho' = 2\rho + R + \delta'.$$

Let $E = M \setminus D$, and suppose we are given $\Omega' \subseteq M'$, with compact closure and smooth boundary.

We first assume that $\Omega' \supseteq B(y'; \rho')$, with $d(y', D') < R$. Then, $B(y'; \rho') \supseteq D'$, which implies $\partial\Omega' \subseteq E$. Set $\Omega = (\Omega' \setminus D') \cup D$. Then, $\partial\Omega = \partial\Omega'$, and Ω contains a disk in M with inradius ρ, which implies

$$A(\partial\Omega') \geq I_{\nu,\rho}(M)V(\Omega)^{1-1/\nu}.$$

We obtain

$$\begin{aligned} A(\partial\Omega') &\geq I_{\nu,\rho}(M)\{V(D) + V(\Omega' \cap E)\}^{1-1/\nu} \\ &\geq \min\{1, V(D)/V(D')\}^{1-1/\nu} I_{\nu,\rho}(M)\{V(D') + V(\Omega' \cap E)\}^{1-1/\nu} \\ &= \min\{1, V(D)/V(D')\}^{1-1/\nu} I_{\nu,\rho}(M)V(\Omega')^{1-1/\nu}. \end{aligned}$$

We now assume $d(y', D') \geq R$. Then, $B(y'; R) \subseteq E$, which implies

$$V(\Omega')^{1-1/\nu} \geq V(y'; R)^{1-1/\nu} \geq \alpha A(\Gamma). \tag{VIII.3.20}$$

Assume first that $V(\Omega' \cap E) \geq \frac{1}{2}V(\Omega')$. Then,

$$\begin{aligned} A(\partial\Omega') &\geq A(\partial\Omega' \cap E) \geq A(\partial(\Omega' \cap E)) - A(\Gamma) \\ &\geq I_{\nu,\rho}(M)V(\Omega' \cap E)^{1-1/\nu} - \frac{1}{\alpha}V(\Omega')^{1-1/\nu} \end{aligned}$$

since $R \geq \rho$. So,

$$A(\partial\Omega') \geq \left\{\frac{I_{\nu,\rho}(M)}{2^{1-1/\nu}} - \frac{1}{\alpha}\right\} V(\Omega')^{1-1/\nu}.$$

Therefore, we pick α at the very outset to also be greater than or equal to $I_{\nu,\rho}(M)/2^{2-1/\nu}$. We obtain

$$A(\partial\Omega') \geq I_{\nu,\rho}(M)/2^{2-1/\nu}V(\Omega')^{1-1/\nu}.$$

The final situation to consider is, therefore, $d(y', D') \geq R$ and $V(\Omega' \cap D') \geq \frac{1}{2}V(\Omega')$. We still have (VIII.3.20); then,

$$\begin{aligned} A(\partial\Omega') &\geq A(\partial\Omega' \cap D') \\ &\geq A(\partial(\Omega' \cap D')) - A(\Gamma) \\ &\geq I_\infty(D')V(\Omega' \cap D') - \frac{1}{\alpha}V(\Omega')^{1-1/\nu} \\ &\geq I_\infty(D')\frac{V(\Omega')}{2} - \frac{1}{\alpha}V(\Omega')^{1-1/\nu} \\ &\geq \left\{\frac{I_\infty(D')V(\Omega')^{1/\nu}}{2} - \frac{1}{\alpha}\right\} V(\Omega')^{1-1/\nu} \\ &\geq \left\{\frac{I_\infty(D')\{\alpha A(\Gamma)\}^{1/(\nu-1)}}{2} - \frac{1}{\alpha}\right\} V(\Omega')^{1-1/\nu}. \end{aligned}$$

In addition to the above, pick, at the outset, α sufficiently large so that

$$\frac{I_\infty(D')\{\alpha A(\Gamma)\}^{1/(\nu-1)}}{2} - \frac{1}{\alpha} \geq \frac{I_\infty(D')A(\Gamma)^{1/(\nu-1)}}{2}.$$

This will then imply the proposition. ■

Remark VIII.3.4. The proof is not valid for $\nu = \infty$, since it uses a *uniform* bound from below for $V(x;r)$. See the previous remark.

Remark VIII.3.5. In our example above, $M = M_0 \times \mathbb{R}^k$, where M_0 is compact, we will have (see below) $I_{\nu,\rho}(M) > 0$ if and only if $1 \leq \nu \leq k$. Our proof will use the discretizations of Riemannian manifolds (see the following section).

§VIII.4. Discretizations and Isoperimetry

We refer the reader to the discussions and notation of §IV.4, wherein we introduced the basic notion of discretization of Riemannian manifolds. Here, we continue the story, starting with the definition of boundaries of subgraphs. To this end, we denote the collection of oriented edges of the connected graph $\mathbf{G}$ by $\mathcal{G}_e$. The oriented edge from ξ to η will be denoted by $[\xi, \eta]$; and when we wish to consider the unoriented edge connecting ξ and η, we denote it by $[\xi \sim \eta]$.

Any finite subset $\mathcal{K}$ in $\mathcal{G}$ determines a finite subgraph $\mathbf{K}$ of $\mathbf{G}$, for which one can describe a variety of suitable definitions for its boundary. Our definition will be that the *boundary of* $\mathbf{K}$, $\partial\mathbf{K}$, will be the subset of $\mathcal{G}_e$ consisting of those oriented edges which connect points of $\mathcal{K}$ to the complement of $\mathcal{K}$ in $\mathcal{G}$.

We define the *area measure* $d\mathsf{A}$ *on* $\mathcal{G}_e$ to be the counting measure for the oriented edges. Thus, for any finite subset of vertices, the area of its boundary will be equal to the number of edges in the boundary.

Definition. For any $\nu \geq 1$ and any finite subgraph $\mathbf{K}$ in $\mathbf{G}$, the *isoperimetric quotient* of $\mathbf{K}$ is defined by

$$\mathfrak{I}_\nu(\mathbf{K}) = \frac{\mathsf{A}(\partial\mathbf{K})}{\mathsf{V}(\mathbf{K})^{1-1/\nu}};$$

and for any $\nu \geq 1$, the *isoperimetric constant* $\mathsf{I}_\nu(\mathbf{G})$ is defined the infimum of $\mathfrak{I}_\nu(\mathbf{K})$, where $\mathbf{K}$ varies over all finite subgraphs of $\mathbf{G}$.

Another definition of the boundary of a finite subgraph $\mathbf{K}$ of $\mathbf{G}$ is given by

$$\partial\mathbf{K} = \{\xi \in \mathcal{G} : \mathsf{d}(\xi, \mathbf{K}) = 1\}.$$

Thus, by this definition, $\partial\mathbf{K}$ is a subset of vertices in the complement of $\mathbf{K}$. Its area is defined to be its cardinality. When $\mathbf{G}$ has bounded geometry, the two choices of area functions

$$\mathbf{K} \mapsto \mathsf{A}(\partial\mathbf{K}),$$

as functions on the collection of subgraphs $\mathbf{K}$ of $\mathbf{G}$, are commensurate each with respect to the other in the sense that the quotient of the two functions are bounded uniformly away from 0 and ∞.

Therefore, when $\mathbf{G}$ has bounded geometry, we will work with the counting measure for the volume of $\mathbf{K}$, and the second definition of $\partial\mathbf{K}$ with counting measure for its area – despite the fact that the theorems are formulated with respect to the original notions of volume, boundary, and area.

Lemma VIII.4.1. *If* $\mathbf{G}$, $\mathbf{F}$ *are roughly isometric graphs, both with bounded geometry, then* $\mathsf{I}_\nu(\mathbf{G}) > 0$ *if and only if* $\mathsf{I}_\nu(\mathbf{F}) > 0$.

Proof. Given any finite $\mathcal{K} \subseteq \mathcal{G}$, we wish to find $\mathcal{J} \subseteq \mathcal{F}$ such that

$$\frac{\text{card}\,\partial\mathcal{J}}{(\text{card}\,\mathcal{J})^{1-1/\nu}} \leq \text{const.}\frac{\text{card}\,\partial\mathcal{K}}{(\text{card}\,\mathcal{K})^{1-1/\nu}},$$

where the constant is independent of $\mathcal{K}$ and $\mathcal{J}$. This will then imply

$$I_\nu(\mathcal{F}) \leq \text{const.}\frac{\text{card}\,\partial\mathcal{K}}{(\text{card}\,\mathcal{K})^{1-1/\nu}}$$

for all finite $\mathcal{K} \subseteq \mathcal{G}$, which implies

$$I_\nu(\mathcal{F}) \leq \text{const.}I_\nu(\mathcal{G}).$$

By switching the roles of $\mathcal{G}$ and $\mathcal{F}$, one obtains the theorem.

So, we are given $\mathcal{K} \subseteq \mathcal{G}$ and we wish to pick $\mathcal{J} \subseteq \mathcal{F}$. Let $\phi : \mathcal{G} \to \mathcal{F}$ be the rough isometry. Then there exists a smallest nonnegative integer k such that ϕ is $(k+1)$–full. Then pick

$$\mathcal{J} := \{\eta \in \mathcal{F} : \mathsf{d}(\eta, \phi(\mathcal{K})) \leq k\}.$$

By Proposition IV.4.2, there exists $\mu \geq 1$ such that

$$\text{card}\,\mathcal{J} \geq \text{card}\,\phi(\mathcal{K}) \geq \mu^{-1}\text{card}\,\mathcal{K}.$$

So, it remains to bound card $\partial\mathcal{J}$ from above, in terms of card $\partial\mathcal{K}$.

Given $\eta \in \partial\mathcal{J}$, then $\mathsf{d}(\eta, \mathcal{J}) = 1$ (we are working with the second definition of the boundary), which implies by the triangle inequality that $\mathsf{d}(\eta, \phi(\mathcal{K})) \leq k+1$ — so $\mathsf{d}(\eta, \phi(\mathcal{K})) = k+1$.

Also, because ϕ is $(k+1)$–full, there exists $\xi \in \mathcal{G}$ such that

$$\text{(VIII.4.1)} \qquad \mathsf{d}(\eta, \phi(\xi)) \leq k.$$

The triangle inequality then implies

$$1 \leq \mathsf{d}(\phi(\mathcal{K}), \phi(\xi)) \leq 2k + 1,$$

which implies (i) that $\phi(\xi) \notin \phi(\mathcal{K})$ — so $\xi \notin \mathcal{K}$. Also, there exist $a \geq 1$ and $b \geq 0$ such that

$$a^{-1}\mathsf{d}(\xi, \mathcal{K}) - b \leq \mathsf{d}(\phi(\xi), \phi(\mathcal{K})),$$

which implies (ii) $1 \leq \mathsf{d}(\xi, \mathcal{K}) \leq a\{2k + 1 + b\} := \sigma + 1$, which implies

$$\text{(VIII.4.2)} \qquad \mathsf{d}(\xi, \partial\mathcal{K}) \leq \sigma.$$

In summary, (VIII.4.1), (VIII.4.2) imply

$$\partial\mathcal{J} \subseteq \beta(\phi(\beta(\partial\mathcal{K}; \sigma)); k),$$

which implies, by Proposition IV.4.2,

$$\operatorname{card} \partial\mathcal{J} \leq 2\mathbf{m_F}^k \operatorname{card} \phi(\beta(\partial\mathcal{K}; \sigma)) \leq 2\mathbf{m_F}^k \operatorname{card} \beta(\partial\mathcal{K}; \sigma) \leq 4\mathbf{m_F}^k \mathbf{m_G}^\sigma \operatorname{card} \partial\mathcal{K},$$

which implies the claim. ∎

Theorem VIII.4.1. *Let M be a complete Riemannian manifold with bounded geometry. Then, for any $\nu \geq 1$, we have $I_{\nu,\rho}(M) > 0$ if and only if $\mathsf{I}_\nu(\mathbf{G}) > 0$, for any discretization $\mathbf{G}$ of M.*

Proof. First, given $I_{\nu,\rho}(M) > 0$.

By Corollary IV.4.1 and the previous lemma, we may work with *any* discretization. Therefore, we consider a discretization $\mathbf{G}$ of M with separation constant $\epsilon > 0$ and covering radius $R = \rho$. We wish to derive the existence of positive constants such that given any $\mathcal{K} \subseteq \mathcal{G}$, we may find $\Omega \subseteq M$ of inradius $\geq \rho$ for which

$$\text{(VIII.4.3)} \qquad A(\partial\Omega) \leq \text{const.} \operatorname{card} \partial\mathcal{K},$$

and

$$\text{(VIII.4.4)} \qquad V(\Omega) \geq \text{const.} \operatorname{card} \mathcal{K}.$$

We proceed as follows: Given a finite subset $\mathcal{K}$, set

$$\Omega := \bigcup_{\xi \in \mathcal{K}} B(\xi; R).$$

Then,

$$\sum_{\xi\in\mathcal{K}} V(\xi;R) \le \mathbf{M}_{\epsilon,R}V(\cup_{\xi\in\mathcal{K}}B(\xi;R)) = \mathbf{M}_{\epsilon,R}V(\Omega),$$

where $\mathbf{M}_{\epsilon,R}$ is an upper bound (depending on ϵ, R, and the lower bound of the Ricci curvature) of the maximum number of ϵ–separated points in a disk of radius R (see Remark IV.3). So, for

$$V_R := \inf_{x\in M} V(x;R) > 0$$

(the positivity of V_R follows from Croke's inequality (Corollary VIII.1.1)), we have

$$V_R \operatorname{card}\mathcal{K} \le \mathbf{M}_{\epsilon,R}V(\Omega),$$

which implies (VIII.4.4).

For the upper bound of $A(\partial\Omega)$, we note that

$$\partial\Omega \subseteq \bigcup_{\xi\in\partial(\mathcal{G}\setminus\mathcal{K})} S(\xi;R).$$

Indeed, if $x\in\partial\Omega$, then $\mathsf{d}(x,\xi)\ge R$ for all $\xi\in\mathcal{K}$, and there exists $\xi_0\in\mathcal{K}$ such that $x\in S(\xi_0;R)$. But there must exist $\xi'\in\mathcal{G}$ such that $\mathsf{d}(x,\xi')<R$, which implies $\xi'\notin\mathcal{K}$. Then, $\mathsf{d}(\xi_0,\xi')<2R$, which implies $\xi_0\in\mathsf{N}(\xi')$. So, $\xi_0\in\partial(\mathcal{G}\setminus\mathcal{K})$, which is the claim.

Therefore,

$$A(\partial\Omega) \le A_\kappa(R)\operatorname{card}\partial(\mathcal{G}\setminus\mathcal{K}) \le \mathbf{m}A_\kappa(R)\operatorname{card}\partial\mathcal{K},$$

which implies (VIII.4.3). So, we have the "only if" claim of the theorem.

For the "if" claim, we again note that we may work with any discretization. Therefore, assume that we are given the graph $\mathbf{G}$, for which $\mathsf{I}_\nu(\mathbf{G})>0$, with covering radius $R=\rho<\operatorname{inj}M/2$. We want to consider small ρ so that we may be able to restrict Ω to smooth hypersurfaces – to apply Buser's isoperimetric inequality.[4]

Suppose we are given Ω, with compact closure, smooth boundary, and inradius greater than ρ. Set

$$\mathcal{K}_0 := \{\xi\in\mathcal{G} : V(\Omega\cap B(\xi;\rho)) > V(\xi;\rho)/2\},$$

$$\mathcal{K}_1 := \{\xi\in\mathcal{G} : 0 < V(\Omega\cap B(\xi;\rho)) \le V(\xi;\rho)/2\}.$$

[4] We made no such fuss in the "only if" part of the argument, since Federer–Fleming (Theorem VIII.3.2 and its argument) and the proof of Proposition VIII.3.1 show that Ω above is admissible for the variational problem defining $I_{\nu,\rho}$.

So, both $\mathcal{K}_0$ and $\mathcal{K}_1$ are contained in $B(\Omega; \rho)$. Then, for at least one of $j = 0, 1$, we have

$$\text{(VIII.4.5)} \qquad \frac{V(\Omega)}{2} \le V(\Omega \cap \cup_{\xi \in \mathcal{K}_j} B(\xi; \rho)).$$

If (VIII.4.5) is valid for $j = 1$, then we have directly from Buser's inequality

$$\begin{aligned} \frac{V(\Omega)}{2} &\le \sum_{\xi \in \mathcal{K}_1} V(\Omega \cap B(\xi; \rho)) \le \text{const.} \sum_{\xi \in \mathcal{K}_1} A(\partial\Omega \cap B(\xi; \rho)) \\ &\le \text{const.} \mathbf{M}_{\epsilon,\rho} A(\partial\Omega) \end{aligned}$$

(without any hypothesis on $\mathsf{I}_\nu(\mathbf{G})$), which implies

$$A(\partial\Omega) \ge \text{const.} V(\Omega) = \text{const.} V(\Omega)^{1/\nu} V(\Omega)^{1-1/\nu} \ge \text{const.} V(\Omega)^{1-1/\nu},$$

since Ω contains a disk of radius ρ, which, by Croke's estimate has volume uniformly bounded from below. So, we must consider the case when (VIII.4.5) is valid only for $j = 0$.

First,

$$\frac{V(\Omega)}{2} \le \sum_{\eta \in \mathcal{K}_0} V(\Omega \cap B(\eta; \rho)) \le V_\kappa(\rho) \operatorname{card} \mathcal{K}_0.$$

Therefore, it suffices to give a lower bound of $A(\partial\Omega)$ by a multiple of card $\partial\mathcal{K}_0$ – the multiple independent of the choice of $\mathcal{K}_0$.

Set

$$H := \{x \in M : \; V(x; \rho)/2 = V(\Omega \cap B(x; \rho))\}.$$

(So, $H \subseteq B(\Omega; \rho)$.) To each $\xi \in \partial\mathcal{K}_0$, there exists $\eta \in \mathsf{N}(\xi)$, $\eta \in \mathcal{K}_0$; we have, of course,

$$d(\xi, \eta) < 3\rho.$$

By definition,

$$V(\Omega \cap B(\eta; \rho)) > V(\eta; \rho)/2, \qquad V(\Omega \cap B(\xi; \rho)) \le V(\xi; \rho)/2,$$

which implies the minimizing geodesic connecting ξ to η contains an element $\zeta \in H$, which implies

$$\partial\mathcal{K}_0 \subseteq B(H; 3\rho),$$

which implies

$$\bigcup_{\xi \in \partial\mathcal{K}_0} B(\xi; \rho) \subseteq B(H; 4\rho).$$

Now let Q be a maximal 2ρ–separated subset of H. Thus,

$$\bigcup_{\xi\in\partial\mathcal{K}_0} B(\xi;\rho) \subseteq B(Q;6\rho),$$

which implies

$$\begin{aligned}
V_\rho \operatorname{card}\partial\mathcal{K}_0 &\le \sum_{\xi\in\partial\mathcal{K}_0} V(\xi;\rho)\\
&\le \mathbf{M}_{\epsilon,\rho}\sum_{\zeta\in Q} V(\zeta;6\rho)\\
&\le \mathbf{M}_{\epsilon,\rho}\text{const.}\sum_{\zeta\in Q} V(\zeta;\rho)\\
&= 2\mathbf{M}_{\epsilon,\rho}\text{const.}\sum_{\zeta\in Q} V(\Omega\cap B(\zeta;\rho))\\
&\le 2\mathbf{M}_{\epsilon,\rho}\text{const.}\sum_{\zeta\in Q} A(\partial\Omega\cap B(\zeta;\rho))\\
&\le 2{\mathbf{M}_{\epsilon,\rho}}^2\text{const.}\,A(\partial\Omega),
\end{aligned}$$

the third inequality uses the Gromov comparison theorem (Theorem III.4.5); the following equality follows from the definition of $H \supseteq Q$; and the fourth inequality uses Buser's inequality. ∎

When $\nu \ge n = \dim M$, then we have the stronger:

Theorem VIII.4.2. (M. Kanai (1985)) *Let M have bounded geometry. Then, for any $\nu \ge n$, we have $I_\nu(M) > 0$ if and only if $\mathsf{I}_\nu(\mathbf{G}) > 0$, for any discretization $\mathbf{G}$ of M.*

Proof. The "only if" is precisely as above.

So we assume that $\mathsf{I}_\nu(\mathbf{G}) > 0$. Suppose we are given Ω, with compact closure, smooth boundary, with no assumption on the inradius. As above, we set

$$\mathcal{K}_0 := \{\xi \in \mathcal{G} : V(\Omega\cap B(\xi;\rho)) > V(\xi;\rho)/2\},$$

$$\mathcal{K}_1 := \{\xi \in \mathcal{G} : 0 < V(\Omega\cap B(\xi;\rho)) \le V(\xi;\rho)/2\}.$$

Again, both $\mathcal{K}_0$ and $\mathcal{K}_1$ are contained in $B(\Omega;\rho)$, and for at least one of $j = 0, 1$, we have (VIII.4.5):

$$\frac{V(\Omega)}{2} \le V(\Omega \cap \cup_{\xi\in\mathcal{K}_j} B(\xi;\rho)).$$

If (VIII.4.5) is valid for $j = 0$, then we argue as above – for the only place we invoked the hypothesis, of the inradius uniformly bounded away from 0, was when (VIII.4.5) is valid only for $j = 1$. We therefore adjust the argument for (VIII.4.5) valid only for $j = 1$.

Lemma VIII.4.2. *There exists a constant $j_\nu > 0$ such that*

$$A(\partial\Omega \cap B(x;\rho)) \geq j_\nu V(\Omega \cap B(x;\rho))^{1-1/\nu}.$$

for all $x \in \mathcal{K}_1$.

Proof of Theorem. Assume the lemma is valid. Then, Minkowski's inequality implies

$$\sum_{\xi\in\mathcal{K}_1} V(\Omega \cap B(\xi;\rho)) \leq \left\{\sum_{\xi\in\mathcal{K}_1} V(\Omega \cap B(\xi;\rho))^{1-1/\nu}\right\}^{\nu/(\nu-1)}.$$

Therefore, (VIII.4.5) and the lemma imply

$$\begin{aligned}\frac{V(\Omega)}{2} &\leq V(\Omega \cap \sum_{\xi\in\mathcal{K}_1} B(\xi;\rho))\\ &\leq \sum_{\xi\in\mathcal{K}_1} V(\Omega \cap B(\xi;\rho))\\ &\leq \text{const.} A(\partial\Omega)^{\nu/(\nu-1)},\end{aligned}$$

which implies the claim. So, it remains to prove the lemma.

Proof of Lemma 6. Let $D = \Omega \cap B(\xi;\rho)$; then,

$$V(D) \leq \frac{V(\xi;\rho)}{2} \leq \frac{V_\kappa(\rho)}{2}.$$

Therefore, $\nu > n$ implies

$$V(D)^{1-1/\nu} \leq \left\{\frac{V_\kappa(\rho)}{2}\right\}^{1/n-1/\nu} V(D)^{1-1/n};$$

so it suffices to prove

$$V(D)^{1-1/n} \leq \text{const.} A(\partial D \cap B(x;\rho)).$$

Well, since $\rho < \operatorname{inj} M/2$, we have, by Croke's isoperimetric inequality (VIII.1.2),

$$V(D)^{1-1/n} \leq \text{const.} A(\partial D) = \text{const.}\{A(\partial D \cap B(x;\rho)) + A(\partial D \cap S(x;\rho))\}.$$

So, we want to show

$$A(\partial D \cap S(x;\rho)) \leq \text{const.} A(\partial D \cap B(x;\rho)).$$

Consider geodesic spherical coordinates centered at x. Let C_x denote the subset of S_x for which

$$\exp \rho \mathsf{C}_x = \partial D \cap S(x;\rho).$$

For each $\xi \in \mathsf{C}_x$, let

$$\sigma(\xi) = \sup\{t \geq 0 : \exp t\xi \in \partial D \cap B(x;\rho)\}.$$

Note that, if $\sigma(\xi) < \rho$, then the geodesic segment from $\exp \sigma(\xi)\xi$ to $\exp \rho\xi$ is contained in D. The Bishop–Gromov theorem (Proposition III.4.1) implies

$$\begin{aligned}
\int_{\sigma(\xi)}^{\rho} \sqrt{\mathbf{g}}(s;\xi)\,ds &\geq \frac{V_\kappa(\rho) - V_\kappa(\sigma(\xi))}{A_\kappa(\rho)} \sqrt{\mathbf{g}}(\rho;\xi) \\
&= \frac{V_\kappa(\rho)}{A_\kappa(\rho)} \sqrt{\mathbf{g}}(\rho;\xi) - \frac{V_\kappa(\sigma(\xi))}{A_\kappa(\rho)} \sqrt{\mathbf{g}}(\rho;\xi) \\
&\geq \frac{V_\kappa(\rho)}{A_\kappa(\rho)} \sqrt{\mathbf{g}}(\rho;\xi) - \frac{V_\kappa(\sigma(\xi))}{A_\kappa(\sigma(\xi))} \sqrt{\mathbf{g}}(\sigma(\xi);\xi) \\
&\geq \frac{V_\kappa(\rho)}{A_\kappa(\rho)} \{\sqrt{\mathbf{g}}(\rho;\xi) - \sqrt{\mathbf{g}}(\sigma(\xi);\xi)\},
\end{aligned}$$

which implies

$$\begin{aligned}
V(D) &\geq \frac{V_\kappa(\rho)}{A_\kappa(\rho)} \int_{\mathsf{C}_x} \{\sqrt{\mathbf{g}}(\rho;\xi) - \sqrt{\mathbf{g}}(\sigma(\xi);\xi)\}\, d\mu_x(\xi) \\
&\geq \frac{V_\kappa(\rho)}{A_\kappa(\rho)} \{A(\partial D \cap S(x;\rho)) - A(\partial D \cap B(x;\rho))\},
\end{aligned}$$

which implies, by Buser's inequality (Theorem VIII.2.1),

$$\begin{aligned}
A(\partial D \cap S(x;\rho)) &\leq \frac{A_\kappa(\rho)}{V_\kappa(\rho)} V(D) + A(\partial D \cap B(x;\rho)) \\
&\leq \text{const.} A(\partial D \cap B(x;\rho)) + A(\partial D \cap B(x;\rho)) \\
&= \text{const.} A(\partial D \cap B(x;\rho)),
\end{aligned}$$

which implies the claim. ∎

§VIII.5. Notes and Exercises

Analytic Isoperimetric Inequalities

Exercise VIII.1. Prove, if $I_\nu(M) > 0$ for some given $\nu > 2$, that for any function ϕ in $C_c^\infty(M)$, we have

$$\|\text{grad}\,\phi\|_2 \geq \frac{\nu-2}{2(\nu-1)} I_\nu(M)\|\phi\|_{2\nu/(\nu-2)}.$$

Exercise VIII.2. Prove, if $I_\nu(M) > 0$ for a given $\nu \geq 2$, that there exists a positive const.$_\nu$ such that

$$||\phi||_2{}^{2+4/\nu} \leq \text{const.}_\nu ||\text{grad}\,\phi||_2{}^2 ||\phi||_1{}^{4/\nu}$$

for all $\phi \in C_c^\infty(M)$.

Exercise VIII.3. (J. Cheeger (1970b)) Prove, if $I_\infty(M) > 0$, that for any function ϕ in $C_c^\infty(M)$, we have

$$\|\text{grad}\,\phi\|_2 \geq \frac{1}{2} I_\infty(M)\|\phi\|_2.$$

In particular, the fundamental tone of M, $\lambda^*(M)$, satisfies

$$\lambda^*(M) \geq \frac{1}{4} I_\infty{}^2(M).$$

Furthermore (H. P. McKean (1970); also see Pinsky (1978)), if M is connected, complete, simply connected, with all sectional curvatures less than or equal to $\kappa < 0$, then

$$\lambda^*(M) \geq \frac{-(n-1)^2\kappa}{4}.$$

The inequality is sharp in the sense that we have equality for M the n–dimensional hyperbolic space of constant sectional curvature κ. See Pinsky (1978), Gage (1980).

The Compact Case

Let M be compact Riemannian, $n \geq 1$ the dimension of M. Then, as noted in Remark VIII.3.1, all isoperimetric constants vanish. Alternatively, by considering the function $f \equiv 1$ on M, one has that all the Sobolev constants of M vanish. Nevertheless, one can adjust the definitions as follows (here one only needs the isoperimetric dimensions n and ∞):

Definition. Define the *isoperimetric constant* $\mathfrak{I}_n(M)$ by

$$\mathfrak{I}_n(M) = \inf_{\Gamma} \frac{A(\Gamma)}{\min\{V(D_1), V(D_2)\}^{1-1/n}},$$

where Γ varies over compact $(n-1)$–dimensional submanifolds of M that divide M into two disjoint open submanifolds D_1, D_2 of M. Define *Cheeger's constant* $\mathfrak{I}_\infty(M)$ by

$$\mathfrak{I}_\infty(M) = \inf_{\Gamma} \frac{A(\Gamma)}{\min\{V(D_1), V(D_2)\}},$$

where Γ varies over compact $(n-1)$–submanifolds of M as described above.

Exercise VIII.4. Prove the analogue of Theorem VIII.3.1 (Yau (1975)), that in the definition of $\mathfrak{I}_\nu(M)$, $\nu = n, \infty$, it suffices to assume that the open submanifolds D_1 and D_2 are connected.

Definition. Define the *Sobolev constant of* M, $\mathfrak{s}_n(M)$, by

$$\mathfrak{s}_n(M) = \inf_f \frac{\|\operatorname{grad} f\|_1}{\inf_\alpha \|f - \alpha\|_{n/(n-1)}},$$

where α varies over $\mathbb{R}$, and f over $C^\infty(M)$.

Exercise VIII.5. Prove (the analogue of the Federer–Fleming theorem (Federer–Fleming (1960), Federer (1969)):

$$\mathfrak{I}_n(M) \le \mathfrak{s}_n(M) \le 2\mathfrak{I}_n(M).$$

Exercise VIII.6. Prove Cheeger's (1970b) inequality:

$$\lambda_2 \ge \frac{\mathfrak{I}_\infty{}^2(M)}{4},$$

where λ_2 is the second (i.e., the lowest nonzero) eigenvalue of the closed eigenvalue problem on M.

Note VIII.1. The Cheeger inequality for compact Riemannian manifolds has been used extensively, especially for surfaces of (constant) negative curvature. For the possibility of constructing metrics on a given compact Riemannian manifold so that the inequality is sharp, and for many other matters, see the survey of Buser (1980). For lower bounds *without* the employment of Cheeger's inequality, see Dodziuk–Randol (1986). For a more recent survey of applications of Cheeger's inequality, see Buser (1992).

In contrast to the McKean result in Exercise VIII.3, it is possible that a *compact* Riemannian manifold of constant negative sectional curvature κ satisfies $\lambda_2 < -(n-1)^2\kappa/4$ (Randol (1974)). This has led to consideration of the phenomenon of "small eigenvalues." See Buser (1992, Chapter 8).

Buser's Isoperimetric Inequality

Note VIII.2. Buser (1982) gave an upper bound of $\lambda_2(M)$ in terms of the Cheeger constant, namely, if compact M of dimension n has Ricci curvature bounded from below by $(n-1)\kappa$, $\kappa \leq 0$, then

$$\lambda_2(M) \leq c(n)\{\mathfrak{I}_\infty(M)\sqrt{-\kappa} + \mathfrak{I}_\infty{}^2(M)\},$$

where $c(n)$ is a constant depending only on n. He gave two proofs, the second using Theorem VIII.2.1.

For M complete noncompact, the result reads as:

$$\lambda^*(M) \leq c(n)\mathfrak{I}_\infty(M)\sqrt{-\kappa}.$$

See Ancona (1990) and Canary (1992) for the analytic argument.

Note VIII.3. Also, for M compact n–dimensional with Ricci curvature bounded below by $(n-1)\kappa$, Buser (1982) also has derived, from his isoperimetric inequality, lower bounds on eigenvalues $\lambda_\ell(M)$ that complement those of Exercise III.41.

IX
Comparison and Finiteness Theorems

In this chapter, we introduce one of the most powerful theorems in Riemannian geometry: H. E. Rauch's comparison theorem. It allows for direct comparison of the growth of Jacobi fields in a given Riemannian manifold M with those in a simply connected space form of constant sectional curvature in both cases, where the constant sectional curvature is an upper or lower bound of the sectional curvatures along the geodesic under consideration in M. The case where the curvature is bounded from above is quite elementary; and for the curvature bounded from below, we have already dealt with the weaker conjugate point (Bonnet–Myers) and volume (Bishop) comparison theorems (in those cases lower bounds on the Ricci curvature sufficed). So, now we turn to the strongest version, the one discussing the Jacobi fields themselves. (See the preliminary discussion in Notes II.10–II.11.)

The major applications we consider here are (i) the Heintze–Karcher volume comparison theorem for the volume of tubular neighborhoods of submanifolds of arbitrary codimension, (ii) the Alexandrov–Toponogov triangle comparison theorems, and (iii) Cheeger's finiteness theorem. Our applications are only a sample. One has, at least, a whole panoply of "sphere theorems" (see §IX.9), which were the initial major application of the Jacobi field comparison theorems – the original program of Rauch. And, most recently, one has M. Gromov's convergence theorems for Riemannian manifolds (see §IX.9).

We first list some small, but necessary, preliminaries.

§IX.1. Preliminaries

We fix our perennial Riemannian manifold M.

1. Fix $p \in M$, its tangent space M_p, and the curvature tensor R acting on M_p. Set

$$R_\alpha(u, v)w = \alpha\{\langle u, w\rangle v - \langle v, w\rangle u\},$$

the curvature tensor associated to constant sectional curvature α (Proposition II.3.1 and Exercise II.2). Suppose all sectional curvatures $\mathcal{K}$ at p satisfy

$$\kappa \le \mathcal{K} \le \delta. \tag{IX.1.1}$$

Then, by Exercise II.1(e), we have

$$|R(v,u)v - R_{(\kappa+\delta)/2}(v,u)v| \le \frac{\delta-\kappa}{2}|u||v|^2. \tag{IX.1.2}$$

2. Parallel translation and curvature. Recall from Exercise IV.2, that if $v : [0,1]\times[0,1] \to M \in D^1$ is a homotopy with fixed endpoints

$$p = v(0,s), \qquad q = v(1,s),$$

X a vector field along v such that

$$X(0,s) = X_0 \in M_p, \qquad \nabla_t X = 0,$$

then

$$|X(1,1) - X(1,0)| \le \frac{4}{3}\{\sup |X|\}\Lambda \int_0^1 ds \int_0^1 |\partial_t v \wedge \partial_s v|\, dt.$$

where $\Lambda = \sup |\mathcal{K}|$. Note that, by Exercise IV.1, the double integral is the area of the surface spanned by the homotopy.

3. Reparameterization of geodesics. Given a geodesic $\gamma = \gamma(t)$, and a Jacobi field $Y = Y(t)$ along γ, that is,

$$\nabla_t{}^2 Y + R(\gamma', Y)\gamma' = 0,$$

consider the reparameterization

$$\gamma(t) = \omega(\alpha t), \qquad Y(t) = Z(\alpha t)$$

of the geodesic, and set $s = \alpha t$. Then,

$$\gamma'(t) = \alpha\omega'(s), \qquad (\nabla_t Y)(t) = \alpha(\nabla_s Z)(s),$$

and the Jacobi equation for Z along ω becomes

$$\nabla_s{}^2 Z + R(\omega', Z)\omega' = 0.$$

In particular, if s is arc length along the geodesic, all the sectional curvatures along the geodesic are equal to κ, and η is any Jacobi field along ω satisfying $\eta(0) = 0$, then

$$\eta(s) = \mathbf{S}_\kappa(s)E(s),$$

where $E(s)$ is a parallel vector field along ω. Therefore, if our Jacobi field satisfies $Y(0) = 0$ then we have

$$Y(t) = \alpha \mathbf{S}_{\kappa\alpha^2}(t)\mathfrak{E}(\alpha t), \qquad (\nabla_t Y)(t) = \alpha \mathbf{C}_{\kappa\alpha^2}(t)\mathfrak{E}(\alpha t),$$

where $\mathfrak{E}$ is a parallel vector field along ω, which implies

$$\frac{\langle Y, \nabla_t Y\rangle}{\langle Y, Y\rangle}(t) = \frac{\mathbf{C}_{\kappa\alpha^2}}{\mathbf{S}_{\kappa\alpha^2}}(t).$$

4. Note that for any vector function $\eta(t)$, whenever $\eta \neq 0$, we have

$$|\eta|' = \frac{\langle \eta', \eta\rangle}{\langle \eta, \eta\rangle};$$

and when $\eta(t_0) = 0$ we have, for the right-hand derivative,

$$|\eta|'_+(t_0) = |\eta'|(t_0).$$

§IX.2. H. E. Rauch's Comparison Theorem

For convenience, we separate the analytic aspects from the geometric aspects. First, the analytic.

We fix a real inner product space V of finite dimension N. For any $\beta > 0$, let

$$\Upsilon = \{X : [0, \beta] \to V \in D^1 : X(0) = 0\}, \qquad \Upsilon_0 = \{X \in \Upsilon : X(\beta) = 0\}.$$

We are given the family $\mathcal{R}(t) : V \to V, t \in [0, \beta]$, of self-adjoint linear maps of V to V, with which we associate the *index form*

$$I(X, Y) = \int_0^\beta \{\langle X', Y'\rangle - \langle \mathcal{R}(t)X, Y\rangle\}\, dt$$

and the *Jacobi equation*

$$\eta'' + \mathcal{R}(t)\eta = 0. \tag{IX.2.1}$$

Of course, we have in mind the identification, via parallel translation, of the subspaces $\gamma'(t)^\perp$ along a geodesic $\gamma(t)$. See §III.1.

Lemma IX.2.1. *Assume I is positive definite on Υ_0, and let $\eta \in \Upsilon$ be a non-trivial solution of* (IX.2.1)*. Then, for any $X \in \Upsilon$ satisfying*

$$X(\beta) = \eta(\beta)$$

we have

$$I(X, X) \geq I(\eta, \eta) = \langle \eta', \eta\rangle(\beta),$$

with equality if and only if $X = \eta$.

Proof. We have

$$\begin{aligned} 0 &\leq I(X-\eta, X-\eta) \\ &= I(X,X) - 2I(X,\eta) + I(\eta,\eta) \\ &= I(X,X) - 2\langle X, \eta' \rangle(\beta) + \langle \eta', \eta \rangle(\beta) \\ &= I(X,X) - \langle \eta', \eta \rangle(\beta) \\ &= I(X,X) - I(\eta,\eta), \end{aligned}$$

which implies the claim. ■

Theorem IX.2.1. (H. E. Rauch (1951)) *Assume*

$$\mathcal{R}(t) \leq \delta I$$

for all $t \in [0, \beta]$. *If* η *is a solution to the Jacobi equation* (IX.2.1), *then the function* $|\eta|$ *satisfies the differential inequality*

$$|\eta|'' + \delta|\eta| \geq 0. \tag{IX.2.2}$$

on $[0, \beta)$. *Furthermore, if* ψ *denotes the solution on* $[0, \beta]$ *of the initial value problem*

$$\psi'' + \delta\psi = 0, \qquad \psi(0) = |\eta|(0), \quad \psi'(0) = |\eta|'(0),$$

and ψ *does not vanish on* $(0, \beta)$, *then*

$$\{|\eta|/\psi\}' \geq 0, \tag{IX.2.3}$$

$$|\eta| \geq \psi, \tag{IX.2.4}$$

on $(0, \beta)$.

We have equality in (IX.2.3) *at* $t_0 \in (0, \beta)$ *if and only if* $\mathcal{R}(t)\eta = \delta\eta$ *on all of* $[0, t_0]$, *and there exists a constant vector* E *for which* $\eta(t) = \psi(t)E$ *on all of* $[0, t_0]$.

The theorem is merely a restatement of Theorem II.6.4, with the geometric data deleted.

Theorem IX.2.2. (H. E. Rauch (1951)) *Assume*

$$\mathcal{R}(t) \geq \kappa I$$

on all of $[0, \beta]$, *and assume that Jacobi's equation* (IX.2.1) *has no points in* $(0, \beta)$ *conjugate to* $t = 0$, *that is, assume that for any nontrivial solution* η *to*

Jacobi's equation satisfying $\eta(0) = 0$, *we have* η *nonvanishing on all of* $(0, \beta)$. *Then, for any nontrivial solution* η *to Jacobi's equation satisfying*

$$\eta(0) = 0$$

we have

$$\frac{|\eta|'}{|\eta|} = \frac{\langle \eta', \eta \rangle}{\langle \eta, \eta \rangle} \le \frac{\mathbf{C}_\kappa}{\mathbf{S}_\kappa}. \tag{IX.2.5}$$

Therefore,

$$\{|\eta|/\mathbf{S}_\kappa\}' \le 0, \tag{IX.2.6}$$

$$|\eta| \le |\eta'(0)|\mathbf{S}_\kappa, \tag{IX.2.7}$$

on $(0, \beta)$.[1]

We have equality in (IX.2.6) *at* $t_0 \in (0, \beta)$ *if and only if*

$$\mathcal{R}(t)\eta = \kappa\eta$$

on all of $[0, t_0]$, *and there exists a constant vector* E *for which*

$$\eta(t) = \mathbf{S}_\kappa(t)E$$

on all of $[0, t_0]$.

Proof. Since $t = 0$ has no conjugate points in $(0, \beta)$, the argument of the Bonnet–Myers theorem (Theorem II.6.1) implies that $\beta \le \pi/\sqrt{\kappa}$ – so, we are guaranteed that $\mathbf{S}_\kappa > 0$ on all of $(0, \beta)$.

Fix $t \in (0, \beta)$. Since $t = 0$ has no conjugate points in $(0, \beta)$, the analytic version of the Jacobi criteria (Theorem II.5.4) implies that the index form for the interval $[0, t]$ is positive definite on Υ_0. Then for *any* $X \in D^1([0, t])$ satisfying

$$X(0) = 0, \qquad X(t) = \eta(t)$$

we have, by Lemma IX.2.1,

$$\langle \eta, \eta' \rangle(t) = \int_0^t |\eta'|^2 - \langle \mathcal{R}\eta, \eta \rangle \le \int_0^t |X'|^2 - \langle \mathcal{R}X, X \rangle \le \int_0^t |X'|^2 - \kappa|X|^2.$$

Pick the *specific* X given by

$$X(s) = \frac{\mathbf{S}_\kappa(s)}{\mathbf{S}_\kappa(t)}\eta(t), \qquad s \in [0, t].$$

[1] See the notes and exercises "On the Rauch theorem" in §II.9.

Then, one has explicitly

$$\int_0^t |X'|^2 - \kappa |X|^2 = \frac{\mathbf{C}_\kappa(t)}{\mathbf{S}_\kappa(t)} |\eta(t)|^2,$$

which implies (IX.2.5). The rest of the claims follow easily. ■

Theorem IX.2.3. (The geometric Rauch theorem) *Given M, $p \in M$, $\xi \in \mathsf{S}_p$, with sectional curvatures satisfying* (IX.1.1) *along γ_ξ; assume that γ_ξ has no points conjugate to p along $\gamma|(0, t]$. Then,*

$$\frac{\mathbf{S}_\delta(t)}{t} \leq \frac{|(\exp_p)_{*|t\xi} \mathfrak{I}_{t\xi} v|}{|v|} \leq \frac{\mathbf{S}_\kappa(t)}{t} \tag{IX.2.8}$$

for all $v \in \xi^\perp$.

Remark IX.2.1. Of course, if one is given only one of the inequalities of (IX.1.1), then one only has the corresponding inequality in (IX.2.8).

Also, if $\kappa < 0$, then the upper bound in (IX.2.8) is valid for all $v \in M_p$. Similarly, if $\delta > 0$, then the lower bound is valid for all $v \in M_p$.

§IX.3. Comparison Theorems with Initial Submanifolds

In this section, we prove the appropriate version of Bishop's theorem (Theorem III.6.1) for Fermi coordinates based on a submanifold with codimension greater than 1. First, we recall the setting and basic background from §III.6.

Let M be our given n–dimensional Riemannian manifold, $\mathfrak{M}$ be a connected k–dimensional submanifold of M, $0 \leq k < n$. We first recall, from §II.2, that the second fundamental form of $\mathfrak{M}$ in M is, at each point $p \in \mathfrak{M}$, a vector-valued symmetric bilinear form $\mathfrak{B} : \mathfrak{M}_p \times \mathfrak{M}_p \to \mathfrak{M}_p{}^\perp$, given by

$$\mathfrak{B}(\xi, \eta) = (\nabla_\xi Y)^N,$$

where Y is any extension of η to a tangent vector field on $\mathfrak{M}$, ∇ denotes the Levi-Civita connection of the Riemannian metric on M, and the superscript N denotes projection onto $\mathfrak{M}_p{}^\perp$. To every vector $v \in \mathfrak{M}_p{}^\perp$, one has the real-valued bilinear form $\mathfrak{b}_v(\xi, \eta) = \langle \mathfrak{B}(\xi, \eta), v \rangle$, and Weingarten map $\mathfrak{A}^v : \mathfrak{M}_p \to \mathfrak{M}_p$ given by $\langle \mathfrak{A}^v \xi, \eta \rangle = \mathfrak{b}_v(\xi, \eta)$ for $\xi, \eta \in \mathfrak{M}_p$ – so,

$$\mathfrak{A}^v \xi = -(\nabla_\xi V)^T,$$

where V is an extension of v to a normal vector field on $\mathfrak{M}$, and the superscript T denotes projection onto $\mathfrak{M}_p$.

For $p \in \mathfrak{M}$, we let $\nu\mathsf{S}_p$ denote the *normal* unit tangent sphere at p, that is,

$$\nu\mathsf{S}_p = \mathsf{S}_p \cap \mathfrak{M}_p{}^{\perp}.$$

Fix $\xi \in \nu\mathsf{S}_p$, and set $\gamma = \gamma_\xi$. We let $\mathfrak{T}$ denote the collection of *transverse vector fields X along the geodesic* γ, that is, those vector fields X along γ for which X is pointwise orthogonal to γ, with initial data

$$X(0) \in \mathfrak{M}_p, \qquad \{(\nabla_t X)(0) + \mathfrak{A}^\xi X(0)\} \in \mathfrak{M}_p{}^{\perp}.$$

The collection of transverse Jacobi fields along γ is an $(n-1)$–dimensional vector space.

If the sectional curvatures along γ are all equal to κ, and the Weingarten map of ξ, $\mathfrak{A}^\xi$, is given by

$$\mathfrak{A}^\xi = \lambda I, \tag{IX.3.1}$$

then the collection of transverse Jacobi fields along γ, pointwise orthogonal to γ, are given as sums of the vector fields:

$$Z(t) = (\mathbf{C}_\kappa - \lambda\mathbf{S}_\kappa)(t)\tau_t\zeta, \; \zeta \in \mathfrak{M}_p, \quad \text{and} \quad Y(t) = \mathbf{S}_\kappa(t)\tau_t\eta, \; \eta \in \xi^\perp \cap \mathfrak{M}_p^\perp,$$

where τ_t denotes parallel translation along γ from p to $\gamma(t)$.

Definition. A point $\gamma(t)$ is said to be *focal to $\mathfrak{M}$ along γ* if there exists a nontrivial transverse Jacobi field Y such that $Y(t) = 0$.

Let Υ denote the collection of vector fields X along γ, pointwise orthogonal to γ, for which $X(0) \in \mathfrak{M}_p$, and let Υ_β consist of those elements of Υ which vanish at $t = \beta$. On Υ, define the index form by

$$I_\beta(X_1, X_2) = -\mathfrak{b}_\xi(X_1(0), X_2(0)) + \int_0^\beta \langle\nabla_t X_1, \nabla_t X_2\rangle - \langle R(\gamma', X_1)\gamma', X_2\rangle\, dt.$$

The Jacobi criteria (Theorems II.5.4 and II.5.5) on the positivity of the index form (on Υ_β) and the nonexistence of focal points remain valid in this setting. In particular, one has the corresponding version of Lemma IX.2.1 above, namely, that if $\mathfrak{M}$ has no focal points along $\gamma|[0, \beta]$, then for any transverse Jacobi field η along γ and any vector field X in Υ satisfying

$$X(\beta) = \eta(\beta)$$

we have

$$I_\beta(X, X) \geq I_\beta(\eta, \eta) = \langle\eta, \eta'\rangle(\beta),$$

with equality if and only if $X = \eta$ on $[0, \beta]$.

Now let $\mathrm{Exp} = \exp|\nu\mathfrak{M}$, where $\nu\mathfrak{M}$ denotes the normal bundle of $\mathfrak{M}$ in M, with natural projection π_ν; also let $\nu\mathsf{S}\mathfrak{M} = \nu\mathfrak{M} \cap \mathsf{S}M$ denote the unit normal bundle of $\mathfrak{M}$. Map $E : [0, +\infty) \times \nu\mathrm{S}\mathfrak{M} \to M$ by

$$E(t, \xi) = \operatorname{Exp} t\xi,$$

so E determines radial coordinates on M, also known as *Fermi coordinates*. Then for $f \in L^1(M)$, we have

(IX.3.2)

$$\int_M f\, dV = \int_{\mathfrak{M}} dV_{\mathfrak{M}}(p) \int_{\nu\mathrm{S}_p} d\mu_{n-k-1,p}(\xi) \int_0^{c_\nu(\xi)} f(\operatorname{Exp} t\xi)\sqrt{\mathbf{g}}(t;\xi)\, dt.$$

In (IX.3.2), $c_\nu(\xi)$ denotes the distance along γ_ξ to the focal cut point of $\mathfrak{M}$ along γ_ξ; $dV_{\mathfrak{M}}$ the (k–dimensional) Riemannian measure of $\mathfrak{M}$; $d\mu_{n-k-1,p}$ the standard $(n-k-1)$–dimensional measure on $\nu\mathrm{S}_p$;

$$\sqrt{\mathbf{g}}(t;\xi) = \det \mathcal{A}(t;\xi),$$

where $\mathcal{A}(t;\xi)$ denotes the matrix solution to Jacobi's equation along γ_ξ, pulled back to $\xi^\perp$, as in §III.1:

$$\mathcal{A}'' + \mathcal{R}\mathcal{A} = 0,$$

subject to the initial conditions

$$\mathcal{A}(0;\xi)|\mathfrak{M}_p = I, \qquad \mathcal{A}'(0;\xi)|\mathfrak{M}_p = -\mathfrak{A}^\xi,$$
$$\mathcal{A}(0;\xi)|\mathfrak{M}_p{}^\perp \cap \xi^\perp = 0, \qquad \mathcal{A}'(0;\xi)|\mathfrak{M}_p{}^\perp \cap \xi^\perp = I.$$

Theorem IX.3.1. (F. W. Warner (1966)) *Assume $\mathfrak{M}$ has codimension 1, that is, that $k = n-1$. Also assume that all sectional curvatures along γ_ξ are bounded below by κ, and that*

$$\mathfrak{A}^\xi \geq \lambda I.$$

Let $\beta_{\kappa,\lambda} \in (0, +\infty]$ denote the first positive zero of $(\mathbf{C}_\kappa - \lambda\mathbf{S}_\kappa)(t)$, should such a zero exist; otherwise, set $\beta_{\kappa,\lambda} = +\infty$. Then (by Theorem III.6.1), $\mathfrak{M}$ has a focal point along γ at distance $\beta \leq \beta_{\kappa,\lambda}$. Let $Y(t)$ be a transverse Jacobi field along γ_ξ, orthogonal to γ_ξ. Then,

$$|Y(t)| \leq (\mathbf{C}_\kappa - \lambda\mathbf{S}_\kappa)(t)|Y(0)|$$

on all of $[0, \beta]$. One has equality at $t_0 \in [0, \beta]$ if and only if $Y(t) = (\mathbf{C}_\kappa - \lambda\mathbf{S}_\kappa)\tau_t Y(0)$ on all of $[0, t_0]$, in which case $\mathfrak{A}^\xi Y(0) = \lambda Y(0)$, and $\mathrm{R}(t)\tau_t Y(0) = \kappa\tau_t Y(0)$ all of $[0, t_0]$.

Proof. The argument is the same for the original Rauch theorem (Theorem IX.2.2), except that one must adjust the argument for the new initial data. It goes as follows (of course, the reader has to check why it does not work in higher codimension – also see Remark III.6.1):

Fix $t \in (0, \beta)$. Then, for *any* $X \in \Upsilon$ satisfying $X(t) = Y(t)$, we have

$$\begin{aligned}\langle Y, \nabla_t Y\rangle(t) &= -\mathfrak{b}_\xi(Y(0), Y(0)) + \int_0^t \{|\nabla_s Y|^2 - \langle \mathsf{R}(s)Y, Y\rangle\}\, ds \\ &\le -\mathfrak{b}_\xi(X(0), X(0)) + \int_0^t \{|\nabla_s X|^2 - \langle \mathsf{R}(s)X, X\rangle\}\, ds \\ &\le -\lambda|X(0)|^2 + \int_0^t \{|\nabla_s X|^2 - \kappa|X|^2\}\, ds.\end{aligned}$$

Pick the *specific* X given by

$$X(s) = \frac{(\mathbf{C}_\kappa - \lambda\mathbf{S}_\kappa)(s)}{(\mathbf{C}_\kappa - \lambda\mathbf{S}_\kappa)(t)}\tau_{s-t}Y(t), \qquad s \in [0, t].$$

Then, one has explicitly

$$-\lambda|X(0)|^2 + \int_0^t \{|\nabla_s X|^2 - \kappa|X|^2\}\, ds = \langle X, \nabla_t X\rangle(t) = \frac{(\mathbf{C}_\kappa - \lambda\mathbf{S}_\kappa)'(t)}{(\mathbf{C}_\kappa - \lambda\mathbf{S}_\kappa)(t)}|Y(t)|^2.$$

Thus,

$$\langle Y, \nabla_t Y\rangle(t) \le \frac{(\mathbf{C}_\kappa - \lambda\mathbf{S}_\kappa)'(t)}{(\mathbf{C}_\kappa - \lambda\mathbf{S}_\kappa)(t)}|Y(t)|^2,$$

which implies the theorem. ∎

Definition. Given $p \in M, \xi \in \mathsf{S}_p$, we say that $\gamma_\xi(\beta)$ *is a focal point of p* along γ_ξ if $\gamma_\xi(\beta)$ is a focal point of the submanifold $\mathfrak{M} = \exp \xi^\perp$ along γ_ξ. Thus, the transverse Jacobi fields under consideration satisfy $\nabla_t Y(0) = 0$.

A geometric application of this last theorem goes as follows: Given M of dimension n, with sectional curvatures bounded below by the constant κ, consider $\mathbb{M}^2_\kappa$ (the 2–dimensional space form of constant curvature κ). Fix $p \in M, \overline{p} \in \mathbb{M}^2_\kappa$, and unit speed geodesics $\gamma : [0, \ell] \to M, \overline{\gamma} : [0, \ell] \to \mathbb{M}^2_\kappa$. Consider parallel vector fields $E, \overline{E}$ along $\gamma, \overline{\gamma}$, respectively, satisfying

$$|E| = |\overline{E}|, \qquad \langle E, \gamma'\rangle = \langle \overline{E}, \overline{\gamma}'\rangle.$$

Let $\phi : [0, \ell] \to \mathbb{R} \in D^1$, and consider the paths

$$\omega(\epsilon) = \exp \phi(\epsilon)E(\epsilon), \qquad \overline{\omega}(\epsilon) = \exp \phi(\epsilon)\overline{E}(\epsilon).$$

Corollary IX.3.1. (M. Berger (1962)) *Assume that for every $\epsilon \in [0, \ell]$, $\gamma(\epsilon)$ has no focal points along the geodesic $t \mapsto \exp tE(\epsilon), t \in [0, \phi(\epsilon)]$ – in* *particular,* $\phi(\epsilon) \le \pi/2\sqrt{\kappa}$. *Then,*

$$\ell(\omega) \le \ell(\overline{\omega}).$$

Proof. Consider the geodesic variations

$$v(t,\epsilon) = \exp t\phi(\epsilon)E(\epsilon), \qquad \overline{v}(t,\epsilon) = \exp t\phi(\epsilon)\overline{E}(\epsilon)$$

in $M, \mathbb{M}_\kappa^2$, respectively. Then,

$$\ell(\omega) = \int_0^\ell |\partial_\epsilon v|(1,\epsilon)\,d\epsilon, \qquad \ell(\overline{\omega}) = \int_0^\ell |\partial_\epsilon \overline{v}|(1,\epsilon)\,d\epsilon.$$

As usual, $\partial_\epsilon v$ is a Jacobi field along the geodesic $t \mapsto v(t,\epsilon)$, for each fixed ϵ. To estimate its length from above one has to decompose it into its projections onto, and perpendicular to, the geodesic; take into account that t is not necessarily arc length – so use **3** of §IX.1; and use Theorem IX.3.1 – with appropriate initial conditions. ■

For codimension greater than or equal to 1, we do have the volume comparison theorem:

Theorem IX.3.2. (E. Heintze & H. Karcher (1978)) *Let H denote the mean curvature vector of $\mathfrak{M}$ in M. Assume that all sectional curvatures along γ_ξ are bounded below by κ. Let τ denote the first positive zero of*

$$\{\mathbf{C}_\kappa - \frac{\langle H,\xi\rangle}{k}\mathbf{S}_\kappa\}(t)$$

(should such a zero exist; otherwise, set $\tau = +\infty$). Then, $\mathfrak{M}$ has a focal point along γ_ξ at distance $c_f(\xi) \le \tau$, and

$$\det \mathcal{A}(t;\xi) \le \{\mathbf{C}_\kappa - \frac{\langle H,\xi\rangle}{k}\mathbf{S}_\kappa\}^k(t)\mathbf{S}_\kappa{}^{n-k-1}(t) \tag{IX.3.3}$$

on all of $[0, c_f(\xi)]$.

Corollary IX.3.2. *Let M be compact, all sectional curvatures of M bounded below by the constant κ. Let $d(M)$ denote the diameter of M. Then for any simple closed geodesic γ in M with length $\ell(\gamma)$ we have*

$$V(M) \le \frac{\mathbf{c_{n-2}}}{n-1}\ell(\gamma)\mathbf{S}_\kappa{}^{n-1}(d(M)). \tag{IX.3.4}$$

Proof of the Heintze–Karcher Theorem. Since ξ is fixed, in what follows, we simply write $\mathcal{A}(t)$ for $\mathcal{A}(t;\xi)$.

Our method, as usual, is to study the logarithmic derivative of det $\mathcal{A}(t)$. To this end, fix $r > 0$ for which det $\mathcal{A}(t) > 0$ on $(0, r)$, fix $s \in (0, r)$, and set

$$\mathcal{B} = \mathcal{A}^*\mathcal{A}, \qquad \mathcal{C}(t) = \mathcal{B}(t)\mathcal{B}(s)^{-1}.$$

Then (the prime denotes differentiation with respect to t),

$$\frac{1}{2}\frac{(\det \mathcal{A})'}{\det \mathcal{A}}(s) = \frac{(\det \mathcal{C})'}{\det \mathcal{C}}(s) = \sum_{j=1}^{n-1} \langle X_j, X_j{}'\rangle(s) = \sum_{j=1}^{n-1} I_s(X_j, X_j),$$

where $\{X_j : j = 1, \ldots, n-1\}$ are transverse Jacobi fields along γ_ξ such that $\{X_j(s) : \ j = 1, \ldots, n-1\}$ is an orthonormal basis of $\gamma_\xi{}'(s)^\perp$.

Now our differential equation lives in $\xi^\perp$. Consider the solution $\mathcal{A}_\kappa(t)$ to the matrix differential equation in $\xi^\perp$

$$\mathcal{A}_\kappa{}'' + \kappa\mathcal{A}_\kappa = 0, \tag{IX.3.5}$$

with the *same* initial conditions as $\mathcal{A}(t)$. Then, the vector solutions of the associated vector differential equation

$$Y'' + \kappa Y = 0 \tag{IX.3.6}$$

are given by $Y = \mathcal{A}_\kappa\eta$, where $\eta \in \xi^\perp$, and Y is transverse to $\mathfrak{M}_p$. One also has the associated index form

$$I_{\kappa,\beta}(X_1, X_2) = -\mathfrak{b}_\xi(X_1(0), X_2(0)) + \int_0^\beta \langle X_1{}', X_2{}'\rangle - \kappa\langle X_1, X_2\rangle\, dt,$$

for any $\beta > 0$.

Pick transverse solutions $\{\overline{X}_j : \ j = 1, \ldots, n-1\}$ of (IX.3.6) such that

$$\overline{X}_j(s) = X_j(s)$$

for all $j = 1, \ldots, n-1$. Then,

$$\sum_{j=1}^{n-1} I_s(X_j, X_j) \le \sum_{j=1}^{n-1} I_s(\overline{X}_j, \overline{X}_j) \le \sum_{j=1}^{n-1} I_{\kappa,s}(\overline{X}_j, \overline{X}_j) = \frac{1}{2}\frac{(\det \mathcal{A}_\kappa)'}{\det \mathcal{A}_\kappa}(s).$$

Since s is arbitrary in $(0, r)$ we have

$$\det \mathcal{A} \le \det \mathcal{A}_\kappa$$

on all of $[0, r]$.

Let $\{e_\epsilon : \epsilon = 1, \ldots, k\}$ be an orthonormal basis of $\mathfrak{M}_p$ consisting of eigenvectors of the Weingarten map $\mathfrak{A}^\xi$, with respective eigenvalues λ_ϵ. And let

$\{e_\delta : \delta = k+1, \ldots, n-1\}$ be an orthonormal basis of $\xi^\perp \cap \mathfrak{M}_p{}^\perp$. Then,

$$\mathcal{A}_\kappa(t)e_\epsilon = \{\mathbf{C}_\kappa - \lambda_\epsilon \mathbf{S}_\kappa\}(t)e_\epsilon, \quad \epsilon = 1, \ldots, k,$$
$$\mathcal{A}_\kappa(t)e_\delta = \mathbf{S}_\kappa(t)e_\delta, \quad \delta = k+1, \ldots, n-1,$$

which implies

$$\det \mathcal{A}_\kappa = \mathbf{S}_\kappa{}^{n-k-1} \prod_{\epsilon=1}^{k} \{\mathbf{C}_\kappa - \lambda_\epsilon \mathbf{S}_\kappa\} \le \{\mathbf{C}_\kappa - \frac{\sum_\epsilon \lambda_\epsilon}{k} \mathbf{S}_\kappa\}^k \mathbf{S}_\kappa{}^{n-k-1},$$

by the arithmetic–geometric mean inequality, which is (IX.3.3). ∎

Remark IX.3.1. Of course, we may read (IX.3.4) as a lower bound on the length of any simple closed geodesic in M in terms of the volume, diameter, and lower bound on the sectional curvature, of M. By Klingenberg's theorem (Theorem III.2.4), we obtain:

Theorem IX.3.3. (J. Cheeger (1970b)) *Let M be compact n–dimensional, all sectional curvatures of M bounded in absolute value from above by Λ,* diam M *bounded from above by D, and $V(M)$ bounded from below by V. Then,*

$$\text{inj}\, M \ge \min \left\{ \frac{\pi}{\sqrt{\Lambda}}, \frac{n-1}{2\mathbf{c}_{n-2}} \frac{V}{\mathbf{S}_{-\Lambda}{}^{n-1}(D)} \right\} := \mathfrak{c}(n, V, D, \Lambda)$$

§IX.4. Refinements of the Rauch Theorem

The work of this section is from Karcher (1977, Proposition A6) and Buser–Karcher (1981, pp. 97ff). It consists of a close study of analytic comparison theorems, without the use of the index form.

Let V be a real inner product space of finite dimension N. We are given the family $\mathcal{R}(t) : V \to V$ of self-adjoint linear maps of V to V, to which we associate the Jacobi differential equation (IX.2.1). We assume

$$\kappa I \le \mathcal{R}(t) \le \delta I, \qquad \text{(IX.4.1)}$$

where $\kappa \le \delta$ are given constants. Set

$$\Lambda := \max\{|\kappa|, |\delta|\}.$$

Let ϵ denote some parameter usually, but not necessarily, in (κ, δ), and set

$$\lambda := \max\{\delta - \epsilon, \epsilon - \kappa\}.$$

We consider the Jacobi differential equations

$$\eta'' + \mathcal{R}(t)\eta = 0, \qquad E'' + \epsilon E = 0,$$

with initial data

$$E(0) = \eta(0), \qquad E'(0) = \eta'(0).$$

We also consider the scalar initial value problem:

$$\sigma'' + (\epsilon - \lambda)\sigma = \lambda|E|, \qquad \sigma(0) = \sigma'(0) = 0.$$

Theorem IX.4.1. *For any $t > 0$ for which $\mathbf{S}_\epsilon|(0, t) > 0$ we have $|\eta - E| \le \sigma$ on all of $[0, t]$.*

Proof. Let P denote a unit vector in V. Then,

$$\langle \eta - E, P\rangle'' + \epsilon\langle \eta - E, P\rangle = \langle \epsilon\eta - \mathcal{R}(t)\eta, P\rangle \le \lambda|\eta|$$

by the self-adjointness of $\mathcal{R}$, and the bounds on $\mathcal{R}$. So $\phi := \langle \eta - E, P\rangle$ satisfies

$$\phi'' + \epsilon\phi \le \lambda|\eta|.$$

Let μ satisfy the scalar initial value problem:

$$\mu'' + \epsilon\mu = \lambda|\eta|, \qquad \mu(0) = \mu'(0) = 0.$$

Then, the standard Sturmian argument (Exercise II.22) implies $\phi \le \mu$ for all such choices of P. Therefore,

$$|\eta - E| \le \mu, \quad \Rightarrow \quad |\eta| \le |E| + \mu,$$

which implies

$$\mu'' + (\epsilon - \lambda)\mu \le \lambda|E| = \sigma'' + (\epsilon - \lambda)\sigma.$$

Again, the Sturmian argument implies $\mu \le \sigma$, which implies

$$|\eta - E| \le \mu \le \sigma,$$

which is the claim. ■

Corollary IX.4.1. *Assume, in addition to the above, that* (i) *the vectors $\eta(0)$ and $\eta'(0)$ are linearly dependent, and* (ii) *the function*

$$f_\epsilon := |\eta(0)|\mathbf{C}_\epsilon + |\eta|'(0)\mathbf{S}_\epsilon$$

is positive on all of $(0, t)$. Then,

$$f_\epsilon = |E|, \qquad |E|^{-1}E = \text{const.}, \qquad \sigma = f_{\epsilon-\lambda} - f_\epsilon$$

on all of $(0, t)$.

Proof. By the differential equation for E, one has constant vectors P and Q such that

$$E = \mathbf{C}_\epsilon P + \mathbf{S}_\epsilon Q.$$

Since $\eta(0)$ and $\eta'(0)$ are linearly dependent, we have the existence of a constant unit vector e, and real constants α and β, such that $E = (\alpha \mathbf{C}_\epsilon + \beta \mathbf{S}_\epsilon)e$, which implies $\alpha = |E(0)| = |\eta(0)|$ and $\beta = |E|'(0) = |\eta|'(0)$, which implies

$$E = f_\epsilon e.$$

The rest is easy. ∎

Corollary IX.4.2. *We also have*

$$|\eta - E| \leq f_{\epsilon-\lambda} - f_\epsilon, \tag{IX.4.2}$$

and

$$|\eta| \leq f_\kappa. \tag{IX.4.3}$$

Proof. The inequality (IX.4.2) follows directly from the theorem; and the inequality (IX.4.3) comes from the theorem, with the specific choice of $\epsilon = (\kappa + \delta)/2$. ∎

Remark IX.4.1. If, in the above corollary, we have $\eta(0) = 0$, then (IX.4.3) is the estimate from Rauch's theorem (Theorem IX.2.1) on the interval

$$(0, \pi/\sqrt{(\kappa+\delta)/2}) \subseteq (0, \pi/\sqrt{\kappa}).$$

Thus, although this argument is extremely elementary, it requires the *upper* bound on $\mathcal{R}(t)$, and the restriction to a smaller interval than the one given by the Rauch theorem.

We also note, more generally, that by lowering ϵ one obtains a larger interval on which $\mathbf{S}_\epsilon > 0$. The price one pays is that the right hand side of (IX.4.2), for example, increases with respect to decreasing ϵ – so the inequality becomes weaker as ϵ decreases.

Corollary IX.4.3. *If we* also *have* $\eta(0) = 0$, *then*

$$|\eta(s) - s\eta'(0)| \leq |\eta'(0)|\{\mathbf{S}_{-\Lambda}(s) - s\}$$

for all $s > 0$.

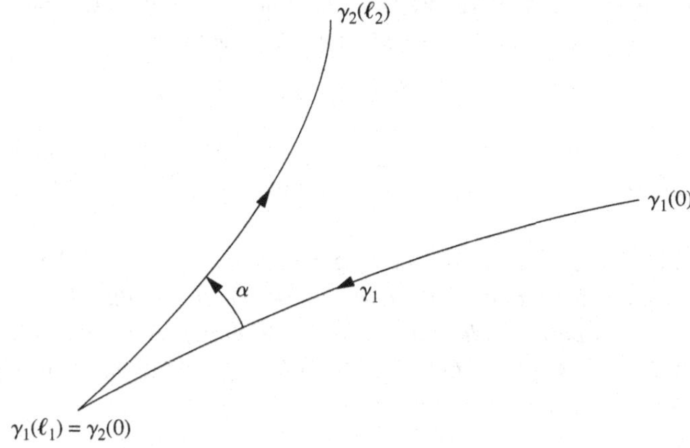

Figure IX.1. A geodesic hinge.

Proof. Here, we pick $\epsilon = 0$, which implies $\lambda = \Lambda$, which implies

$$E = s\eta'(0), \qquad \sigma = |\eta'(0)|\{\mathbf{S}_{-\Lambda}(s) - s\},$$

which implies the claim. ∎

§IX.5. Triangle Comparison Theorems

By "triangle comparison theorems," we mean global forms of the Rauch comparison theorem. Such theorems were first proved by A. D. Alexandrov (1948) and V. A. Toponogov (1959). The theorems are presented in detail, with increasing simplifications of the earliest arguments, in Berger (1962), Cheeger–Ebin (1975), Gromoll–Klingenberg–Meyer (1968), Grove (1987), Karcher (1989) and Klingenberg (1982). For completeness, we sketch the proof here, following Grove (1987).

Definition.

A *geodesic hinge* $(\gamma_1, \gamma_2, \alpha)$ in a Riemannian manifold is a configuration of two unit speed geodesics $\gamma_i : [0, \ell_i] \to M$ which meet at $p = \gamma_1(\ell_1) = \gamma_2(0)$ with oriented angle $\angle(-\gamma_1{}'(\ell_1), \gamma_2{}'(0)) = \alpha$. (See Figure IX.1.)

Let $(\gamma_1, \gamma_2, \alpha)$ be a hinge in M, and consider the unit parallel field E along γ_1 determined by $E(\ell_1) = \gamma_2{}'(0)$. We say that the hinge $(\gamma_1, \gamma_2, \alpha)$ is *thin* if,

(i) for the corresponding hinge $(\overline{\gamma}_1, \overline{\gamma}_2, \alpha)$ and $\overline{E}$ in $\mathbb{M}^2_\kappa$, a minimal path $\overline{c}(\epsilon)$ from $\overline{\gamma}_1(0)$ to $\overline{\gamma}_2(\ell_2)$ is given by $\overline{c}(\epsilon) = \exp \phi(\epsilon)\overline{E}(\epsilon)$ for some function $\phi : [0, \ell_1] \to \mathbb{R}$, and

(ii) for each $\epsilon \in [0, \ell_1]$, $\gamma_1(\epsilon)$ has no focal points (as first defined in §IX.3 above) along the geodesic $t \mapsto \exp t\phi(\epsilon)E(\epsilon)$, $t \in [0, 1]$.

A *geodesic triangle* in a Riemannian manifold M is a configuration of three unit speed geodesics $\gamma_i : [0, \ell_i] \to M$ (the *sides*) such that

$$\gamma_i(\ell_i) = \gamma_{i+1}(0), \qquad \ell_i + \ell_{i+1} \geq \ell_{i+2},$$

indices taken modulo 3. The points $p_i = \gamma_{i+2}(0)$ are called the *vertices* of the triangle and $\alpha_i = \angle(-\gamma_{i+1}'(\ell_{i+1}), \gamma_{i+2}'(0))$ the *corresponding angles.*

Theorem IX.5.1. (Alexandrov–Toponogov distance comparison theorem) *Let M be a complete Riemannian manifold with sectional curvature $\mathcal{K} \geq \kappa$.*

Let $(\gamma_1, \gamma_2, \alpha)$ be a geodesic hinge in M. Suppose γ_1 is minimal, and if $\kappa > 0$ suppose $\ell_2 \leq \pi/\sqrt{\kappa}$. Let $(\overline{\gamma}_1, \overline{\gamma}_2, \alpha)$ be a geodesic hinge in $\mathbb{M}^2_\kappa$ such that $\ell(\overline{\gamma}_i) = \ell_i$. Then,

$$d(\gamma_1(0), \gamma_2(\ell_2)) \leq d(\overline{\gamma}_1(0), \overline{\gamma}_2(\ell_2)).$$

Theorem IX.5.2. (Alexandrov–Toponogov angle comparison theorem) *Let M be a complete Riemannian manifold with sectional curvature $\mathcal{K} \geq \kappa$.*

Let $(\gamma_1, \gamma_2, \gamma_3)$ be a geodesic triangle in M. Suppose γ_1, γ_3 are minimal, and if $\kappa > 0$ suppose $\ell_2 \leq \pi/\sqrt{\kappa}$. Then there exists a geodesic triangle $(\overline{\gamma}_1, \overline{\gamma}_2, \overline{\gamma}_3)$ in $\mathbb{M}^2_\kappa$ such that

$$\ell(\overline{\gamma}_i) = \ell_i \ \ \forall i, \qquad \overline{\alpha}_1 \leq \alpha_1, \qquad \overline{\alpha}_3 \leq \alpha_3.$$

Except in the case $\kappa > 0$, and $\ell_i = \pi/\sqrt{\kappa}$ for some i, the triangle $(\overline{\gamma}_1, \overline{\gamma}_2, \overline{\gamma}_3)$ is uniquely determined.

Lemma IX.5.1. *A geodesic triangle $(\overline{\gamma}_1, \overline{\gamma}_2, \overline{\gamma}_3)$ in $\mathbb{M}^2_\kappa$, with side lengths ℓ_i, is uniquely determined (up to congruence) by the triplet (ℓ_1, ℓ_2, ℓ_3), unless $\ell_i = \pi/\sqrt{\kappa}$ for some i.*

Moreover, if we fix the lengths ℓ_1, ℓ_2, and the geodesic $\overline{\gamma}_1$, and consider the hinge $(\overline{\gamma}_1, \overline{\gamma}_2, \alpha)$ in $\mathbb{M}^2_\kappa$, with $\phi(\alpha) = d(\overline{\gamma}_1(0), \overline{\gamma}_2(\ell_2))$, then $\alpha \mapsto \phi(\alpha)$ is strictly increasing on $[0, \pi]$, except when $\kappa > 0$ and $\ell_i = \pi/\sqrt{\kappa}$ for at least one i, in which case ϕ is constant.

Proof. Assume $\kappa > 0$. Note that $\phi(0) = |\ell_1 - \ell_2|$, and $\phi(\pi) = \min\{\ell_1 + \ell_2, (2\pi/\sqrt{\kappa}) - (\ell_1 + \ell_2)\}$. Also, if $\phi(\alpha) = \pi/\sqrt{\kappa}$ for some α, then one easily sees that $\alpha = \pi$. Therefore, for all κ, we consider $\alpha < \pi$.

Then there exists a unique minimal geodesic $\overline{c}_\alpha$ from $\overline{\gamma}_1(0)$ to $\overline{\gamma}_2(\ell_2)$. One can easily deduce the result from the Law of Cosines (see Note II.4, when $\kappa \neq 0$). ∎

Corollary IX.5.1. *Theorems* IX.5.1 *and* IX.5.2 *are equivalent.*

Proof of Theorems IX.5.1 and IX.5.2. For the moment, we refer to Theorems IX.5.1 and IX.5.2 as the `hinge theorem` and the `angle theorem`, respectively. We refer to Lemma IX.5.1 as the `monotonicity lemma`.

Whenever we invoke restrictions on lengths in terms of $\sqrt{\kappa}$, we are in the situation of $\kappa > 0$.

Step 1. Given a thin hinge $(\gamma_1, \gamma_2, \alpha)$, consider the path $c(\epsilon) = \exp \phi(\epsilon)E(\epsilon)$ in M from $\gamma_1(0)$ to $\gamma_2(\ell_2)$, as determined above. Then, Corollary IX.3.1 implies

$$d(\gamma_1(0), \gamma_2(\ell_2)) \leq L(c) \leq L(\overline{c}) = d(\overline{\gamma}_1(0), \overline{\gamma}_2(\ell_2)),$$

which implies the hinge theorem for thin hinges. (Note that $\mathcal{K} \geq \kappa$ implies $\ell_2 < \pi/2\sqrt{\kappa}$.) The ultimate proof of the theorem will then be to reduce the case of arbitrary hinges to thin hinges.

Step 2. Consider the subclass $\mathcal{H}$ of hinges $(\gamma_1, \gamma_2, \alpha)$ in M for which

$$\ell_2 < \frac{\pi}{\sqrt{\kappa}}, \qquad \max_t d(\gamma_1(0), \gamma_2(t)) := d_2 < \frac{\pi}{\sqrt{\kappa}}.$$

The set of all minimal geodesics from $\gamma_1(0)$ to $\gamma_2(t)$, $t \in [0, \ell_2]$ is compact (in the topology of the unit sphere of $M_{\gamma_1(0)}$). Then there exists a subdivision $\{0 = t_0 < t_1 < \cdots < t_k = \ell_2\}$ such that

(i) $\gamma_2|[t_i, t_{i+1}]$ is minimizing, and

(ii) the minimal geodesics $\gamma_{3,i}$ from $\gamma_1(0)$ to $\gamma_2(t_i)$, $i = 0, \ldots k$, are such that the hinges

$$(\gamma_{3,i}, \gamma_2|[t_i, t_{i+1}], \alpha_i), \qquad (\gamma_{3,i}, {\gamma_2}^{-1}|[t_i, t_{i-1}], \beta_i),$$

are thin – in particular $|t_i - t_{i-1}| < \pi/2\sqrt{\kappa}$ (see Fig. IX.2).

Step 3. Note that, in $\mathbb{M}_\kappa$, $\ell_2 \leq \ell_1 + d(\overline{\gamma}_1(0), \overline{\gamma}_2(\ell_2))$; so, if $\ell_2 \geq \ell_1 + d(\gamma_1(0), \gamma_2(\ell_2))$ in M, then the hinge theorem is certainly valid. Therefore, we may also assume[2] that

$$\ell_2 \leq \ell_1 + d(\gamma_1(0), \gamma_2(\ell_2)).$$

One can easily check that all triangle inequalities are valid for all the triangles $(\gamma_{3,i}, \gamma_2|[t_i, t_j], {\gamma_{3,j}}^{-1})$, $i < j$.

Step 4. Now use an induction that oscillates between the hinge theorem and the angle theorem.

First, the hinge theorem is valid for the hinges

$$(\gamma_{3,0}, \gamma_2|[t_0, t_1], \alpha_0), \quad \text{and} \quad (\gamma_{3,1}, {\gamma_2}^{-1}|[t_1, t_0], \beta_1),$$

which implies, by the monotonicity lemma, that the angle theorem is valid for the triangle $(\gamma_{3,0}, \gamma_2|[t_0, t_1], {\gamma_{3,1}}^{-1})$.

[2] The point is even though $\gamma_2|[0, \ell_2]$ might not minimize the distance from $\gamma_2(0)$ to $\gamma_2(\ell_2)$ in M.

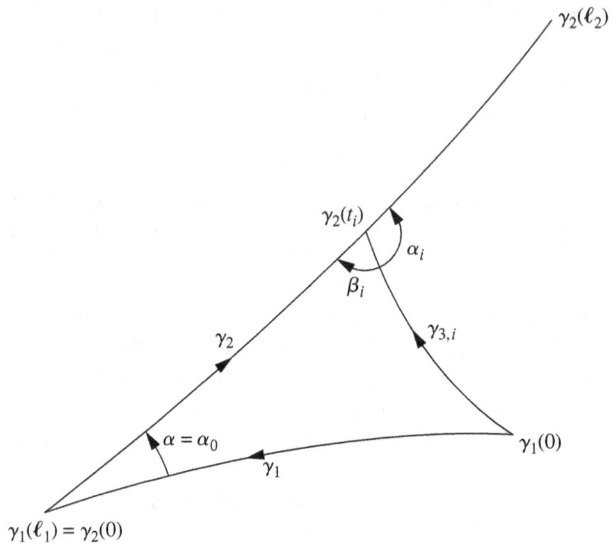

Figure IX.2. For the Alexandrov–Toponogov theorem.

So, if $\overline{\beta_1}$ is the angle in $\mathbb{M}_\kappa$ corresponding to β_1 in M, then $\overline{\beta_1} \leq \beta_1$. Therefore, for the supplementary angles α_1 and $\overline{\alpha_1}$ and of β_1 and $\overline{\beta_1}$, respectively, we have $\alpha_1 \leq \overline{\alpha_1}$. But the hinge theorem is valid for the hinge $(\gamma_{3,1}, \gamma_2|[t_1, t_2], \alpha_0)$; the monotonicity lemma then implies that the hinge theorem is valid for the hinge $(\gamma_{3,0}, \gamma_2|[t_0, t_2], \alpha_0)$.

We now have the hinge theorem for the hinges

$$(\gamma_{3,0}, \gamma_2|[t_0, t_2], \alpha_0) \quad \text{and} \quad (\gamma_{3,2}, {\gamma_2}^{-1}|[t_2, t_1], \beta_2).$$

The monotonicity lemma implies the validity of angle theorem to the larger triangle $(\gamma_{3,0}, \gamma_2|[t_0, t_2], {\gamma_{3,2}}^{-1})$. Now one can extend the hinge theorem to the hinge $(\gamma_{3,0}, \gamma_2|[t_0, t_3], \alpha_0)$ as above.

By continuing the argument, one has the theorem for hinges in the class $\mathcal{H}$.

Step 5. To obtain the general theorem for *all* hinges, apply a limit argument to hinges (first on α and then on the lengths ℓ_i) in $\mathcal{H}$. ∎

Remark IX.5.1. Suppose in the angle theorem we have equality, that is, we have $\ell_2 < \pi/\sqrt{\kappa}$, $0 < \alpha < \pi$, and

$$d(\gamma_1(0), \gamma_2(\ell_2)) = d(\overline{\gamma}_1(0), \overline{\gamma}_2(\ell_2)).$$

Then, one has an isometric totally geodesic imbedding (see Exercise II.3) of the triangular surface in $\mathbb{M}^2_\kappa$ determined by $(\overline{\gamma}_1, \overline{\gamma}_2, \alpha)$ to M, which maps

$(\overline{\gamma}_1, \overline{\gamma}_2, \alpha)$ to $(\gamma_1, \gamma_2, \alpha)$ to M. Moreover, the image of the minimal geodesics in $\mathbb{M}^2_\kappa$ from $\overline{\gamma}_1(0)$ to $\overline{\gamma}_2(t)$ are minimal geodesics in M from $\gamma_1(0)$ to $\gamma_2(t)$.

§IX.6. Convexity

In the general Riemannian setting, there are many notions of convexity, all of which coincide in $\mathbb{R}^n$.

Definition. Let M be a complete Riemannian manifold. A set A in M is:

• *weakly convex* if for any $p, q \in A$ there exists a geodesic $\gamma_{pq} \subseteq A$ such that γ_{pq} is the unique minimizer in A connecting p to q;

• *convex* if for any $p, q \in A$ there exists a geodesic $\gamma_{pq} \subseteq A$ such that γ_{pq} is the unique minimizer in M connecting p to q;

• *strongly convex* if for any $p, q \in A$ there exists a geodesic $\gamma_{pq} \subseteq A$ such that γ_{pq} is the unique minimizer in M connecting p to q, and γ_{pq} is the only geodesic contained in A joining p to q.

Calculations Associated with Convexity

Given a point $p \in M$, and a path $\Omega(t)$ in M_p. Set

$$\omega(t) = \exp_p \Omega(t), \qquad v(s,t) = \exp_p s\Omega(t), \qquad \phi(t) = \frac{1}{2}|\Omega(t)|^2.$$

Of course, $|\Omega(t)|$ is the length of the geodesic $\gamma_t(s) = v(s,t)$ from p to $\Omega(t)$. Then,

$$\phi'(t) = \langle \partial_s v, \partial_t v\rangle_{s=1},$$

by Gauss' lemma (Lemma I.6.1) or the first variation of arc length (Theorem II.4.1). If ω is a geodesic, then

$$\phi''(t) = \langle \nabla_t \partial_s v, \partial_t v\rangle_{s=1} = \langle \nabla_s \partial_t v, \partial_t v\rangle_{s=1}.$$

Write

$$\partial_t v = s\alpha \partial_s v + Y_t.$$

Then, Y_t is the Jacobi field along the geodesic $\gamma_t(s) = v(s,t)$, pointwise orthogonal to γ_t, and satisfying $Y_t(0) = 0$, which implies

$$\nabla_s \partial_t v = \alpha \partial_s v + \nabla_s Y_t.$$

One concludes

$$\phi''(t) = \alpha^2 |\Omega(t)|^2 + \langle \nabla_s Y_t, Y_t\rangle_{s=1}.$$

If, in addition, one assumes that

$$\mathcal{K} \leq \delta \tag{IX.6.1}$$

on M, then

$$\phi'' \geq \alpha^2|\Omega|^2 + \frac{\mathbf{C}_{\delta|\Omega|^2}}{\mathbf{S}_{\delta|\Omega|^2}}(1)|Y_t|^2(1) \geq \min\left\{1, \frac{\mathbf{C}_{\delta|\Omega|^2}}{\mathbf{S}_{\delta|\Omega|^2}}(1)\right\}|\omega'|^2, \tag{IX.6.2}$$

when $|\Omega| < \pi/\sqrt{\delta}$. Of course, this is only useful when $|\Omega| < \pi/2\sqrt{\delta}$.

Applications

Theorem IX.6.1. *Assume* (IX.6.1) *on M, and set*

$$r_1 = \min\left\{\frac{\operatorname{inj} M}{2}, \frac{\pi}{2\sqrt{\delta}}\right\}.$$

Then, $B(x;r_1)$ is strongly convex.

Proof. By replacing δ by $\delta + \epsilon$, $\epsilon > 0$ arbitrarily small, we may assume that the inequality in (IX.6.1) is *strict* inequality.

If $p, q \in B(x;r_1)$, then any geodesic in $B(x;r_1)$ joining p to q must be a unique minimizer. Therefore, for any minimizer in M joining p to q, it suffices to show that it is completely contained in $B(x;r_1)$.

Consider such a minimizer. Then, the geodesic triangle xpq has length $< 4r_1$. Therefore, if $\delta > 0$, no two points on the path xpq have distance $\pi/\sqrt{\delta}$ one from the other. By comparison to the triangle $x_\delta p_\delta q_\delta$ in $\mathbb{M}^2_\delta$ with the same corresponding sides, the angles at p and q are strictly less than the respective angles at p_δ and q_δ. (See Corollary IX.5.1 and Exercise IX.1 in §IX.9.)

Parameterize both geodesic segments γ_{pq} and $\gamma_{p_\delta q_\delta}$ from 0 at p (respectively, p_δ) to 1 at q (respectively, q_δ). Then, the first variation of arc length formula (Theorem II.4.1) implies there exists $\rho > 0$ for which we have

$$d(x, \gamma_{pq}(s)) < d(x_\delta, \gamma_{p_\delta q_\delta}(s)) \tag{IX.6.3}$$

for all $s \in (0, \rho) \cup (1 - \rho, 1)$. But, this very same argument now implies that (IX.6.3) is valid for all $s \in (0, 1)$. Therefore,

$$\begin{aligned} d(x, \gamma_{pq}(s)) &< d(x_\delta, \gamma_{p_\delta q_\delta}(s)) < \max\{d(x_\delta, p_\delta), d(x_\delta, q_\delta)\} \\ &= \max\{d(x, p), d(x, q)\} \end{aligned}$$

for all $s \in (0, 1)$, which is the claim. ∎

Theorem IX.6.2. *Assume* (IX.6.1) *on* M, *and let*

$$r_2 = \min\left\{\operatorname{inj} M, \frac{\pi}{2\sqrt{\delta}}\right\}.$$

Then, $\overline{B(x;r_2)}$ *is weakly convex.*

Proof. (Buser–Karcher (1981, p. 102)) Given p and q in $B(x;r_2)$. Then, the Arzela–Ascoli theorem implies (see Exercises IV.24 and IV.25) that there exists a shortest *path* $\omega(t) \subseteq \overline{B(x;r_2)}$ connecting p to q. (Of course, any subsegment of this path contained in the interior $B(x;r_2)$ is geodesic.) We assume $d(x,p) \geq d(x,q)$, and show that

$$\omega \subseteq \overline{B(x;d(x,p))}.$$

This will then imply that ω is a minimizing geodesic in $B(x;r_2)$.

To this end, we write

$$\omega(t) = \exp_x \Omega(t),$$

set

$$\tilde{\Omega}(t) = \frac{d(x,p)}{\max\{d(x,p), |\Omega(t)|\}}\Omega(t),$$

$$\tilde{\omega}(t) = \exp_x \tilde{\Omega}(t) \subseteq \overline{B(x;d(x,p))} \subseteq B(x;r_2).$$

Then, $\tilde{\omega}$ is a path from p to q; and (IX.6.1), the estimate on r_2, and the Rauch theorem imply

$$\ell(\tilde{\omega}) \leq \ell(\omega).$$

We conclude that the shortest connection from p to q in $\overline{B(x;r_2)}$ is contained in $\overline{B(x;d(x,p))}$, and is therefore a geodesic.

We assumed thus far that p and q were in the interior of $\overline{B(x;r_2)}$. If we are given that p and q are arbitrary in $\overline{B(x;r_2)}$, then we may approach p and q from the *interior* $B(x;r_2)$ to obtain the existence of a minimizing geodesic connecting p to q, which is completely contained in $\overline{B(x;r_2)}$. So, it remains to show that this geodesic is unique in $\overline{B(x;r_2)}$.

First, minimizers from p to q must have length $\leq 2r_2 < \pi/\sqrt{\delta}$, which implies p and q have no conjugate points on γ_{pq} (to be precise, one should also say γ_{qp}). Therefore, consider the collection of points $\{(p,q) \in \overline{B(x;r_2)} \times \overline{B(x;r_2)}\}$ for which p is joined to q by more than one $\overline{B(x;r_2)}$–minimizing geodesic. Consider the pair with minimal distance. Then, the Klingenberg argument (Theorem III.2.4) implies there exists a closed geodesic γ in $\overline{B(x;r_2)}$. But, the function

$t \mapsto d(x, \gamma(t))^2/2$ can have no maximum, which implies a contradiction, which implies the uniqueness of the minimizer. ∎

Definition. For any $x \in M$, we define $\operatorname{conv} x$, the *convexity radius of* x, by

$$\operatorname{conv} x = \sup\{\rho : B(x; r) \text{ convex for all } r < \rho\}.$$

Lemma IX.6.1. *Given* $x, y \in M$ *such that*

$$d(x, y) := R < 2 \min\{\operatorname{conv} x, \operatorname{conv} y\}.$$

Then, $B(x; R/2) \cap B(y; R/2) = \emptyset$.

Proof. Let γ be a unit speed geodesic from x to y, with midpoint $z_0 = \gamma(R/2)$. If $z \in B(x; R/2) \cap B(y; R/2)$, then join z_0 to z by a *unique* minimizer

$$\omega \subseteq \overline{B(x; R/2)} \cap \overline{B(y; R/2)}.$$

Then, $\omega(t)$ may be lifted to respective paths $\Omega_x(t)$ and $\Omega_y(t)$ in M_x and M_y, with associated functions of distance $\phi_x(t)$, $\phi_y(t)$ are described in our calculations above. The formula for the first variation of arc length (Theorem II.4.1) implies

$$\lim_{t \downarrow 0} \phi_x{}'(t) = \lim_{t \downarrow 0} \phi_y{}'(t) = 0,$$

and the hypothesis on convexity implies $\phi_x{}'' \geq 0, \quad \phi_y{}'' \geq 0$, which implies $\phi_x \geq R/2, \quad \phi_y \geq R/2$, which is a contradiction. ∎

Proposition IX.6.1. (M. Berger (1976)) *We always have*

$$\operatorname{conv} M \leq \frac{\operatorname{inj} M}{2}.$$

Proof. If not, that is, if $\operatorname{conv} M > (1/2)\operatorname{inj} M$, then there exist $p, q \in M$ such that p and q are cut points one to the other, and

$$d(p, q) < 2\operatorname{conv} M.$$

Let γ be *a* minimizing unit speed geodesic from p to q, z the midpoint of γ, and consider $B(z; \operatorname{conv} M)$. Then, both p and q are in $B(z; \operatorname{conv} M)$, which implies that γ is the unique minimizer from p to q, which implies (see §III.2) p and q are conjugate along γ.

Let $t_0 = d(p, q)$, and set $\gamma(0) = p, \gamma(t_0) = q$. Then, for small $t - t_0 > 0$, we have $d(p, \gamma(t)) < t$, which determines a unique minimizing unit speed geodesic γ_t from p to $\gamma(t)$, with $\gamma_t \to \gamma$ as $t \downarrow t_0$. Consider $B(p; t/2)$ and $B(\gamma(t); t/2)$.

Both closures are convex, which implies by the above lemma that they are disjoint (indeed: $t/2 < (\text{conv } M)/2$). On the other hand, their intersection must contain the point $\gamma_t(d(p, \gamma(t))/2)$, which is a contradiction. ∎

§IX.7. Center of Mass

Here, we follow Karcher (1977).

In $\mathbb{R}^n$, one has the following typical situation. One is given a subset A of $\mathbb{R}^n$, with positive finite measure, and one considers the vector field on $\mathbb{R}^n$ given by

$$v(x) = \frac{1}{V(A)} \int_A (x - a)\, dV(a).$$

More generally, for any probability measure $\mathfrak{m}$ on A one can consider the vector field

$$v(x) = \int_A (x - a)\, d\mathfrak{m}(a).$$

The usual definition of *center of mass of* $(A, \mathfrak{m})$ is that point in $\mathbb{R}^n$ solving the equation $v(x) = 0$. Needless to say, the solution x is unique. Note that for

$$\mathcal{E}(x) = \frac{1}{2} \int_A |x - a|^2\, d\mathfrak{m}(a),$$

we have

$$(\operatorname{grad} \mathcal{E})(x) = v(x).$$

To consider a center of mass in a general Riemannian manifold, we start with a given probability space $(A, \mathfrak{m})$, a mapping $f : A \to M$, and consider the function $\mathcal{E}$ on M given by

$$\mathcal{E}(x) = \frac{1}{2} \int_A d^2(x, f(a))\, d\mathfrak{m}(a).$$

Then ask when does $\mathcal{E}$ have a unique minimum.

First consider

$$\mathcal{E}_a(x) := \frac{1}{2} d^2(x, f(a)).$$

Assume $f(A) \subseteq B$, where B is a weakly convex subset of M. Then to each $x, y \in B$, we have well-defined $\exp_x{}^{-1} y$. One immediately sees that, for all $x \in B$ we have,

$$(\operatorname{grad} \mathcal{E}_a)(x) = -\exp_x{}^{-1} f(a).$$

Proposition IX.7.1. *Assume* (IX.6.1) *on M, and* $\operatorname{diam} B \leq \pi/2\sqrt{\delta}$. *Then $\mathcal{E}$ has a unique minimum $\mathcal{C}$ in B.*

Proof. Let $\omega(t)$ be a geodesic in B, and $\phi(t) = \mathcal{E}_a(\omega(t))$. Then, the hypotheses imply that $\phi'' > 0$, which implies (Exercises II.9 and II.20(a)) the claim. ∎

Proposition IX.7.2. *Assume* (IX.6.1) *is valid on M, and let $B = B(p; r_0)$, where*

$$r_0 < \min\left\{\frac{\operatorname{inj} M}{2}, \frac{\pi}{4\sqrt{\delta}}\right\}.$$

Then, for $x \in B$, we have

$$|\operatorname{grad} \mathcal{E}|(x) \geq d(x, \mathcal{C}) \frac{\mathbf{C}_{2\delta r_0}}{\mathbf{S}_{2\delta r_0}}(1).$$

Proof. Let $\gamma : [0, 1] \to M$ denote the minimizing geodesic from $\mathcal{C}$ to x. Then,

$$\begin{aligned} |\operatorname{grad} \mathcal{E}|(x)|\gamma'|(1) &\geq \langle(\operatorname{grad} \mathcal{E})(x), \gamma'(1)\rangle \\ &= \left\{\frac{d}{dt}\mathcal{E}(\gamma(t))\right\}_{t=1} \\ &= \int_0^1 \frac{d^2}{dt^2}(\mathcal{E} \circ \gamma)(t)\, dt \\ &= \int_A d\mathfrak{m}(a) \int_0^1 \frac{d^2}{dt^2}(\mathcal{E}_a \circ \gamma)(t)\, dt \\ &\geq \frac{\mathbf{C}_{2\delta r_0}}{\mathbf{S}_{2\delta r_0}}(1) d^2(x, \mathcal{C}), \end{aligned}$$

by (IX.6.2), which implies the claim. ∎

§IX.8. Cheeger's Finiteness Theorem

Since the dimension n of the manifolds under consideration never changes, we write $\mathbb{B}(r)$ for $\mathbb{B}^n(r)$, the n–disk of radius r in $\mathbb{R}^n$.

The heart of the matter is contained in the following lemma.

Lemma IX.8.1. (S. Peters (1984)) *We are given two compact n–dimensional Riemannian manifolds M and $\overline{M}$, $n \geq 2$, whose sectional curvatures satisfy*

$$|\mathcal{K}_M|, |\mathcal{K}_{\overline{M}}| \leq \Lambda,$$

with injectivity radii satisfying

$$\operatorname{inj} M, \operatorname{inj} \overline{M} \geq \iota.$$

Fix R such that

$$0 < R < \frac{\iota}{4}.$$

We consider respective discretizations of M, $\overline{M}$, *with the same number of elements equal to* N, *and both having separation distance and covering radius equal to* R (see §IV.4).

In M: *To each* $z_i, i = 1, \ldots, N$ *of the discretization, associate a linear isometry*

$$u_i : \mathbb{R}^n \to M_{z_i},$$

and Riemann normal coordinates

$$\phi_i = \exp_{z_i} \circ u_i$$

on D_{z_i}, *the set about* z_i *inside the cut locus of* z_i (see §III.2). *Also, for* $d(z_i, z_j) < \iota$, *let*

$$P_{ij} : M_{z_i} \to M_{z_j}$$

denote parallel translation along the minimizing geodesic connecting z_i *to* z_j.

In $\overline{M}$: *Consider the corresponding data* $\overline{z}_i, \overline{u}_i, \overline{\phi}_i, \overline{P}_{ij}, i, j = 1, \ldots, N$.

Then, $R = R(n, \Lambda, \iota)$ *may be chosen sufficiently small so that there exist constants*

$$\epsilon_0 = \epsilon_0(R), \qquad \epsilon_1 = \epsilon_1(R)$$

for which M *and* $\overline{M}$ *are diffeomorphic whenever*

$$|{\phi_j}^{-1}\phi_i - {\overline{\phi}_j}^{-1}\overline{\phi}_i| < \frac{2\epsilon_0}{3} \tag{IX.8.1}$$

on $\mathbb{B}(3R)$, *and*

$$|{u_j}^{-1}P_{ij}u_i - {\overline{u}_j}^{-1}\overline{P}_{ij}\overline{u}_i| < \epsilon_1, \tag{IX.8.2}$$

on $\mathbb{R}^n$, *for all* i, j. (See Fig. IX.3.)

The proof of the lemma is quite involved, and we shall break it into a number of steps. But first we show how to use it to derive Cheeger's finiteness theorem.

Theorem IX.8.1. (J. Cheeger (1970a)) *Given real numbers* $n, d, V, \Lambda > 0$. *Then there exist only finitely many diffeomorphism classes of compact* n–*dimensional Riemannian manifolds satisfying*

$$\operatorname{diam} M \leq d, \qquad V(M) \geq V, \qquad |\mathcal{K}| \leq \Lambda. \tag{IX.8.3}$$

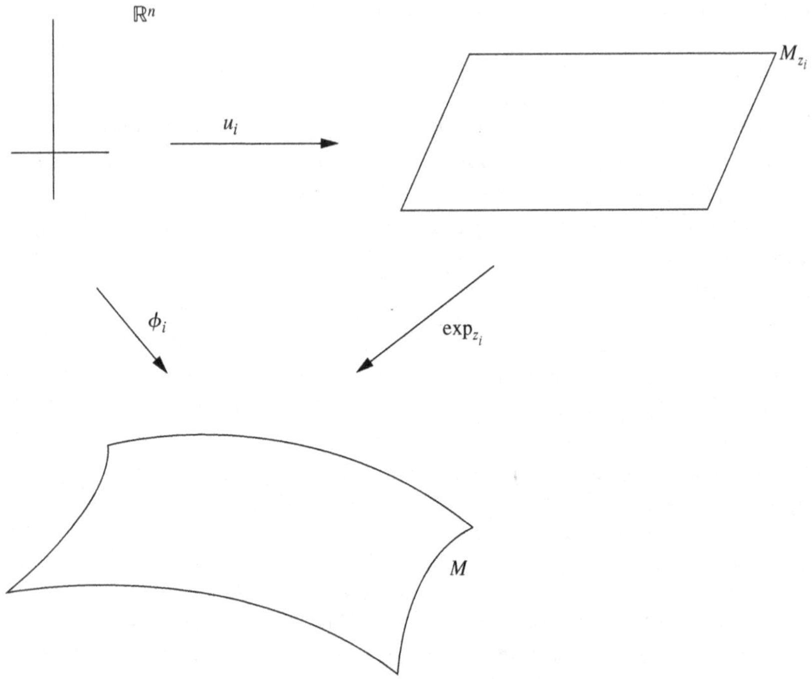

Figure IX.3. Indexing the Riemann normal coordinate systems.

Proof. It suffices to show that given any infinite sequence M_α of compact n–dimensional Riemannian manifolds satisfying (IX.8.3), then it is possible to pick an infinite subsequence for which any two of the manifolds are diffeomorphic. To actually pick this subsequence, we proceed as follows:

By Theorem IX.3.3, there exists ι, depending only on n, d, V, Λ, such that

$$\operatorname{inj} M_\alpha \geq \iota$$

for all α, and the disk $B(x_\alpha; \iota/2)$ is convex for all α, $x_\alpha \in M_\alpha$.

Fix $R < \iota/4$. Pick a typical M in the sequence $\{M_\alpha\}$. Pick a discretization of M with separation constant and covering radius both equal to R. Then, N the number of elements of the discretization satisfies (see Lemma IV.4.1)

$$N \leq \frac{V_{-\Lambda}(\operatorname{diam} M)}{V_\Lambda(R/2)}.$$

Now the Arzela–Ascoli theorem implies that if we consider the collection $\mathcal{H}_{K_1}$ of imbeddings $\mathbf{F} : \mathbb{B}(R) \to \mathbb{B}(R)$ for which

$$\sup_{\mathbb{B}(R)} \{|\mathbf{F}| + |\operatorname{grad} \mathbf{F}|\} \leq K_1,$$

then $\mathcal{H}_{K_1}$ is totally bounded in the sense that, given any $\epsilon_0 > 0$, there exists a subset $\mathcal{H}$ consisting of finitely many elements of $\mathcal{H}_{K_1}$ such that any $\mathbf{F} \in \mathcal{H}_{K_1}$ has C^0–distance (i.e., the distance between $\mathbf{F}_1$ and $\mathbf{F}_2$ is $\sup |\mathbf{F}_1 - \mathbf{F}_2|$) from $\mathcal{H}$ strictly less than $\epsilon_0/3$.

Note that the Rauch comparison theorem implies there exists a K_1 for which the mappings ${\phi_j}^{-1}\phi_i$ of any discretization of M, as described in Peters' lemma, are all elements of $\mathcal{H}_{K_1}$, for all i, j.

Also, the mappings ${u_j}^{-1}R_{ij}u_i$ of any discretization of M, as described in Peters' lemma, are all elements of $\mathcal{O}(n)$, the orthogonal group of $\mathbb{R}^n$, for all i, j. Since $\mathcal{O}(n)$ is compact, given any $\epsilon_1 > 0$, there exists a finite covering of $\mathcal{O}(n)$ by disks of radius $\epsilon_1/2$. (See Exercise I.22(a) for the Riemannian metric on $\mathcal{O}(n)$.)

We are now ready to finish the proof. Since, in each of the conditions below, there are only finitely many distinct possibilities, we have for our sequence M_α of Riemannian manifolds, a subsequence – also called M_α – satisfying:

(i) $N_\alpha = N \leq N_0$ for all α, that is, all discretizations under consideration have the same number of elements;

(ii) the network of overlaps match, more precisely, $\phi_i^\alpha(\mathbb{B}(R)) \cap \phi_j^\alpha(\mathbb{B}(R)) \neq \emptyset$ if and only if $\phi_i^\beta(\mathbb{B}(R)) \cap \phi_j^\beta(\mathbb{B}(R)) \neq \emptyset$ for all $\alpha, \beta; i, j$;

(iii) for each i, j, the full collection $\{(\phi_j^\alpha)^{-1}\phi_i^\alpha\}_\alpha$ all belong to the same $\epsilon_0/3$–disk in $\mathcal{H}_{K_1}$;

(iv) for each i, j, the full collection $\{(u_j^\alpha)^{-1}P_{ij}^\alpha u_i^\alpha\}_\alpha$ all belong to the same $\epsilon_1/2$–disk in $\mathcal{O}(n)$.

Then, for sufficiently small choices of ϵ_0, ϵ_1, we may use Peters' lemma to guarantee that all $\mathbb{M}_\alpha$ are diffeomorphic. ∎

Proof of Peters' Lemma. The idea behind the proposed diffeomorphism is as follows: The collection of maps

$$F_i = \overline{\phi}_i {\phi_i}^{-1}$$

is a collection of *locally* defined diffeomorphisms from M to $\overline{M}$. The conditions (IX.8.1) imply that for any $p \in M$ one has a neighborhood $U(p)$ such that the different images $\{F_i(U(p))\}_{i=1}^N$ are pointwise sufficiently close one to the other to allow the construction of an "average image" through the center of mass construction of the previous section.

Step 1. Given any $p \in B(z_i; R) \cap B(z_j; R)$, we want an upper bound on $d(F_i(p), F_j(p))$. We start with some

$$R < \frac{\iota}{4}.$$

Of course, for any such $R > 0$, there exists $\epsilon_0 > 0$ for which (IX.8.1) is valid on $\mathbb{B}(3R)$. Then, $p \in B(z_i; R) \cap B(z_j; R)$ implies

$$d(z_i, z_j) < 2R < \iota,$$

which implies, since $\phi_j{}^{-1}\phi_i$ and $\overline{\phi}_j{}^{-1}\overline{\phi}_i$ will both map $\mathbb{B}(R)$ to $\mathbb{B}(3R)$, for all i, j, that

$$d(\phi_j{}^{-1}(p), \overline{\phi}_j{}^{-1}F_i(p)) = |\{\phi_j{}^{-1}\phi_i - \overline{\phi}_j{}^{-1}\overline{\phi}_i\}\phi_i{}^{-1}(p)| < \frac{2\epsilon_0}{3}.$$

That is,

$$d(\phi_j{}^{-1}(p), \overline{\phi}_j{}^{-1}F_i(p)) < \frac{2\epsilon_0}{3}.$$

Now, $\phi_j{}^{-1}(p), \overline{\phi}_j{}^{-1}(p) \in \mathbb{R}^n$, and $\overline{u}_j\phi_j{}^{-1}(p), \overline{u}_j\overline{\phi}_j{}^{-1}F_i(p) \in \overline{M}_{\overline{z}_j}$. So, we must know the largest possible expansion in distance to which two points are subject under $\exp_{\overline{z}_j}$. Well, if $\rho \leq \iota$ and $F_j(p), F_i(p) \in B(\overline{z}_j; \rho)$, then by (IX.2.8) (see Remark IX.1), we have

$$d(F_j(p), F_i(p)) < \frac{\mathbf{S}_{-\Lambda}(\rho)}{\rho}\frac{2\epsilon_0}{3},$$

so we must find a good estimate of ρ for which $F_j(p), F_i(p) \in B(\overline{z}_j; \rho)$. Clearly, $d(F_j(p), \overline{z}_j) < R$. What about $d(F_i(p), \overline{z}_j)$? Well,

$$\begin{aligned} d(F_i(p), \overline{z}_j) &\leq d(F_i(p), F_i(z_i)) + d(F_i(z_i), \overline{z}_j) \\ &= d(p, z_i) + d(\overline{z}_i, \overline{z}_j) \\ &\leq R + d(\overline{z}_i, F_i(z_j)) + d(F_i(z_j), \overline{z}_j) \\ &< 3R + d(F_i(z_j), \overline{z}_j). \end{aligned}$$

Then, by applying (IX.8.1) to $\zeta = \phi_i{}^{-1}(z_j)$, we obtain

$$d(F_i(z_j), \overline{z}_j) = |\overline{\phi}_j{}^{-1}F_i(z_j)| = |\{\phi_j{}^{-1}\phi_i - \overline{\phi}_j{}^{-1}\overline{\phi}_i\}(\zeta)| < \frac{2\epsilon_0}{3}.$$

So, we want $\rho = 3R + 2\epsilon_0/3$. Thus, we shall require of ϵ_0 that $\epsilon_0 \leq R$. Then, for

$$\epsilon_0 \leq R < \frac{\iota}{4},$$

such that (IX.8.1) is valid on $\mathbb{B}(3R)$ for all i, j, we have

$$d(F_j(p), F_i(p)) < \frac{\mathbf{S}_{-\Lambda}(4R)}{4R}\frac{2\epsilon_0}{3}.$$

Pick

$$R_0 > 0: \quad \frac{\mathbf{S}_{-\Lambda}(4R_0)}{4R_0} = \frac{5}{4}, \qquad R_1 = \min\{R_0, \iota/5\}.$$

Then, for $\epsilon_0 \le R \le R_1$, the validity of (IX.8.1) on $\mathbb{B}(3R)$ implies

$$d(F_i(p), F_j(p)) < \epsilon_0$$

for $p \in B(z_i; R) \cap B(z_j; R)$. Henceforth, $R \le R_1$.

Step 2. We now let

$$\epsilon_0 = \min\{\pi/2\sqrt{\Lambda}, R\}$$

and assume (IX.8.1) is valid on $\mathbb{B}(3R)$. Then,

$$\epsilon_0 < \operatorname{conv} M, \ \operatorname{conv} \overline{M}.$$

Now, for any $p \in M$, there exists $\overline{q}(p) \in \overline{M}$ for which $F_i(p) \in B(\overline{q}(p); \epsilon_0)$ for all those i for which $p \in B(z_i; R)$. So, we apply the center of mass construction to the set $\{F_i(p)\}$ as follows:

Fix $\eta : [0, +\infty) \to [0, 1] \in C_c^\infty$ such that

$$|\eta'| \le 4, \qquad \eta = \begin{cases} 1 & r \le 1/2 \\ 0 & r \ge 1 \end{cases}.$$

For every i, consider

$$\eta_i(p) = \eta\left(\frac{d(p, z_i)}{2R}\right), \qquad \psi_i(p) = \frac{\eta_i(p)}{\sum_j \eta_j(p)}.$$

For every $p \in M$, consider the measure space $(A_p, \mathfrak{m}_p)$, where $\mathfrak{m}_p$ is the measure on A_p, given by

$$A_p := \{F_i(p) \in \overline{M} : \ p \in B(z_i; R)\}, \qquad \mathfrak{m}_p(F_i(p)) = \psi_i(p).$$

The map f (in the definition of the center of mass) from A_p to $\overline{M}$ is the inclusion $\{F_i(p)\}_{i=1}^N \hookrightarrow \overline{M}$. The energy function $\mathcal{E}$ on $\overline{M}$ is given by

$$\mathcal{E}_p(\overline{x}) = \frac{1}{2} \sum_{i=1}^N d^2(\overline{x}, F_i(p))\psi_i(p),$$

with gradient vector field

$$(\operatorname{grad} \mathcal{E}_p)_{|\overline{x}} = -\sum_{i=1}^N \psi_i(p) \exp_{\overline{x}}{}^{-1} F_i(p).$$

We therefore define $F(p)$ to be the unique minimum point in $B(\overline{q}(p); \epsilon_0)$ of the function $\mathcal{E}_p$, that is, *the* solution in $B(\overline{q}(p); \epsilon_0)$ of

$$\sum_{i=1}^{N} \eta_i(p) \exp_{\overline{x}}{}^{-1} F_i(p) = 0.$$

Write

$$v(p; \overline{x}) := \sum_{i=1}^{N} \eta_i(p) \exp_{\overline{x}}{}^{-1} F_i(p).$$

To solve

(IX.8.4) $$v(p; \overline{x}) = 0$$

for $\overline{x} = \overline{x}(p)$, we may use the implicit function theorem, since Hess $\mathcal{E}_p$ is nonsingular. Furthermore, $\overline{x}$ is differentiable with respect to p.

Step 3. The next thing to do is to show that $\overline{x}(p)$ has maximal rank. Differentiate (IX.8.4) with respect to p_j (local coordinates for p). We obtain

$$\nabla_{\partial p_j} v + \sum_k \frac{\partial \overline{x}^k}{\partial p_j} \nabla_{\partial \overline{x}^k} v = 0.$$

We already know that the matrix (relative to some coordinate system about $\overline{x}$) $\nabla_{\partial \overline{x}^k} v$ is nonsingular. So, it suffices to show that $\nabla_{\partial p_j} v$ is nonsingular.

Consider a path ω in M with

$$p = \omega(0), \qquad \xi = \omega'(0).$$

Then, we wish to show that the covariant derivative $\nabla_t v$ of v along the map $t \mapsto \overline{x}(\omega(t))$ does not vanish at $t = 0$. First, we have

$$\begin{aligned} \nabla_\xi v &= \{\nabla_t \, v_{|\omega(t)}\}_{|t=0} \\ &= \sum_i \left\{ \frac{d(\eta_i(\omega(t)))}{dt} \exp_{\overline{x}}{}^{-1} F_i(p) + \eta_i(p)(\exp_{\overline{x}}{}^{-1})_{*|F_i(p)} \frac{d(F_i(\omega(t))}{dt} \right\}_{|t=0}. \end{aligned}$$

For convenience, write, for $q \in B(p; r)$

$$\overline{x}_i(q) = \exp_{\overline{x}}{}^{-1} F_i(q), \qquad \overline{x} = F(p).$$

Then, we may write

$$\nabla_\xi v = \sum_i \left\{ (d\eta_{i|p} \cdot \xi) \overline{x}_i(p) + \eta_i(p)(\overline{x}_i)_{*|p} \cdot \xi \right\},$$

so

$$|\nabla_\xi v| \geq \left| \sum_i \eta_i(p)(\overline{x}_i)_{*|p} \cdot \xi \right| - \left| \sum_i (d\eta_{i|p} \cdot \xi) \overline{x}_i(p) \right|.$$

Start with estimating

$$\left|\sum_i (d\eta_{i|p} \cdot \xi)\overline{x}_i(p)\right|$$

from above. Of course,

$$|\overline{x}_i(p)| = d(F(p), F_i(p)) < \epsilon_0$$

(since $F(p)$ is the minimum point of $\mathcal{E}_p$), and

$$|d\eta_{i|p} \cdot \xi| \leq 4|\xi|/R,$$

which implies

$$|(d\eta_{i|p} \cdot \xi)\overline{x}_i(p)| \leq \frac{4\epsilon_0}{R}|\xi|;$$

so we must now give an upper bound on the number of $d\eta_i$ that do not vanish at p, that is, an upper bound for the cardinality $N'(p)$ of

$$G_p := \{z_i \,:\, i = 1, \ldots, N\} \cap \{B(p; R) \setminus B(p; R/2)\} \subseteq B(p; R).$$

The Bishop–Gromov argument (see Lemma IV.4) gives the upper bound –

$$N'(p) \leq \frac{V_{-\Lambda}(5R/2)}{V_{-\Lambda}(R/2)} \leq \text{const.}_{\Lambda,\iota};$$

we conclude

$$\left|\sum_i (d\eta_{i|p} \cdot \xi)\overline{x}_i(p)\right| \leq \frac{4\epsilon_0}{R}\frac{V_{-\Lambda}(5R/2)}{V_{-\Lambda}(R/2)}|\xi|.$$

Step 4. The hard part is to estimate

$$\left|\sum_i \eta_i(p)(\overline{x}_i)_{*|p} \cdot \xi\right|$$

from below. We are evaluating at $v = 0$, so

$$\overline{x}_i(p) = \exp_{F(p)}{}^{-1} F_i(p) = \exp_{F(p)}{}^{-1} \circ \overline{\phi}_i \circ {\phi_i}^{-1}(p),$$

which implies directly from the geometric Rauch theorem (Theorem IX.2.3), that for each $|(\overline{x}_i)_{*|p} \cdot \xi|$, we have

$$|(\overline{x}_i)_{*|p} \cdot \xi| \geq \frac{|\overline{x}_i(p)|}{\mathbf{S}_{-\Lambda}(|\overline{x}_i(p)|)}\frac{\mathbf{S}_{\Lambda}(d(p, z_i))}{\mathbf{S}_{-\Lambda}(d(p, z_i))}|\xi|.$$

So, we will want

$$\epsilon_0 \leq R \leq R_2 := \min\{R_1, \sup \rho\},$$

where the sup is taken over all ρ for which

$$\frac{\alpha}{\mathbf{S}_{-\Lambda}(\alpha)}\frac{\mathbf{S}_{\Lambda}(\beta)}{\mathbf{S}_{-\Lambda}(\beta)} \geq 1 - \Lambda\rho^2 \qquad \forall\alpha, \beta \in (0, \rho).$$

This will then imply

$$|(\overline{x}_i)_{*|p} \cdot \xi| \geq 1 - \Lambda R^2.$$

We denote the elements of G_p by $z_{i'}$, $i' = 1, \ldots, N'(p)$, and the associated maps by $\overline{x}_{i'}$. Since we have such an estimate for each individual $(\overline{x}_i)_{*|p} \cdot \xi$, we now want to show that the collection of points

$$\{(\overline{x}_{i'})_{*|p} \cdot \xi\}_{i'=1}^{N'(p)} \subseteq \overline{M}_{F(p)}$$

are sufficiently close each to the other so that no significant cancellation effects can apply to

$$\sum_{i'} \eta_{i'}(p)(\overline{x}_{i'})_{*|p} \cdot \xi.$$

Step 5. We prepare here for Step 6, in which we replace the linearized exponential map $(\overline{x}_{i'})_{*|p}$ by parallel translation. So, we first compare the two. For $\xi \in \mathsf{S}_p$, we have

$$|\{(\exp_p)_{*|t\xi} \circ \mathfrak{I}_{t\xi} - \tau_{p,\gamma_\xi(t)}\}\eta| \leq \left\{\frac{\mathbf{S}_{-\Lambda}(t)}{t} - 1\right\}|\eta|,$$

where $\tau_{p,\gamma_\xi(t)}$ denotes parallel translation along the minimizing geodesic γ_ξ from p to $\gamma_\xi(t)$. That is, we have $\eta \in \xi^\perp \mapsto Y_\eta$, and the above estimate is for $|t^{-1}Y_\eta(t) - \tau_{p,\gamma_\xi(t)}\eta|$ (see Theorem II.16 and Corollary IX.4.3). To go in the opposite direction, we are given $v \in \gamma_\xi(t)^\perp \mapsto \eta \in \xi^\perp$: $v = t^{-1}Y_\eta(t)$ and estimate $|\eta - \tau_{p,\gamma_\xi(t)}{}^{-1}v|$. Well,

$$\begin{aligned}
|\eta - \tau_{p,\gamma_\xi(t)}{}^{-1}v|\frac{\mathbf{S}_\Lambda(t)}{t} &\leq |t^{-1}Y_{\eta - \tau_{p,\gamma_\xi(t)}{}^{-1}v}(t)| \\
&= |t^{-1}Y_\eta(t) - t^{-1}Y_{\tau_{p,\gamma_\xi(t)}{}^{-1}v}(t)| \\
&= |v - t^{-1}Y_{\tau_{p,\gamma_\xi(t)}{}^{-1}v}(t)| \\
&\leq \left\{\frac{\mathbf{S}_{-\Lambda}(t)}{t} - 1\right\}|v|,
\end{aligned}$$

which implies

$$|\eta - \tau_{p,\gamma_\xi(t)}{}^{-1}v| \leq \frac{t}{\mathbf{S}_\Lambda(t)}\left\{\frac{\mathbf{S}_{-\Lambda}(t)}{t} - 1\right\}|v|,$$

that is,

$$|\{\Im_{t\xi}{}^{-1}\circ(\exp_p{}^{-1})_{*|t\xi} - \tau_{p,\gamma_\xi(t)}{}^{-1}\}v| \le \frac{t}{\mathbf{S}_\Lambda(t)}\left\{\frac{\mathbf{S}_{-\Lambda}(t)}{t} - 1\right\}|v|.$$

Step 6. Now, we want to replace

$$(\overline{x}_i)_{*|p} = (\exp_{F(p)}{}^{-1})_{*|F_i(p)}\circ(\overline{\phi}_i)_{*|\phi_i{}^{-1}(p)}\circ(\phi_i{}^{-1})_{*|p}.$$

by

$$\tau_{F_i(p),F(p)}\circ\tau_{\overline{z}_i,F_i(p)}\circ\overline{u}_i\circ u_i{}^{-1}\circ\tau_{p,z_i}.$$

In what follows, we let

$$t = d(p, z_i) = d(F_i(p), \overline{z}_i) < R, \qquad \alpha = d(F_i(p), F(p)) < \epsilon_0.$$

Given

$$\begin{aligned}
v \in M_p \mapsto \tau_{p,z_i}v \in M_{z_i}, &\quad \eta \in M_{z_i} : t^{-1}Y_\eta(t) = v \\
&\mapsto \overline{v} := \overline{u}_i\circ u_i{}^{-1}(\tau_{p,z_i}v), \quad \overline{\eta} := \overline{u}_i\circ u_i{}^{-1}(\eta) \in \overline{M}_{\overline{z}_i} \\
&\mapsto \tau_{\overline{z}_i,F_i(p)}\overline{v}, \quad t^{-1}\overline{Y}_{\overline{\eta}}(t) \in \overline{M}_{F_i(p)} \\
&\mapsto \tau_{F_i(p),F(p)}\circ\tau_{\overline{z}_i,F_i(p)}\overline{v} \in \overline{M}_{F(p)},\ \zeta_i \in \overline{M}_{F(p)} : \alpha^{-1}\overline{Y}_{\zeta_i}(\alpha) = t^{-1}\overline{Y}_{\overline{\eta}}(t).
\end{aligned}$$

Therefore, if we set

$$\zeta_i = (\overline{x}_i)_{*|p}v,$$

then, recalling that $\overline{u}_i\circ u_i{}^{-1}$ is an isometry, we have

$$\begin{aligned}
|\tau_{F_i(p),F(p)}\circ\tau_{\overline{z}_i,F_i(p)}\overline{v} - \zeta_i| &\le |\tau_{\overline{z}_i,F_i(p)}\overline{v} - t^{-1}\overline{Y}_{\overline{\eta}}(t)| + |\tau_{F_i(p),F(p)}(t^{-1}\overline{Y}_{\overline{\eta}}(t)) - \zeta_i| \\
&\le |\tau_{p,z_i}v - \eta| + |\tau_{\overline{z}_i,F_i(p)}\overline{\eta} - t^{-1}\overline{Y}_{\overline{\eta}}(t)| \\
&\quad + |\tau_{F_i(p),F(p)}(t^{-1}\overline{Y}_{\overline{\eta}}(t)) - \zeta_i| \\
&\le \frac{t}{\mathbf{S}_\Lambda(t)}\left\{\frac{\mathbf{S}_{-\Lambda}(t)}{t} - 1\right\}|v| + \left\{\frac{\mathbf{S}_{-\Lambda}(t)}{t} - 1\right\}|\eta| \\
&\quad + \frac{\alpha}{\mathbf{S}_\Lambda(\alpha)}\left\{\frac{\mathbf{S}_{-\Lambda}(\alpha)}{\alpha} - 1\right\}|t^{-1}\overline{Y}_{\overline{\eta}}(t)| \\
&\le \frac{t}{\mathbf{S}_\Lambda(t)}\left[2\left\{\frac{\mathbf{S}_{-\Lambda}(t)}{t} - 1\right\}\right. \\
&\quad \left. + \frac{\alpha}{\mathbf{S}_\Lambda(\alpha)}\left\{\frac{\mathbf{S}_{-\Lambda}(\alpha)}{\alpha} - 1\right\}\frac{\mathbf{S}_{-\Lambda}(t)}{t}\right]|v|.
\end{aligned}$$

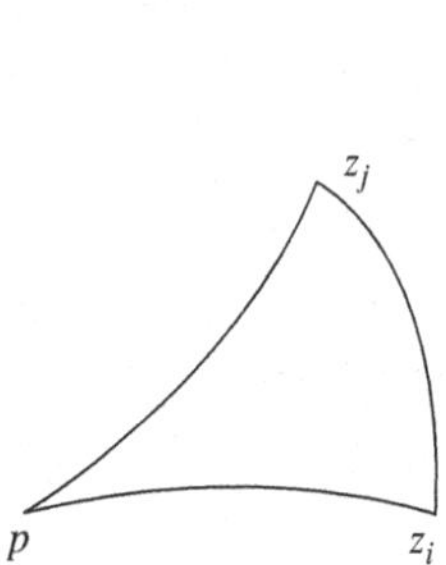

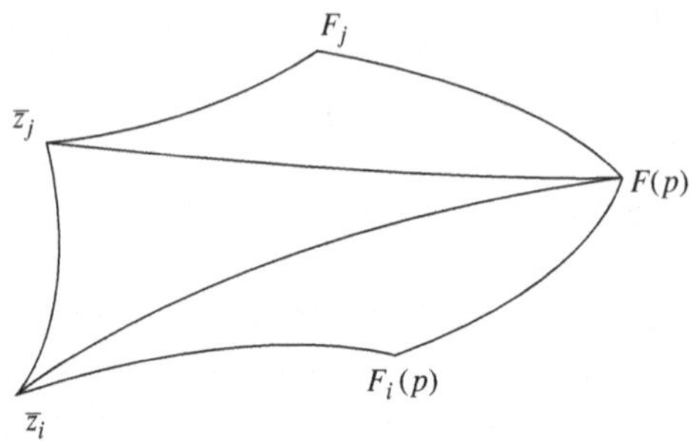

Figure IX.4. For **Step 7**.

So, we restrict ϵ_o and R to

$$\epsilon_0 \leq R \leq R_3 := \min\{R_2, \sup \rho\},$$

where the sup is taken over all ρ for which

$$\frac{t}{\mathbf{S}_\Lambda(t)}\left[2\left\{\frac{\mathbf{S}_{-\Lambda}(t)}{t} - 1\right\} + \frac{\alpha}{\mathbf{S}_\Lambda(\alpha)}\left\{\frac{\mathbf{S}_{-\Lambda}(\alpha)}{\alpha} - 1\right\}\frac{\mathbf{S}_{-\Lambda}(t)}{t}\right] \leq \frac{\Lambda\rho^2}{2}$$

for all $\alpha, t \in (0, \rho)$.

Step 7. Now that we have replaced $(\overline{x}_i)_{*|p}$ with $\tau_{F_i(p),F(p)}\circ\tau_{\overline{z}_i,F_i(p)}\circ\overline{u}_i {u_i}^{-1}\circ \tau_{p,z_i}$, we must therefore compare

$$\tau_{F_i(p),F(p)}\circ\tau_{\overline{z}_i,F_i(p)}\circ\overline{u}_i {u_i}^{-1}\circ\tau_{p,z_i} \quad \text{with} \quad \tau_{F_j(p),F(p)}\circ\tau_{\overline{z}_j,F_j(p)}\circ\overline{u}_j {u_j}^{-1}\circ\tau_{p,z_j}.$$

Then, by **3** of §XI.1, the difference is bounded above by the sum of the areas of the triangles pz_iz_j, $\overline{z}_iF_i(p)F(p)$, $\overline{z}_iF(p)\overline{z}_j$, and $\overline{z}_jF(p)F_j(p)$, added to the difference between the parallel translations P_{ij}, $\overline{P}_{ij}$, as described in (IX.8.2). (See Fig. IX.4.) Since all the sides of the triangles have lengths less than or equal to $3R$, we will further restrict R so that

$$R \leq R_4 := \min\{R_3, \pi/6\sqrt{\Lambda}\}.$$

This will then imply the existence of a constant $\text{const.}_{\Lambda,\iota} > 0$ such that the sum of the areas of the triangles is less that or equal to $\text{const.}_{\Lambda,\iota}R^2$. (We leave it to the reader to estimate the area of each triangle.)

Step 8. We therefore conclude that

$$|\{(\overline{x}_i)_{*|p} - \{(\overline{x}_j)_{*|p}\}\cdot\xi| \leq \{\text{const.}_{\Lambda,\iota}R^2 + \epsilon_1\}|\xi|;$$

so pick $\epsilon_1 = \text{const.}_{\Lambda,\iota} R^2$. Then,

$$\begin{aligned}
\left|\sum_i \eta_i(p)(\overline{x}_i)_{*|p} \cdot \xi\right| \\
&= \left|\sum_i \eta_i(p)(\overline{x}_{j_o})_{*|p} \cdot \xi + \sum_i \eta_i(p)\{(\overline{x}_i)_{*|p} - (\overline{x}_{j_o})_{*|p}\} \cdot \xi\right| \\
&\geq |(\overline{x}_{j_o})_{*|p} \cdot \xi| - \sup_i |\{(\overline{x}_i)_{*|p} - (\overline{x}_{j_o})_{*|p}\} \cdot \xi| \\
&\geq \{1 - \text{const.}_{\Lambda,\iota} R^2\}|\xi|,
\end{aligned}$$

which implies

$$\left|\sum_i \left\{(d\eta_{i|p} \cdot \xi)\overline{x}_i(p) + \eta_i(p)(\overline{x}_i)_{*|p} \cdot \xi\right\}\right| \geq \{1 - \text{const.}_{\Lambda,\iota}(R^2 + \epsilon_0/R)\}|\xi|.$$

We may therefore choose R, ϵ_0 sufficiently small to imply that F has maximal rank.

But if F has maximal rank, then F is a covering (Theorem IV.IV.1.3). This implies that the similarly constructed map $\overline{F} : \overline{M} \to M$ is a covering. Therefore, the mappings $F \circ \overline{F}, \overline{F} \circ F$ are coverings. But both $F \circ \overline{F}, \overline{F} \circ F$ map every point to a convex neighborhood of itself, in which case they are homotopic to the identity. This implies that both $F, \overline{F}$ are diffeomorphisms. ∎

§IX.9. Notes and Exercises

The Rauch Comparison and Sphere Theorems

Note IX.1. Our proof of Rauch's theorem (Theorem IX.2.1) followed his original argument, wherein he used the index form on vector fields along the geodesic. A proof avoiding use of the index form, using instead the matrix Riccati equation, was first given in Karcher (1977). (See the discussion in §III.8.) A proto-version of this approach was given by L. W. Green (1954, Lemma 2.1, 1958, Lemma 3). More subtle details are treated in Eschenburg–Heintze (1990).

Note IX.2. Let M be a simply connected compact Riemannian manifold of positive sectional curvatures. We say that M is *δ–pinched*, $0 < \delta \leq 1$, if

$$\min \mathcal{K} = \delta \sup \mathcal{K}$$

where $\mathcal{K}$ varies over all sectional curvatures on M. H. E. Rauch (1951) in his seminal work first proved that there exists a $\delta_0 > 0$ such that any δ–pinched Riemannian manifold, with $\delta \geq \delta_0$, is homeomorphic to a sphere. His first estimate of δ_0 was $\delta_0 \sim 3/4$. The most ambitious choice δ_0 is: any value of $\delta_0 > 1/4$; for the Riemannian symmetric spaces of strictly positive curvature, which are not homeomorphic to spheres, have pinching equal to $1/4$. The theorem was further improved until M. Berger (1960) proved the theorem in even dimensions for $\delta > 1/4$, and W. Klingenberg (1961) proved the theorem for $\delta > 1/4$ in odd dimensions. See the presentations in Cheeger–Ebin (1975), Gromoll–Klingenberg–Meyer (1968), Grove (1987), Karcher (1989) and Klingenberg (1982). Also see Eschenburg (1986), and the recent elementary argument in do Carmo (1992, Chapter 13). Surveys of the subsequent developments of the sphere theorems in Riemannian geometry are given in Berger (1985), Grove–Petersen (1997), Petersen (1997, 1998, 1999), Sakai (1984), and Shiohama (2000).

Diameter Sphere Theorems

One first recalls the Bonnet–Myers theorem that if M is Riemannian complete with Ricci curvatures bounded below by $(n-1)\kappa$, where κ is a positive constant, then M is compact with diameter less than or equal to $\pi/\sqrt{\kappa}$. The Toponogov–Cheng theorem then states that if, in addition, the diameter of M is maximal, that is equal to $\pi/\sqrt{\kappa}$, then M is isometric to the sphere of constant sectional curvature κ. The following theorem gives a different type of pinching result:

Theorem. (Grove–Shiohama (1977); see also Grove (1987, pp. 205–207)) *Let M be a complete Riemannian manifold with sectional curvatures $\mathcal{K} \geq \kappa$, $\kappa > 0$ a constant, and* $\operatorname{diam} M > \pi/2\sqrt{\kappa}$. *Then, M is homeomorphic to a sphere.*

Note IX.3. Of course, $\pi/2\sqrt{\kappa}$ is sharp, since the real projective space of constant sectional curvature $\kappa > 0$ provides a counterexample.

The theorem cannot be simply extended to Ricci curvature, without additional hypotheses (see Anderson (1990a) and Otsu (1991)). For some positive results, see also Shiohama (1983) and Eschenburg (1991).

The Alexandrov–Toponogov Comparison Theorems

Exercise IX.1. We give an easy Toponogov theorem for curvature bounded from above. Given a complete Riemannian manifold M, and geodesic hinge $(\gamma_1, \gamma_2, \alpha)$ at p in M. Assume

$$\mathcal{K}_{|B(p;\,\mathrm{inj}\,p)} \leq \delta, \qquad \sum_{i=1}^{3} \ell_i < \min\left\{2\,\mathrm{inj}\,p, \frac{2\pi}{\sqrt{\delta}}\right\},$$

where ℓ_3 denotes the distance between the endpoints of γ_1, γ_2. Let $(\overline{\gamma}_1, \overline{\gamma}_2, \alpha)$ be a geodesic hinge in $\mathbb{M}^2_\kappa$ such that $L(\overline{\gamma}_i) = \ell_i$. Prove

$$d(\gamma_1(0), \gamma_2(\ell_2)) \geq d(\overline{\gamma}_1(0), \overline{\gamma}_2(\ell_2)).$$

Characterize the case of equality.

Busemann Functions and Halfspaces

We are given a complete noncompact Riemannian manifold.

Exercise IX.2. Let $\gamma : [0, +\infty) \to M$ be a ray (Exercise I.5) in M. To every $t \in [0, +\infty)$, associate the function $\phi_t : M \to \mathbb{R}$ given by

$$\phi_t(x) = t - d(x, \gamma(t)).$$

Prove:

(a) $|\phi_t(x) - \phi_t(y)| \leq d(x, y)$ for all $t, x, y \in M$.

(b) $|\phi_t(x)| \leq d(x, \gamma(0))$ for all $t, x \in M$.

(c) $t > s \Rightarrow \phi_t(x) \geq \phi_s(x)$ for all $t > s, x \in M$. One therefore has the existence of a limit function, *the Busemann function of* γ,

$$\phi_\gamma(x) = \lim_{t \uparrow +\infty} \phi_t(x).$$

(d) For any $s > 0$, let γ_s denote the geodesic given by $\gamma_s(t) = \gamma(s + t)$. Show that $\phi_{\gamma_s}(x) = \phi_\gamma(x) - s$ for all $s > 0, x \in M$.

(e) Consider the subset

$$B_\gamma = \bigcup_{t>0} B(\gamma(t); t)$$

in M. Show that $\phi_\gamma(x) = s$ for all $s > 0, x \in \partial B_{\gamma_s}$.

Definition. A subset A of any Riemannian manifold is *totally convex* if for any $p, q \in A$ and γ_{pq} *any* geodesic in M connecting p to q then $\gamma_{pq} \subseteq A$.

Exercise IX.3. Let M be as above, with nonnegative sectional curvature on all of M. For any ray γ in M, let H_γ denote the *half-space* given by

$$H_\gamma = M \setminus B_\gamma,$$

where B_γ is given above. Prove that H_γ is totally convex.

Note IX.4. One can continue from here to the study of the Cheeger–Gromoll theory of manifolds of nonnegative curvature. See the references in Note II.8. Also, see the nice treatment in Grove (1987, pp. 208ff).

Note IX.5. Continue with the assumption that all sectional curvatures are nonnegative. Check that, inside the cut locus of $\gamma(t)$, the function ϕ_t has a nonnegative Hessian – which suggests a convexity of ϕ_γ. This lies behind the result of the above exercise. When we are only given the nonnegativity of the Ricci curvature, this "nonnegativity of the Hessian" translates to a weak form of the nonnegativity of the Laplacian of ϕ_γ – enough to prove that ϕ_γ is subharmonic. See Besse (1987, pp. 171ff), and the proof there of the splitting theorems of Cohn-Vossen (1936), Toponogov (1964), and Cheeger–Gromoll (1971), following simplifications by Eschenburg–Heintze (1984).

Now assume M is complete, simply connected of nonpositive curvature. It is common to refer to M as a *Hadamard–Cartan* manifold. Note that all geodesics are *lines,* that is, they minimize distance between any two of their points.

Exercise IX.4.

(a) We know that for any $q \in M$ the distance function $r_q(x) = d(q, x)$ is a convex function on M. Show that, for any ray γ, the Busemann function ϕ_γ is *concave,* that is, $-\phi_\gamma$ is convex. Show, therefore, that B_γ is convex.

(b) Show that for:

(i) any $p \in M, r > 0$, and $x \notin B(p;r)$ there exist points $y_1, y_2 \in S(p;r)$ such that $|r_x(y_1) - r_x(y_2)| = 2r$.

(ii) the Busemann function ϕ_γ of the ray γ, $r > 0$, and any $p \in M$, there exist $y_1, y_2 \in S(p;r)$ such that $|\phi_\gamma(y_1) - \phi_\gamma(y_2)| = 2r$. Furthermore, the choice of two points y_1, y_2 is unique.

(c) Show that ϕ_γ is C^1, with $|\operatorname{grad} \phi_\gamma| = 1$ on all of M.

Exercise IX.5. Show that if $\phi : M \to \mathbb{R}$ is a concave C^1 function with $|\operatorname{grad} \phi| = 1$ on all of the Hadamard–Cartan manifold M, then ϕ is the Busemann function of a geodesic in M.

Note IX.6. For extended discussion see Ballman–Gromov–Schroeder (1985) and Jost (1997).

Finiteness Theorems

Note IX.7. A short elegant argument was given by A. Weinstein (1967) to obtain the following finiteness theorem for homotopy type.

Theorem. *For any even number $2n$ and any constant $\delta > 0$, there are only finitely many homotopy types of $2n$–dimensional, δ–pinched manifolds.*

A more general theorem is:

Theorem (Grove–Peterson (1988)). *Given real numbers* $n, d, V, \kappa > 0$. *Then there exist only finitely many homotopy types of compact n–dimensional Riemannian manifolds satisfying*

$$\operatorname{diam} M \leq d, \qquad V(M) \geq V, \qquad \mathcal{K} \geq \kappa.$$

In fact, the number of homotopy types is bounded above by a constant depending only n, $V^{-1}d^n$, and κd^2. For a survey of these and other directions in finiteness theorems, see Cheeger (1991).

Convergence Theorems for Riemannian Manifolds

The pinching and finiteness theorems have since led to convergence theorems of M. Gromov (1981) for Riemannian metrics. We mention here some of the definitions and results in this theory. See Grove (1987, pp. 214ff) for introductory arguments.

Definition. Given metric spaces X and Y, we define their *Lipschitz distance* $d_L(X, Y)$ by

$$d_L(X, Y) = \inf\{|\ln \operatorname{dil} f| + |\ln \operatorname{dil} f^{-1}|\},$$

where f varies over Lipschitz homeomorphisms $f : X \to Y$. Should no Lipschitz homeomorphisms exist, then we define $d_L(X, Y) = +\infty$.

Exercise IX.6. If X and Y are compact, with $d_L(X, Y) = 0$, then X and Y are isometric.

Exercise IX.7. Show that, on the space $\mathfrak{X}$ of compact metric spaces, the function d_L is a distance metric.

Theorem. (A. Shikata (1966)) *For all integers* $n \geq 2$, *there exists a positive* $\epsilon = \epsilon(n)$ *such that any two n–dimensional compact Riemannian manifolds* M, $\overline{M}$ *satisfying*

$$d_L(M, \overline{M}) < \epsilon$$

are diffeomorphic.

See the refinements in Karcher (1977) and the application to the differentiable pinching problem in Shikata (1967).

Definition. Fix a metric space Z. Then, for sets $A, B \subseteq Z$ define their *Hausdorff distance* $d_H^Z(C, D)$ by

$$d_H^Z(C, D) = \inf\{\epsilon > 0 : [C]_\epsilon \supseteq D \quad [D]_\epsilon \supseteq C\},$$

where $[C]_\epsilon = \{q \in Z : d(q, C) < \epsilon\}$, and similarly for D_ϵ.

For two fixed metric spaces X and Y, define their *Hausdorff distance* $d_H(X, Y)$ by

$$d_H(X, Y) = \inf\{d_H^Z(f(X), g(Y))\},$$

where Z varies over *all* metric spaces, and (f, g) vary over all isometric immersions

$$f : X \to Z, \qquad g : Y \to Z.$$

Exercise IX.8. When X and Y are compact then $d_H(X, Y) < +\infty$.

The following exercise is evocative of S. Peters' approach to Cheeger's finiteness theorem:

Exercise IX.9.

(a) Suppose we are given a sequence of compact metric spaces $X_j \in \mathfrak{X}$ converging, relative to d_H, to $X \in \mathfrak{X}$. Then, for any discretization $\mathbf{G}$ of X (see §IV.4), there exist discretizations $\mathbf{G}_j$ of X_j such that $d_L(\mathbf{G}_j, \mathbf{G}) \to 0$ as $j \to \infty$.

(b) Suppose we are given the collection $\{X_j : j = 1, \ldots\} \subseteq \mathfrak{X}$, and $X \in \mathfrak{X}$ such that

$$\sup\{\text{diam } X_j, \text{diam } X\} < +\infty;$$

and suppose that for every $R > 0$ there exist discretizations $\mathbf{G}_j$ of X_j, $\mathbf{G}$ of X, of covering radius R, such that $d_L(\mathbf{G}_j, \mathbf{G}) \to 0$ as $j \to \infty$. Then, $d_H(X_j, X) \to 0$, as $j \to \infty$.

(c) Show that for $X, Y \in \mathfrak{X}$, we have $d_H(X, Y) = 0 \Rightarrow d_L(X, Y) = 0$.

The two Gromov convergence theorems are as follows:

Theorem. *Given constants $\kappa \in \mathbb{R}$ and $d > 0$. Then, the set of all compact n–dimensional Riemannian manifolds satisfying* $\text{Ric} \geq (n-1)\kappa$ *and* $\text{diam } M \leq d$ *is precompact in the Hausdorff metric on $\mathfrak{X}$.*

Theorem. *Given real numbers* $n, d, V, \Lambda > 0$. *With respect to the Lipschitz distance topology, the collection of compact* n*–dimensional Riemannian manifolds satisfying*

$$\operatorname{diam} M \leq d, \qquad V(M) \geq V, \qquad |\mathcal{K}| \leq \Lambda$$

is relatively compact in the larger class of n*–dimensional* $C^{1,1}$*–manifolds with* C^0*–Riemannian metrics.*

One can find a proof of the first theorem in Grove (1987, p. 218); and a proof of the second theorem in Peters (1987). The latter is built on Peters' Lemma IX.8.1. In fact, he proves a stronger result: Given any sequence of compact n–dimensional Riemannian manifolds satisfying the conditions on diameter, volume, and sectional curvature, as above, there exists a subsequence converging with respect to the Lipschitz topology to an n–dimensional differentiable manifold M with metric of Hölder class $C^{1,\alpha}$, $0 < \alpha < 1$.

Other versions of similar approaches can be found in Greene–Wu (1988) and Kasue (1989). Recent treatments can be found in Gromov (1999), Petersen (1997), and Shiohama–Shioya–Tanaka (2003).

Hints and Sketches for Exercises

Hints and Sketches: Chapter I

Exercise I.2. Hint: Use the first Bianchi identity.

Exercise I.3. Sketch: If the Christoffel symbols of a chart vanish at p, then certainly $T = 0$ at p. So, consider the converse. Assume $T = 0$ at p. Given a chart $x : U \to \mathbb{R}^n$, $p \in U$, with Christoffel symbols $\Gamma_{jk}{}^{\ell}$ symmetric in j, k, define

$$y^{\ell}(q) = x^{\ell}(q) - x^{\ell}(p) + \frac{1}{2}\sum_{j,k} \Gamma_{jk}{}^{\ell}(x(p))(x^j(q) - x^j(p))(x^k(q) - x^k(p)).$$

Check that there exists a domain V in U, such that $(y|V) : V \to \mathbb{R}^n$ is a chart for which the Christoffel symbols vanish at p.

Exercise I.4. Sketch: Since grad f has constant length, one certainly has grad $f \perp \nabla_{\text{grad}\, f}\text{grad}\, f$. Therefore, given $p \in M$, $\xi \in M_p$, $\xi \perp (\text{grad}\, f)_{|p}$, it suffices to show $\xi \perp \nabla_{(\text{grad}\, f)_{|p}}\text{grad}\, f$. Let $\omega(t)$ be the integral curve of grad f satisfying $\omega(0) = p$, and X the parallel vector field along ω satisfying $X(0) = \xi$. Then, one easily shows that $\langle \nabla_{(\text{grad}\, f)_{|p}}\text{grad}\, f, \xi \rangle = \langle \nabla_t \omega', X \rangle_{|t=0}$ vanishes.

Exercise I.5. Sketch: One can easily check that for such ψ one has

$$|\psi(x) - \psi(y)| \le d(x, y).$$

So, the issue is to show that, among these functions, we may choose ψ so that $|\psi(x) - \psi(y)|$ is arbitrarily close to $d(x, y)$. To this end, consider the function

$$\psi(z) = \min \{d(z, x), d(y, x)\}.$$

Then, ψ is constant (actually equal to $d(x, y)$) outside $B(x; d(x, y))$, and

$$|\psi(x) - \psi(y)| = d(x, y).$$

Even though ψ is not C^∞, it is uniformly Lipschitz, in that

$$|\psi(z) - \psi(w)| \le d(z, w)$$

for all $z, w \in M$. Now, one requires an argument that ψ may be approximated by C^∞ functions ψ_n for which $|\text{grad}\, \psi_n| \leq 1$.

Exercise I.7. Sketch:

(i) First show it is always possible to reparameterize any f in the length structure with the new parameter proportional to arc length, that is, one can reparameterize f to $\gamma : [0, 1] \to X$ so that

$$\ell(\gamma|[0, t]) = \ell(\gamma)t.$$

(ii) Fix the interval $I = [0, 1]$. Assume we are given a sequence $\gamma_n : I \to X \in C^0$, parameterized proportional to arc length, such that $\gamma_n(0) = x$, $\gamma_n(1) = y$ for all $n = 1, 2, \ldots,$ and $\ell(\gamma_n) \downarrow d(x, y)$ as $n \to \infty$. Show that the sequence of mappings (γ_n) is equicontinuous. Since $\overline{B(x; 2d(x, y))}$ is compact, the Arzela–Ascoli Theorem implies that (γ_n) converges uniformly to $\gamma : I \to X \in C^0$ connecting x to y.

(iii) Now verify that

$$\ell(\gamma) = d(x, y).$$

Exercise I.9. Sketch: We want to show

$$\ell_d(\omega) = \int_a^b |\omega'|\, dt$$

for any D^1 path $\omega : [a, b] \to M$. One easily has from the definition of ℓ_d that

$$\ell_d(\omega) \leq \int_a^b |\omega'|\, dt;$$

so the real issue is the opposite inequality. The argument is as follows:

One proves that given any compact K in M and any real $\lambda > 1$, there exists (see Riemann normal coordinates below, in §II.8) a finite cover of K, $\{U_1, \ldots, U_k\}$, with charts $x_j : U_j \to \mathbb{R}^n$ such that

$$\lambda^{-1} \leq \frac{|\xi|}{|\xi|_{\mathbb{R}^n}} \leq \lambda$$

(where $|\xi|_{\mathbb{R}^n}$ denotes the standard norm on $\mathbb{R}^n$) for all $\xi \in TU_j$, $j = 1, \ldots, k$, and

$$\lambda^{-1} \leq \frac{d(p, q)}{|x_j(p) - x_j(q)|} \leq \lambda$$

for all $p, q \in U_j$, $j = 1, \ldots, k$. From this, it is easy to prove that

$$\ell_d(\omega) \geq \lambda^{-2} \int_a^b |\omega'|\, dt,$$

for all $\lambda > 1$.

Exercise I.11(d). Hint: Consider $0 = d^2 F$.

Exercise I.13(b). Hint: Use the properties of the Levi-Civita connection to show

$$\langle [X, Y], T \rangle = \langle X, \nabla_Y T \rangle - \langle Y, \nabla_X T \rangle.$$

Exercise I.15(b). Hint: Use Exercise I.13(b) and the properties of the Levi-Civita connection to show that $\Gamma_{rs}{}^{\alpha}$ is skew-symmetric with respect to r, s.

Exercise I.21(b). Hint: From

$$\operatorname{Ad} \exp t\xi = e^{t \operatorname{ad} \xi}, \qquad (\operatorname{ad} \xi)(\eta) = [\xi, \eta],$$

we have

$$[\xi, \eta] = \left.\frac{d}{dt}\right|_{t=0} \left.\frac{d}{ds}\right|_{s=0} \exp t\xi \exp s\eta \exp -t\xi.$$

Now that we know the 1–parameter subgroups explicitly, one can check that

$$[\xi, \eta] = \xi\eta - \eta\xi.$$

Exercise I.21(c). Hint: Use Exercise I.19.

Exercise I.21(d). Hint: Use Exercise I.20 to show that the space has constant curvature 4. Check that the geodesics have length equal to π. Then construct a map from the 2–sphere in $\mathbb{R}^3$ of radius $1/2$ to G/H. Check that it is an isometry. (See §IV.1. Also see Thurston (1997, §2.7).)

Hints and Sketches: Chapter II

Exercise II.2. Hint: Use the second Bianchi identity (Exercise I.2).

Exercise II.5. Hint: Since the surface M is compact, it is contained in some open 3–disk, $\mathbb{B}(o; R)$. Now, keep $o \in \mathbb{R}^3$ fixed, and decrease R towards 0, and consider what happens when $R = r_o := \inf\{r : M \subset B(o; r)\}$.

Exercise II.7. Sketch: Use moving frames as in Exercise II.10 to show $\omega_j{}^m = \lambda(\mathbf{x})\omega^j$, for $j = 1, \ldots, m-1$. Now, calculate $d\omega_j{}^m$ two ways, and show thereby that $d\lambda = 0$ on M. Conclude that $e_m = -\lambda\mathbf{x} + \mathbf{q}_o$, for some constant vector $\mathbf{q}_o$, which implies that $|\mathbf{x} - \mathbf{q}_o/\lambda| = \text{const}$.

Exercise II.8. Hint: Use moving frames.

Exercise II.13. Hint: Use (II.3.6).

Exercise II.15(a). Hint: For convenience, assume $\tau > 0$. First, check that the Riemannian metric on M_o is given by

$$ds^2 = |dx|^2 - \frac{(x \cdot dx)^2}{|x|^2 + \rho^2}.$$

Then, introduce spherical coordinates

$$x = r\xi \qquad r > 0, \ \ \xi \in \mathbb{S}^{n-1},$$

and show

$$ds^2 = \frac{\rho^2 (dr)^2}{r^2 + \rho^2} + r^2 |d\xi|^2.$$

Finally, set

$$r = \rho \sinh t/\rho,$$

and substitute.

Exercise II.16. Hint: The first argument consists of noting that if an M–geodesic lies in the surface, and passes through a point in the surface, then the second fundamental form at that point cannot be definite.

The second argument consists of noting that, for any Jacobi field Y along an M–geodesic lying in the surface, we have

$$|Y|'' + \langle R(\gamma', Y)\gamma', Y\rangle \geq 0,$$

see the beginning of the proof of Theorem II.6.4.

Exercise II.18. Hint: First use Theorem II.7.1 to derive the existence of ϵ in $(0, 1)$, $r > \beta$, such that $\exp|\mathcal{C}_{\epsilon,r}(\xi)$ is a diffeomorphism, and use the arguments of Theorem I.6.2.

Exercise II.20(b). Hint: Use Theorem II.6.2 and the previous exercise.

Exercise II.20(c). Hint: Adapt the derivation of the second variation of arc length.

Exercise II.21. Hint: Use a linear isometry

$$\iota : (M_1)_{\gamma_1(0)} \to (M_2)_{\gamma_2(0)}$$

and parallel translation along the respective geodesics to identify vector fields along γ_1 with vector fields along γ_2. Now compare the respective index forms and use the Jacobi criteria.

Exercise II.25. Hint: Use Theorem II.8.1.

Exercise II.26. Sketch: Recall that Exercise II.17 shows that points along γ, conjugate to p along γ, are isolated. So, pick $t_o < t_1 < t_2$ so that $[t_o, t_2]$ has only $\gamma(t_1)$ conjugate to p along γ.

One considers both geodesic polar coordinates *and* Riemann normal coordinates based at at p. Let $e_1 = \xi$, and e_2 orthonormal to e_1. The polar coordinates are given by

$$v(t, \theta) = \exp_p t\{\cos\theta + \sin\theta\},$$

and the normal coordinates by

$$\zeta = \exp \sum_{j=1}^{2} \zeta^j e_j \quad \Rightarrow \quad \mathbf{n}(\exp \zeta) = (\zeta^1, \zeta^2).$$

So, we want to solve the equation $\zeta^2(t, \theta) = 0$ for some function $t = \varphi(\theta)$ in a neighborhood of $(t_1, 0)$.

Use Theorem II.5.1 to show that there exists a nonzero constant σ such that the Taylor expansion of $\zeta^2(t, \theta)$ in a neighborhood of $(t_1, 0)$ is given by

$$\zeta^2(t, \theta) = (t - t_1)\theta\sigma + o(|t - t_1|) + o(|\theta|)h(t),$$

where $h(t)$ is bounded. Then show that $\zeta^2(t, \theta) = 0$, can be solved for some function $t = \varphi(\theta)$ in a neighborhood of $(t_1, 0)$.

Exercise II.27. Sketch: Again, recall that Exercise II.17 shows that points along γ, conjugate to p along γ, are isolated. So pick $t_o < t_1 < t_2$ so that $[t_o, t_2]$ has only $\gamma(t_1)$ conjugate to p along γ.

Pick a nonzero Jacobi field $Y(t)$ along γ satisfying: $Y(0) = Y(t_1) = 0$ and set

$$Y_1(t) = \begin{cases} Y(t) & t \in [0, t_1] \\ 0 & t \in [t_1, \beta] \end{cases}.$$

As in the proof of Theorem II.5.5, let $Z(t)$ be a differentiable vector field along γ satisfying

$$Z|[0, t_o] \cup [t_2, b] = 0, \qquad Z(t_1) = -(\nabla_t Y)(t_1),$$

and

$$X = Y_1 + \lambda\varphi Z.$$

Assume the theorem is false. So, we assume there is a neighborhood U of $t_1\xi$ on which $\exp_p$ is one-to-one. Then, the geodesic $s \mapsto \exp_{\gamma(t_o)} sY(t_o)$ has a lift to a path $\overline{\delta}(s)$ in M_p, that is,

$$\delta(s) := \exp_{\gamma(t_o)} sY(t_o) = \exp_p \overline{\delta}(s).$$

Therefore, consider the variation $v(t, s)$ of $\gamma(t)$ given by

$$v(t, s) = \begin{cases} \exp_p t\overline{\delta}(s)/t_o & t \in [0, t_0] \\ \exp_{\gamma(t)} sX(t) & t \in [t_0, \beta] \end{cases}.$$

First show that $\partial_s v(t, 0) = X(t)$.

Then, use the argument of Theorem II.5.5 to show that, if $|s|$ sufficiently small, the length of $\omega_s(t) := v(t, s)$ is strictly less that β.

Next, note that, for s sufficiently small, $\omega_s|[t_o, t_2] \subset \exp(\mathsf{U})$, which implies ω_s has a lift $\overline{\omega}_s$ to M_p. Now use the *argument* of Exercise II.18 to show the length of ω_s is greater than or equal to β – a contradiction.

Hints and Sketches: Chapter III

Exercise III.3. Hint: Given that M is locally symmetric, use the explicit knowledge of the Jacobi fields along any geodesic to show that the geodesic symmetry is a local isometry. Conversely, given that the geodesic symmetry is a local isometry, show that M satisfies the criterion of Exercise III.2(a).

Exercise III.5(a). Hint: First show that group property is valid along $\gamma(t)$. Then use Exercise II.13.

Exercise III.5(b). Hint: Pick the 2–section along the geodesic with the lowest curvature, assumed to be negative. Let ξ be a unit vector in this 2–section perpendicular to the geodesic, and consider the 1–parameter group of isometries Γ_ξ. Then, by differentiating with respect to the group parameter, the group will determine a periodic Jacobi field along the geodesic, which will contradict the specific knowledge of the Jacobi field as unbounded.

Exercise III.5(c). Hint: Use Klingenberg's lemma (Theorem III.2.4), and Exercises III.4(b) and III.5(b).

Exercise III.9. Hint: First show that the matrix Wronskian of two solutions of (III.1.4) is constant, that is, if $\mathcal{C}(t)$, $\mathcal{D}(t)$ are solutions of (III.1.4), then

$$\mathcal{C}'^*\mathcal{D} - \mathcal{C}^*\mathcal{D}' = \text{const.}$$

Then pick appropriate $\mathcal{C}$ and $\mathcal{D}$.

Exercise III.10(b). Hint: Use Exercise III.9.

Exercise III.12(c). Hint: One requires Sard's theorem.

Exercise III.12(h). Hint: Integrate the previous formula, that is, (g), over the unit sphere.

Exercise III.14. Sketch: One easily verifies that the contribution of $d \circ \mathrm{i}(\xi) + \mathrm{i}(\xi) \circ d$ to the integral is 0. So, one must deal exclusively with the contribution of $d \circ \mathrm{i}(\eta) + \mathrm{i}(\eta) \circ d$.

First check that

$$A_k'(\epsilon) = \int_{\mathfrak{M}} \phi_\epsilon{}^*((\mathrm{i}(\eta) \circ d)\, \omega^1 \wedge \cdots \wedge \omega^k).$$

(Hint: $\mathrm{i}(\eta)\omega^r = 0$ for all r.)

Next, check that, when restricted to $\phi_\epsilon(\mathfrak{M})$, one has

$$(\mathrm{i}(\eta) \circ d)\, \omega^s = -\sum_r (\omega_r{}^s(\eta))\omega^r + \sum_\alpha \eta^\alpha \omega_\alpha{}^s,$$

where $\{\omega_A{}^B\}$ are the connection forms of the coframe $\{\omega^A\}$ (as described in §I.8 and §II.2), and

$$\eta^\alpha = \langle \eta, e_\alpha \rangle,$$

which implies (when restricted to $\phi_\epsilon(\mathfrak{M})$)

$$\begin{aligned}(\mathrm{i}(\eta)\circ d)\,\omega^1\wedge\cdots\wedge\omega^k &= \sum_r \omega^1\wedge\cdots\wedge((\mathrm{i}(\eta)\circ d)\,\omega^r)\wedge\cdots\wedge\omega^k\\ &= \sum_{r,\alpha}\eta^\alpha\,\omega^1\wedge\cdots\wedge{\omega_\alpha}^r\wedge\cdots\wedge\omega^k\\ &= -\langle\eta, H\rangle\,\omega^1\wedge\cdots\wedge\omega^k.\end{aligned}$$

Exercise III.17. Sketch: On the image of the normal bundle of $\mathfrak{M}$ under the exponential map Exp , first define the vector field Z locally on a neighborhood of $\mathfrak{M}$ by

$$Z_{|(w,\epsilon)} = \frac{f(w)}{\sqrt{\mathbf{g}}(\epsilon; w)}\partial_\epsilon,$$

where ∂_ϵ is the velocity vector of the geodesic $\gamma_\nu(\epsilon)$. Then extend Z smoothly to all of M so that the support of Z is contained in Exp $(\mathfrak{M}\times(-\epsilon_o, \epsilon_o))$. Let Σ_ϵ denote the flow of Z. Finally, let $\phi(w, \epsilon)$ denotes the solution to the initial-value problem

$$\frac{\partial\phi}{\partial\epsilon} = \frac{f(w)}{\sqrt{\mathbf{g}}(\phi; w)}, \qquad \phi(w; 0) = 0.$$

Verify that (i) $\Sigma_\epsilon(\mathfrak{M})$ is the boundary of $\Sigma_e p(\Omega)$; and (ii) for $w\in\mathfrak{M}$, we have

$$\Sigma_\epsilon(w) = \mathrm{Exp}\,(w, \phi(w, \epsilon)).$$

Now prove that

$$\frac{\partial\{V(\Sigma_\epsilon(\Omega))\}}{\partial\epsilon} = \int_\Gamma f\,dA_\Gamma = 0.$$

Exercise III.20. Hint: Pick a ray $\gamma : [0, +\infty)\to M$ from x to ∞, $|\gamma'| = 1$. Let $x_k = \gamma(k)$. First show that

$$V(x_k; k-1) \geq \left\{\frac{k-1}{k+1}\right\}^n V(x_k; k+1),$$

which implies (why?)

$$V(x; 2k) \geq V(x_k; k-1) \geq \frac{(k-1)^n}{(k+1)^n - (k-1)^n} V(x; 1),$$

which implies the claim.

Exercise III.29(a). Hint: Use Exercise II.10.

Exercise III.29(b). Hint: Use Exercise III.29(a), with Green's formula.

Exercise III.30. Hint: See Exercise II.10.

Exercise III.32. Solution: We start from

$$A_k{}'(\epsilon) = \int_{\phi_\epsilon(\mathfrak{M})} \sum_{r,\alpha} \eta^\alpha\, \omega^1 \wedge \cdots \wedge \omega_\alpha{}^r \wedge \cdots \wedge \omega^k$$

(see the sketch for Exercise III.14). Then,

$$A_k{}''(\epsilon) = \int_{\phi_\epsilon(\mathfrak{M})} (\mathrm{i}(\eta) \circ d) \left\{ \sum_{r,\alpha} \eta^\alpha\, \omega^1 \wedge \cdots \wedge \omega_\alpha{}^r \wedge \cdots \wedge \omega^k \right\}$$

(again, there is no contribution from ξ).

First, one has

$$\begin{aligned}
(1) \quad & d \sum_{r,\alpha} \eta^\alpha \omega^1 \wedge \cdots \wedge \omega_\alpha{}^r \wedge \cdots \wedge \omega^k \\
&= \sum_{r,\alpha} d\eta^\alpha \wedge \omega^1 \wedge \cdots \wedge \omega_\alpha{}^r \wedge \cdots \wedge \omega^k \\
&+ \sum_{\alpha,r} \sum_{s<r} (-1)^{s-1} \eta^\alpha \omega^1 \wedge \cdots \wedge (\omega^r \wedge \omega_r{}^s)_{\mathsf{s}} \wedge \cdots \wedge \omega_\alpha{}^r \wedge \cdots \wedge \omega^k \\
&+ \sum_{\alpha,\beta,r} \sum_{s<r} (-1)^{s-1} \eta^\alpha \omega^1 \wedge \cdots \wedge \left(\omega^\beta \wedge \omega_\beta{}^s\right)_{\mathsf{s}} \wedge \cdots \wedge \omega_\alpha{}^r \wedge \cdots \wedge \omega^k \\
&+ \sum_{j,r,\alpha} (-1)^{r-1} \eta^\alpha \omega^1 \wedge \cdots \wedge \left(\omega_\alpha{}^j \wedge \omega_j{}^r\right)_{\mathsf{r}} \wedge \cdots \wedge \omega^k \\
&+ \sum_{r,\alpha,\beta} (-1)^{r-1} \eta^\alpha \omega^1 \wedge \cdots \wedge \left(\omega_\alpha{}^\beta \wedge \omega_\beta{}^r\right)_{\mathsf{r}} \wedge \cdots \wedge \omega^k \\
&- \sum_{r,\alpha} (-1)^{r-1} \eta^\alpha \omega^1 \wedge \cdots \wedge \Omega_\alpha{}^r \wedge \cdots \wedge \omega^k \\
&+ \sum_{\alpha,r} \sum_{s>r} (-1)^{s-1} \eta^\alpha \omega^1 \wedge \cdots \wedge \omega_\alpha{}^r \wedge \cdots \wedge (\omega^r \wedge \omega_r{}^s)_{\mathsf{s}} \wedge \cdots \wedge \omega^k \\
&+ \sum_{\alpha,\beta,r} \sum_{s>r} (-1)^{s-1} \eta^\alpha \omega^1 \wedge \cdots \wedge \omega_\alpha{}^r \wedge \cdots \wedge \left(\omega^\beta \wedge \omega_\beta{}^s\right)_{\mathsf{s}} \wedge \cdots \wedge \omega^k.
\end{aligned}$$

For lines 3, 5, and 8 of (1), we have

$$\begin{aligned}
& \sum_{\alpha,r} \sum_{s<r} (-1)^{s-1} \eta^\alpha \omega^1 \wedge \cdots \wedge (\omega^r \wedge \omega_r{}^s)_{\mathsf{s}} \wedge \cdots \wedge (\omega_\alpha{}^r)_{\mathsf{r}} \wedge \cdots \wedge \omega^k \\
&\quad + \sum_{j,r,\alpha} (-1)^{r-1} \eta^\alpha \omega^1 \wedge \cdots \wedge \left(\omega_\alpha{}^j \wedge \omega_j{}^r\right)_{\mathsf{r}} \wedge \cdots \wedge \omega^k \\
&\quad + \sum_{\alpha,r} \sum_{s>r} (-1)^{s-1} \eta^\alpha \omega^1 \wedge \cdots \wedge (\omega_\alpha{}^r)_{\mathsf{r}} \wedge \cdots \wedge (\omega^r \wedge \omega_r{}^s)_{\mathsf{s}} \wedge \cdots \wedge \omega^k \\
&= \sum_{\alpha,r} \sum_{s<r} (-1)^s \eta^\alpha \omega^1 \wedge \cdots \wedge (\omega_\alpha{}^r \wedge \omega_r{}^s)_{\mathsf{s}} \wedge \cdots \wedge \omega^k \\
&\quad + \sum_{s,r,\alpha} (-1)^{s-1} \eta^\alpha \omega^1 \wedge \cdots \wedge (\omega_\alpha{}^r \wedge \omega_r{}^s)_{\mathsf{s}} \wedge \cdots \wedge \omega^k \\
&\quad + \sum_{\alpha,r} \sum_{s>r} (-1)^s \eta^\alpha \omega^1 \wedge \cdots \wedge (\omega_\alpha{}^r \wedge \omega_r{}^s)_{\mathsf{s}} \wedge \cdots \wedge \omega^k \\
&= 0.
\end{aligned}$$

For lines 4 and 9 of (1), we note again that $\omega^\alpha|\phi_\epsilon(\mathfrak{M}) = 0$. Therefore, one calculates to obtain

$$\Bigg\{ \mathrm{i}(\eta)\Bigg(\sum_{\alpha,\beta,r}\sum_{s<r}(-1)^{s-1}\eta^\alpha\omega^1\wedge\cdots\wedge\left(\omega^\beta\wedge\omega_\beta{}^s\right)_{\mathsf{s}}\wedge\cdots\wedge\omega_\alpha{}^r\wedge\cdots\wedge\omega^k$$

$$+\sum_{\alpha,\beta,r}\sum_{s>r}(-1)^{s-1}\eta^\alpha\omega^1\wedge\cdots\wedge\omega_\alpha{}^r\wedge\cdots\wedge\left(\omega^\beta\wedge\omega_\beta{}^s\right)_{\mathsf{s}}\wedge\cdots\wedge\omega^k\Bigg)\Bigg\}_{|\phi_\epsilon(\mathfrak{M})}$$

$$=\Bigg\{\sum_{\alpha,\beta}\sum_{s<r}\eta^\alpha\eta^\beta\omega^1\wedge\cdots\wedge\omega_\beta{}^s\wedge\cdots\wedge\omega_\alpha{}^r\wedge\cdots\wedge\omega^k$$

$$+\sum_{\alpha,\beta}\sum_{s>r}\eta^\alpha\eta^\beta\omega^1\wedge\cdots\wedge\omega_\alpha{}^r\wedge\cdots\wedge\omega_\beta{}^s\wedge\cdots\wedge\omega^k\Bigg\}_{|\Phi_\epsilon(\mathfrak{M})}$$

$$=\sum_{\alpha,\beta,j,\ell}\sum_{s<r}\eta^\alpha\eta^\beta h_{sj}{}^\beta h_{r\ell}{}^\alpha\omega^1\wedge\cdots\wedge\left(\omega^j\right)_{\mathsf{s}}\wedge\cdots\wedge\left(\omega^\ell\right)_{\mathsf{r}}\wedge\cdots\wedge\omega^k$$

$$+\sum_{\alpha,\beta,j,\ell}\sum_{s>r}\eta^\alpha\eta^\beta h_{sj}{}^\beta h_{r\ell}{}^\alpha\omega^1\wedge\cdots\wedge\left(\omega^\ell\right)_{\mathsf{r}}\wedge\cdots\wedge\left(\omega^j\right)_{\mathsf{s}}\wedge\cdots\wedge\omega^k$$

$$=\sum_{\alpha,\beta}\sum_{s<r}\eta^\alpha\eta^\beta h_{ss}{}^\beta h_{rr}{}^\alpha\omega^1\wedge\cdots\wedge\omega^k$$

$$+\sum_{\alpha,\beta}\sum_{s<r}\eta^\alpha\eta^\beta h_{sr}{}^\beta h_{rs}{}^\alpha\omega^1\wedge\cdots\wedge\omega^k$$

$$+\sum_{\alpha,\beta}\sum_{s>r}\eta^\alpha\eta^\beta h_{ss}{}^\beta h_{rr}{}^\alpha\omega^1\wedge\cdots\wedge(\omega^r)_{\mathsf{s}}\wedge\cdots\wedge(\omega^s)_{\mathsf{r}}\wedge\cdots\wedge\omega^k$$

$$+\sum_{\alpha,\beta}\sum_{s>r}\eta^\alpha\eta^\beta h_{sr}{}^\beta h_{rs}{}^\alpha\omega^1\wedge\cdots\wedge(\omega^s)_{\mathsf{r}}\wedge\cdots\wedge(\omega^r)_{\mathsf{s}}\wedge\cdots\wedge\omega^k$$

$$=\sum_{\alpha,\beta}\sum_{s\neq r}\eta^\alpha\eta^\beta h_{ss}{}^\beta h_{rr}{}^\alpha\omega^1\wedge\cdots\wedge\omega^k$$

$$-\sum_{\alpha,\beta}\sum_{s\neq r}\eta^\alpha\eta^\beta h_{sr}{}^\beta h_{rs}{}^\alpha\omega^1\wedge\cdots\wedge\omega^k$$

$$=\sum_{\alpha,\beta,s,r}\left\{\eta^\alpha\eta^\beta h_{rr}{}^\beta h_{ss}{}^\alpha-\eta^\alpha\eta^\beta h_{sr}{}^\beta h_{rs}{}^\alpha\right\}\omega^1\wedge\cdots\wedge\omega^k$$

$$=\left\{\langle\eta,H\rangle^2-\|\mathfrak{B}\|_{\mathsf{gs}}{}^2\right\}\omega^1\wedge\cdots\wedge\omega^k.$$

For line 7 of (1), we first have

$$\Omega_A{}^B(X,Y)=\langle R(X,Y)e_A,e_B\rangle,$$

which implies

$$\Omega_A{}^B=\frac{1}{2}\langle R(e_C,e_D)e_A,e_B\rangle\,\omega^C\wedge\omega^D;$$

therefore

$$\mathrm{i}(Z)\Omega_A{}^B = \langle R(Z, e_C)e_A, e_B\rangle\, \omega^C.$$

Then, $\mathrm{i}(\eta)$ applied to line 7, and then restricted to $\phi_\epsilon(\mathfrak{M})$, is given by

$$\left\{-\sum_{r,\alpha}(-1)^{r-1}\eta^\alpha\omega^1\wedge\cdots\wedge\mathrm{i}(\eta)(\Omega_\alpha{}^r)\wedge\cdots\wedge\omega^k\right\}_{\phi_\epsilon(\mathfrak{M})}$$
$$= -\sum_r\langle R(\eta, e_r)\eta, e_r\rangle\,\omega^1\wedge\cdots\wedge\omega^k$$
$$= -\mathfrak{Ric}(\eta,\eta)\,\omega^1\wedge\cdots\wedge\omega^k.$$

It remains to consider line 2 and 6 of (1), that is, we must study

$$\mathrm{i}(\eta)\cdot\left(\sum_{r,\alpha} d\eta^\alpha\wedge\omega^1\wedge\cdots\wedge\omega_\alpha{}^r\wedge\cdots\wedge\omega^k\right.$$
$$\left.+\sum_{r,\alpha,\beta}(-1)^{r-1}\eta^\alpha\omega^1\wedge\cdots\wedge\left(\omega_\alpha{}^\beta\wedge\omega_\beta{}^r\right)_{\mathrm{r}}\wedge\cdots\wedge\omega^k\right)$$
$$= \mathrm{i}(\eta)\left(\sum_{r,\alpha,\beta}\{d\eta^\alpha+\eta^\beta\omega_\beta{}^\alpha\}\wedge\omega^1\wedge\cdots\wedge\omega_\alpha{}^r\wedge\cdots\wedge\omega^k\right)$$
$$= \sum_{r,\alpha,\beta}\{d\eta^\alpha+\eta^\beta\omega_\beta{}^\alpha\}(\eta)\omega^1\wedge\cdots\wedge\omega_\alpha{}^r\wedge\cdots\wedge\omega^k$$
$$+\sum_{r,\alpha,\beta}(-1)^r\{d\eta^\alpha+\eta^\beta\omega_\beta{}^\alpha\}\wedge\omega^1\wedge\cdots\wedge\omega_\alpha{}^r(\eta)\wedge\cdots\wedge\omega^k.$$

If we restrict to $\phi_\epsilon(\mathfrak{M})$, then

$$\sum_{r,\alpha,\beta}\{d\eta^\alpha+\eta^\beta\omega_\beta{}^\alpha\}(\eta)\wedge\omega^1\wedge\cdots\wedge\omega_\alpha{}^r\wedge\cdots\wedge\omega^k$$
$$= -\sum_{r,\alpha}(\mathcal{D}\eta^\alpha)(\eta)h_{rr}{}^\alpha\omega^1\wedge\cdots\wedge\omega^k$$
$$= -\langle\nabla_\eta\eta, H\rangle\omega^1\wedge\cdots\wedge\omega^k,$$

where $\mathcal{D}$ denotes the Levi-Civita connection in the normal bundle of $\mathfrak{M}$.

Finally,

$$\omega_\alpha{}^r(\eta) = \langle\nabla_\eta e_\alpha, e_r\rangle$$
$$= -\langle e_\alpha, \nabla_\eta e_r\rangle$$
$$= -\langle e_\alpha, \nabla_{e_r}\eta + [\eta, e_r]\rangle.$$

But, since η is the projection of $\Phi_*(\partial_\epsilon^{\mathfrak{M}})$ into the normal bundle of $\mathfrak{M}$ in M, we have $[\eta, e_r]$ is tangent to $\Phi(\mathfrak{M}\times\{\epsilon\})$, which implies

$$\omega_\alpha{}^r(\eta) = -\langle e_\alpha, \nabla_{e_r}\eta\rangle = -\langle e_\alpha, \mathcal{D}_{e_r}\eta\rangle.$$

One then has

$$\left\{\sum_{r,\alpha,\beta}(-1)^r\{d\eta^\alpha+\eta^\beta\omega_\beta{}^\alpha\}\wedge\omega^1\wedge\cdots\wedge\omega_\alpha{}^r(\eta)\wedge\cdots\wedge\omega^k\right\}_{\phi_\epsilon(\mathfrak{M})}$$
$$=\left\{\sum_{r,\alpha,\beta}(-1)^{r-1}\langle\mathcal{D}_{e_r}\eta,e_\alpha\rangle\{d\eta^\alpha+\eta^\beta\omega_\beta{}^\alpha\}\wedge\omega^1\wedge\cdots\wedge\widehat{\omega^r}\wedge\cdots\wedge\omega^k\right\}_{\phi_\epsilon(\mathfrak{M})}$$

But for

$$\mathcal{D}\eta=\sum_{\alpha,j}\eta^\alpha{}_{;j}e_\alpha\otimes\omega^j$$

we have

$$\eta^\alpha{}_{;j}=\langle\mathcal{D}_{e_j}\eta,e_\alpha\rangle,$$

which implies

$$\left\{\{d\eta^\alpha+\sum_\beta\eta^\beta\omega_\beta{}^\alpha\}\wedge\omega^1\wedge\cdots\wedge\widehat{\omega^r}\wedge\cdots\wedge\omega^k\right\}_{\phi_\epsilon(\mathfrak{M})}$$
$$=\sum_j\eta^\alpha{}_{;j}\omega^j\wedge\omega^1\wedge\cdots\wedge\widehat{\omega^r}\wedge\cdots\wedge\omega^k=(-1)^{r-1}\eta^\alpha{}_{;r}\omega^1\wedge\cdots\wedge\omega^k,$$

which implies

$$\left\{\sum_{r,\alpha,\beta}(-1)^r\{d\eta^\alpha+\eta^\beta\omega_\beta{}^\alpha\}\wedge\omega^1\wedge\cdots\wedge\omega_\alpha{}^r(\eta)\wedge\cdots\wedge\omega^k\right\}_{\phi_\epsilon(\mathfrak{M})}$$
$$=|\mathcal{D}\eta|^2\,\omega^1\wedge\cdots\wedge\omega^k.$$

Exercise III.34. Sketch: Check that if $Z=\xi+\eta$ (ξ tangent to $\Sigma_\epsilon(\Omega)$), and η normal to $\Sigma_\epsilon(\Omega)$)) is the vector field of a variation of D, with flow Σ_ϵ, then

$$\frac{\partial\{V(\Sigma_\epsilon(\Omega))\}}{\partial\epsilon}=\int_{\Sigma_\epsilon(\partial D)}\langle\eta,\nu\rangle\,dA,$$

and

$$\frac{\partial^2\{V(\Sigma_\epsilon(\Omega))\}}{\partial\epsilon^2}=\int_{\Sigma_\epsilon(\partial D)}\left\{\langle\nabla_\eta\eta,\nu\rangle-H\langle\eta,\nu\rangle^2\right\}dA.$$

Now use the fact that the variation is volume preserving, and $H=$ const. when $\epsilon=0$.

Exercise III.35. Hint: Consider the function

$$f(\mathbf{x})=H(\mathbf{x}\cdot\mathbf{n})+n-1.$$

Check that $\int_M f\,dA = 0$ and show that

$$0 \le -\int_M f\{\Delta f + f\|\mathfrak{B}\|^2\}\,dA$$
$$= -\int_M (n-1)\left\{-H^2 + (n-1)\|\mathfrak{B}\|^2\right\}\,dA \le 0,$$

which implies M is everywhere umbilic, which implies the claim.

Exercise III.36. Hint: Show that a nontrivial function

$$f = \sum_{j=1}^{k} \alpha_j \phi_j$$

exists, where $\phi_1, \ldots, \phi_k$ are orthonormal, and each ϕ_j is an eigenfunction of λ_j, such that f is orthogonal to $v_1, \ldots, v_{k-1}$ in $L^2(M)$. Then, use the argument of the proof of Rayleigh's theorem.

Exercise III.38. Hint: Let ψ_j be a Dirichlet eigenfunction on Ω_j of $\lambda(\Omega_j)$, with $L^2(\Omega_j)$–norm equal to 1. Extend ψ_j to vanish on $M \setminus \Omega_j$.

(i) Show that among all functions f on M of the form

$$f = \sum_{j=1}^{k} \alpha_j \psi_j,$$

there is at least one that is $L^2(M)$–orthogonal to $\{\phi_1, \ldots, \phi_{k-1}\}$. Thus,

$$\lambda_k(M) \le \mathcal{D}[f, f]/\|f\|^2.$$

(ii) Show that

$$\mathcal{D}[f, f] \le \left\{\sup_j \lambda(\Omega_j)\right\} \|f\|^2.$$

Exercise III.41. Hint: Let $\mathfrak{v}_\kappa$ denote the inverse function of V_κ (so $\mathfrak{v}_\kappa(c)$ is the radius of the disk in $\mathbb{M}_\kappa$ whose volume is c). Prove, using Exercises III.36 and III.38,

$$\lambda_\ell(M) \le \lambda_\kappa\left(\frac{1}{2}\mathfrak{v}_\kappa\left(\frac{V(M)}{\ell}\right)\right).$$

Then, prove, using Exercise III.37,

$$\lambda_\kappa\left(\frac{1}{2}\mathfrak{v}_\kappa\left(\frac{V(M)}{\ell}\right)\right) \le c(n, \kappa)\left\{\frac{\ell}{V(M)}\right\}^{2/n}.$$

Exercise III.43(a). Hint: Let $\mathbf{x} : M^2 \to \mathbb{R}^3$ denote the imbedding, and $u : M \to \mathbb{R}$ given by $u(\mathbf{x}) = |\mathbf{x}|^2$. Recall that (i) $|\operatorname{grad} \mathbf{x}|^2 = 2$, (ii) $\Delta \mathbf{x} = 0$. Then show that $\Delta u = 4$ on all of M^2. To prove that $\lambda^*(\Omega) > 1/(4R^2)$ is to prove that

$$\iint_\Omega |\operatorname{grad} \phi|^2\,dA \ge \frac{1}{4R^2}\iint_\Omega \phi^2\,dA$$

for any C^∞ function with compact support in Ω. Prove the estimate by considering

$$\iint_\Omega \phi^2 \Delta u \, dA,$$

and applying, to it, Green's theorem and the Cauchy–Schwarz inequality.

Exercise III.43(b). Hint: Given any $B(o; R)$, where o is a fixed point in M, and $\alpha \in (0, 1)$, consider the function ϕ on $B(o; R)$ defined by

$$\phi(x) = \begin{cases} 1 & d(x, o) \le \alpha R, \\ (R - d(x, o))/(R - \alpha R) & \alpha R \le d(x, o) \le R. \\ 0 & d(x, o) \ge R \end{cases}$$

Show that the assumption

$$\iint_{B(o;R)} |\mathrm{grad}\, \phi|^2 \, dV \ge \text{const.} \iint_{B(o;R)} \phi^2 \, dV$$

leads to the inequality

$$V(o; R) \ge \{\text{const.}(1 - \alpha^2)R^2 + 1\} V(\alpha R).$$

Now show that any polynomial estimate

$$V(o; R) \le \text{const.} r^\ell \qquad \text{for all } R \gg 1,$$

for any fixed $\ell > 0$ leads to a contradiction.

Exercise III.44. Hint: Use moving frames, as in Exercise I.11. The argument can be simplified by assuming, at any fixed point p, that one may choose the frame field $\{e_j\}$ to satisfy $[e_j, e_k]_{|p} = 0$ for all j, k.

(i) Write, as in Exercise I.11,

$$\nabla F = dF = \sum_j F_j e_j, \qquad \nabla\nabla F = \sum_{jl} F_{jl}\, \omega^l \otimes \omega^j,$$

and

$$\nabla\nabla\nabla F = \sum_{jlk} F_{jlk}\, \omega^k \otimes \omega^l \otimes \omega^j,$$

where

$$dF_{jl} - \sum_r \omega_j{}^r F_{rl} - \sum_r \omega_l{}^r F_{jr} = \sum_k F_{jlk} \omega^k.$$

Use the symmetry of F_{jl} to show

$$F_{jlk} = F_{lkj} + \langle R(e_j, e_k) \mathrm{grad}\, F, e_l \rangle.$$

(ii) Show

$$\sum_k F_{kkj} = \langle \mathrm{grad}\, \Delta F, e_j \rangle.$$

(iii) Consider

$$G = \frac{1}{2}|dF|^2, \qquad dG = \sum_i G_i\,\omega^i,$$

and show

$$G_i = \sum_j F_j F_{ji},$$
$$\nabla\nabla G = \sum_{jkl} \{F_{jk}F_{jl} + F_j F_{jlk}\}\,\omega^l \otimes e_k.$$

(iv) Now use all of the above to derive the formula.

Hints and Sketches: Chapter IV

Exercise IV.2(b). Hint: Use (a) to first show that

$$|X(1,1) - X(1,0)| \le \int_0^1 ds \int_0^1 |R(\partial_s v, \partial_t v)X|\,dt,$$

and then use Exercise II.1(f).

Exercise IV.3(e). Hint: Use (IV.6.1).

Exercise IV.4. Sketch: If there is an isometry Φ of M_o, so that $\Gamma_2 = \Phi^{-1}\Gamma_1\Phi$, then map $\phi : M_2 \to M_1$ by

$$\phi(x) = \pi_1 \circ \Phi \circ {\pi_2}^{-1}[x].$$

If $\overline{x} \in {\pi_2}^{-1}[x]$ and $\gamma_2 \in \Gamma_2$ then there exists $\gamma_1 \in \Gamma_1$ such that

$$\Phi(\gamma_2 \cdot \overline{x}) = \gamma_1 \cdot \Phi(\overline{x})$$

which implies ϕ is a well-defined isometry.

If $\phi : M_1 \to M_2$ is an isometry, then $\pi_1 \circ \phi$ is a covering of M_2 by M_o. The universal property of the covering $\pi_2 : M_o \to M_2$ implies that $\pi_1 \circ \phi$ factors through $\Phi : M_o \to M_o$, that is, there exists a Φ such that $\pi_1 \circ \phi = \Phi \circ \pi_2$, which implies the claim.

Exercise IV.5(b). Sketch: One proves the result by induction on n.

For $n = 1$: Let Γ be a nontrivial discrete subgroup of $\mathbb{R}$. Then, for any $r > 0$, the interval $(-r, r)$ has at most a finite number of elements of Γ, which implies Γ has an element $u \neq 0$ closest to the origin. This implies $\{ku : k \in \mathbb{Z}\} \subseteq \Gamma$. If given any $v \in \Gamma$, then there exists $k \in \mathbb{Z}$ such that $v \in [ku, (k+1)u)$, which implies $|v - ku| < |u|$, which implies $v - ku = 0$, which implies the claim for $n = 1$.

Assume the result is true for $\mathbb{R}^{n-1}$, $n \ge 1$, and let Γ be a nontrivial discrete subgroup of $\mathbb{R}^n$. Again, there is an element $u \neq 0$ closest to the origin. So $\Gamma \cap \mathbb{R}u = \{ku : k \in \mathbb{Z}\}$. Consider the projection

$$\pi : \mathbb{R}^n \to \mathbb{R}^n/\mathbb{R}u;$$

Here is the argument to show $\pi(\Gamma)$ is discrete in $\mathbb{R}^n/\mathbb{R}u$ (this will imply the result):

Assume one has a sequence $v_j \in \Gamma$ such that $\pi(v_j) \to 0$, that is, $v_j \to 0 \pmod{\mathbb{R}u}$. Then there exists a sequence r_j in $\mathbb{R}$ such that

$$v_j - r_j u \to 0 \quad \text{in } \mathbb{R}^n.$$

Write each r_j as

$$r_j = \rho_j + k_j, \qquad k_j \in \mathbb{Z},\ |\rho_j| \leq 1/2.$$

Then,

$$|v_j - k_j u - \rho_j u| < |u|/2$$

for sufficiently large j, which implies $|v_j - k_j u| < |u|$, which contradicts the fact that u is closest to the origin. Therefore, $\rho_j = 0$ for all but a finite number of j, which implies the result.

Exercise IV.5(c). Sketch: Clearly, if $T \in \mathcal{E}(n)$ has a line ℓ on which it acts as a translation, then T maps every hyperplane perpendicular to ℓ to a hyperplane perpendicular to ℓ. This implies that T has no fixed points.

If $T = Ax + a$ has no fixed points, then decompose

$$\mathbb{R}^n = V \oplus W, \qquad V = \{x \in \mathbb{R}^n : Ax = x\},\ W = V^{\perp},$$

and write

$$a = b + c, \quad x = y + z, \qquad b, y \in V,\ c, z \in W.$$

Then, $A - I|W : W \to W$ is an isomorphism. Check that this implies that $b \neq 0$ (so V is nontrivial) and that a line ℓ is given by

$$\ell(t) = tb - \{(A - I)|W\}^{-1}c.$$

Exercise IV.12(a). Hint: Use the Hadamard–Cartan theorem (Theorem IV.1.3).

Exercise IV.12(b). Hint: Pick up where Note II.11 leaves off. Also, see Note II.5.

Exercise IV.13(b). Hint: Let $\omega'_{|x}$ denote the velocity vector of ω at x; it suffices to show that

$$\gamma_* \cdot \omega'_{|x} = \omega'_{|\gamma \cdot x}.$$

Use the triangle inequality.

Exercise IV.14. Hint: Let $\omega_o : \mathbb{R} \to M$ be an axis of $\gamma_o \in \Gamma_o$. For any other $\gamma \in \Gamma_o$ and $t \in \mathbb{R}$, consider $\gamma_o(\gamma \circ \omega(t))$.

Exercise IV.16. Sketch: (Check the notations, definitions, and results of §§III.3–5.) First, fix $r > 0$ and $x \in M$, and consider that part of $S(x; r)$ in D_x (that is, the inside of the cut locus) which intersects F. The idea is to show the desired inequality for *this* area for all r and then to use Proposition III.5.1. We will use the fact that, since M covers a compact, its Ricci curvature is bounded uniformly from below by a constant, say, κ.

Proceed as follows: Let $\mathcal{F}$ denote the union of all translates of F by Γ, which neighbor F (i.e., whose closures intersect the closure of F), and let $\delta > 0$ denote the distance from F to $M \setminus \mathcal{F}$. Let $\mathsf{DF}(r)$ denote the collection of unit tangent vectors in $\mathsf{D}_x(r)$, such that $\exp r\xi \in F$. Now, verify

$$V(\mathcal{F}) \geq \int_{\mathsf{DF}(r)} d\mu_x(\xi) \int_{r-\delta}^{r} \sqrt{\mathbf{g}}(t;\xi)\,dt \geq \delta \frac{\mathbf{S}_\kappa(r-\delta)}{\mathbf{S}_\kappa(r)} \int_{\mathsf{DF}(r)} \sqrt{\mathbf{g}}(r;\xi)\,d\mu_x(\xi),$$

which implies the claim.

Exercise IV.20. Hint:

(i) Assume M is not simply connected. Then, M possesses a nontrivial free homotopy class, with attendant shortest closed geodesic Γ. Prove that Γ possesses a periodic parallel vector field.

(ii) Use the second variation of arc length to obtain a contradiction.

Exercise IV.24. Hint: Here, one must use the fact that a homotopy class in the space of continuous maps of $\mathbb{S}^1$ to X, $C^0(\mathbb{S}^1, X)$, is an open set in $C^0(\mathbb{S}^1, X)$ (with topology of uniform convergence).

Exercise IV.26. Sketch: Use the isomorphism of Γ with $\pi_1(M_o, x_o)$. Given any homotopy class $\alpha \in \pi_1(M_o, x_o)$, fix $\epsilon > 0$, and let α be represented by the continuous path $\gamma : [0, 1] \to M$ with $\gamma(0) = \gamma(1) = x_o$.

(i) Show that $[0, 1]$ may be subdivided into subintervals $[t_{j-1}, t_j]$, $j = 1, \ldots, N$, so that

$$d(\gamma(t_{j-1}), \gamma(t_j)) < \epsilon$$

for all j.

(ii) Connect x_o to every $\gamma(t_j)$, $j = 1, \ldots, N-1$, by a minimizing geodesic, and prove that α is generated by homotopy classes each of whose minimizing loops have length less than $2d(M) + \epsilon$.

(iii) Now use Exercise IV.21, with fixed basepoint, to prove (IV.6.3).

Exercise IV.27. Hint: The argument is a variant of the argument of Lemma IV.3.2.

Hints and Sketches: Chapter V

Exercise V.1. Hint: Note that the degree of T is given by the integral of the geodesic curvature of the imbedding with respect to arc length. So, the first step is to reparameterize the imbeddings with respect to multiples of arc length (show that this can be done through smooth homotopies), smoothly deform through convex combinations the curvature function of Γ_0 to the curvature function of Γ_1, and then check that this homotopy can be "integrated" to a homotopy of Γ_0 to Γ_1.

Exercise V.3.

(a) Hint: Use Stokes' theorem on $d\omega_1{}^2$.

(b) Hint: Use Exercise III.19.

Hints and Sketches: Chapter VI

Exercise VI.4(a). Sketch: Indeed, for the second claim, we have

$$\begin{aligned}\iint_D F^2\,dV &= \int_0^{r_0} \psi^2(r)A_\kappa(r)\,dr \\ &= -\int_0^T \psi^2(\mathfrak{r}(t))A_\kappa(\mathfrak{r}(t))\mathfrak{r}'(t)\,dt \\ &= -\int_0^T t^2 V'(t)\,dt \\ &= \iint_\Omega f^2\,dV,\end{aligned}$$

by the coarea formula.

Exercise VI.4(b). Sketch: Use the coarea formula and the Cauchy–Schwarz inequality:

$$\begin{aligned}\iint_\Omega |\text{grad}\, f|^2\,dV &= \int_0^T dt \int_{\Gamma(t)} |\text{grad}\, f|\,dA_t \\ &\geq \int_0^T A^2(t)\left\{\iint_{\Gamma(t)} |\text{grad}\, f|^{-1}\,dA_t\right\}^{-1} dt.\end{aligned}$$

Exercise VI.4(c). Sketch: Use the isoperimetric inequality:

$$\begin{aligned}\int_0^T A^2(t)\left\{\iint_{\Gamma(t)} |\text{grad}\, f|^{-1}\,dA_t\right\}^{-1} dt &= -\int_0^T A^2(t)(V'(t))^{-1}\,dt \\ &= -\int_0^T A^2(t)\{A_\kappa(\mathfrak{r}(t))\mathfrak{r}'(t)\}^{-1}\,dt \\ &\geq -\int_0^T A_\kappa(\mathfrak{r}(t))(\mathfrak{r}'(t))^{-1}\,dt.\end{aligned}$$

Exercise VI.4(d). Sketch: Use $|\text{grad}\, F|^2(\mathfrak{r}(t)) = (\mathfrak{r}'(t))^{-2}$ to show

$$\begin{aligned}\iint_D |\text{grad}\, F|^2\,dV &= \int_0^{r_0} |\text{grad}\, F|^2(r)A_\kappa(r)\,dr \\ &= -\int_0^T (\mathfrak{r}'(t))^{-2}A_\kappa(\mathfrak{r}(t))\mathfrak{r}'(t)\,dt \\ &= -\int_0^T A_\kappa(\mathfrak{r}(t))(\mathfrak{r}'(t))^{-1}\,dt.\end{aligned}$$

Exercise VI.6(a). Hint: Define $\mathfrak{r}(t)$ by $V(t) = \beta V_\kappa(\mathfrak{r}(t))$.

Exercise VI.6(b). Hint: Given the equality, let Ω be a nodal domain (see Exercise VI.4 above) of the eigenfunction ϕ of $\lambda_2(M)$, with volume less than or equal to half the volume of M. Then show

$$n\kappa = \lambda^*(\Omega) \geq \lambda_\kappa(D) \geq n\kappa,$$

using Exercise III.46.

Hints and Sketches: Chapter VII

Exercise VII.2(d). Hint: Use Parts (b) and (c) of this exercise.

Exercise VII.2(e). Hint: Assume such a submanifold N exists. Show, using (d) of this exercise, that $\int_N \Theta^{n-1} \neq 0$. Also show, on the other hand, that Θ^{n-1} is exact, which implies its integral over N is 0.

Exercise VII.5. Hint: First show that the Jacobi field $J_{\mathfrak{x}}(t)$ associated with the geodesic variation $\Omega(t, \epsilon)$ satisfies the initial conditions

$$J_{\mathfrak{x}}(0) = \pi_*\mathfrak{x}, \qquad (\nabla_t J_{\mathfrak{x}})(0) = \mathcal{T}\mathfrak{x};$$

then show that the Jacobi field along $\gamma_\xi(t)$ associated with the geodesic variation given by

$$\Sigma(t, \epsilon) = \pi \circ \Phi_t Z(\epsilon)$$

has the same initial conditions. Therefore, they are the very same.

Exercise VII.7. Hint: Use the Rauch comparison theorem (Theorem II.6.4) to show that, for $\eta \in \xi^\perp, \xi \in \mathsf{S}M$, we have $|\mathcal{D}_\xi^+(t)\eta| \to +\infty$ as $t \uparrow +\infty$, and $|\mathcal{D}_\xi^+(t)\eta|$ remains bounded as $t \downarrow -\infty$.

Exercise VII.8. Hint: Use the formula for the derivative of determinants, and the invariance of the Liouville measure under the geodesic flow.

Exercise VII.9. Hint: For an arbitrary orthonormal basis $\{e_1, \ldots, e_{n-1}\}$ of $\xi^\perp$, we have

$$\begin{aligned} -\sigma(\xi) &= \operatorname{tr} {K_\xi}^{1/2} \\ &= \sum_j \langle {K_\xi}^{1/2} e_j, e_j \rangle \\ &\leq \sum_j |{K_\xi}^{1/2} e_j| \\ &= \sum_j \langle {K_\xi}^{1/2} e_j, {K_\xi}^{1/2} e_j \rangle^{1/2} \\ &= \sum_j \langle K_\xi e_j, e_j \rangle^{1/2}. \end{aligned}$$

Exercise VII.10. Hint: Multiply Riccati's equation on the right by U^{-1}, take the trace, and the use Exercise VII.8.

Exercise VII.12. Hint: Equality in Exercise VII.11(b) implies

$$\beta \int \operatorname{tr} U = \int K U^{-1} = \int \operatorname{tr} U,$$

by the exercise above; so, $\beta = 1$. Since $\beta = 1$, we have, with equality in Exercise VII.11(a),

$$\operatorname{tr} U = \operatorname{tr} K U^{-1} = k^2(\xi) \operatorname{tr} U,$$

which implies $k = 1$ on all of $\mathsf{S}M$. So, $K = U^2$, which implies U is parallel along the geodesic flow, which implies K is parallel along the geodesic flow.

Exercise VII.14. Sketch: Use Lord Rayleigh's characterization of eigenvalues (Theorem III.9.2), that is, show

$$\int_\Omega |\operatorname{grad} f|^2 \, dV \geq \pi^2 n \inf_{x \in \Omega} \frac{1}{\mathbf{c_{n-1}}} \int_{\mathsf{S}_x} \frac{d\mu_x(\xi)}{\delta^2(\xi)} \int_\Omega |f|^2 \, dV.$$

To carry it out, first use Lemma VII.3.1, applied to the bilinear form

$$(\xi, \eta) \mapsto \langle \nabla f, \xi \rangle \langle \nabla f, \eta \rangle,$$

to show

$$\int_\Omega |\operatorname{grad} f|^2 \, dV = \frac{n}{\mathbf{c_{n-1}}} \int_{\mathsf{S}\Omega} \langle (\operatorname{grad} f) \circ \pi, \xi \rangle \, d\mu(\xi).$$

Note that along $\mathsf{S}^+\Omega$ one has $\delta(\xi) = \tau(\xi)$, which implies by Santalo's formula that this last integral is equal to

$$\frac{n}{\mathbf{c_{n-1}}} \int_{\mathsf{S}^+\partial\Omega} \langle \xi, \overline{\nu}_{\pi(\xi)} \rangle \, d\sigma(\xi) \int_0^{\delta(\xi)} (f \circ \gamma_\xi{}'(t))^2 \, dt.$$

Now use the fixed-endpoint version of Wirtinger's inequality (Exercise III.42) to estimate this last integral from below by

$$\frac{n}{\mathbf{c_{n-1}}} \int_{\mathsf{S}^+\partial\Omega} \langle \xi, \overline{\nu}_{\pi(\xi)} \rangle \, d\sigma(\xi) \int_0^{\delta(\xi)} \frac{\pi}{\delta^2(\xi)} (f \circ \gamma_\xi(t))^2 \, dt.$$

Note that $\delta(\Phi_t \xi)$ is constant with respect to t. Now finish off the argument.

Hints and Sketches: Chapter VIII

Exercise VIII.1. Sketch: Set $f = |\phi|^{2(\nu-1)/(\nu-2)}$; $p = 2(\nu-1)/(\nu-2)$. So, $|df| = p|\phi|^{p-1}|\operatorname{grad}|\phi|| = p|\phi|^{p-1}|\operatorname{grad}\phi|$, which implies, by the Cauchy–Schwarz inequality, that

$$\int_M |df| \, dV \leq p \|\operatorname{grad}\phi\|_2 \|\phi\|_{2(p-1)}{}^{p-1},$$

which easily implies the claim.

Exercises VIII.2. Sketch: Start with Exercise VIII.1 for $\nu > 2$. Next one shows Moser (1964, p. 116)),

$$\int \phi^{2+4/\nu}\, dV \le \text{const.} \|\text{grad}\,\phi\|_2{}^2 \|\phi\|_2{}^{4/\nu}$$

for all $\nu \ge 2$. Indeed, if $\nu > 2$, set $f = |\phi|^2$, $p = \nu/(\nu - 2)$ and $g = |\phi|^{4/\nu}$, $q = \nu/2$, and use Hölder's inequality. On the other hand, if $\nu = 2$, one has directly

$$\|\phi^2\|_2 \le \text{const.} \|\text{grad}\,(\phi^2)\|_1 \le \text{const.} \|\phi\|_2 \|\text{grad}\,\phi\|_2.$$

Square both sides.

To obtain the claimed inequality (see Cheng–Li (1981)), simply apply Hölder's inequality to $f = |\phi|^{4/(\nu+4)}$, $p = (\nu + 4)/4$ and $g = |\phi|^{(2\nu+4)/(\nu+4)}$, $q = (\nu + 4)/\nu$.

Exercises VIII.4–VIII.6. Hint: See Chavel (1984, pp. 109–112).

Hints and Sketches: Chapter IX

Exercise IX.1. Hint: See Exercise II.24 and Note II.11.

Exercise IX.3. Sketch: If H_γ is not totally convex, then there exists geodesic $\omega : [0, 1] \to M$ such that $\omega(0), \omega(1) \in H_\gamma$, and there exists $s \in (0, 1)$ such that $q = \omega(s) \in B_\gamma$.

(i) Show that there exist $t_0 > 0$, $\epsilon > 0$ such that

$$t_0 - \epsilon = d(q, \gamma(t_0)), \qquad d(q, \gamma(t)) \le t - \epsilon$$

for all $t \ge t_0$.

(ii) To each $t \ge t_0$, associate a point ω_{s_t} on ω closest to $\gamma(t)$; construct an appropriate hinge at ω_{s_t} with which one can conclude, using the fact that $\omega(0) \in H_\gamma$, and using the Toponogov–Alexandrov theorem, that

$$t^2 \le d(\gamma(t), \omega(0))^2 \le \ell(\omega)^2 + (t - \epsilon)^2$$

for all $t \ge t_0$, which is impossible for large t.

Exercise IX.4(b). Hint: In a Hadamard–Cartan manifold, the function $r_q(x) = d(q, x)$ is C^∞ with gradient equal to 1. The same applies to ϕ_t. Now, let $t \uparrow +\infty$.

Exercise IX.4(c). Hint: The idea is to first pick a candidate vector field for grad ϕ_γ, and then to verify that it is indeed the gradient.

Given $p \in M$, pick y_1, y_2 as in (b–ii). Show that we may assume

$$\phi_\gamma(y_1) = \phi_\gamma(p) + r, \qquad \phi_\gamma(y_2) = \phi_\gamma(p) - r.$$

Pick the unit vector field $\xi_{|p}$ to be the initial vector field of the unit speed geodesic connecting p to y_1. Show that the geodesic passes through y_2. Then, show that, for any unit speed geodesic $\omega(s)$, satisfying $\omega(0) = p$, we have

$$r - d(\omega(s), y_1) \le \phi_\gamma(\omega(s)) \le -r + d(\omega(s), y_2).$$

Now use the first variation formula (Theorem II.4.1) to show that the function $\phi_\gamma \circ \omega$ has a derivative at $s = 0$ equal to $\langle \xi, \omega'(0) \rangle$.

Exercise IX.5. Hint: First show that the integral curves of grad ϕ are geodesics (see Exercise I.4 – one does not need the C^∞ hypothesis here). Pick $x \in M$ so that $\phi(x) = 0$, $\gamma = \gamma_{(\mathrm{grad}\,\phi)_{|x}}$, that is, γ is the unit speed geodesic emanating from x with initial velocity vector equal to the gradient of ϕ at x. This is the candidate for the Busemann function; that is, show $\phi = \phi_\gamma$. Use $|\mathrm{grad}\,\phi| = 1$ on all M to show that $\phi_\gamma \le \phi$. Use the concavity of ϕ and ϕ_γ to show they are in fact equal.

Exercise IX.6. Hint: For any $\epsilon > 0$, there exists a Lipschitz homeomorphism $f_\epsilon : X \to Y$ such that

$$|\ln \mathrm{dil}\, f_\epsilon| + |\ln \mathrm{dil}\, {f_\epsilon}^{-1}| < \epsilon.$$

Show that one can use the Arzela–Ascoli theorem to obtain a sequence $\epsilon_j \to 0$ for which one has the uniform limit $f = \lim f_{\epsilon_j}$, as $j \to \infty$. Then verify that f is an isometry.

Exercise IX.8. Hint: The idea is to construct a metric space Z_0 for which there are isometric immersions of X and Y. Pick Z_0 to be the disjoint union of X and Y. The metric? For two points in X (resp. Y), keep the distance as before. This will provide for an isometric immersion. But first, a genuine metric. What will the distance be between points $x \in X$, $y \in Y$? Keep in mind the safest way to guarantee the triangle inequality.

Exercise IX.9. Hint: See Grove (1987, pp. 215–217).

Bibliography

Accola, R. D. M. (1960). Differential and extremal lengths on Riemann surfaces. *Proc. Natl. Acad. Sci. USA* **46**, 540–543.

Adams, C. and F. Morgan (1999). Isoperimetric curves on hyperbolic surfaces. *Proc. Am. Math. Soc.* **127**, 1347–1356.

Adams, R. A. (1975). *Sobolev Spaces*. New York: Academic Press.

Ahlfors, L. V. (1973). *Conformal Invariants*. New York: McGraw-Hill.

Ahlfors, L. V. and A. Beurling (1950). Conformal invariants and function-theoretic null-sets. *Acta Math.* **83**, 101–129.

Alexandrov, A. D. (1948). *The Intrinsic Geometry of Convex Surfaces*. Moscow–Leningrad: Gosudarstv. Izdat Tehn–Teor. Lit. (Russian). German transl., Die Innere Geometrie der konvexen Flächen. Berlin: Akademie Verlag, 1955.

Alexandrov, A. D. (1962). A characteristic property of spheres. *Ann. di Math. Pura Appl.* **58**, 303–315.

Allendoerfer, C. B. (1940). The Euler number of a Riemannian manifold. *Am. J. Math.* **62**, 243–248.

Allendoerfer, C. B. and A. Weil (1943). The Gauss–Bonnet theorem for Riemannian polyhedra. *Trans. Am. Math. Soc.* **53**, 101–129.

Almgren, F. (1976). *Existence and Regularity Almost Everywhere of Solutions to Elliptic Variational Problems with Constraints*, Volume 4 of *Memoir AMS*. Providence, RI: American Mathematical Society.

Almgren, F. (1986). Optimal isoperimetric inequalities. *Ind. U. Math. J.* **35**, 451–547.

Ambrose, W. (1956). Parallel translation of Riemannian curvature. *Ann. Math.* **64**, 337–363.

Ambrose, W. (1961). The index theorem in Riemannian geometry. *Ann. Math.* **73**, 49–86.

Ancona, A. (1990). Théorie du potentiel sur les graphes et les variétés. In A. Ancona, D. Geman, and N. Ikeda (Eds.), *École d'Été de Probabilités de Saint-Flour XVIII–1988*, pp. 5–112. Berlin: Springer Verlag.

Anderson, M. T. (1990a). Metrics of positive Ricci curvature. *Manusc. Math.* **68**, 405–415.

Anderson, M. T. (1990b). On the topology of complete manifolds of non-negative Ricci curvature. *Topology* **29**, 41–55.

Angenent, S. B. (1992). Shrinking doughnuts. *Progr. Nonlin. Diff. Eq. Appl.* **7**, 71–111.

Anosov, D. V. (1967). Geodesic flows on Riemannian manifolds of negative curvature. *Proc. Steklov Math. Inst.* **90**. Engl. transl. Providence, RI: American Mathematical Society.

Arnold, V. I. (1980). *Mathematical Methods of Classical Mechanics*. New York: Springer Verlag.

Aronszajn, N. (1957). A unique continuation theorem for solutions of partial differential equations or inequalities of second order. *J. Math. Pure Appl.* **36**, 237–249.

Aubin, T. (1976). Problèmes isopérimétriques et espaces de Sobolev. *J. Diff. Geom.* **11**, 573–598.

Aubin, T. (1982). *Nonlinear Analysis on Manifolds. Monge–Ampère Equations*. New York: Springer Verlag.

Baernstein, A. (2004). *Symmetrization in Analysis*. In preparation.

Baernstein, A. and B. A. Taylor (1976). Spherical arrangements, subharmonic functions, and *-functions in n–space. *Duke Math. J.* **43**, 245–268.

Ballman, W., M. Brin, and K. Burns (1987). On surfaces with no conjugate points. *J. Diff. Geom.* **25**, 249–273.

Ballman, W., M. L. Gromov, and V. Schroeder (1985). *Manifolds of Nonpositive Curvature*, Volume 61 of *Progress in Math.* Boston: Birkhäuser.

Barbosa, J. L. and A. Colares (1986). *Minimal Surfaces in* $\mathbb{R}^3$, Volume 1195 of *Lecture Notes Math.* Berlin: Springer Verlag.

Barbosa, J. L. and M. do Carmo (1976). On the size of a stable minimal surface in $\mathbb{R}^3$. *Am. J. Math.* **98**, 515–528.

Barbosa, J. L. and M. do Carmo (1984). Stability of hypersurfaces with constant mean curvature. *Math. Zeit.* **185**, 339–353.

Barta, J. (1937). Sur la vibration fundamentale d'une membrane. *C. R. Acad. Sci. Paris* **204**, 472–473.

Basmajian, A. (1972). Generalizing the hyperbolic collar lemma. *Bull. Am. Math. Soc.* **27**, 154–158.

Bavard, C. (1984). Le rayon d'injectivité des surfaces à courbure majoree. *J. Diff. Geom.* **20**, 137–142.

Bavard, C. and P. Pansu (1986). Sur le volume minimal de R^2. *Ann. École Norm. Sup. Paris* **19**, 479–490.

Beardon, A. F. (1983). *The Geometry of Discrete Groups*. New York: Springer Verlag.

Beckenbach, E. F. and T. Radó (1933). Subharmonic functions and surfaces of negative curvature. *Trans. Am. Math. Soc.* **35**, 662–674.

Bedford, T., M. Keane, and C. Series (Eds.) (1991). *Ergodic Theory, Symbolic Dynamics and Hyperbolic Spaces*. Oxford: Oxford University Press.

Bendetti, R. and C. Petronio (1994). *Lectures on Hyperbolic Geometry*. New York: Springer Verlag.

Benjamimi, I. and J. Cao (1996). A new isoperimetric theorem for surfaces of variable curvature. *Duke Math. J.* **85**, 359–396.

Benjamini, Y. (1983). On two-point symmetrization. Texas Functional Analysis Seminar. University of Texas. Preprint.

Bérard, P. H. (1986). *Spectral Geometry: Direct and Inverse Problems*, Volume 1207 of *Lecture Notes Math.* Berlin: Springer Verlag.

Bérard, P. H. and D. Meyer (1982). Inégalités isopérimétrique et applications. *Ann. Sci. École Norm. Sup., Paris* **15**, 531–542.

Berger, M. (1960). Les variétés riemannienes (1/4)–pincées. *Ann. Scuola Norm. Sup. Pisa* **14**, 161–170.

Berger, M. (1962). An extension of Rauch's metric comparison theorem and some applications. *Ill. J. Math.* **6**, 700–712.

Berger, M. (1965). *Lectures on Geodesics in Riemannian Geometry*. Bombay: Tata Institute of Fund. Research.

Berger, M. (1972a). Du côté de chez Pu. *Ann. Sci. École Norm. Sup., Paris* **4**, 1–44.

Berger, M. (1972b). A l'ombre de Loewner. *Ann. Sci. École Norm. Sup., Paris* **4**, 241–260.

Berger, M. (1976). Some relations between volume, injectivity radius, and convexity radius in Riemannian geometry. In M. Cahen and M. Flato (Eds.), *Differential Geometry and Relativity*, pp. 33–42. Dordrecht-Holland: D. Reidel Publishing Co.

Berger, M. (1980). Une borne inférieure pour le volume d'une variété riemannienes en fonction du rayon d'injectivité. *Ann. Inst. Fourier, Grenoble* **30**, 259–265.

Berger, M. (1985). H. E. Rauch géomètre différentiel. In I. Chavel and H. Farkas (Eds.), *Differential Geometry and Complex Analysis. H. E. Rauch Memorial Volume*, pp. 1–13. Berlin: Springer Verlag.

Berger, M. (1987). *Geometry*. Berlin: Springer Verlag.

Berger, M. (2000). *Riemannian Geometry during the Second Half of the Twentieth Century*. University Lecture Series. Providence, RI: American Mathematical Society.

Berger, M. (2003). *A Panoramic View of Riemannian Geometry*. Berlin: Springer Verlag.

Berger, M., P. Gauduchon, and E. Mazet (1974). *Le Spectre d'une Variété Riemanniene*, Volume 194 of *Lecture Notes Math.* Berlin: Springer Verlag.

Berger, M. and B. Gostiaux (1988). *Differential Geometry: Manifolds, Curves, and Surfaces*. New York: Springer Verlag.

Berger, M. and J. L. Kazdan (1980). A Sturm–Liouville inequality with applications to an isoperimetric inequality for volume, injectivity radius, and to wiedersehns manifolds. In E. Beckenbach (Ed.), *General Inequalities*, Volume II, pp. 367–377. Basel: Birkhäuser.

Bernstein, F. (1905). Über die isoperimetrische Eigenschaft des Kreises auf der Kugeloberfläche und in der Ebene. *Math. Ann.* **60**, 117–136.

Besse, A. (1987). *Einstein Manifolds*. Berlin: Springer Verlag.

Besse, A. L. (1978). *Manifolds All of Whose Geodesics Are Closed*. Berlin: Springer Verlag.

Bishop, R. L. (1977). Decomposition of cut loci. *Proc. Am. Math. Soc.* **65**, 133–136.

Bishop, R. L. and R. Crittenden (1964). *Geometry of Manifolds*. New York: Academic Press.

Blaschke, W. (1967). *Vorlesungen über Differential Geometrie* (4th ed.). New York: Chelsea Publishing Company.

Blatter, C. (1961). Über extremallänge auf geschlossenen Flächen. *Comment. Math. Helv.* **35**, 153–168.

Bol, G. (1941). Isoperimetrische Ungleichung für Bereiche auf Flächen. *Jber. Deutsch. Math.-Verein.* **51**, 219–257.

Bolza, G. (1909). *Vorlesungen über Variationsrechnung*. New York: Chelsea Publishing Company, 1962. Reprint of first edition.

Bombieri, E. (Ed.) (1983). *Seminar on Minimal Submanifolds*, Volume 103 of *Ann. Math. Studies*. Princeton University Press.

Bonnet, O. (1848). Mémoire sur la théorie générales des surfaces. *J. l'École Polytechnique* **32**, 1–148.

Bonnet, O. (1855). Sur quelques propriétés des lignes géodésiques. *C. R. Acad. Sci. Paris* **40**, 1311–1313.

Bourguignon, J. P. and H. Karcher (1978). Curvature operators: pinching estimates and geometric examples. *Ann. Sci. École Norm. Sup., Paris* **11**, 71–92.

Burago, D., Y. D. Burago, and S. Ivanov (2001). *A Course in Metric Geometry*. Providence, RI: American Mathematical Society.

Burago, D. and S. Ivanov (1994). Riemannian tori without conjugate points are flat. *Geom. Fcnl. Anal.* **3**, 259–269.

Burago, Y. D. (1978). The radius of injectivity on the surfaces whose curvature is bounded from above. *Ukrain. Geom. Sb.* **21**, 10–14. (Russ.).

Burago, Y. D. and V. A. Zalgaller (1988). *Geometric Inequalities*. Berlin: Springer Verlag. Original Russian edition: *Geometricheskie neravenstva*, Leningrad 1980.

Buser, P. (1978). The collar theorem and examples. *Manusc. Math.* **25**, 349–357.

Buser, P. (1980). On Cheeger's inequality $\lambda_1 \geq h^2/4$. In *Geometry of the Laplace Operator*, Volume 36 of *Proc. Symp. Pure Math.*, pp. 29–77. Providence, RI: American Mathematical Society.

Buser, P. (1982). A note on the isoperimetric constant. *Ann. Sci. École Norm. Sup., Paris* **15**, 213–230.

Buser, P. (1992). *Geometry and Spectra of Compact Riemann Surfaces*. Basel: Birkhäuser Verlag.

Buser, P., J. Conway, P. Doyle, and K.-D. Semmler (1994). Some planar isospectral domains. *Int'l. Math. Res. Notices* **9**, 167–192.

Buser, P. and H. Karcher (1981). *Gromov's almost flat manifolds*, Volume 81 of *Astérique*. Paris: Société Mathématiques de France.

Cabré, X. (2000). Equcions en derivades parcials, geometria i control estocastic. *Butl. Soc. Catal. Mat.* **15**, 7–27.

Cabré, X. (2003). The isoperimetric inequality and the principal eigenvalue via the ABP method. Preprint.

Calabi, E. (1967). On Ricci curvature and geodesics. *Duke Math. J.* **34**, 667–676.

Canary, R. D. (1992). On the Laplacian and the geometry of hyperbolic 3–manifolds. *J. Diff. Geom.*, 349–367.

Carathéodory, C. and E. Study (1909). Zwei Beweise des Satzes dass der Kreis unter alle Figuren gleichen Umgangs den grössten Inhalt hat. *Math. Ann.* **68**, 133–144.

Carleman, T. (1921). Zur theorie der Minimalflächen. *Math. Zeit.* **9**, 154–160.

Cartan, E. (1927). La géometrie des groupes des transformations. *J. Math. Pure Appl.* **6**, 1–119.

Cartan, E. (1946). *Leçons sur la Géométrie des Espaces de Riemann* (2nd ed.). Paris: Gauthier–Villars.

Charlap, L. S. (1986). *Bieberbach Groups and Flat Manifolds*. New York: Springer Verlag.

Chavel, I. (1967). Isotropic Jacobi fields, and Jacobi's equations on Riemannian homogeneous spaces. *Comment. Math. Helv.* **44**, 237–248.

Chavel, I. (1970). On Riemannian symmetric spaces of rank one. *Adv. Math.* **4**, 236–264.

Chavel, I. (1972). *Riemannian Symmetric Spaces of Rank One*, Volume 5 of *Lecture Notes Pure Appl. Math.* New York: Marcel Dekker, Inc.

Chavel, I. (1978). On A. Hurwitz' method in isoperimetric inequalities. *Proc. Am. Math. Soc.* **71**, 275–279.

Chavel, I. (1984). *Eigenvalues in Riemannian Geometry*. New York: Academic Press.

Chavel, I. (2001). *Isoperimetric Inequalities: Differential Geometric and Analytic Perspectives*. New York: Cambridge University Press.

Chavel, I. and E. A. Feldman (1978a). Cylinders on surfaces. *Comment. Math. Helv.* **53**, 439–447.

Chavel, I. and E. A. Feldman (1978b). Spectra of domains in compact manifolds. *J. Fcnl. Anal.* **30**, 275–289.

Chavel, I. and E. A. Feldman (1980). Isoperimetric inequalities in curved surfaces. *Adv. Math.* **37**, 83–98.

Chavel, I. and E. A. Feldman (1988). Spectra of manifolds less a small domain. *Duke Math. J.* **56**, 339–414.

Chavel, I. and E. A. Feldman (1991). Modified isoperimetric constants, and large time heat diffusion in Riemannian manifolds. *Duke Math. J.* **64**, 473–499.

Cheeger, J. (1970a). Finiteness theorems for Riemannian manifolds. *Am. J. Math.* **92**, 61–74.

Cheeger, J. (1970b). A lower bound for the smallest eigenvalue of the Laplacian. In *Problems in Analysis*, pp. 195–199. Princeton, NJ: Princeton University Press.

Cheeger, J. (1991). Critical points of distance functions and applications to geometry. In P. de Bartolomeis and F. Tricerri (Eds.), *Geometric Topology: Recent Developments*, Volume 1504 of *Lecture Notes Math.*, pp. 1–38. Berlin: Springer-Verlag.

Cheeger, J. and D. Ebin (1975). *Comparison Theorems in Riemannian Geometry.* Amsterdam: North Holland Publishing Company.

Cheeger, J. and D. Gromoll (1971). The splitting theorem for manifolds of nonnegative Ricci curvature. *J. Diff. Geom.* **6**, 119–128.

Cheeger, J. and D. Gromoll (1972). On the structure of complete manifolds of nonnegative curvature. *Ann. Math.* **96**, 413–443.

Cheeger, J., M. Gromov, and M. Taylor (1982). Finite propagation speed, kernel estimates for functions of the Laplace operator, and the geometry of complete Riemannian manifolds. *J. Diff. Geom.* **17**, 15–54.

Cheng, S. Y. (1975). Eigenvalue comparison theorems and its geometric applications. *Math. Zeit.* **143**, 289–297.

Cheng, S. Y. and P. Li (1981). Heat kernel estimates and lower bounds of eigenvalues. *Comment. Math. Helv.* **56**, 327–338.

Chern, S. S. (1944). A simple intrinsic proof of the Gauss–Bonnet formula for closed Riemannian manifolds. *Ann. Math.* **45**, 747–752.

Chern, S. S. (1968). *Minimal Submanifolds in a Riemannian Manifold* (Lecture Notes). University of Kansas.

Chern, S. S. (Ed.) (1989). *Studies in Global Geometry and Analysis* (2nd ed.), Volume 27 of *Studies in Math.* Washington, DC: Mathematical Association of America.

Choe, J. (1990). The isoperimetric inequality for a minimal surface with radially connected boundary. *Ann. Scuola Norm. Sup. Pisa* **17**, 583–593.

Cohn-Vossen, S. (1936). Totalkrümmung und geodetische Linien auf einfachzussamenhängenden offnen vollständigen Fläschenstücken. *Mat. Sbornik* **43**, 139–163.

Costa, C. J. (1984). Example of a complete minimal immersion in $\mathbb{R}^3$ of genus one and three ends. *Bol. Sci. Brasil. Mat.* **15**, 47–54.

Courant, R. and D. Hilbert (Vol. I, 1953; Vol. II, 1967). *Methods of Mathematical Physics.* New York: Wiley (Interscience).

Courtois, G. (1987). Comportement d'une variétés riemanniene compacte sous perturbation topologique par excision d'un domaine. Thèse, Grenoble, France.

Courtois, G. (1995). Spectrum of manifolds with holes. *J. Fcnl. Anal.* **134**, 194–221.

Croke, C. B. (1980). Some isoperimetric inequalities and eigenvalue estimates. *Ann. Sci. École Norm. Sup., Paris* **13**, 419–435.

Croke, C. B. (1982). Poincaré's problem and the length of the shortest closed geodesic on a convex hypersurface. *J. Diff. Geom.* **17**, 595–634.

Croke, C. B. (1984). A sharp 4–dimensional isoperimetric inequality. *Comment. Math. Helv.* **59**, 187–192.

Croke, C. B. and A. Derdziński (1987). A lower bound for λ_1 on manifolds with boundary. *Comment. Math. Helv.* **62**, 106–121.

Croke, C. B. and H. Karcher (1988). Volumes of small balls on open manifolds: lower bounds and examples. *Trans. Am. Math. Soc.* **309**, 753–762.

Damek, E. and F. Ricci (1992). A class of nonsymmetric harmonic spaces. *Bull. Am. Math. Soc.* **27**, 139–142.

Darboux, G. (1894). *Leçons sur la Théorie Générale des Surfaces* (3rd ed.). New York: Chelsea Publishing Company, 1972. Reprint of 1st ed. Paris: Gauthier–Villars.

Darboux, G. (1898). *Leçons sur les Systemes Orthogonaux et les Coordonées Curviligne*, Volume I. Paris: Gauthier–Villars.

Davies, E. B. (1984). Some norm bounds and quadratic form inequalities for Schrödinger operators, II. *J. Operator Theory* **12**, 177–196.

Davies, E. B. (1987). Explicit constants for Gaussian upper bounds on heat kernels. *Am. J. Math.* **109**, 319–334.

Davies, E. B. (1989). *Heat Kernels and Spectral Theory*. Cambridge: Cambridge University Press.

Davies, E. B. and Y. Safarov (Eds.) (1999). *Spectral Theory and Geometry*. London Math. Soc., Lecture Note Series. Cambridge: Cambridge University Press.

de Rham, G. (1952). Sur la reducibilité d'un espace de Riemann. *Comment. Math. Helv.* **26**, 328–344.

de Rham, G. (1955). *Differentiable Manifolds*. Berlin: Springer Verlag, 1984. Engl. transl. of Paris: Hermann, 1955 edition.

Dierkes, U., S. Hildebrandt, A. Kuester, and O. Wohlrab (1992). *Minimal Surfaces*. New York: Springer Verlag.

do Carmo, M. (1976). *Differential Geometry of Curves and Surfaces*. New Jersey: Prentice Hall.

do Carmo, M. (1989). Hypersurfaces of constant mean curvature. In F. Carreras, O. Gil-Medrano, and A. Naveira (Eds.), *Differential Geometry, Proceedings, Peñiscola 1988*, Volume 1410 of *Lecture Notes Math.*, pp. 128–144. Berlin: Springer Verlag.

do Carmo, M. (1992). *Riemannian Geometry*. Boston: Birkhauser.

Dodziuk, J. and B. Randol (1986). Lower bounds for λ_1 on a finite-volume hyperbolic manifold. *J. Diff. Geom.* **24**, 133–139.

Dombrowski, P. (1962). On the geometry of the tangent bundle. *J. Reine Angew. Math.* **210**, 73–88.

Eberlein, P. (1973). When is a geodesic flow of Anosov type? I. *J. Diff. Geom.* **8**, 437–463.

Eberlein, P. and B. O'Neill (1973). Visibility manifolds. *Pac. J. Math.* **46**, 45–109.

Efimov, N. V. (1964). Generation of singularities on surfaces of negative curvature (Russian). *Mat. Sbornik* **64**, 286–320.

Efremovič, V. (1953). The proximity geometry of Riemannian manifolds (Russian). *Uspehi Mat. Nauk.* **8**, 189.

Epstein, D. B. A. (Ed.) (1987). *Analytic and Geometric Aspects of Hyperbolic Space*. Cambridge: Cambridge University Press.

Eschenburg, J.-H. (1986). Local convexity and nonnegative curvature–Gromov's proof of the sphere theorem. *Invent. Math.* **84**, 507–522.

Eschenburg, J.-H. (1987). Comparison theorems and hypersurfaces. *Manusc. Math.* **59**, 295–323.

Eschenburg, J.-H. (1991). Diameter, volume, and topology for positive Ricci curvature. *J. Diff. Geom.* **33**, 743–747.

Eschenburg, J.-H. and E. Heintze (1984). An elementary proof of the Cheeger–Gromoll splitting theorem. *Ann. Global Anal. Geom.* **2**, 141–151.

Eschenburg, J.-H. and E. Heintze (1990). Comparison theory for Riccati equations. *Manusc. Math.* **68**, 209–214.

Faber, C. (1923). Beweiss, dass unter allen homogenen Membrane von gleicher Fläche und gleicher Spannung die kreisförmige die tiefsten Grundton gibt. *Sitzungber. Bayer Akad. Wiss., Math.–Phys., Munich*, 169–172.
Farkas, H. and I. Kra (1980). *Riemann Surfaces*. New York: Springer Verlag.
Federer, H. (1969). *Geometric Measure Theory*. New York: Springer Verlag.
Federer, H. and W. H. Fleming (1960). Normal integral currents. *Ann. Math.* **72**, 458–520.
Fenchel, W. (1940). On total curvatures of Riemannian manifolds I. *J. Lond. Math. Soc.* **15**, 15–22.
Fenchel, W. (1989). *Elementary Geometry in Hyperbolic Space*. Berlin: W. de Gruyter.
Fiala, F. (1940). Le problème des isopérimètres sur les surfaces ouvertes à courbure positive. *Comment. Math. Helv.* **13**, 293–346.
Fischer-Colberie, D. and R. Schoen (1980). The structure of complete stable minimal surfaces in 3–manifolds of nonnegative scalar curvature. *Comm. Pure Appl. Math.* **33**, 199–211.
Flanders, H. (1963). *Differential Forms*. New York: Academic Press.
Frankel, T. (1961). Manifolds with positive curvature. *Ann. Math.* **11**, 165–174.
Gage, M. (1980). Upper bounds for the first eigenvalue of the Laplace–Beltrami operator. *Ind. U. Math. J.* **29**, 897–912.
Gage, M. (1983). An isoperimetric inequality with applications to curve shortening. *Duke Math. J.* **50**, 1225–1229.
Gage, M. (1984). Curve shortening makes convex curves circular. *Invent. Math.* **76**, 357–364.
Gage, M. and R. Hamilton (1986). The heat equation shrinking plane convex curves. *J. Diff. Geom.* **23**, 69–96.
Gallot, S., D. Hulin, and J. Lafontaine (1987). *Riemannian Geometry*. Berlin: Springer Verlag.
Gardner, R. J. (2002). The Brunn–Minkowski inequality. *Bull. Am. Math. Soc.* **39**, 355–405.
Gauss, C. (1825). In *New General Investigations of Curved Surfaces*, pp. 79–114. Hewlett, NY: Raven Press, 1965. Reprint of 1902 English translation by A. Hiltebeitel & J. Moorehead.
Gauss, C. (1827). In *General Investigations of Curved Surfaces*, pp. 3–78. Hewlett, NY: Raven Press, 1965. Reprint of 1902 English translation by A. Hiltebeitel & J. Moorehead.
Gilbarg, D. and N. S. Trudinger (1977). *Elliptic Partial Differential Equations of Second Order*. Berlin: Springer Verlag.
Goldstein, H. (1950). *Classical Mechanics*. Reading, MA: Addison-Wesley.
Gordon, C. (2000). In *Handbook of Differential Geometry*, pp. 3–78. Amsterdam: North-Holland.
Gordon, C. and D. Webb (1996). You can't hear the shape of a drum. *Am. Scientist*, 46–53.
Gordon, C., D. Webb, and S. Wolpert (1992a). Isospectral plane domains and surfaces via Riemannian orbifolds. *Invent. Math.* **110**, 1–22.
Gordon, C., D. Webb, and S. Wolpert (1992b). One can't hear the shape of a drum. *Bull. Am. Math. Soc.* **27**, 134–138.
Gray, A. (1990). *Tubes*. Reading, MA: Addison-Wesley.
Gray, A. (1998). *Modern Differential Geometry of Curves and Surfaces with Mathematica* (2nd ed.). Berlin: Springer Verlag.
Grayson, M. (1987). The heat equation shrinks embedded plane curves to round points. *J. Diff. Geom.* **26**, 285–314.

Grayson, M. (1989a). A short note on the evolution of a surface by its mean curvature. *Duke Math. J.* **58**, 555–558.

Grayson, M. (1989b). Shortening emdedded curves. *Ann. Math.* **129**, 71–111.

Green, L. and R. Gulliver (1985). Planes without conjugate points. *J. Diff. Geom.* **22**, 43–47.

Green, L. W. (1954). Surfaces without conjugate points. *Trans. Am. Math. Soc.* **76**, 529–546.

Green, L. W. (1958). A theorem of E. Hopf. *Mich. Math. J.* **5**, 31–34.

Green, L. W. (1963). Aufwiedersehnsflächen. *Ann. Math.* **78**, 289–299.

Greene, R. E. and H. Wu (1988). Lipschitz convergence of riemannian manifolds. *Pac. J. Math.* **131**, 119–141.

Gribkov, I. (1980). The incorrectness of Schur's theorem. *Soviet Math. Doklady Transl.* **21**, 922–925.

Grimaldi, R. and P. Pansu (1994). Sur la régularité de la fonction croissance d'une variété riemanniene. *Geom. Dedicata* **50**, 301–307.

Gromoll, D., W. Klingenberg, and W. Meyer (1968). *Riemannsche Geometrie im Grossen*, Volume 55 of *Lecture Notes Math.* Berlin: Springer Verlag.

Gromoll, D. and W. Meyer (1969). On complete open manifolds of positive curvature. *Ann. Math.* **90**, 75–90.

Gromov, M. (1980). Paul Levy's isoperimetric inequality. Preprint.

Gromov, M. (1981). *Structures métriques pour les Variétés Riemannienes*. Paris: Cedic/Ferdnand Nathan.

Gromov, M. (1983). Filling Riemannian manifolds. *J. Diff. Geom.* **18**, 1–147.

Gromov, M. (1986). Isoperimetric inequalities in Riemannian manifolds. In *Asymptotic Theory of Finite Dimensional Normed Spaces*, Volume 1200 of *Lecture Notes Math.*, pp. 114–129. Berlin: Springer Verlag.

Gromov, M. (1996). Systoles and isosystolic inequalities. In A. L. Besse (Ed.), *Actes de la Table Rondes de Géométrie Differentielle* (*Luminy, 1992*), Volume 1 of *Sémin. Congr.*, pp. 291–362. Paris: Société Mathématiques de France.

Gromov, M. (1999). *Metric Structure for Riemannian and Non-Riemannian Spaces*. Boston: Birkhauser. Appendices by M. Katz, P. Pansu, S. Semmes (Eds.). J. La Fontaine & P. Pansu. Engl. trans.: S. M. Bates.

Grossman, N. (1967). The volume of a totally geodesic hypersurface in a pinched manifold. *Pacific J. Math.* **23**, 257–262.

Grove, K. (1987). Metric differential geometry. In V. Hansen (Ed.), *Differential Geometry*, Volume 1263 of *Lecture Notes Math.*, pp. 171–227. Berlin: Springer-Verlag.

Grove, K. and P. Petersen (Eds.) (1997). *Comparison Geometry*, Volume 30 of *MSRI Publications*. New York: Cambridge University Press.

Grove, K. and K. Shiohama (1977). A generalized sphere theorem. *Ann. Math.* **106**, 201–211.

Grove, K. and K. Shiohama (1988). Bounding homotopy types by geometry. *Ann. Math.* **128**, 195–206.

Günther, P. (1960). Einige Sätze über das Volumenelement eines Riemannschen Raumes. *Publ. Math. Deberecen* **7**, 78–93.

Hadamard, J. (1898). Les surfaces à courbures opposées et leurs géodésiques. *J. Math. Pures Appl.* **4**, 27–73.

Hamilton, R. (1982). Three-manifolds with positive ricci curvature. *J. Diff. Geom.* **3017**, 255–306.

Hartman, P. (1964). Geodesic parallel coordinates in the large. *Am. J. Math.* **86**, 705–727.

Hass, J. and F. Morgan (1996). Geodesics and soap bubbles in surfaces. *Math. Zeit.* **223**, 185–196.

Hebda, J. (1982). Some lower bounds for the area of surfaces. *Invent. Math.* **65**, 485–490.

Heintze, E. and H. Karcher (1978). A general comparison theorem with applications to volume estimates for submanifolds. *Ann. Sci. École Norm. Sup., Paris* **11**, 451–470.

Helgason, S. (1959). Differential operators on homogeneous spaces. *Acta Math.* **102**, 239–299.

Helgason, S. (1962). *Differential Geometry and Symmetric Spaces*. New York: Academic Press.

Hicks, N. (1965). *Notes on Differential Geometry*. New York: Van Nostrand.

Hilbert, D. (1901). Über Flächen von constanter Gausscher Krümmung. *Trans. Am. Math. Soc.* **2**, 87–99.

Hilbert, D. and S. Cohn-Vossen (1952). *Geometry and the Imagination*. New York: Chelsea Publishing Co.

Hirsch, M. (1976). *Differential Topology*. New York: Springer Verlag.

Hoffman, D. and W. Meeks III (1985). A complete embedded minimal surface in $\mathbb{R}^3$ with genus one and three ends. *J. Diff. Geom.* **21**, 109–127.

Hoffman, D. and H. Karcher (1997). Complete embedded minimal surfaces of finite total curvature. In R. Osserman (Ed.), *Geometry V: Minimal Surfaces*, Encyclopedia of Mathematical Science. Volume 90. New York: Springer Verlag. 5–93.

Hoffman, D. and R. Osserman (1980). The geometry of the generalized Gauss map. *Mem. Am. Math. Soc.* **28**.

Holopainen, I. (1994). Rough isometries and p–harmonic functions with finite Dirichlet integral. *Revista Math. Iberoamer.* **217**, 459–477.

Hopf, E. (1948). Closed surfaces without conjugate points. *Proc. Natl. Acad. Sci., USA* **34**, 47–51.

Hopf, H. (1925). Zum Clifford–Kleinschen Raumproblem. *Math. Ann.* **95**, 313–339.

Hopf, H. (1926a). Abbildungsklassen n–dimensionaler Mannigfaltigkeiten. *Math. Ann.* **96**, 209–224.

Hopf, H. (1926b). Vektorfelden in n–dimensionaler Mannigfaltigkeiten. *Math. Ann.* **96**, 225–250.

Hopf, H. (1935). Über die Drehung der Tangenten und Sehnen ebener Kurven. *Comp. Math.* **2**, 50–62.

Hopf, H. (1946). In *Differential Geometry in the Large*, Volume 1000 of *Lecture Notes Math.*, pp. 1–75. Berlin: Springer Verlag, 1983. Reprint of Lecture Notes from Stanford University 1946.

Hopf, H. (1956). In *Differential Geometry in the Large*, Volume 1000 of *Lecture Notes Math.*, pp. 77–184. Berlin: Springer Verlag, 1983. Reprint of Lecture Notes from New York University 1956.

Hopf, H. and W. Rinow (1931). Über den Begriff der vollständigen differential–geometrischen Flächen. *Comment. Math. Helv.* **3**, 209–225.

Howards, H., M. Hutchings, and F. Morgan (1999). The isoperimetric problem on surfaces. *Am. Math. Monthly* **106**, 430–439.

Hsiang, W. Y. (1991). Isoperimetric regions and soap bubbles. In *do Carmo Symposium*, Volume 52 of *Pitman Monographs Surveys Pure & Appl. Math.*, pp. 229–240. Essex: Longman Scientific & Technical.

Huber, A. (1954). On the isoperimetric inequality on surfaces of variable curvature. *Ann. Math.* **60**, 237–247.

Huisken, G. (1984). Flow by mean curvature of convex surfaces into spheres. *J. Diff. Geom.* **20**, 237–266.
Huisken, G. (1985). Ricci deformation of the metric on a Riemannian manifold. *J. Diff. Geom.* **21**, 47–62.
Hurwitz, A. (1901). Sur le problème des isopérimétres. *C. R. Acad. Sci. Paris* **132**, 401–403.
Jacobi, C. F. (1836). Zur Theorie der Variationsrechnung und der Differentialgleichungen. *J. Reine Angew. Math.* **17**, 68–82. Reprinted: *Gesammelte Werke* IV. 2nd ed. pp. 41–55. New York: Chelsea, 1969.
Jellet, J. H. (1853). Sur la surface dont la courbure moyenne est constante. *J. Math. Pure Appl.* **18**, 163–167.
Jost, J. (1995). *Riemannian Geometry and Geometric Analysis*. Berlin: Springer Verlag.
Jost, J. (1997). *Nonpositive Curvature: Geometric and Analytic Aspects*. Basel: Birkhauser Verlag.
Kac, M. (1966). Can one hear the shape of a drum? *Am. Math. Monthly* **73**, 1–23.
Kanai, M. (1985). Rough isometries, and combinatorial approximations of geometries of noncompact Riemannian manifolds. *J. Math. Soc. Japan* **37**, 391–413.
Karcher, H. (1977). Riemannian center of mass and mollifier smoothing. *Comm. Pure Appl. Math.* **30**, 509–541.
Karcher, H. (1989). Riemannian comparison constructions. In *Studies in Global Geometry and Analysis* (2nd ed.), Volume 27 of *Studies in Math.* Washington, DC: Mathematical Association of America.
Kasue, A. (1989). A convergence theorem for riemannian manifolds and some applications. *Nagoya Math. J.* **114**, 21–51.
Katz, M. (1983). The filling radius of two-point homogeneous spaces. *J. Diff. Geom.* **18**, 505–511.
Katz, M. (2005). Spectral Geometry and Topology. Preprint.
Kawohl, B. (1985). *Rearrangements and Convexity of Level Sets in PDE*, Volume 1150 of *Lecture Notes Math.* Berlin: Springer Verlag.
Kazdan, J. L. (1978). An inequality arising in geometry. In Besse (1978). Appendix E, 243–246.
Kazdan, J. L. (1982). An isoperimetric inequality and wiedersehns manifolds. In Yau (1982). 143–157.
Keen, L. (1974). Collars in Riemann surfaces. In *Discontinuous Groups Riemann Surfaces*, pp. 263–268. Princeton: *Annals of Mathematics Studies*.
Killing, W. (1891). Über die Clifford–Kleinschen Raumformen. *Math. Ann.* **39**, 257ff.
Killing, W. (1893). *Einfuhrung in die Grundlagen der Geometrie*. Munster: Paderborn.
Kleiner, B. (1992). An isoperimetric comparison theorem. *Invent. Math.* **108**, 37–47.
Klingenberg, W. (1959). Contributions to Riemannian geometry in the large. *Ann. Math.* **69**, 654–666.
Klingenberg, W. (1961). Über Riemannsche Mannigfaltigkeiten mit positiver Krümmung. *Comment. Math. Helv.* **35**, 47–54.
Klingenberg, W. (1976). *A Course in Differential Geometry*. New York: Springer Verlag.
Klingenberg, W. (1978). *Lectures on Closed Geodesics*. Berlin: Springer Verlag.
Klingenberg, W. (1982). *Riemannian Geometry*. Berlin: Walter de Gruyter.
Kobayashi, S. (1958). Fixed points of isometries. *Nagoya Math. J.* **13**, 63–68.
Kobayashi, S. and K. Nomizu (Vol. I, 1963; Vol. II, 1969). *Foundations of Differential Geometry*. New York: Interscience Publishers.
Krahn, E. (1925). Über eine von Rayleigh formulierte Minimaleigenschafte des Kreises. *Math. Ann.* **94**, 97–100.

Kröger, P. (2004). An extension of Günther's volume comparison theorem. *Math. Ann.* **329**, 593–596.

Lang, S. (1995). *Differentiable and Riemannian Manifolds*. New York: Springer Verlag.

Laugwitz, D. (1966). *Differential and Riemannian Geometry*. New York: Academic Press.

Lawson, H. B. (1980). *Lectures on Minimal Submanifolds* (2nd ed.). Berkeley, CA: Publish or Perish.

Lee, J. (1997). *Riemannian Manifolds*. New York: Springer Verlag.

Levi-Civita, T. (1929). *Absolute Differential Calculus*. London: Blackie & Sons. First published in Rome, 1925, in Italian.

Li, P., R. Schoen, and S. T. Yau (1984). On the isoperimetric inequality for minimal surfaces. *Ann. Sc. Norm. Sup. Pisa* **11**, 237–244.

Lichnerowicz, A. (1958). *Geometries des Groupes des Transformationes*. Paris: Dunod.

Lieb, E. H. and M. Loss (1996). *Analysis*. Providence, RI: American Mathematical Society.

Manning, A. (1979). Topological entropy for geodesic flows. *Ann. Math.* **110**, 567–573.

Massey, W. S. (1967). *Algebraic Topology: An Introduction*. New York: Harcourt, Brace, and World, Inc.

McKean, H. P. (1970). An upper bound for the spectrum of Δ on a manifold of negative curvature. *J. Diff. Geom.* **4**, 359–366.

Meschkowski, H. (1964). *Noneuclidean Geometry* (2nd ed.). New York: Academic Press.

Milnor, J. (1963). *Morse Theory*, Volume 51 of *Ann. Math. Studies*. Princeton, NJ: Princeton University Press.

Milnor, J. (1964). Eigenvalues of the Laplace operator on certain manifolds. *Proc. Natl. Acad. Sci., USA* **51**, 542.

Milnor, J. (1965). *Topology from the Differentiable Viewpoint*. Charlottesville, VA: University Press of Virginia.

Milnor, J. (1968). A note on curvature and the fundamental group. *J. Diff. Geom.* **2**, 1–7.

Milnor, T. K. (1972). Efimov's theorem about complete immersed surfaces of negative curvature. *Adv. Math.* **8**, 474–543.

Morgan, F. (1988). *Geometric Measure Theory: A Beginner's Guide*. Boston: Academic Press.

Morgan, F. (1992). *Riemannian Geometry: A Beginner's Guide*. Boston: Jones and Bartlett.

Morgan, F., M. Hutchings, and H. Howards (2000). The isoperimetric problem on surfaces of revolution of decreasing gauss curvature. *Trans. AMS* **352**, 4889–4909.

Morse, M. (1930). A generalization of the Sturm separation and comparison theorems. *Math. Ann.* **103**, 52–69.

Moser, J. (1964). A Harnack inequality for parabolic differential equations. *Comm. Pure Appl. Math.* **17**, 101–134.

Myers, S. B. (1935). Connections between differential geometry and topology. *Duke Math. J.* **1**, 376–391.

Myers, S. B. (1941). Riemannian manifolds with positive mean curvature. *Duke Math. J.* **8**, 401–404.

Myers, S. B. and N. Steenrod (1939). The group of isometries of a Riemannian manifold. *Ann. Math.* **40**, 400–416.

Nagano, T. (1959). Homogeneous sphere bundles and the isotropic Riemannian manifolds. *Nagoya Math. J.* **15**, 29–55.

Narasimhan, R. (1968). *Analysis on Real and Complex Differentiable Manifolds*. Amsterdam: North Holland Publishing Company.

Nitsche, J. C. C. (1989). *Introduction to Minimal Surfaces*. Cambridge: Cambridge University Press.

Nomizu, K. (1954). Invariant affine connections on homogeneous spaces. *Am. J. Math.* **76**, 33–65.

Obata, M. (1962). Certain conditions for a Riemannian manifold to be a sphere. *J. Math. Soc. Japan* **14**, 333–340.

Ohman, D. (1955). Über den Brunn–Minkowskischen Satz. *Comment. Math. Helvitici* **29**, 215–222.

Olver, F. (1974). *Asymptotics and Special Functions*. New York: Academic Press.

O'Neill, B. (1966a). *Elementary Differential Geometry*. New York: Academic Press.

O'Neill, B. (1966b). The fundamental equations of a submersion. *Mich. Math. J.* **13**, 459–469.

O'Neill, B. (1983). *Semi–Riemannian Geometry*. New York: Academic Press.

Oprea, J. (1997). *Differential Geometry and its Applications*. Upper Saddle River, NJ: Prentice Hall.

Osserman, R. (1978). The isoperimetric inequality. *Bull. Am. Math. Soc.* **84**, 1182–1238.

Osserman, R. (1979). Bonnesen-style inequalities. *Am. Math. Monthly* **86**, 1–29.

Osserman, R. (1980). Minimal surfaces, Gauss maps, total curvature, eigenvalue estimates, and stability. In W.-Y. Hsiang, S. Kobayashi, I. Singer, A. Weinstein, J. Wolf, and H.-H. Wu (Eds.), *The Chern Symposium 1979*, pp. 199–227. New York: Springer Verlag.

Osserman, R. (1986). *A Survey of Minimal Surfaces* (2nd ed.). New York: Dover.

Osserman, R. (1990). Curvature in the eighties. *Am. Math. Monthly* **97**, 731–756.

Osserman, R. (Ed.) (1997). *Geometry V: Minimal Surfaces*, Volume 91 of *Encyclopedia of Mathematical Sciences*. New York: Springer Verlag.

Osserman, R. and P. Sarnak (1984). A new curvature invariant and entropy of geodesic flows. *Invent. Math.* **77**, 455–462.

Otsu, Y. (1991). On manifolds of positive Ricci curvature with large diameter. *Math. Zeit.* **264**, 206–255.

Palais, R. S. (1957). The differentiability of isometries. *Proc. Am. Math. Soc.* **8**, 805–807.

Peters, S. (1984). Cheeger's finiteness theorem for diffeomorphism classes of Riemannian manifolds. *J. Reine Angew. Math.* **349**, 77–82.

Peters, S. (1987). Convergence of Riemannian manifolds. *Comp. Math.* **62**, 3–16.

Petersen, P. (1997). In K. Grove and P. Petersen (Eds.), *Convergence Theorems in Riemannian Geometry*, Volume 30 of *MSRI Publications*, pp. 167–202. New York: Cambridge University Press.

Petersen, P. (1998). *Riemannian Geometry*, Volume 171 of *GTM*. New York: Springer Verlag.

Petersen, P. (1999). Aspects of global riemannian Geometry. *Bull. Am. Math. Soc.* **36**, 297–344.

Pinsky, M. A. (1978). The spectrum of the Laplacian on a manifold of negative curvature. I. *J. Diff. Geom.* **13**, 87–91.

Poincaré, H. (1905). Sur les lignes géodésiques des surfaces convexes. *Trans. Am. Math. Soc.* **6**, 237–274.

Pólya, G. and G. Szegö (1951). *Isoperimetric Inequalities in Mathematical Physics*, Volume 27 of *Ann. Math. Studies*. Princeton, NJ: Princeton University Press.

Preissmann, A. (1943). Quelques propriétés globales des espaces de Riemann. *Comment. Math. Helv.* **15**, 175–216.

Pu, P. M. (1962). Some inequalities in certain nonorientable Riemannian manifolds. *Pac. J. Math.* **11**, 55–71.

Randol, B. (1974). Small eigenvalues of the Laplace operator on compact Riemann surfaces. *Bull. Am. Math. Soc.* **80**, 996–1000.

Randol, B. (1979). Cylinders in Riemann surfaces. *Comment. Math. Helv.* **54**, 1–5.

Ratcliffe, J. (1994). *Foundations of Hyperbolic Manifolds*. New York: Springer Verlag.

Rauch, H. E. (1951). A contribution to differential geometry in the large. *Ann. Math.* **54**, 38–55.

Rayleigh, L. J. W. S. (1877). *The Theory of Sound*. New York: Macmillan. Reprinted: Dover Publishing. New York. 1945.

Reid, W. T. (1959). The isoperimetric inequality and associated boundary problems. *J. Math. Mech.* **8**, 571–581.

Riemann, B. (1854). Über die Hypothesen, welche der Geometrie zur Grunde liegen. Habilitationschrift. In *Collected Works of Bernard Riemann*, pp. 272–287. New York: Dover Publishing, Inc., 1953. Reprint of 1892 edition and 1902 supplement. English translation of *Habilitationschrift* in Spivak (1970), Vol. II, 4A 1–4A 20.

Rinow, W. (1961). *Die innere Geometrie der metrischen Räume*. Berlin: Springer Verlag.

Ritoré, M. (2001). The isoperimetric problem in complete surfaces with nonnegative curvature. *J. Geom. Anal.* **11**, 509–517.

Rodrigues, O. (1814–1816). Recherches sur la théorie analytique des lignes et des rayons de courbure des surfaces, et sur la transformation d'une classe d'integrales doubles, qui ont un rapport direct avec les formules de cette théorie. *École Polytechnique Corresp.* **III**, 162–182.

Ros, A. (2005). The isoperimetric problem. *Global Theory of Minimal Surfaces*. Providence, RI: American Mathematical Society, 175–209. Available at: http://www.ugr.es/~aros/isoper.htm.

Ruh, E. A. (1982). Riemannian manifolds with bounded curvature ratios. *J. Diff. Geom.* **17**, 643–653.

Ruse, H. S., A. G. Walker, and T. J. Willmore (1961). *Harmonic Spaces*. Rome: Edz. Cremonese.

Sakai, T. (1984). Comparison and finiteness theorems in Riemannian geometry. In *Geometry of Geodesics and Related Fields*, Volume 3 of *Advanced Studies in Pure Mathmatics*, pp. 125–181.

Sasaki, S. (1958). On the differential geometry of tangent bundles of Riemannian manifolds. *Tôhuku Math. J.* **10**, 338–354.

Sasaki, S. (1962). On the differential geometry of tangent bundles of Riemannian manifolds, II. *Tôhuku Math. J.* **14**, 146–155.

Schmidt, E. (1948). Der Brunn–Minkowskische Satz und sein Spiegel-theorem sowie die isoperimetrische Eigenschaft der Kugel in der euklidischen und nichteuklidischen Geometrie I, II. *Math. Nach.* **1**, 81–157. **2** (1949), 171–244.

Schneider, R. (1972). Konvexe Flächen mit langsam abnehmender Krümmung. *Arch. der Math* **23**, 650–654.

Schönberg, I. J. (1932). Some applications of the calculus of variations to Riemannian geometry. *Ann. Math.* **33**, 485–495.

Schur, F. (1886). Über den Zusammenhang der Räume konstanter Krümmungmasses mit den projektiven Räumen. *Math. Ann.* **27**, 537–567.

Schwarz, H. A. (1884). Beweis des Satzes, dass die Kugel kleinere Oberfläche besizt als jeder andere Körper gleichen Volumens. *Nach. Ges. Wiss. Göttingen*, 1–13. Reprinted: *Gesammelte Werke*. Bronx, NY: Chelsea Publishing Co., 1971 (reprint of 1881–1882 ed.), pp. 327–340.

Shikata, A. (1966). On a distance function on a set of differentiable structures. *Osaka Math. J.* **3**, 65–79.

Shikata, A. (1967). On the differentiable pinching problem. *Osaka Math. J.* **4**, 279–287.

Shiohama, K. (1983). A sphere theorem for manifolds of positive Ricci curvature. *Trans. Am. Math. Soc.* **275**, 811–819.

Shiohama, K. (2000). Sphere theorems. In F. Dillen and L. Verstraelen (Eds.), *Handbook of Differential Geometry*, Volume I, pp. 865–903. Amsterdam: North Holland Publishing Company, Elsevier.

Shiohama, K., T. Shioya, and M. Tanaka (2003). *The Geometry of Total Curvature on Complete Open Surfaces*. Cambridge, UK: Cambridge University Press.

Spivak, M. (1970). *A Comprehensive Introduction to Differential Geometry*. Boston: Publish or Perish.

Stantalo, L. A. (1976). *Integral Geometry and Geometric Probability*. Reading, MA: Addison-Wesley.

Steiner, J. (1838). Einfache Beweise der isoperimetrische Hauptsätze. *J. Reine Angew. Math.* **18**, 281–296. Reprinted: *Gesammelte Werke*. Bronx, NY: Chelsea Publishing Company, 1971 (reprint of 1881–1882 ed.).

Steiner, J. (1842). Über Maximum und Minimum. *Gesammelte Werke*. Bronx, NY: Chelsea Publishing Company, 1971 (reprint of 1881–1882 ed.) Vol. II, 245–308.

Stoker, J. J. (1969). *Differential Geometry*. New York: Wiley (Interscience).

Stredulinsky, E. and W. P. Ziemer (1997). Area minimizing sets subject to a volume constraint in a convex set. *J. Geom. Anal.* **7**, 653–677.

Struik, D. J. (1961). *Lectures on Classical Differential Geometry* (2nd ed.). Reading, MA: Addison–Wesley.

Sturm, J. C. F. (1836). Mémoire sur les équationes differentielles du second ordre. *J. Math. Pure Appl.* **1**, 106–186.

Svarc, A. S. (1955). A volume invariant of coverings. *Dokl. Akd. Nauk SSR* **105**, 32–34.

Synge, J. L. (1934). On the deviation of geodesics and null–geodesics, particularly in relation to the properties of spaces of constant curvature and indefinite line–element. *Ann. Math.* **35**, 705–713.

Synge, J. L. (1936). On the connectivity of spaces of positive curvature. *Q. J. Math.* **7**, 316–320.

Szabo, Z. I. (1990). The Lichnerowicz conjecture on harmonic manifolds. *J. Diff. Geom.* **31**, 1–28.

Szabo, Z. I. (1991). A short toplogical proof for the symmetry of two-point homogeneous spaces. *Invent. Math.* **106**, 61–64.

Thurston, W. P. (1979). *The Geometry and Topology of Three-Manifolds*. Based on author's course 1978–1979. Princeton, NJ: Princeton University Written by B. Floyd & S. Kerchoff; Chapter 7 based on lecture by J. Milnor.

Thurston, W. P. (1997). *Three Dimensional Geometry and Topology*, Volume 1. Princeton, NJ: Princeton University.

Tits, J. (1955). Sur certains classes d'espaces homogènes de groupes de Lie. *Acad. R. Belg. Cl. Sci. Mém. Coll.* **29**.

Toponogov, V. A. (1959). Riemannian spaces having their curvature bounded below by a positive number. *Usphei Math. Nauk.* **14**, 87–135. Engl. trans., *Trans. Am. Math. Soc.* **37** (1964), 291–336.

Toponogov, V. A. (1964). The metric structure of Riemannian spaces with nonnegative curvature which contain straight lines. *Sibirsk Math. Z.* **5**, 1358–1369. Engl. trans., *Trans. Am. Math. Soc.* **2** (1968), 225–239.

Topping, P. (1998). Mean curvature flow and geometric inequalities. *J. Reine Angew. Math.* **503**, 47–61.

Topping, P. (1999). The isoperimetric inequality on a surface. *Manuscr. Math.* **100**, 23–33.

Velling, J. (1999). Existence and uniqueness of complete constant mean curvature surfaces at infinity of $\mathbb{H}^3$. *J. Geom. Anal.* **9**, 457–489.

von Mangolt, H. (1881). Über diejenigen Punkte auf positiv gekrummten Flächen, welche die Eigenschaft haben, dass die von ihnen ausgehended geodätischen Linien nie aufhörn, kürzeste linien zu sien. *J. Reine Angew. Math.* **91**, 23–52.

Wang, H. C. (1952). Two-point homogeneous spaces. *Ann. Math.* **55**, 177–191.

Warner, F. W. (1965). The conjugate locus of a Riemannian manifold. *Am. J. Math.* **87**, 575–604.

Warner, F. W. (1966). Extensions of the Rauch comparison theorem to submanifolds. *Trans. Am. Math. Soc.* **122**, 341–356.

Warner, F. W. (1971). *Foundations of Differentiable Manifolds and Lie Groups.* Glenview, IL: Scott, Foresman, and Co.

Watson, G. N. (1916). A problem in analysis situs. *Proc. Lond. Math. Soc.* **15**, 227–242.

Weil, A. (1926). Sur les surfaces a courbure negative. *C. R. Acad. Sci. Paris* **182**, 1069–1071.

Weinberger, H. F. (1956). An isoperimetric inequality for the n–dimensional free membrane. *J. Rat. Mech. Anal.* **5**, 633–636.

Weinstein, A. (1967). On the homotopy type of positively-pinched manifolds. *Arch. der Math.* **18**, 523–524.

Weinstein, A. (1968). The cut locus and conjugate locus of a Riemannian manifold. *Ann. Math.* **87**, 29–41.

Weinstein, A. (1974). On the volume of manifolds all of whose geodesics are closed. *J. Diff. Geom.* **9**, 513–517.

Weyl, H. (1911). Über die asymptotische Verteilung der Eigenwerte. *Nachr. d. Königl. Ges. d. Wiss. zu Göttingen*, 110–117. Reprinted: *Gesammelte Abhandlungen.* Berlin: Springer Verlag, 1968, Vol. I, pp. 368–375.

Weyl, H. (1912). Der asymptotische Verteilungsgesetz der Eigenwerte linear partieller Differentialgleichungen (mit einer Anwendung auf die Theorie der Hohlraumstrahlung). *Math. Ann.* **71**, 441–479. Reprinted: *Gesammelte Abhandlungen.* Berlin: Springer Verlag, 1968, Vol. I, pp. 393–430.

Weyl, H. (1939). On the volume of tubes. *Am. J. Math.* **61**, 461–472.

White, B. (1991). Existence of smooth embedded surfaces of prescribed genus that minimize parametric even elliptic functionals on 3–manifolds. *J. Diff. Geom.* **33**, 413–433.

Whitehead, J. H. C. (1932). Convex regions in the geometry of paths. *Q. J. Math.* **3**, 33–42.

Whitney, H. (1937). On regular closed curves in the plane. *Comp. Math.* **4**, 276–284.

Wolf, J. A. (1967). *Spaces of Constant Curvature* (5th ed.). New York: McGraw–Hill (Wilmington, DE: Publish or Perish, 1984).

Yang, C. T. (1980). Odd dimensional wiedersehns manifolds are spheres. *J. Diff. Geom.* **15**, 91–96.

Yang, C. T. (1990). On the Blaschke conjecture. In S. Yau (Ed.), *Seminar on Differential Geometry*, Volume 102 of *Ann. Math. Studies*, pp. 159–171. Berlin: Springer Verlag.

Yau, S. T. (1975). Isoperimetric constants and the first eigenvalue of a compact manifold. *Ann. Sci. École Norm. Sup., Paris* **8**, 487–507.

Yau, S. T. (Ed.) (1982). *Seminar on Differential Geometry*, Volume 102 of *Ann. Math. Studies*. Berlin: Springer-Verlag.

Author Index

Subject Index

Printed in the United States
By Bookmasters